Convention on International Civil Aviation

Ruwantissa Abeyratne

Convention on International Civil Aviation

A Commentary

Springer

Ruwantissa Abeyratne
Legal Affairs and External Relations Bureau
International Civil Aviation Organization (ICAO)
Montreal
Quebec
Canada

ISBN 978-3-319-00067-1 ISBN 978-3-319-00068-8 (eBook)
DOI 10.1007/978-3-319-00068-8
Springer Cham Heidelberg New York Dordrecht London

Library of Congress Control Number: 2013944364

Printed on acid-free paper

Springer is part of Springer Science+Business Media (www.springer.com)

Preface

The 70th Anniversary of the signature ceremony of the Convention on International Civil Aviation (Chicago Convention) will fall on 7 December 2014. This day each year also happens to be designated "International Civil Aviation Day" marked by a modest ceremony by the International Civil Aviation Organization (ICAO). In reality, the 7th of December is not the day the Chicago Convention's anniversary should be associated with. It should be the date on which the Chicago Convention entered into force, which was 4 April 1947. However, the ICAO Assembly chose otherwise and, at its 29th Session (Montreal, 22 September–8 October 1992), adopted Resolution A29-1, which declared that each year, starting in 1994, the 7th of December shall be designated "international civil aviation day". This practice is at variance with "The United Nations Day", which is celebrated each year on 24 October—the day on which the United Nations came into existence.

At the time of writing, there were 191 States that had signed or otherwise adhered to the Chicago Convention, which *ipso facto* make them member States of ICAO. However, in 1944, only 52 signatory States (approximately 27 % of the current number) were party to the Convention. Over the years, the Convention has retained its pristine purity with no fundamental amendments or revisions, although a few "cosmetic" revisions have been added. In particular, three amendments entered into force, relating to articles: Article 3 *bis*, Article 83 *bis*, Article 50(a), and Article 56, in 1995–1998.

The Chicago Convention has been an enduring multilateral treaty for the past several decades, showing both resilience and vision. The treaty is far-reaching and can today be taken to apply to aspects of aviation such as security and environmental protection, which are not even explicitly referred to therein. However, The Chicago Convention has been vulnerable to misinterpretation and has often been misquoted by States mostly for political reasons and gains. For example, the provision on State sovereignty over airspace has been used to block useful initiatives on the liberalization of air transport, the imposition of air navigation charges, and other levies on airlines. It is submitted that the Convention should be interpreted to accord with the intent of its forefathers and current exigencies so that it achieves its main objective of serving the needs of the people of the world and not exclusively those of individual businesses and States.

One of the unique characteristics of the Chicago Convention is its wording in various provisions that ascribes specific meaning and purpose to its provisions. To this extent, the Chicago Convention stands out as an international treaty carved out in the early years of international comity after World War 2, having particular diplomatic nuances in its language. Various provisions, depending on their compelling nature, use words that effectively describe the meaning and intent of the treaty. For example, Article 1, on the question of sovereignty, states that the Contracting States "recognize" that each State has complete and exclusive sovereignty over the air space above its territory. Here, the word "recognize" conveys the meaning that the legal recognition of sovereignty of nations has already existed, which is a fact, as sovereignty over national airspace was first referred to in the Paris Convention of 1919. In Articles 2 and 3 that follow, the Convention uses the word "shall" to denote a peremptory rule of law (for example, in Article 3(a)) the Convention stipulates that it "shall" be applicable only to civil aircraft and shall not be applicable to State aircraft).

In Article 3(a) and (b), one sees again the word "recognize", where the Convention provides that Contracting States recognize that every State must refrain from resorting to the use of weapons against civil aircraft in flight and that States also recognize that each State has the right to require aircraft to land at designated airports. However, in Article 3(c), the provision starts with "Every civil aircraft shall comply with an order given in pursuance of paragraph b) of the Article", thus bringing in the mandatory element of compliance.

A slight deviation is seen in Article 4, where the Convention provides that each Contracting State "agrees" not to use civil aviation for any purposes inconsistent with the aims of the Convention. Here, the word "agrees" implies general agreement of States. It is arguable that the particular use of the word leaves a window of opportunity for a State to deviate from its agreement if it is impossible for that State to keep to its agreement. In the following Article, the word "agrees" occurs once again where States are recognized as having agreed to allow non-scheduled flights the right to make technical and non-commercial flights into their territory.

Article 6 deviates from the positive approach of the preceding provisions by saying that each Contracting State shall have the right to refuse cabotage rights or commercial air traffic rights to foreign aircraft between points within their own territory. The use of the words "shall have the right to refuse" is skillfully used to convey the meaning that a State's right to grant cabotage rights already exists.

The discretionary right of a State is explicitly recognized in Article 9, which provides that each Contracting State may, for reasons of military necessity or public safety, restrict or prohibit aircraft in certain circumstances from flying over their territory. The use of the word "may" is clear in its meaning and purpose.

Article 12 carries yet another nuance of language where each Contracting State is required to undertake to adopt certain measures. The word "undertake" implies accountability and responsibility. The difference between the use of the words "agree" and "undertake" brings to bear the clear intent of a treaty carved out many years ago with vision and foresight by its founding fathers.

The above terminology can be compared with the use of the words in Article 17, which states that "aircraft have the nationality of the State in which they are registered". It is to be noted that this provision does not have the peremptory admonition issued by the word "shall", and one could only conclude that the provision conveys that it is a fact taken for granted, that once an aircraft is registered in a particular State it shall *ipso facto* be deemed registered in that State. The following statement in Article 18, that aircraft cannot be validly registered in more than one State, conveys the impossibility of such an exigency. Here, the use of the word "cannot" instead of "shall not" leaves no room for doubt that in this instance the right for dual registration of aircraft did not exist to begin with. This usage is contrasted with the use of the words "shall not", which implies that a right that seemingly exists is taken away.

This book provides a commentary on the Chicago Convention and its various provisions against the backdrop of legal analysis. I was prompted to write this commentary as I had not seen a comparable treatise that explains the Convention, its nuances, and the manner in which the ICAO Assembly and Council have interpreted the Convention. In doing so, I address the main provisions of the Convention that impact civil aviation law. Those provisions, which are self-explanatory and have not been subject to actions of the international aviation community or of ICAO, are not mentioned in the text under separate chapters. The text of the Chicago Convention is attached for ease of reference.

1 July 2013 Ruwantissa Abeyratne

Contents

Chapter VI International Standards and Recommended Practices

Part II The International Civil Aviation Organization

Chapter VII The Organization

Chapter VIII The Assembly

Chapter IX The Council

Preamble

Preamble

WHEREAS the future development of international civil aviation can greatly help to create and preserve friendship and understanding among the nations and peoples of the world, yet its abuse can become a threat to the general security;

WHEREAS it is desirable to avoid friction and to promote that cooperation between nations and peoples upon which the peace of the world depends;

THEREFORE, the undersigned governments having agreed on certain principles and arrangements in order that international civil aviation may be developed in a safe and orderly manner and that international air transport services may be established on the basis of equality of opportunity and operated soundly and economically;

Have accordingly concluded this Convention to that end.

At a glance, one would note that the Preamble to the Convention on International Civil Aviation (Chicago Convention), which sets the tone of the Convention, resonates a message of peace and harmony among nations of the world through aviation. The Preamble, and its *raison d'etre* was invoked at the 15th Session of the Assembly (Montreal, 16 June–22 July 1965) of the International Civil Aviation Organization (ICAO)[1] on the theme of peace when the Assembly adopted

[1]The International Civil Aviation Organization is the United Nations specialized agency dealing with international civil aviation. ICAO was established by the Convention on International Civil Aviation (Chicago Convention), signed at Chicago on 7 December 1944. Fifty two States signed the Chicago Convention on 7 December 1944. The Convention came into force on 4 April 1947, on the thirtieth day after deposit with the Government of the United States. Article 43 of the Convention states that an Organization to be named the International Civil Aviation Organization is formed by the Convention. ICAO is made up of an Assembly, which is the sovereign body of the Organization composed of the entirety of ICAO member (Contracting) States, and a Council which elects its own president. The Assembly, which meets at least once every 3 years, is convened by the Council. The Council is a permanent organ responsible to the Assembly, composed of 36 Contracting States. These 36 Contracting States are selected for representation in the Council in three categories: States of chief importance to air transport; States not otherwise included which make the largest contribution to the provision of facilities for international air navigation; and States not otherwise included whose designation will insure that all the major geographic areas of the world are represented on the Council. Article 47 of the Chicago Convention provides that ICAO enjoys "such legal capacity as may be necessary for the performance of its functions" and goes on to say that "full juridical personality shall be granted to the Organization wherever compatible with the constitution of the laws of the State concerned." The Council has two main subordinate governing bodies, the *Air Navigation Commission* and the *Air Transport Committee. The Air Navigation Commission* is serviced by The Air Navigation Bureau and is responsible for the examination, coordination and planning of all of ICAO's work in the air navigation field. This includes the development and modification of SARPS) contained in the ICAO Annexes (all except Annexes 9 and 17), subject to the final adoption by the ICAO Council. At the time of writing, ICAO had 191 member States.

R. Abeyratne, *Convention on International Civil Aviation*,
DOI 10.1007/978-3-319-00068-8_1, © Springer International Publishing Switzerland 2014

Resolution A15-7 (Condemnation of the Policies of Apartheid and Racial Discrimination of South Africa). The Assembly Resolution went on to say inter alia:

> BEARING IN MIND that the apartheid policies constitute a permanent source of conflict between the nations and peoples of the world; and
>
> RECOGNIZING, furthermore, that the policies of apartheid and racial discrimination are a flagrant violation of the principles enshrined in the Preamble to the Chicago Convention;
>
> THE ASSEMBLY: . . .URGES South Africa to comply with the aims and objectives of the Chicago Convention.

The aims and objectives of the Chicago Convention which are enshrined in Article 44 of the Convention are to develop the principles and techniques of international air navigation and to foster the planning and development of international air transport so as to:

- Insure the safe and orderly growth of international civil aviation throughout the world;
- Encourage the arts of aircraft design and operation for peaceful purposes;
- Encourage the development of airways, airports, and air navigation facilities for international civil aviation;
- Meet the needs of the peoples of the world for safe, regular, efficient and economical air transport;
- Prevent economic waste caused by unreasonable competition;
- Insure that the rights of Contracting States are fully respected and that every Contracting State has a fair opportunity to operate international airlines;
- Avoid discrimination between Contracting States;
- Promote safety of flight in international air navigation;
- Promote generally the development of all aspects of international civil aeronautics.

Another Assembly Resolution followed at the 17 Session of the ICAO Assembly (Montreal, 16–13 June 1970) wherein the Assembly adopted Resolution A 17-1 (Declaration by the Assembly) which stated inter alia:

> WHEREAS international civil air transport helps to create and preserve friendship and understanding among the peoples of the world and promotes commerce between nations;
>
> THE ASSEMBLY ADOPTS THE FOLLOWING DECLARATION: The Assembly of the International Civil Aviation Organization. . .mindful of the principles enunciated in the Convention on International Civil Aviation;
>
> SOLEMNLY requests concerted action on the part of states towards suppressing all acts which jeopardize the safe and orderly development of international civil air transport.

At its next Session (Vienna, 15 June–7 July 1971) the Assembly adopted Resolution A 18-4 (Measures to be taken in pursuance of resolutions 2555 and 2704 of the United Nations General Assembly in relation to South Africa where the Assembly stated:

> THE ASSEMBLY, recalling its condemnation of the apartheid policies in South Africa in Resolution A15-7;
>
> RECOGNIZING the need for maximum co-operation with the United Nations General Assembly in implementing its Resolutions;

> RESOLVES that as long as the Government of South Africa continues to violate the United Nations General Assembly resolutions on apartheid and on the Declaration on the Granting of Independence to Colonial Countries and Peoples;
>
> South Africa will not be invited to attend any meetings convened by ICAO. . . .

Two years later, the ICAO Assembly, at its 19th (Extraordinary) Session (New York, 27 February–2 March 1973) adopted Resolution A19-1 which condemned Israeli action which resulted in the loss of 108 lives.

The same year, the ICAO Assembly, at its 20th (Extraordinary) Session (Rome, 28 August–21 September 1973) adopted Resolution A20-2 (Acts of Unlawful Interference with Civil Aviation) which stated *inter alia*:

> THE ASSEMBLY, MINDFUL that the development of international civil aviation can greatly help to create and preserve friendship and understanding among the nations and peoples of the world, yet it's abuse can become a threat to general security;
>
> CONSCIOUS of the mandate bestowed on the International Civil Aviation Organization to ensure the safe and orderly development of international civil aviation;
>
> REAFFIRMS the important role of the International Civil Aviation Organization to facilitate the resolution of questions which may arise between Contracting States in relation to matters affecting the safe and orderly operation of civil aviation throughout the world.

One gleans four key elements in the Preamble to the Chicago Convention: peace and friendship through aviation; safety; economical and orderly air transport.

The elements of peace and friendship have already been discussed as being reflective of the Assembly Resolutions discussed above. On safety, and its relevance projected by the Preamble to the Convention, the ICAO Council on 4 June 1973 adopted a Resolution which recalled the adoption by the United Nations Security Council of Resolution 262 in 1969 which condemned Israel for its premeditated action against Beirut Civil Airport which resulted in the destruction of thirteen commercial and civil aircraft. The Resolution urged Israel to comply with the aims and objectives of the Chicago Convention.

On the economic side, the Preamble to the Convention featured prominently in ICAO Assembly Resolution A21-28 (International Air Services Transit Agreement) adopted by the 21st Session of the Assembly (Montreal, 24 September–15 October 1974) which quoted the Preamble in part which recognized that one of the objectives of the Chicago Convention was that international air transport services may be operated soundly and economically. In pursuance of this objective, Resolution A21-28 urged Contracting States to become Parties to the International Air Services Transit Agreement which strengthened the operation of international scheduled services and facilitated the achievement of that objective.

In general terms, the Preamble to the Chicago Convention leaves no room for doubt that, being a post war instrument, its overall theme is on aviation and peace, which has its genesis in the Chicago Conference that led to the adoption of the Chicago Convention. The Conference which took place from 1 November to 7 December 1944 was inaugurated with the reading of a message to the Conference from the President of the United States. In his message, President Roosevelt, referring to the Paris Conference of 1919 which was designed to open Europe to air traffic, but unfortunately took years to be effectively implemented, stated:

> I do not believe that the world today can afford to wait several years for its air communications. There is no reason why it should.
>
> Increasingly, the aeroplanes will be in existence. When either the German or Japanese enemy is defeated, transport planes should be available for release from military work in numbers sufficient to make a beginning. When both enemies have been defeated, they should be available in quantity. Every country has its airports and trained pilots; practically every country knows how to organize airlines.
>
> You are fortunate to have before you one of the great lessons of history. Some centuries ago, an attempt was made to build great empires based on domination of great sea areas. The lords of these areas tried to close the areas to some, and to offer access to others, and thereby to enrich themselves and extend their power. This led directly to a number of wars both in the Eastern and Western Hemispheres. We do not need to make that mistake again. I hope you will not dally with the thought of creating great blocs of closed air, thereby tracing in the sky the conditions of future wars. I know you will see to it that the air which God gave everyone shall not become the means of domination over anyone.[2]

Thus, President Roosevelt urged States to eschew protectionism, while encouraging them to avoid dominance over one another. Ever since, the fate of economic regulation of international air transport has become an obdurate dilemma to regulators since they were faced with the question as to how States could avoid dominance by others without protecting themselves. The elusive and delicate balance between the two is still being vigorously sought, as will be seen in discussions to follow in this paper.

The Chairman of the Conference, Adolf A Berle Jr. endorsed the President's comments by observing:

> There are many tasks which our countries have to do together, but in none have they a clearer and plainer common interest than in the work of making the air serviceable to mankind. For the air was given to all; every nation in the world has access to it. To each nation there is now available a means of friendly intercourse with all the world, provided a working basis for that intercourse can be found and maintained.[3]

At the Conference, the United States took the position that the use of the air and the use of the sea were both common in that they were highways given by nature to all men. They were different in that man's use of the air is subject to the sovereignty of nations over which such use is made. The United States was therefore of the opinion that nations ought to arrange among themselves for its use in such manner as would be of the greatest benefit to all humanity, wherever situated. The United States further asserted the rule that each country has a right to maintain sovereignty of the air which is over its lands and its territorial waters. There was no question of alienating or qualifying this sovereignty. This absolute right, according to the United States, had to be qualified by the subscription by States to friendly intercourse between nations and the universal recognition of the natural rights of States

---

[2] *Proceedings of the International Civil aviation Conference, Chicago, Illinois, November 1–December 7 1944, Vol I & II (Washington, D.C.: U.S. Government Printing Office, 1948)* at 42–43.

[3] *Id.* 43.

to communicate and trade with each other. This right could not be derogated by the use of discriminatory measures.[4] The fact that the United States required States to exchange air traffic rights reciprocally is clearly evident in the statement:

> It is therefore the view of the United States, that, without prejudice to full rights of sovereignty, we should work upon the basis of exchange of needed privileges and permissions which friendly nations have a right to expect from each other.[5]

The privilege of communication by air with friendly countries, according to the United States was not a right to wander at will throughout the world. In this respect, it was contended that traffic by air differed materially from traffic by sea, where commerce need have no direct connection with the country from which the ship may have come. The air routes were analogous to railroad lines and the right to connect communication links between States was to establish a steady flow of traffic, thereby opening economic routes between countries. According to the United States, it was too early to go beyond this concept and States should accept the fact that what the Chicago Conference would accomplish was to adopt a Convention that would establish communication between States.[6]

The ICAO Assembly, at its 28th Session of the Assembly (Montreal, 22–26 October 1990) adopted Resolution A28-7 (Aeronautical consequences of the Iraqi invasion of Kuwait) which recalled that the *Convention on International Civil Aviation* is based on the belief that the future development of international civil aviation can greatly help to create and preserve friendship and understanding among the nations and peoples of the world, yet its abuse can become a threat to the general security; and that it is desirable to avoid friction and to promote that cooperation between nations and peoples upon which the peace of the world depends and noted United Nations Security Council condemnation of the invasion of Kuwait and Security Council Resolution 662 which decided that annexation of Kuwait by Iraq has no legal validity and is considered null and void and called upon all States, International Organizations and Specialized Agencies not to recognize that annexation and to refrain from any action or dealing that might be interpreted as an indirect recognition of the annexation. The Resolution also noted further, Security Council Resolution 661, which calls upon all States to take appropriate measures to protect assets of the legitimate Government of Kuwait and its agencies.

The Assembly also noted Security Council Resolution 670 which affirmed that the specialized agencies are required to take such measures as may be necessary to give effect to the terms of Resolution 661 and condemned the violation of the sovereignty of the airspace of Kuwait and the plunder of Kuwait International Airport by Iraqi armed forces including the seizure and removal to Iraq of 15 aircraft of Kuwait Airways and their purported registration by Iraq. The Assembly called upon Iraq to facilitate the early recovery by their owners of foreign registered aircraft stranded at

---

[4] *Id.* at 55.

[5] *Id.* 56.

[6] *Id.* 57.

Kuwait International Airport and declared that the unilateral registration of aircraft of Kuwait Airways by Iraqi aircraft is null and void and called upon the Iraqi government to return the Kuwaiti aircraft to the legitimate Government of Kuwait. It also requested all States in whose territory any of these aircraft are found to hand them over to the legitimate Government of Kuwait and not to supply Iraq, its companies or nationals, whether directly or indirectly, with any spare parts, equipment or supplies or services to enable Iraq to use the aircraft.

The attacks of 11 September 2001 inevitably highlighted the strategic position of civil aviation both as an industry vulnerable to attack and as an integral tool in ensuring peace and security in the world. The modernist view of civil aviation, as it prevailed when the Convention on International Civil Aviation was signed at Chicago on 7 December 1944, was centered on State sovereignty and the widely accepted post-war view that the development of international civil aviation can greatly help to create and preserve friendship and understanding among the nations and peoples of the world, yet its abuse can become a threat to general security. This essentially modernist philosophy focussed on the importance of the State as the ultimate sovereign authority which can overrule considerations of international community welfare if they clashed with the domestic interests of the State. It gave way, in the 1960s and 1970s to a post-modernist era of recognition of the individual as a global citizen whose interests at public international law were considered paramount over considerations of individual State interests.

The 11 September 2001 events led to a new era that now calls for a neo-post modernist approach which admits of social elements and corporate interests being involved with States in an overall effort at securing world peace and security. The role of civil aviation in this process is critical, since it is an integral element of commercial and social interactivity and a tool that could be used by the world community to forge closer interactivity between the people of the world. This discussion will assess the position of civil aviation in a world community embroiled in a neo post modernist approach towards securing world peace and understanding among nations.

Until 11 September 2001, the link between civil aviation and world peace was somewhat conceptual and intellectual. However, when four civilian aircraft on United States domestic services were destroyed by terrorist acts and crews, hundreds of passengers and thousands of innocent victims in buildings located in New York City and Washington DC were killed, civil aviation ceased to be isolated from the world peace efforts and became immediately inextricably linked to overall endeavours of the world community toward achieving peace and economic sustainability.

The significance of peace and security of the world involving civil aviation was signalled by United National Resolution A/RES/421(XIV) which referred to the immediate consequences of the attacks of 11 September 2001 as the closure of civil airports in the United States and disruptions of air services. The Resolution also referred to A/RES/145(V) which concerned the safety of civil aviation in relation to tourism. The new era brought about by the paralysis experienced in terms of world trade brought in both states and their instrumentality together with the private sector to join in finding solutions toward keeping the trade machine of the world functioning.

The pursuit of peace has been inseparable from policy making and dispute settlement in affairs of aviation. Varied and chronologically sequential instances where ICAO was requested by its Contracting States to address contentious issues relating to civil aviation are reflective of the importance of political considerations that underlie such disputes and the relentless search by nations of the world to settle disputes peacefully. Although political contentions may exist between States, which is a natural corollary of Statecraft and international politics, it is not the purview of an international organization to address political motivations of individual States when considering issues referred to it or adjudicating disputes between States. In this regard, ICAO has tread a delicate line between diplomacy and objectivity.

Part I
Air Navigation

Chapter I. General Principles and Application of the Convention

Article 1. Sovereignty
Article 2. Territory
Article 3. Civil and State Aircraft
Article 3 *bis*
Article 4. Misuse of Civil Aviation

Chapter II. Flight Over Territory of Contracting States

Article 5. Right of Non-scheduled Flight
Article 6. Scheduled Air Services
Article 7. Cabotage
Article 8. Pilotless Aircraft
Article 9. Prohibited Areas
Article 10. Landing at Customs Airport
Article 11. Applicability of Air Regulations
Article 12. Rules of the Air
Article 13. Entry and Clearance Regulations
Article 14. Prevention of Spread of Disease
Article 15. Airport and Similar Charges
Article 16. Search of Aircraft

Chapter III. Nationality of Aircraft

Article 17. Nationality of Aircraft
Article 18. Dual Registration
Article 19. National Laws Governing Registration
Article 20. Display of Marks
Article 21. Report of Registrations

Chapter IV. Measures to Facilitate Air Navigation

Article 22. Facilitation of Formalities
Article 23. Customs and Immigration Procedures
Article 24. Customs Duty
Article 25. Aircraft in Distress
Article 26. Investigation of Accidents
Article 27. Exemption from Seizure on Patent Claims
Article 28. Air Navigation Facilities and Standard Systems

Chapter V. Conditions to be Fulfilled with Respect to Aircraft

Article 29. Documents Carried in Aircraft
Article 30. Aircraft Radio Equipment
Article 31. Certificates of Airworthiness
Article 32. Licenses of Personnel
Article 33. Recognition of Certificates and Licenses
Article 34. Journey Log Books
Article 35. Cargo Restrictions
Article 36. Photographic Apparatus

Chapter VI. International Standards and Recommended Practices

Article 37. Adoption of International Standards and Procedures
Article 38. Departures from International Standards and Procedures
Article 39. Endorsement of Certificates and Licenses
Article 40. Validity of Endorsed Certificates and Licenses
Article 41. Recognition of Existing Standards and Airworthiness
Article 42. Recognition of Existing Standards of Competency of Personnel

Article 1
Sovereignty

The contracting States recognize that every State has complete and exclusive sovereignty over the air space above its territory.

Contents

1 Conceptual Aspects

1.1 State

With regard to what a State constitutes at law, The Montevideo Convention of 1933 in its Article 1 provides that a State as a legal person of international law should possess: (a) a permanent population; (b) a defined territory; (c) government; and (d) capacity to enter into relations with the other states.[1] Jurists have argued that this list is not exhaustive and that the four elements themselves have to be further elaborated. Accordingly, it has been said that "a permanent population" should essentially connote a stable community, the absence of which in a given territory would effectively preclude that territory from being designated a State at law. With regard to "defined territory" the acceptable notion is that it must be politically controlled by the stable community mentioned above. By "government" is usually meant a defined legal order that has the three separately identifiable factors of the legislature, judiciary and the executive.

The tripartite legal doctrine of separation of powers, which Baron de Montesquieu propounded through his theory on the division of political power among a legislature, an executive and a judiciary, advocates that the three branches of government (legislative, executive, judicial) exist largely independent of each other, with their own prerogatives, domains of activity, and exercises of control over each other.

[1]Montevideo Convention on the Rights and Duties of States, signed at Montevideo, 26 December 1933. The Convention entered into force on 26 December 1934. See http://www.taiwandocuments.org/montevideo01.htm.

R. Abeyratne, *Convention on International Civil Aviation*,
DOI 10.1007/978-3-319-00068-8_2, © Springer International Publishing Switzerland 2014

According to this philosophy, the legislative body has control of the executive finances, and has judiciary powers.[2] It also has control of the way the judiciary works. The judiciary often has control of laws not being contradictory to the constitution or other laws and it has the power to correct and control the way the executive body exercises its powers (to execute the law). The executive is the arm of government that has sole authority, power and responsibility for the daily administration of the State, and for executing the law of the land.

This separation is essential for ensuring the legal maxim *Omnia praesumuntur rite et solemniter esse acta* (all acts are presumed to have been done rightly and regularly). It would also ensure good governance. Overall public interest in good governance is now a common feature in the modern state, and is not restricted to the academics and practitioners who bore the burden of evaluating governance in the past. The increasing concern and interest in good governance may be attributed to the public being more educated and aware than before, which is now popularly known as "civic literacy", coupled with the proliferation of complex issues that have emerged with globalization and an international awareness that has spread to national boundaries. Therefore, an empirical demonstration of good governance has now become a compelling need that could provide the necessary tools for the public to develop their own desired models of governance which are capable of delivering goods that accord with their expectations.

Essentially, governance, which is critical to the proper running of a State, is a set of responsibilities and practices that are aimed at achieving strategic direction and ensuring that objectives are achieved. Indicators of good governance are: involvement of citizens; accountability of actions of the governing body; transparency; equality in social inclusion (gender, ethnicity, age, religion etc.); ethical conduct; integrity; ability to compete in a global environment; ability to work as partners with other governments or bodies; fair procedures and due process; and respect for the rule of law. A State's adherence to the rule of law is extremely important as a determinant of good governance. It carries the principle that law (as administered by the ordinary courts) is supreme and that all citizens (including members of the government) are equally subject to it and equally entitled to its protection.

As regards "independence" as the fourth feature of a State, it is tied up in the Montevideo Convention to the ability and capacity to enter into relations with other States, and is widely recognized as the decisive factor in the determination of Statehood. Preeminent in this issue is the essential requirement for a State to have

[2]Bijo Francis, a human rights lawyer with the Asian Human Rights CommissionIt states: "It is interesting to note that in India, whenever in India, the state legislatures or the central parliament has tried using parliamentary privilege for unjustifiable reasons against the judiciary, the judiciary has corrected the legislative houses. Additionally, the basic structure doctrine, postulated in Kesavananda Bharati (petitioner) against State of Kerala and others (respondents) [All India Reporter 1973 Supreme Court p. 1461], triumphs the clarion call of the power of judicial review and the limits drawn upon the parliament, even in its legislative authority." See Bijo Francis, Does the Impeachment of the Chief Justice of Sri Lanka Matter, *The Sri Lanka Guardian*, 22 November 2012.

a certain centrality in its functions and protection from interference from other States or entities.

In modern parlance, two other requirements for Statehood have been identified: willingness to observe international law; and a certain degree of civilization.

1.2 Sovereignty

1.2.1 Theoretical Aspects

Recognition by States of their sovereignty over their airspace inevitably presupposes that this rule has already been entrenched in the annals of air law in an earlier instrument. The Convention Relating to the Regulation of Aerial Navigation signed by 26 States on 13 October 1919 established that *the High Contracting Parties recognize that every Power has complete and exclusive sovereignty over the airspace above its territory.*[3]

Even the 1919 Paris Convention merely recognized *sovereignty* of every State over the airspace above its territory. This means that sovereignty over airspace was already an established right at international law.

The concept goes back in time: to Roman times in fact:

> States claimed, held, and in fact exercised sovereignty in the air space above their national territories. . .and that the recognition of an existing territorial airspace of States by the Paris Convention of 1919 was well founded in law and history.[4]

The Roman State adopted an all-embracing approach in ensuring the protection of private and public rights of its citizens. It could not have assumed full jurisdiction to lay down rules for its citizenry unless it exercised rights of sovereignty in airspace as well as on land. The genesis of the concept of sovereignty in airspace is traced to Emperor Justinian's *Corpus Juris Civilis*, where the concept seemed to be an inference from a passage in the Digest.[5] Accordingly, airspace at that time became, at international law a contentious issue when it came to justifying the removal of projections from an adjoining property over a place of burial. Bouve added the view that airspace was new space added to accommodate man's ability to fly.[6] Thus, the right bestowed by the private law maxim *Cujus est solum, ejus est usque ad coelom* was formally entrenched as a absolute right of a person under ancient Roman law. This maxim, which means that a right of land ownership brings with it rights of ownership of airspace above the land, was later found to be unacceptable as an absolute rule. Disparaged by some commentators as the "product of some black letter lawyer",[7] the rule was later adapted to mean that no nation

[3]For an extended discussion on this issue see Milde (2012) at 8–10

[4]Cooper (1968) at 55.

[5]42 tit.24, pr 22 S 4.

[6]Bouve (1930).

[7]Baldwin (1910).

acquired any domain in what was known as navigable airspace until such domain was needed to protect subjacent territory.[8]

The doctrine of sovereignty was introduced to the Western world by the French philosopher Bodin. At a time when political attitudes were in transition from the dominance of the universal church to a universal legal order, Bodin introduced sovereignty as a supreme power over citizens and subjects that was not itself bound by laws. Bodin elaborated that every independent community had to consider that while acknowledging the authority of the law, a State was above the law if it wished to govern successfully. Other jurists who supported the theory of exclusive sovereignty were Hugo Grotius, who maintained that sovereign States were independent of foreign control, and Thomas Hobbes, who said that sovereignty was absolute and its misuse was unthinkable. John Locke attempted to compromise the absolute quality of sovereignty by opining that sovereignty was not absolute and unquestionable in that it was an exchange of social trust between the government and the people. Accordingly, there was an inarticulate premise that a breach of the social trust between the two parties would erode the concept of sovereignty.

The important question is how is sovereignty determined? Both juristic and judicial opinion favor the view that sovereignty of airspace should be determined on the role played by the importance of subjacent airspace in its relation to land and sea. In other words, a symbolic possession of the airspace is necessary in order that States can claim sovereignty over their airspace. Therefore, the concept of sovereignty becomes compatible with the concept of ownership of property with possession by the owner, to the exclusion of others. To determine sovereignty in airspace, three elements would have to be resolved: the use of airspace; the nature of its possession; and the nature of its control to the exclusion of others.

The use of airspace is inextricably linked to the social needs that the airspace in question would subserve. Roscoe Pound envisaged that one of the fundamental bases for the control and use of property was its sociological importance. There is no difficulty in establishing a nexus between the sociological value of territorial land and sea and the protection offered to them by the subjacent air space of a country. Weber and Erlich both contended that the law is not a formal set of rules but a prime method of establishing order in society and accordingly required a person merely to show incontrovertible reason for the need to possess property. The final element—the nature of control of airspace—can be subsumed in modern juristic thought; that the modern interpretation of the concept of sovereignty is not the ability to make war or to exploit others, but to legislate over a given State or community.

Perhaps the most convincing justification for the acceptance of sovereignty in airspace as the fundamental legal norm in air law is seen in Hans Kelsen's pure theory of the law. Kelsen considered that all international laws derived their basis from a *grundnorm* or a basic legal postulate derived purely from law and not from morality. This basic norm was international custom. In this context, the philosophy

---

[8]Cooper (1968) at 29.

of air law is founded on the concept of sovereignty in airspace and would sustain its credibility through this customary concept. The basic idea of sovereignty is then taken to its final conclusion and ultimate justification by Pound when he states that

> Men must be able to assume in civilized society that they may control, for purposes beneficial to themselves, what they have created by their own labor and what they have acquired under existing social and economic order. This is a final postulate of civilized society

By the end of the nineteenth century, the private law concept of absolute ownership of airspace over land was antiquated. The beginning of the twentieth century saw the emergence of States' sovereignty in airspace. The impetus for public international law to take over the issue of rights over airspace was given by the August 1904 aerial incident where Russian guards shot down the German balloon *Tschudi* when it was flying outside Russian territory and two unrelated but similar incidents that occurred in 1908 and 1910 respectively. The French Government hastened to call a conference of European powers in 1910. For the first time, participating States at this conference recognized airspace as belonging to individual States.

Sovereignty in international law is the right to exercise the functions of a State to the exclusion of all other States in regard to a certain area of the world. In international aviation the concept of sovereignty is the fundamental postulate upon which other norms and virtually all air law is based. Post World War II attitudes towards the concept of sovereignty in airspace and the philosophy of air law range between the unlimited public law right of a State to exercise sovereignty over its airspace and the idea of free movement of air traffic. Professor O.J. Lissitzyn analyses the concept of sovereignty in its modern development as having three basic principles: that each State has exclusive sovereignty over its airspace; each State has complete discretion as to the admission of any aircraft into its airspace; and, that airspace over the high seas and other areas not subject to a State's jurisdiction is *res nullius* and is free to the aircraft of all States.

Sovereignty, in its pristine sense, involved independence of a State in regard to a portion of the globe and the right to exercise therein, to the exclusion of any other State, the function of a State. However in the 1960s and 1970s, a shift in focus of international law, later fuelled by the end of the Cold War, impelled legal scholars to view the concept of sovereignty as veering from the normative perspective of exclusivity to an approach accommodating globalization and democratization. Although political theory and social justice may have caused, through such events as the disintegration of the Warsaw Pact; the reunification of Germany; and the emergence of the Commonwealth of Independent States, a paradigm shift from exclusive sovereignty to extended sovereignty, particularly with regard to the exercise of some control over the high seas pertaining to air navigation, it remains to be seen whether a clearly identifiable and distinct legal regime exists in this field.

Sovereignty has two attributes:

- Internal sovereignty, whereby a State exercises its exclusive right and competence to determine the character of its own institutions and to provide for their function. Internal sovereignty also includes the exclusive power of a State to enact its own internal laws and to ensure their respect; and

- External sovereignty, whereby a State freely determines its relations with other States or entities without the restraint or control of another State.

Justice Huber noted in the 1928 *Island of Palmas* case:

> Sovereignty in the relations between States signifies independence. Independence in relation to a portion of the globe is the right to exercise therein, to the exclusion of any other State, the function of a State. The development of the national organization of States during the last few centuries and, as a corollary, the development of international law, have established this principle of exclusive competence of the State in regard to its own territory in such a way as to make it the point of departure in settling most questions that concern international relations. Sovereignty in relation to a portion of the surface of the globe is the legal condition necessary for the inclusion of such portion in the territory of any particular State.[9]

Professor Bin Cheng addresses the principles governing post World War II sovereignty over airspace as enunciated in Article 1 of the Chicago Convention and concludes:

> The now firmly established rule of international law that each State possesses complete and exclusive sovereignty over the airspace above its territory means that international civil aviation today rests on the tacit acquiescence or express agreement of States flown over.[10]

Ian Brownlie, Professor of International Law at Oxford University cites the principle corollaries of the sovereignty and equality of States as:

> a jurisdiction, prima facie exclusive, over a territory and the permanent population living there; a duty of non-intervention in the area of exclusive jurisdiction of other States; and the dependence of obligations arising from customary law and treaties on the consent of the obligor.[11]

A more modern view is that which is taken by Brownlie (cited above) who, in his book *Principles of Public International Law* states that the term sovereignty is synonymous with independence. Article 2.4 of the United Nations Charter exhorts all members of the United Nations to refrain in their international relations from the threat or use of force against the territorial integrity or political independence of any State, or in any other manner inconsistent with the purposes of the United Nations. In keeping with this fundamental premise, the 1965 *Declaration on the Inadmissibility of Intervention in the Domestic Affairs of States* emphasized that no state has the right to intervene, directly or indirectly, for any reason whatsoever, in the internal or external affairs of any other State. The Declaration went on to say that consequently, armed intervention and all other forms of interference or attempted threats against the personality of the State or its political, economic and cultural

---

[9]2RIAA, at pp. 829, at 838 (1928).

[10]Cheng (1962) at 3. See also pages 3–17 for a discussion of the manner in which air rights devolve upon carriers under the sovereignty doctrine of the Chicago Convention.

[11]Brownlie (1990) at 287.

elements, are condemned. This principle was reaffirmed in the 1970 Declaration on Principles of International Law contained in United Nations General Assembly Resolution 2625.

Starke is inclined to stretch the principle of sovereignty to accommodate external involvement by a State in the affairs of another in special circumstances:

> ... "Sovereignty" has a much more restricted meaning today than in the eighteenth and nineteenth centuries when, with the emergence of powerful highly nationalized States, few limits on State autonomy were acknowledged. At the present time there is hardly a State which, in the interests of the international community, has not accepted restrictions on its liberty of action. Thus most States are members of the United Nations and the International Labour Organization (ILO), in relation to which they have undertaken obligations limiting their unfettered discretion in matters of international policy. Therefore, it is probably more accurate today to say that the sovereignty of a State means the residuum of power which it possesses within the confines laid down by international law. It is of interest to note that this conception resembles the doctrine of early writers on international, law, who treated the State as subordinate to the law of nations, then identified as part of the wider "law of nature".[12]

United Nations Secretary General Kofi Annan in defining sovereignty said:

> State sovereignty is being redefined by the forces of globalization and international cooperation. The state is now widely understood to be the servant of its people, not vice versa. At the same time, individual sovereignty –the human rights and fundamental freedoms of each and every individual as enshrined in our Charter– has been enhanced by a renewed consciousness of the right of every individual to control his or her own destiny.[13]

Sovereignty technically precludes intervention by one State in the affairs of another. Starke is inclined to stretch the principle of sovereignty to accommodate external involvement by a State in the affairs of another in special circumstances:

> The principle of non-intervention is part of international law and is based on the recognition of the territorial sovereignty and integrity of States. Intervention is not permitted at international law if such adversely affects the free choice of States made by virtue of State sovereignty. Intervention becomes unacceptable when it restricts free choice of a State.[14]

The above notwithstanding, it is incontrovertible that sovereignty is no longer an absolute concept that would shield States against any internal acts of aggression or irrationality against its citizens. Sovereignty can therefore no longer be accepted in the international for a as seen an absolute protection against interference. It is no longer an absolute right but a charge of responsibility on a State where it e is accountable to both domestic and external constituencies. A Brookings Institute study has recently revealed that in internal conflicts in Africa, sovereign states have often failed to take responsibility for their own citizens' welfare and for the humanitarian consequences of conflict, leaving the victims with no assistance.

[12]Starke (1977) at 106.

[13]http://users.lmi.net/wfanca/pp_annan_on_sov.html.

[14]Starke, Supra Note 12. Ibid.

Therefore, what is needed is a delicate balance between respect for State sovereignty and protection of the citizenry against arbitrary and capricious acts of States.

1.2.2 Practical Aspects

From an aviation perspective, the first official instance of sovereignty of airspace under Article 1 of the Chicago Convention being recognized at ICAO was at its 21st Assembly (held in Montreal from 24 September to 15 October 1974) where the Assembly adopted Resolution A21-7 (the Airport of Jerusalem) where the Assembly recognized that Jerusalem airport lay in the Arab occupied territories and was registered under the jurisdiction of Jordan in ICAO's Middle East Air Navigation Plan. The Assembly, in the context of Article 1 of the Chicago Convention, resolved that all Contracting States to the Convention should take all necessary measures to refrain from operating, or giving permission to any airline to operate any air service, whether scheduled or non-scheduled, to or from Jerusalem airport, unless prior permission is granted pursuant to the relevant article[15] of the Chicago Convention.

At the same session, the Assembly adopted Resolution A21-21 (Consolidated Statement of Continuing Policies and Associated Practices Related Specifically to Air Navigation) Appendix N of which declared that any Contracting State which delegated to another State the Responsibility for providing air traffic services over its territory to another State by mutual agreement did so without delegation of its sovereignty.

At the same session the Assembly (Montreal, 24 September–15 October 1974) adopted Resolution A21-21 (consolidated statement of continuing policies and associated practices related specifically to air navigation) Appendix N of which resolved inter alia that any Contracting State which delegates to another State the responsibility for providing air traffic services within airspace over its territory does so without derogation of its sovereignty.

Encroachment on the sovereignty over the territory of one State by another has been subject to consideration by the ICAO Council on several occasions.

On 24 February 1996, the United States registered private (general aviation) civil aircraft were shot down by Cuban military aircraft, which resulted in the loss of four lives. Consequent upon information received from the United States authorities of the incident, the President of the ICAO Council, on 26 February 1996,

[15]According to Article 5 each contracting State agrees that all aircraft of the other contracting States, being aircraft not engaged in scheduled international air services shall have the right, subject to the observance of the terms of this Convention, to make flights into or in transit non-stop across its territory and to make stops for non-traffic purposes without the necessity of obtaining prior permission, and subject to the right of the State flown over to require landing. Each contracting State nevertheless reserves the right, for reasons of safety of flight, to require aircraft desiring to proceed over regions which are inaccessible or without adequate air navigation facilities to follow prescribed routes, or to obtain special permission for such flights. Article 6 states that no scheduled international air service may be operated over or into the territory of a contracting State, except with the special permission or other authorization of that State, and in accordance with the terms of such permission or authorization.

wrote to the Government of Cuba expressing his deep concern and requesting authentic and authoritative information pertaining to the incidents.[16] Further developments ensued on 27 February 1996 when the United States formally requested that the Council of ICAO consider the matter under Article 54(n) of the Chicago Convention, and, on the same day, the United States Security Council issued a statement through its President deploring the shooting down, by Cuban military aircraft, of the two United States registered aircraft. The Security Council also alluded to Article 3 *bis* of the Chicago Convention and the Montreal Protocol of 1984 which provide that States must refrain from the use of weapons against civil aircraft in flight and must not endanger the lives of persons on board and the safety of aircraft. The Security Council requested the ICAO Council to look into the matter and report to it as soon as possible.[17] For its part, Cuba, in its communications to the President of the Council, chronicled a series of chronological violations by United States registered aircraft. This was followed by a further communication on 28 February 1996 from the Cuban Ministry of Foreign Affairs addressed to the Secretary General of ICAO alluding to a series of violations, which had allegedly increased in number over a 20 month period, of Cuban airspace by civil aircraft registered and based in the United States. The Government of Cuba urged ICAO to carry out an extensive investigation into the violations, repeated over the years, of Cuban airspace by aircraft coming from the United States, including the incidents of 24 February 1996.

The communications received by ICAO with regard to the incidents of 24 February 1996 clearly required the Organization, under Article 54(n) of the Chicago Convention, to investigate two issues:

- The incidents of 24 February 1996, an investigation into which was requested both by the United States and Cuba; and
- Repeated violations of Cuban airspace by aircraft registered and based in and coming from the United States, alleged by Cuba which requested an investigation.

When the abovementioned issues were addressed by the ICAO Council on 6 March 1996, the position taken by the United States was primarily based on Article 3 *bis* of the Chicago Convention, whereby the US claimed that there was a duty incumbent upon every State to refrain from resorting to the use of weapons against civil aircraft in flight. Accordingly, the United States claimed that the Cuban action was a blatant violation of international law and that firing on unarmed, known civil aircraft could never be justified. The United States claimed

[16]Memorandum PRES AK/97 dated 26 February 1996 from President of the Council to Representatives on the Council, Attachment.

[17]S/PRST/1996/9, 27 February 1996 at 35 I.L.M. 493 (1996).

that, consequently, as required at international law, the Cuban Government should pay appropriate compensation to the families of those whose lives were lost.[18]

In response, the Cuban Delegation claimed that Cuba had been a victim of violations of its sovereignty and territorial integrity for many years which involved aircraft coming from the territory of the United States and that, over the past 20 months, as many as 25 such incursions and violations had been detected by Cuba. Cuba also counterclaimed that, in response to the reference by the United States of Article 3 *bis*, there was a stipulation in the Article obliging every civil aircraft to comply with orders of the subjacent State making the State of origin of the aircraft obligated to ensure compliance with such orders. Another argument adduced by Cuba was that paragraph (d) of Article 3 *bis*, that each Contracting State was required to take appropriate measures to prohibit the deliberate use of any civil aircraft registered in that State, *inter alia*, for any purpose inconsistent with the Chicago Convention, was applicable to the instances concerned.

The overall trend in the Council, when the US–Cuba dispute was taken up, was indicative of a consensus that action taken by Cuba was deplorable[19] and, in the words of the United Kingdom which seemingly echoed the general view: "the principle is simple. Weapons must not be used against civil aircraft in international and civil aviation".[20] On the issue of violation of airspace, which was brought up by Cuba, many States voiced the view that there was indeed an obligation on the part of all States to refrain from violating the sovereignty of States, while some States focused their attention on Article 4 of the Convention which requires that civil aviation must not be used for any purpose inconsistent with the aims of the Convention.

The United Nations Security Council adopted Resolution 1067, adopted on 26 July 1996, after noting various statements and resolutions by the President of the Security Council and International Civil Aviation Organization (ICAO) deploring the shooting down of two civilian aircraft by the Cuban Air Force on 24 February 1996, the Council called on Cuba to comply with international obligations relating to aviation, particularly the Convention on International Civil Aviation.

The Security Council recalled the sovereignty that all countries had in the airspace above their territory and territorial waters. In this regard, all countries had to abide by principles, rules and standards in the Convention on International Civil Aviation (Chicago Convention), including rules relating to the interception and non-use of weapons against civil aircraft.

The Resolution noted that the shooting down of the two planes, which were part of the Brothers to the Rescue organisation run by Cuban exiles, was a violation of the principle that no weapons were to be used against civil aircraft in flight and that, when intercepting such aircraft, the lives of those on board not should be

[18]ICAO Doc 9676-C/1118, C-MIN 147/1-16: Council—147th Session, Summary Minutes with Subject Index at 68–71.

[19]*Id.* at 79–92.

[20]*Id.* at 88.

jeopardised. Cuba had argued that the flights were provocative acts in its airspace. Condolences were expressed to the families of the four persons who died as a result of the interception, which was condemned by the Council. All the parties were called to respect international civil aviation laws and procedure, while at the same time reaffirming the right of states to use appropriate measures against aircraft being used for purposes contrary to that of the Chicago Convention.

Due to its inherent complexities, this was clearly one issue which demanded that ICAO's diplomatic fabric be tested to its limits. The wisdom and diplomacy of the President of the Council proved invaluable when he advised the Council of the three alternatives available to Council in its pronouncement: resolution; decision; or conclusion. The President further advised the Council that whether the Council pronounced by resolution, by decision or by conclusion, any one of these would be binding in terms of implementation. Consequently, the President of the Council presented a revised version of the draft Resolutions presented by both the United States and by Cuba, for consideration of the Council. The draft Resolution suggested by the President, while recognizing that the use of weapons against civil aircraft in flight is incompatible with elementary considerations of humanity and the norms governing international behaviour, reaffirmed that States must refrain from the use of weapons against civil aircraft in flight and that, when intercepting aircraft, the lives of persons on board and the safety of the aircraft must not be endangered. For action, the draft Resolution required that the Secretary General initiate an investigation into the shooting down of the aircraft immediately, in particular with reference to the request of the United Nations Security Council Resolution, and that the Report of such investigation should be made available to the Council within 60 days in order to be transmitted to the United Nations Security Council.[21]

As to the relevance of including a reference to Article 3 *bis* in the Resolution, the President of the Council advised that Article 3 *bis* merely recognized a principle of customary international law and there was an addition to the principles embodied in the Convention. As such, it was the President's view that there was no need for the resolution to reaffirm an Article which in effect was an affirmation of the humanitarian principles already incorporated in the text.[22] It is noted that, by effectively precluding the express mention of a principle of customary international law as incorporated into the Chicago Convention, the Council played its ultimate role in diplomacy and political rectitude, by staying within the parameters of its own jurisdiction and avoiding incursions into judgment prior to facts being properly ascertained.

The final Resolution of the ICAO Council, adopted on 27 June 1996 following the Report of the Secretary General, embodies two critical principles. These were that the Council recalled and recognized the principle that every State has complete and exclusive sovereignty over the airspace above its territory and that the territory

---

[21] *Id.* at 102–103.

[22] *Ibid.* at 103.

of a State shall be deemed to be the land areas and territorial waters adjacent thereto; and that States must refrain from the use of weapons against civil aircraft in flight and that, when intercepting civil aircraft, the lives of persons on board and the safety of the aircraft must not be endangered. Integral to the Resolution was also the principle that each Contracting State should ensure that appropriate measures are taken to prohibit the deliberate use of any civil aircraft registered in that State or operated by an operator who has his principal place of business or permanent residence in that State for any purpose inconsistent with the Chicago Convention. The Council's condemnation of the use of weapons against civil aircraft involved the explicit mention of Article 3 *bis* at this advanced stage of the resolution making process, which, when examined from a diplomatic perspective, is seemingly appropriate and purposeful.

The Council Resolution was an example of the comprehensive manner in which the Council addresses issues referred to it under Article 54(n). Additionally, the Resolution masterfully indicates the views of the Council by recognizing that, while on the one hand it should be recognized that all states have complete and exclusive sovereignty over the air space above their territories and that such sovereignty should not be encroached upon, on the other hand States do not have the right to use weapons against aircraft endangering the lives of those on board, no matter what the circumstances.

In the consideration of ICAO's role as a specialized agency of the United Nations which is from time to time called upon to address contentious issues at the request of its Contracting States, it is inevitable that some determination must be made on whether ICAO should refrain from transgressing the parameters of international politics within its diplomatic efforts. The US–Cuba issue was clearly one where the ICAO Council traversed the diplomatic rope with a balanced sense of purpose and dedication to its role. The duality of sovereignty and protection of its territory by a State balanced well with the somewhat peremptory admonition that whatever the rights of a State may be, the use of weaponry could not be condoned under any circumstance.

The extent to which ICAO will be exposed politically in issues addressed by the Council is perhaps best illustrated by the consideration of the Council, in 1988 of the Iran Air incident. This concerned the shooting down of an Iran Air Airbus A300 (IR655) carrying commercial passengers on a scheduled flight from Bandar-Abbas (Iran) to Dubai. The aircraft was brought down by the *U.S.S. Vincennes* over the Persian Gulf, resulting in the death of all 290 persons on board the aircraft. The incident, which occurred on 3 July 1988, was considered by the Council at several of its meetings, notably on 7 December 1988 when the Council adopted its decision. The Council decision, while recalling the event of 3 July 1988, acknowledged the fact finding investigation report of the Secretary General of ICAO, and urged all States to take all necessary action for the safety of navigation of civil aircraft, particularly by assuring effective coordination of civil and military activities. The Resolution went on to refer to the fundamental principle of general international law that States must refrain from resorting to the use of weapons against civil aircraft and urge States to ratify Article 3 *bis* as soon as possible if they had already not done so.

One of the emergent facts about the ICAO Council which became clear was the Council's resolve to address its deliberations to purely technical issues pertaining to the incident, while stringently avoiding political issues and pitfalls. This is certainly true of all incidents discussed above, where the Council restricted its scope to technical issues as applicable to the principles embodied in the Chicago Convention.

Take concerted action towards suppressing all acts which jeopardized the sage and orderly development of international air transport. In this context, the most forceful example of ICAO's role can be seen in Resolution A20-2—*Acts of Unlawful Interference with Civil Aviation*, adopted in March 1973 by the Assembly, which reaffirmed ICAO's role as facilitating the resolution of questions which may arise between Contracting States in matters affecting the safe and orderly operation of civil aviation throughout the world.[23]

On 1 September 1983, the President of the Council of ICAO received a communiqué from the Minister of Foreign Affairs of the Republic of Korea that Flight KE 007 which was being carried out by a Korean Airlines Boeing 747 passenger airliner had disappeared off the radar screens after it took off from Anchorage, Alaska on 31 August 1983 bound for Seoul. The Minister requested ICAO's assistance with regard to ensuring the safety of the passengers, crew and aircraft.[24] The diplomatic response of the President was instantaneous and immediate, containing a message to the Minister of Civil Aviation of the USSR. It stated that information had been received by ICAO that an aircraft may have possibly landed in Soviet territory and that ICAO was confident that the Soviet authorities were rendering all assistance to persons and property concerned.[25]

As an initial response to the incident, the ICAO Council met in extraordinary session on 15 and 16 September 1983 at the request of the Government of the Republic of Korea and the Government of Canada, and adopted a resolution which averred to the fact that a Korean Air Lines civil aircraft was destroyed on September 1, 1983 by Soviet military aircraft. The Council, by Resolution, expressed its deepest sympathy to the families bereaved in this tragic incident; and reaffirmed the principle that States, when intercepting civil aircraft, should not use weapons against them. *Inter alia*, the Resolution also deplored the destruction of an aircraft in commercial international service resulting in the loss of 269 innocent lives and recognized that such use of armed force against international civil aviation is incompatible with the norms governing international behaviour and elementary considerations of humanity and with the rules, Standards and Recommended Practices enshrined in the Chicago Convention and its Annexes. The Council

---

[23]For the resolutions quoted above, see Doc 8900/2 *Repertory—Guide to the Convention on International Civil Aviation*, Second Edition, 1977 at pp. 1–3.

[24]Memorandum dated 2 September 1983 from President of the Council to the Representatives on the Council, Attachment 1.

[25]*Id.* Attachment 2.

directed the Secretary General to institute an investigation to determine the facts and technical aspects relating to the flight and destruction of the aircraft and to provide an interim report to the Council within 30 days of the adoption of this Resolution and a complete report during the 110th Session of the Council. All parties were requested to cooperate fully in the investigation.

The issue was further discussed under the auspices of ICAO at the 24th (Extraordinary) Session of the ICAO Assembly which met at Montreal from 20 September to 7 October 1983 with the participation of 131 Contracting States. In the general discussion, much attention focused on the tragedy of the Korean Airlines flight 007 and on the resolutions of the Extraordinary Session of the Council. The Assembly adopted Resolution A24-5 which, while endorsing Council action taken so far, urged all Member States to cooperate fully in their implementation.

During the Assembly, the Delegation of Canada presented a proposal for a new Convention on the Interception of Civil Aircraft[26] and the Assembly referred the proposal to the Council of ICAO for further study on the understanding that the Council was empowered to consider the inclusion of this item into the General Work Programme of the Legal Committee.

At its 138th Session, the Council examined the interim report of the ICAO investigative team into the KAL 007 incident and progress made in collecting facts regarding the shooting down of the aircraft. The Council noted the excellent cooperation provided to the ICAO investigative team by the Contracting States concerned and noted that a final report on the ICAO investigation would be placed by the Secretary General before the Council at its 139th Session.

The completed report of the Secretary General was presented to the Council during its 139th Session[27] and the Council closed the matter of KAL 007 on 14 June 1993. From a diplomatic perspective, and irrespective of the findings of the Report—which are not relevant to this work—it must be noted that the outcome of the Report and discussions that ensued in the Council endorsed the usefulness of the Council. As reflected in the Statement issued in Council by the Republic of Korea:

> The Council must once again make it clear to the world that, while reaffirming the principle of prohibition of the use of arms against civil aircraft, it unreservedly condemns the destruction of a civilian aircraft simply because it strayed into the airspace of another country.[28]

The role of the ICAO Council was aptly brought to bear by the United Kingdom during the Council's deliberations on KAL 007, which was supported by several other States, that the Council should not seek to endorse the conclusions and recommendations in the Report since it was not a tribunal seeking to reach a

[26] A24-WP/85.

[27] See C-WP/9781 Appendix for the Secretary General's Report.

[28] ICAO Doc 9615-C/1110, C-MIN 139/1-17: Council—139th Session, Summary Minutes with Subject Index at 69.

judgment on the facts.[29] The significance of the Council's role as a diplomatic tool in international civil aviation is borne out by the Summary of the President of the Council which formed the substance of the Council Resolution which followed and which, *inter alia*, expressed appreciation for the full cooperation extended to the fact finding mission by the authorities of all the States concerned. The President appealed to all Contracting States to ratify Article 3 *bis* to the Chicago Convention which approved the fundamental principle of general international law that States must refrain from resorting to the use of weapons against civil aircraft.

The KAL 007 investigation and the ICAO approach to the issue of dispute resolution was clearly a reiteration of the position taken by the Council in its earlier determination of the Libya–Israel dispute in 1973. The incident concerned the shooting down of a Libyan Airlines Boeing 727 aircraft by Israeli fighter aircraft on 21 February 1973 over Israeli occupied Sinai territory. One hundred and ten persons were killed in the incident and the Boeing 727 aircraft involved was completely destroyed. As an immediate response, the ICAO Council convened the 19th Session (Extraordinary) of the Assembly, at which speakers generally condemned the act of destruction. An investigation was called for and the Assembly proceeded to adopt Resolution A19-1 which stated that the Assembly, having considered the item concerning the Libyan civil aircraft shot down on 21 February 1973 by Israeli fighters over the occupied Egyptian territory of Sinai, condemned the Israeli action which resulted in the loss of innocent lives. Convinced that such an action adversely affects and jeopardizes the safety of international civil aviation and therefore, emphasizing the urgency of undertaking an immediate investigation, the Assembly directed the Council to instruct the Secretary General to institute an investigation in order to undertake fact findings and report to the Council. The Assembly also called upon all parties involved to cooperate fully in the investigation.[30]

Consequently, the Secretary General of ICAO presented his report,[31] which was in effect a report of the Secretariat investigative team containing, *inter alia*, a draft resolution[32] developed by numerous ICAO Contracting States. Pursuant to sustained discussion in Council, the Representatives on Council agreed upon a Resolution which was adopted by the Council. The Resolution, while recalling United Nations Security Council Resolution 262 of 1969, which condemned Israel for premeditated action against Beirut International Airport resulting in the destruction of 13 commercial and civil aircraft, expressed its deep conviction and belief that such acts constitute a serious danger against the safety of international civil aviation, and recognized that such an attitude is a flagrant violation of the principles enshrined in the Chicago Convention.

---

[29]*Id.* at 72.

[30]ICAO Doc 9061, Chapter II, note 60.

[31]C-WP/5764, Attachment.

[32]C-WP/5792 at p. 33.

The above statement of the ICAO Council truly typifies the quintessentially diplomatic approach taken by ICAO on contentious issues between ICAO Contracting States. If one analyses the first part of the Council Resolution as given above, it is difficult not to note that the Council has skilfully restated an already adopted resolution of the United Nations, ensuring that, while avoiding being judgmental, it nonetheless conveys to the international aviation community its position on the issue at hand.

In the second part of the Resolution, the Council proved to be even more dexterous, in courageously taking a stand by strongly condemning the Israeli action which resulted in the destruction of the Libyan civil aircraft and the loss of 108 innocent lives, and urging Israel to comply with the aims and objectives of the Chicago Convention. The mastery of the Council, in encompassing into a single resolution compelling precedent established by a United Nations resolution together with its own resolute position, is diplomacy at its most astute. The dexterity of the Council in this instance must not be mistaken for tendentiousness nor deviousness as the Council Resolution is clearly forthright and direct.

The extent to which ICAO will be exposed politically in issues addressed by the Council is perhaps best illustrated by the consideration of the Council, in 1988 of the Iran Air incident. This concerned the shooting down of an Iran Air Airbus A300 (IR655) carrying commercial passengers on a scheduled flight from Bandar-Abbas (Iran) to Dubai. The aircraft was brought down by the *U.S.S. Vincennes* over the Persian Gulf, resulting in the death of all 290 persons on board the aircraft. The incident, which occurred on 3 July 1988, was considered by the Council at several of its meetings, notably on 7 December 1988 when the Council adopted its decision. The Council decision, while recalling the event of 3 July 1988, acknowledged the fact finding investigation report of the Secretary General of ICAO, and urged all States to take all necessary action for the safety of navigation of civil aircraft, particularly by assuring effective coordination of civil and military activities. The Resolution went on to refer to the fundamental principle of general international law—that States must refrain from resorting to the use of weapons against civil aircraft—and urged States to ratify Article 3 *bis* as soon as possible if they had already not done so.

States have a responsibility to respect the sovereignty of other nations. Apart from the direct attribution of responsibility to a State, particularly in instances where a State might be guilty of a breach of treaty provisions, or violate the territorial sovereignty of another State, there are instances where an act could be imputed to a State.[33] Imputability or attribution depends upon the link that exists

[33]There are some examples of imputability, for example the incident in 1955 when an Israeli civil aircraft belonging to the national carrier El Al was shot down by Bulgarian fighter planes, and the consequent acceptance of liability by the USSR for death and injury caused which resulted in the payment of compensation to the victims and their families. See 91 *ILR* 287. Another example concerns the finding of the International Court of Justice that responsibility could have been be imputed to the United States in the *Nicaragua* case, where mines were laid in Nicaraguan waters and attacks were perpetrated on Nicaraguan ports, oil installations and a naval base by persons

between the State and the legal person or persons actually responsible for the act in question. The legal possibility of imposing liability upon a State wherever an official could be linked to that State encourages a State to be more cautious of its responsibility in controlling those responsible for carrying out tasks for which the State could be ultimately held responsible. In the same context, the responsibility of placing mines was attributed to Albania in the *Corfu Channel* case since the court attributed to Albania the responsibility, since Albania was known to have knowledge of the placement of mines although it did not know who exactly carried out the act. It is arguable that, in view of the responsibility imposed upon a State by the Chicago Convention on the provision of air navigation services, the principles of immutability in State responsibility could be applied to an instance of an act or omission of a public or private official providing air navigation services.

The sense of international responsibility that the United Nations ascribed to itself had reached a heady stage at this point, where the role of international law in international human conduct was perceived to be primary and above the authority of States. In its Report to the General Assembly, the International Law Commission recommended a draft provision which required that:

> Every State has the duty to conduct its relations with other States in accordance with international law and with the principle that the sovereignty of each State is subject to the supremacy of international law.[34]

This principle, which forms a cornerstone of international conduct by States, provides the basis for strengthening international comity and regulating the conduct of States both internally—within their territories—and externally, towards other States. States are effectively precluded by this principle of pursuing their own interests untrammelled and with disregard to principles established by international law.

Generally under legal theory, each State is sovereign and equal and the term sovereignty may be used as a synonym for independence. However, in modern parlance, with the rapid growth in telecommunications and global competition and rivalries, no State can be entirely sovereign to the exclusion of others. Today, the words "sovereignty" and "intervention" tend to be interlinked in practice.

If Starke is right, and sovereignty is the residuum of power within the parameters prescribed by international law, and most States circumscribe their actions in the interests of the international community, then no State has the moral right to reject the collective will of the international community with regard to its internal acts if they jeopardise the populace within that State. At the same time, the international

identified as agents of the United States. See *Nicaragua* v. *the United States*, ICJ Reports 1986, 14. Also, 76 *ILR* 349. There was also the instance when the Secretary General of the United Nations mediated a settlement in which a sum, *inter alia* of $ 7 million was awarded to New Zealand for the violation of its sovereignty when a New Zealand vessel was destroyed by French agents in New Zealand. See the *Rainbow Warrior* case, 81 *AJIL*, 1987 at 325. Also in 74 *ILR* at 241.

[34] *Report of the International Law Commission to the General Assembly on the Work of the 1st Session, A/CN.4/13*, June 9 1949, at 21.

community cannot expect to claim the right to arbitrarily intervene in the internal acts of a State unless there is overwhelming consensus within that community. This is a delicate balance of diplomacy best left to seasoned diplomats.

1.2.3 Territoriality

As mentioned earlier in the Preface, although the Chicago Convention does not mention aviation and environmental protection (which was not a subject of concern at all in 1944) the Convention has been dragged into a spat that has developed with the enforcement by the European Union of its extension to aviation of the European Emissions Trading Scheme. Many States who have opposed this scheme do so on the basis that it erodes the protection afforded to sovereign States by Article 1.

In 2003, The European Union, by Directive 2003/87/EC[35] amended the European Union's Emissions Trading Scheme (EU ETS) to include the aviation sector in the scheme. Initially, in 2011, only flights between EU airports were to be included in the Scheme. From 2012 this was extended to all flights arriving at or departing from an EU airport. This means that the EU-ETS is applicable to any airline and its flight from anywhere in the world that is destined to Europe and vice versa. This would mean that the Scheme would be applicable, for example, to an American carrier's flight from New York to London all the way, covering inter alia emissions released over American airspace right through the flight over other territorial airspace before entering European airspace. Detractors of EU-ETS claim that this is extra territorial application of European law. Another contention is that EU has not made it clear as to where the funds collected under the Scheme would go. Brian Havel and John Mulligan, in a meticulously analyzed and well-reasoned article have addressed the issue of extra-territoriality—the bone of contention with all opponents to the EU-ETS as follows:

> The planned extension of the European Union emissions trading scheme to non-EU carriers has provoked vociferous opposition and prompted debate among scholars and practitioners of international and environmental law. US airlines challenged the relevant EU directive in the English High Court, arguing that it is incompatible with international law... the Advocate General and the Court, in their juristic pronouncements regarding the novel legal issues in this case, exalt political concerns over adherence to traditional understandings of fundamental principles of international aviation law, notably sovereignty, extraterritoriality, and the applicability of the Chicago Convention...[36]

[35]The European Union Emission Trading Scheme (EU ETS) is the largest multi-national, greenhouse gas emissions trading scheme in the world and is a main pillar of EU climate policy. Under the Scheme, each participating country has a National Allocation Plan (NAP) specifying caps on greenhouse gas emissions for individual power plants and other large point sources. Each facility gets a maximum amount of emission "allowances" for a particular period (e.g. 2005–2007). To comply, facilities can either reduce their emissions or purchase allowances from facilities with an excess of allowances. Progressively tightening caps are foreseen for each new period, forcing overall reductions in emissions.

[36]See generally, Havel and Mulligan (2012) at 3–33.

Extra territorial jurisdiction is exercised when a State (or in this case a community of States) seeks to apply its laws outside its territory in such a manner as may cause conflicts with other States.[37] It can be justified by the invocation of the effects doctrine or the "effects theory" which goes beyond the principles of sovereignty. This theory relates to a situation where a State assumes jurisdiction beyond its territorial limits on the ground that the behaviour of a party is adversely affecting the interests of that State by producing "effects" within its territory. It does not matter whether all the conduct and practices take place in another State or whether part of the conduct is within the State adopting the legislation. In the latter instance, the conduct of the party would come under the "objective territorial principle" where part of the offence takes place within the jurisdiction. In the case of aircraft engine emissions, the applicable principles would come under both headings as trans-boundary pollution of the environment by an aircraft which flies into Europe may involve the emissions of gases in one State that could cross boundaries and affect Europe.

The effects doctrine has been robustly applied in the United States particularly in the field of antitrust legislation.[38] Judicial recognition of the principle lay in the premise that any State may impose liabilities, even upon persons not within its allegiance, for conduct outside its borders that has consequences within its borders which the State reprehends.[39] This blanket principle was later toned down within the United States to acknowledge growing international protests against the wide ranging and arbitrary manner in which the principle could be applied. The modification involved the need to prove intentional conduct and the fact that the effect should be substantial for the doctrine to be applied.[40] In addition, courts began to insist on a jurisdictional rule of reason that involved consideration of interests of other nations and the nature of relationship between the US and the actors concerned. It is also noteworthy that the *Third Restatement of Foreign Relations Law* provides that a State may exercise jurisdiction based on the effects in the State when the effect or intended effect is substantial and the exercise of jurisdiction is reasonable.[41] Reasonableness is based on the extent the enacting State limited its

[37]There is a general common law presumption against the extra territorial application of legislation. See the House of Lords decision in *Holmes* v. *Bangladesh Biman Corporation* [1989] AC 1112 at 1126; 87 ILR 365 at 369. Also, *Air India* v. *Wiggins* [1980] 1WLR 815 t 819; 77 ILR 276 at 27.

[38]See The *US Sherman Antitrust Act 1896* 15 USC paras 1ff.

[39]*US*. v. *Aluminium Company of America*, 148 F.2d 416 (1945).

[40]*Timberlane Lumber Company* v. *Bank of America* 549 F.2d 597 (1976); 66 ILR, 270. Also, *Mannington Mills* v. *Congoleum Corporation*, 595 F.2d 1287 (1979); 66 ILR, 487.

[41]The Third Restatement constitutes a comprehensive revision of the earlier (1965) Restatement, covering many more subjects, and reflecting important developments in the intervening decades. This Restatement consists of international law as it applies to the United States, and domestic law that has substantial impact on the foreign relations of the United States or has other important international consequences.

jurisdiction so as to obviate conflict with the jurisdiction of the State affected to the extent possible.

The 1984 case of *Laker Airways* v. *Sabena*[42] held that once law was declared applicable it could not be subject to qualification or ignored by virtue of comity.[43] However, changes could be effected through diplomatic negotiations. The United States Supreme Court ruled in 1993 that US legislation (in this case the *Sherman Act*) could apply to foreign conduct that was meant to produce some substantial effect in the United States.[44] Extra territorial application of laws can be effectively rendered destitute of effect by blocking legislation[45] which a State can enact to preclude the application of a foreign law to citizens of that State.

In several instances, the United States has controlled or influenced activities occurring outside its borders which are calculated to harm the environment. For example: Congress passed a law prohibiting persons and vessels subject to the jurisdiction of the United States from "taking" (killing or injuring) marine mammals on the high seas; The EPA issued subpoenas to American companies demanding information on the use and release of chemicals from companies operating in Mexico, with a view to curbing pollution from the New River in Mexico from flowing into the United States; and Congress passed a law banning the import of ivory from countries that did not have an elephant protection programme, so that the numbers of elephants in Africa and Asia would not decrease due to poaching.

The United States has also used trade and investment measures to influence the conduct of other States. For example, during the 1990s, Congress drew a link between the human rights record of China with most-favoured nations treatment of the World Trade Organization. There have also been instances where goods from States are banned from importation to the United States unless that State complies with certain standards set in U.S. law. Conversely US exports are banned from import into those countries.

In every instance of extra territorial jurisdiction, there are two issues to be considered: the first is whether the State or group of States has the authority to

[42] 731 F.2d 909 (1984).

[43] Comity, at law, refers to legal reciprocity where one jurisdiction will extend certain courtesies to other nations, particularly by recognizing the validity and effect of their executive, legislative and judicial acts. The term refers to the idea that courts should not act in a way that demeans the jurisdiction, laws, or judicial decisions of another country. It is especially important in the application of principles of public international law. Part of the presumption of comity is that other nations will reciprocate the courtesy shown to them.

[44] *Hartford Fiore Insurance Company* v. *California*, 113 S. Ct. 2891 (1993) per Souter J.

[45] The most common instance of blocking legislation concerns the prevention of private information being demanded and obtained from nationals of a State by another State. Several countries have enacted so-called "blocking legislation." Blocking legislation mandates the confidentiality of information and documents and attempts to block foreign efforts to obtain evidence from residents of the enacting jurisdiction. It is often enacted by countries seeking to foster banking and financial industries, such as Switzerland, the Bahamas, Panama and Vanuatu. It generally prohibits residents of those countries and corporations doing business there disclosing confidential business information about others doing business there.

exercise extra territorial jurisdiction; and the second is, whether the exercise of that authority reasonable (taking into consideration the law concerned and the potential foreign policy conflicts).

1.2.4 Air Space and Outer Space

The Permanent Court of International Justice, when requested for a definition of "air space" in the 1933 *Eastern Greenland's Case,*[46] was of the view that the natural meaning of the term was its geographical meaning. The most fundamental assumption that one could reach from this conclusion is that air space is essentially geophysical, meaning that it is space where air is found. Simplistically put, "air space" has been considered as going upwards into space from the territorial boundaries of a State and downwards to the centre of the Earth, in the shape of an inverted cone. This theory, advanced mathematically, in terms of space where air is found, would encompass the atmosphere, which has is layered into components starting from the troposphere (from sea level to about 10 km); the stratosphere (from about 10 to 40 km up); the ionosphere (from about 40 to 375 km); and the exosphere (from 375 to 20,000 km). Based on this methodology, a sub-orbital flight, which goes up to about 62.5 miles (100 km) above the landmass of the Earth, would hover somewhere in the lower level of the ionosphere, prompting the conclusion that it is a space flight traversing outer space, while others would maintain that the vehicle does not leave the Earth's atmosphere and therefore is airborne.

The United Nations Committee on the Peaceful Uses of Outer Space (UNCOPUOS), which is the UN forum where technical and legal aspects of space activities with global impact are considered, has discussed the issue of the definition and delimitation of outer space from 1962 and no definite conclusion has been reached so far in this regard. In this connection, it is of interest to note that the Legal Subcommittee of UNCOPUOS, through its Working Group on Matters Relating to the Definition and Delimitation of Outer Space, has been considering possible legal issues with regard to aerospace objects. A questionnaire thereon has been circulated to all U.N. Member States. A compilation of the replies received so far and an analytical summary of such replies, as well as a historical summary on the consideration of the question on the definition and delimitation of outer space, may be found on the OOSA website.[47]

As debated for decades in the framework of UNCOPUOS, it may be questioned whether the vertical limit of airspace would be critical to determine the scope of applicability of air law as opposed to international space law conventions (spatialist approach), or whether the type of activities at issue would determine which law should apply (functionalist approach) to sub orbital flights. The latter school of thought submits that flights which would be passing merely in transit through (sub)

---

[46]PCIJ Series A/B, No. 53, at pp. 53ff.

[47]www.oosa.unvienna.org/index.html.

orbital space in the course of an earth-to-earth transportation would be in air space and therefore remain subject to principles of air law.

A sub-orbital flight is a flight up to a very high altitude which does not involve sending the vehicle into orbit. 'sub-orbital trajectory', which a sub orbital flight would follow, is defined in the legislation of the United States as

> The intentional flight path of a launch vehicle, re-entry vehicle, or any portion thereof, whose vacuum instantaneous impact point does not leave the surface of the Earth.

In 2004, SpaceShipOne was the first private vehicle to complete two sub-orbital flights within 2 weeks carrying weight equivalent to three human adults up to about 62.5 miles (100 km) to win the Ansari X Prize. It was carried during 1 h by an aeroplane up to nearly 50,000 feet (9.5 miles) from where it was released into a glide and then propelled vertically for 80 s by a rocket motor to an altitude of more than 62 miles at apogee, reaching a speed over Mach 3. Then falling back to return to earth, it re-entered the atmosphere and glided during 15–20 min before landing back on the runway of departure.

SpaceShipOne, strictly speaking, does not operate as an aeroplane or even as an aircraft during the ballistic portion of the flight while it is not supported by the reactions of the air, even though some degree of aerodynamic control exists throughout the trajectory from launch altitude until the craft enters the upper reaches of the atmosphere where the air density is no longer sufficient for aerodynamic flight. After apogee, during re-entry into the atmosphere the vehicle transitions to unpowered aerodynamic (gliding) flight for the return to earth. Consequently, depending upon some design and operational aspects, it could be considered operating as an aircraft in flight during this latter portion of the journey.

Therefore, such vehicles could fulfil the principal elements in the definition of aircraft and be used as such during a portion of their flights, but they offer some characteristics of a rocket as well. It is likely that other vehicles engaged in the future in such sub-orbital flights would similarly be of an hybrid nature, taking into account that developments to come may lead to a range of designs, some of which could be more clearly classified as aircraft. Should sub-orbital vehicles be considered (primarily) as aircraft, when engaged in international air navigation, consequences would follow under the Chicago Convention, mainly in terms of registration, airworthiness certification, pilot licensing and operational requirements (unless they are otherwise classified as State aircraft under Article 3 of the Convention).

Plans have been announced by Virgin Galactic for the development of a fleet of five sub-orbital vehicles to carry paying passengers, six per vehicle; it plans that the first of these will be ready for commercial operations in 2008 at the earliest. There are indications that at least one other company is planning to offer rival sub-orbital flights.

Manned and unmanned sub-orbital flights have been undertaken to test spacecraft and launch vehicles intended for later orbital flight, but some vehicles have been designed exclusively to reach space sub-orbitally: manned vehicles such as the X-15 and SpaceShipOne, and unmanned ones such as ICBMs and sounding rockets.

Sub-orbital tourist flights will initially focus on attaining the altitude required to qualify as reaching space. The flight path will probably be either vertical or very steep, with the spacecraft landing back at its take-off site.

The spacecraft will probably shut off its engines well before reaching maximum altitude, and then coast up to its highest point. During a few minutes, from the point when the engines are shut off to the point where the craft begins to slow its descent for landing, the passengers will experience.

A suborbital flight is known to be the next generation of commercial passenger travel. At the present time flight testing of commercial reusable launch vehicles (RLVs) is underway, making the availability of frequent suborbital flight closer than ever. As earlier mentioned sub orbital flights are a considered missions that fly out of the atmosphere but does not reach speeds needed to sustain continuous orbiting of the earth. They allow passengers to look down at the brilliant curvature of the earth as they would from orbit.

One must not confuse a sub orbital flight with a space flight which is a flight *into* or *through* space. The craft which undertakes a spaceflight is called a spacecraft. It is often thought that orbital spaceflights are spaceflights and sub-orbital spaceflights are less than actual spaceflights. This is not entirely accurate as both orbital and sub-orbital spaceflights are true spaceflights.

The term *orbit* can be used in two ways: it can mean a trajectory in general, or it can mean a closed trajectory. The terms *sub-orbital* and *orbital spaceflights* refer to the latter: an orbital spaceflight is one which completes an orbit fully around the central body.

From the above discussion the conclusions that could be drawn are that for a flight from Earth to be a spaceflight, the spacecraft has to ascend from Earth and at the very least go past the edge of space. The edge of space is, for the purpose of space flight, often accepted to lie at a height of 100 km (62 miles) above mean sea level. Any flight that goes higher than that is by definition a spaceflight. Although space begins where the Earth's atmosphere ends, the atmosphere fades out gradually so the precise boundary is difficult to ascertain. Therefore one could argue that there is a need to accept the fact that vehicles which would effect earth-to-earth connections through sub-orbital space could incorporate the constitutive elements of aircraft and fly as such at least during descending phase while gliding. However, rocket-propelled vehicles could be considered as not falling under the classification of aircraft.

From a spatialist viewpoint, there is no clear indication in international law on the delimitation between airspace and outer space which would permit to conclude on the applicability of either air law or space law to sub-orbital flights. On the other hand, it might be argued from a functionalist viewpoint that air law would prevail since airspace would be the main centre of activities of sub-orbital vehicles in the course of an earth-to-earth transportation, any crossing of outer space being brief and only incidental to the flight. The United Nations Committee on the Peaceful Uses of Outer Space (UNCOPUOS), and more particularly its Legal Subcommittee, is considering the question of possible legal issues with regard to aerospace objects but no final conclusion has been reached yet.

The ICAO Assembly, at its 29th Session (Montreal, 22 September–8 October 1992), adopted Resolution A29-11 by which ICAO was recognized as responsible for stating the position of international civil aviation on all related outer space matters.

1.2.5 Drones and Sovereignty

Drones are essentially pilotless aircraft (a detailed discussion of which will follow under Article 8) which are usually called unmanned aircraft systems. Remotely piloted aircraft system (RPAS) operations are spreading beyond the original military applications, towards other State non-military operations (e.g. police, coast guard and similar), but also into civil aviation. However, these aircraft are mostly, at present, used in military attacks and therefore do not strictly form part of the discussion in this book. However they are not without their influence on civil aviation as seen in Article 3 (c) which provides that no State (military, my parenthesis) aircraft of a Contracting State shall fly over the territory of another State or land thereon without authorization by special agreement or otherwise, and in accordance with the terms thereof.

Those responsible for operating drones such as the remotely-piloted Predator MQ-1 or Reaper MQ-9 aircraft do not seek nor receive such permission usually. *The Economist* reports:

> Laden with sophisticated sensors and carrying Hellfire missiles and laser-guided bombs, they patrol the skies above Afghanistan, launch lethally accurate strikes against terrorists in the tribal areas of Pakistan, Yemen and Somalia and have helped NATO turn the tide against Muammar Qaddafi's forces in Libya. Even calling them Unmanned Aerial Vehicles (RPASs) or Unmanned Aerial Systems (RPAS) is slightly misleading. There may not be a man in the cockpit, but each Reaper, a bigger, deadlier version of the Predator, requires more than 180 people to keep it flying. A pilot is always at the controls (albeit from a base that might be 7,500 miles, or 12,000km, away); and another officer operates its sensors and cameras.[48]

It is also recorded that under the current US Presidency of Barack Obama, drone strikes on terrorists in Pakistan's tribal areas has risen tenfold, which now amounts to one in every 4 days as against during the Presidency of George W. Bush which amounted to one in every 40 days. And the frequency is growing. At the Presidential Debates in October 2012 between President Obama and Governor Romney, both agreed on the necessity of continuing drone attacks in the war against terrorism, which meant that even if Governor Romney won the election, the drone attacks over sovereign airspace would have continued. John Brennan, Mr Obama's counter-terrorism chief, has explicitly asserted that that as America gradually withdraws its forces in Afghanistan over the next 3 years, there will reduction in drone strikes, which, he claims, are partly responsible for al-Qaeda's terror campaign being curbed.

[48] *The Economist*, 8th October 2011.

This brings to bear an interesting point. Article 2.4 of the United Nations Charter states unequivocally:

> All member States shall refrain from in their international relations from the threat or use of force against the territorial integrity or political independence of any State, or in any other manner inconsistent with the Purposes of the United Nations.[49]

Article 51 further qualifies this provision when it says:

> Nothing in the present Charter shall impair the inherent right of individual or collective self-defence if an armed attack occurs against a Member of the United Nations, until the Security Council has taken measures necessary to maintain international peace and security. Measures taken by Members in the exercise of this right of self-defence shall be immediately reported to the Security Council and shall not in any way affect the authority and responsibility of the Security Council under the present Charter to take at any time such action as it deems necessary in order to maintain or restore international peace and security.

Of course after the events of 11 September 2001 the concept of sovereignty as we have known above took a new twist with the invasions of Afghanistan and Iraq. Following the events of 11 September 2001, where civilian commercial aircraft were used as weapons of mass destruction. Air Defence Identification Zones (ADIZ) gained prominence as a tool with which a State could control and be prepared for aircraft approaching their territory. An Air Defence Identification Zone (ADIZ) is an area in airspace over land or water which may not be over the sovereign territory of a State in which ready identification, location and control of all aircraft is required in the interest of national security.[50] ADIZ must not be

[49]Article 2 of the UN Charter in 2.1 states that the United Nations is based on the sovereign equality of all its Members. Article 1 of the Charter gives the purposes of the United Nations as:

1. To maintain international peace and security, and to that end: to take effective collective measures for the prevention and removal of threats to the peace, and for the suppression of acts of aggression or other breaches of the peace, and to bring about by peaceful means, and in conformity with the principles of justice and international law, adjustment or settlement of international disputes or situations which might lead to a breach of the peace;
2. To develop friendly relations among nations based on respect for the principle of equal rights and self-determination of peoples, and to take other appropriate measures to strengthen universal peace;
3. To achieve international co-operation in solving international problems of an economic, social, cultural, or humanitarian character, and in promoting and encouraging respect for human rights and for fundamental freedoms for all without distinction as to race, sex, language, or religion; and
4. To be a centre for harmonizing the actions of nations in the attainment of these common ends.

[50]A similar definition is found in the United States Federal Regulations. See 14 C.F.R. S. 99.3 (2009). The United States has four ADIZs: The Contiguous US ADIZ; Alaska ADIZ; Guam ADIZ; and Hawai ADIZ. In the United States, ADIZ applies only to commercial aircraft intending to enter United States airspace. The United States does not recognize the right of a coastal nation to apply its ADIZ procedures to foreign aircraft not intending to enter national airspace nor does the United States apply its ADIZ procedures to foreign aircraft not intending to enter U.S. airspace. Accordingly, U.S. military aircraft not intending to enter national airspace should not identify themselves or otherwise comply with ADIZ procedures established by other nations, unless the

confused with Flight Information Regions (FIRs) which are areas established for the facilitation of airspace and air traffic management. FIRs generally involve a subjacent State which has undertaken responsibility for providing air traffic control services.[51] The main purpose of establishing an ADIZ is to properly identify all approaching aircraft for security purposes so that they could, prior to entry into national airspace, satisfy certain local entry requirements.[52] Although there is no overwhelming evidence, either from a scholastic or legislative perspective that lends legal legitimacy to the establishment of ADIZs, such a concept has never been challenged as being inconsistent with existing law.[53]

It has been argued that ADIZs came into prominence as a security tool in air navigation as a corollary to the events of 11 September 2001 where aircraft were used as weapons of mass destruction.[54] Norway and the United Kingdom, India, Pakistan and Canada (CADIZ) are some countries which maintain ADIZs as well as the United States.[55] If an analogy from maritime law and practice were to be applied to ADIZ, one could cite the *United Nations Convention on the Law of the Sea* (UNCLOS)[56] which was signed by the Parties on December 10, 1982 and entered into force on November 16, 1994 after receiving 60 ratifications or accessions. UNCLOS divides the seas into zones over which States have varying degrees of rights and controls. The territorial sea, which is exclusively controlled by the State, is the first zone which extends 12 nautical miles from the coast or coastal baselines. The territorial sea is open to all vessels to enjoy the right of innocent passage. Beyond the territorial zone comes the contiguous zone of another 12

United States has specifically agreed to do so. See U.S. Navy's *Commander's Handbook on the Law of Naval Operations*. Also see Williams (2007) at 95–96.

[51]States may delegate such responsibility to another State or States without abdicating their sovereignty. See Annex 11 to the Chicago Convention (Air Traffic Services) which provides that Flight information service is provided to aircraft operating in controlled airspace and to others known to the air traffic services units. The information includes significant meteorological (SIGMET) information, changes in the serviceability of navigation aids and in the condition of aerodromes and associated facilities and any other information likely to affect safety. Flights operated by Instrument Flight Rules (IFR) receive, in addition, information on weather conditions at departure, destination and alternate aerodromes, collision hazards to aircraft operating outside of control areas and control zones and, for flight over water, available information on surface vessels. Flights operated by Visual Flight Rules (VFR) receive information on weather conditions which would make visual flight impractical. Annex 11 also contains specifications for operational flight information service (OFIS) broadcasts, including automated terminal information service (ATIS) broadcasts. See Franklin (2007) at 426.

[52]See Petras (2010) at 62–63.

[53]See McDougal et al. (1963), at 306–311 where the author suggests that if for security reasons States have certain claims on those who enter their sovereign territories, such claims may not be inconsistent with the principles of international law.

[54]See Dutton (2009) at 691.

[55]*Ibid.*

[56]Law of the Sea, Official Text of the United Nations Convention on the Law of the Sea with Indexes and Annex, Final Act of the Third United Nations Conference on the Law of the Sea, United Nations: 1983.

nautical miles followed by the exclusive economic zone of 200 nautical miles from the coastal baseline. In Article 76 titled "Definition of the Continental Shelf" UNCLOS provides that a States Party may extend its continental margin beyond the 200 nautical mile Exclusive Economic Zone (EEZ) if certain criteria are fulfilled.

It is an incontrovertible principle of international maritime law that international navigation, however founded and whatever the right of innocent passage may be, often conflicts with a State's desire to protect itself from activities that infringe its sovereignty, resource rights or more importantly, internal security. In such instances jurisdiction inevitably vests with the coastal State.[57] Turning to ADIZs, it must be emphasized that a State can by no means arrogate to itself territorial sovereignty over an ADIZ *ipso facto*. Neither can a State interfere with a State's exercise of legitimate rights of navigation over the high seas.[58]

Broadly speaking, ADIZ requirements are those that sovereign States require aircraft to comply with if they are to be permitted to enter sovereign airspace. Therefore ADIZs requirements act as conditions precedent that are calculated to ensure the protection of that State. The justification for ADIZ lies theoretically in the precautionary principle which asserts that the absence of empirical or scientific evidence should not preclude States from taking action to prevent a harm before it occurs.[59] The evolution of the principle in international law, particularly in the field of environmental protection, began in the early 1980s[60] although there is evidence that it was domestically popular in Europe in the 1930s in the German socio-legal tradition, centering on the concept of good household management. In German the concept is *Vorsorgeprinzip*, which translates into English as *precaution principle*. One commentator has added the thought provoking comment that in today's political sphere, the precautionary principle enjoys a wide, unprecedented recognition[61]:

> The precautionary principle has become of such tremendous importance because in many cases, the scientific establishment of cause and effect is a difficult task sometimes approaching a fruitless investigation of infinite series of events.[62]

[57]See Posner and Sykes (2010) at 577.

[58]In the United States context, see Restatement 3d, Foreign Relations Law, American Law Institute, at S. 521. The Restatement is persRPASive law in the United States. See Cardozo (1924) at 9.

[59]The precautionary principle (a moral and political concept) states that if an action or policy might cause severe or irreversible harm to the public, in the absence of a scientific consensus that harm would not ensue, the burden of proof falls on those who would advocate taking the action. The precautionary principle is most often applied in the context of the impact of human actions on the environment and human health where the consequences of actions may be unpredictable.

[60]For a discussion of the emergence of the precautionary principle see Scott Lafranchi, Surveying the Precautionary Principle' Ongoing Global Development 32 *B.C. ENVTL. AFF. L. REV.* 2005 at 678.

[61]Marr (2003) at 3.

[62]*Id.* at 6.

For the precautionary principle to apply, States must take measures according to their capabilities and they must be cost effective. Also, threats that are responded to must be both serious and irreversible. The precautionary principle is usually applied through a structured approach to the analysis of risk, which comprises three elements: risk assessment; risk management; and risk communication. The principle is particularly relevant to the management of risk. It is based on the presupposition that potentially dangerous effects from a particular process or phenomenon have been identified and that scientific evaluation does not guarantee that the risk could be averted.

There are instances where a State can be defended for invoking preventive action based on the overarching principle of social contract by which the citizens charge the State with the responsibility of ensuring their security. Social Contract describes a broad class of philosophical theories whose subject is the implied agreements by which people form nations and maintain social order. Social contract theory provides the rationale behind the historically important notion that legitimate state authority must be derived from the consent of the governed, which, in other words means that a democratic State is precluded from enacting draconian laws against the civil liberty of citizens unless with the consent of the people. The first modern philosopher to articulate a detailed contract theory was Thomas Hobbes (1588–1679), who contended that people in a state of nature ceded their individual rights to create sovereignty, retained by the state, in return for their protection and a more functional society, so social contract evolves out of pragmatic self-interest. Hobbes named the state *Leviathan*, thus pointing to the artifice involved in the social contract.

Alan Dershowitz, Professor of Law at Harvard University, asserts that

> there is a desperate need in the world for a coherent and widely accepted jurisprudence of preemption and prevention, in the context of both self-defence and defense of others.[63]

Of course, here Dershowitz is referring to the international scene, but it would not be wrong to ascribe this principle to the national level when there is a dire need to control anarchy and insecurity of a nation. However, the bottom line for any preventive jurisprudence in the domestic context is the social contract theory where State authority must be derived from the people. There must be a preventive jurisprudence in place governing the acts of the executive and law enforcement officers. Preventive acts must never be *ad hoc*, or decided at the whim of the law enforcer.

Preemption[64] and prevention[65] are necessary elements in today's political and military fabric, where legal legitimacy is ascribed to actions of States which act swiftly to avoid harm and protect its citizenry. This is often accomplished

---

[63] *Preemption—A Sword that Cuts Both Ways,* Norton: New York, 2006, at 11.

[64] Preemption is when an act, which is potentially harmful to a State and is imminent, is effectively precluded by military or other action.

[65] Prevention is when an act, which is potentially harmful to a State and is inevitable, is effectively precluded by military or other action.

bypassing rigid dogma and entrenched rules based on the precautionary principle and on the maxim *necessitat non habet legem* (necessity has no law or rules). Another is *Inter arma enim silent leges* is a maxim attributed to Cicero, which translates as "In times of war, the laws are silent". In the twenty-first century, this maxim, which was purported to address the growing mob violence and thuggery of Cicero's time, has taken on a different and a more complex dimension, extending from the idealistic synergy between the executive and the judiciary in instances of civil strife, to the overall power, called "prerogative" or "discretion" of the sovereign, to act for the public good and the role of the judiciary as the guardian of the rule of law.

The enduring conflict between executive power and the rule of law is at the heart of this maxim. In modern usage it has become a watchword for the erosion of civil liberties during internal and external strife. The implication of Cicero's aphorism is that freedoms, such as the right of free passage through a territory of a State, are subservient to a nation's self-defence from enemies within or without.

The "state of exception" or "abnormal times" is considered by some political scientists to call for legal justification for a State to be uncontrolled.[66] This theory justifies the sovereign, as guardian of the Constitution, in its extra judicial response to all exceptions to dangers within the political and legal spectrum, on the basis of the sovereign's exclusive capability of identifying the enemy and the threat it poses to the State. However, in common law States, this extremist view is blended harmoniously with the essential philosophy of the Rule of Law, which is the foundation of civil liberty and order, and the underlying constitutional principle requiring government to be conducted according to law, thus making all public officers answerable for their acts in the ordinary courts. Common law jurisdictions such as the United States, Canada and the United Kingdom take pride in their long tradition of parliamentary democracy which would effectively preclude arbitrary acts of the Executive in curbing civil liberties guaranteed by the law. This principle is embodied in the dissenting judgment of Lord Atkin in *Liversidge* v. *Anderson,* to which courts pay frequent lip service, that:

> amidst the clash of arms, the laws are not silent. They may be changed, but they speak the same language in war as in peace.[67]

Another issue is the judicial interpretation of the extent to which a State is justified in suspending an existing legal order to protect its citizens. Against the backdrop of the *Bush Doctrine*[68] which followed the events of 11 September 2001 and the justification of the invasion of Afghanistan, in spite of Article 2.4 of the

[66]Schmitt (1988) at 5.

[67]*Liversidge* v. *Anderson* [1942] A.C. 206 at 244.

[68]The Bush Doctrine is attributed to the modern notion of preventive war and the justification that the United States had the right to secure itself against countries that harbor or give aid to terrorist groups, which was used to justify the 2001invasion of Afghanistan.

Charter of the United Nations.[69] Of special relevance in this regard is the judicial examination by the United States Supreme Court of the imprisonment of an Arab–American immigrant, 3 months after the terrorist attacks of 9/11. Justice Scalia stated:

> Many think it not only inevitable but entirely proper that liberty give way to security in times of national crisis that, at the extremes of military exigency, *inter arma silent leges*. Whatever the general merits of the view that war silences law or modulates its voice, that view has no place in the interpretation and application of a Constitution designed precisely to confront war and, in a manner that accords with democratic principles, to accommodate it.[70]

This statement, which strongly supports the Cicero maxim, reinforces the legal legitimacy of a statement made earlier by Chief Justice Rehnquist who opined:

> It is neither desirable nor is it remotely likely that civil liberty will occupy as favored a position in wartime as it does in peacetime. But it is both desirable and likely that more careful attention will be paid by the courts to the basis for the government's claims of necessity as a basis for curtailing civil liberty…the laws will thus not be silent in time of war, but they will speak with a somewhat different voice.[71]

Modern perspectives of the sovereign prerogative and conflicting principles of juridification of war bring to bear the important distinction between the ultimate question as to whether the judiciary is able in every circumstance, to impose the rule of law or whether, under the state of exception and the principles of guardianship, a sovereign can suspend the legal order in order to protect its citizens. This dilemma was well expounded by Lord Shaw in 1917 when His Lordship stated:

> The basic danger is found in an especial degree whenever the law is not same for all, but the selection of the victim is left to the plenary discretion whether of a tyrant, a committee, a bureaucracy or any other depository of despotic power…it is a poison to the commonwealth.[72]

The attacks of 11 September 2001 inevitably highlighted the strategic position of civil aviation both as an industry vulnerable to attack and as an integral tool in ensuring peace and security in the world. The modernist view of civil aviation, as it prevailed when the Chicago Convention[73] was signed at Chicago on 7 December 1944, was centred on State sovereignty and the widely accepted post-war view that the development of international civil aviation can greatly help to create and preserve friendship and understanding among the nations and peoples of the

---

[69]Article 2.4 provides: "All members of the United Nations must refrain in their international relations from the threat or use of force against the territorial integrity or political independence of any State, or in any other manner inconsistent with the purposes of the United Nations. See *Charter of the United Nations and Statute of the International Court of Justice*, United Nations: New York.

[70]*Hamdi* v. *Rumsfeld*, 542 US 5047 (2004).

[71]Rehnquist (1998) at 5.

[72]*R*. v. *Halliday*, [1917] A.C. at 292.

[73]Preamble *supra* note 1.

world, yet its abuse can become a threat to general security.[74] This essentially modernist philosophy focussed on the importance of the State as the ultimate sovereign authority which can overrule considerations of international community welfare if they clashed with the domestic interests of the State. It gave way, in the 1960s and 1970s to a post-modernist era of recognition of the individual as a global citizen whose interests at public international law were considered paramount over considerations of individual State interests.

The 11 September 2001 events led to a new era that now calls for a neo-post modernist approach which admits of social elements and corporate interests being involved with States in an overall effort at securing world peace and security. The role of civil aviation in this process is critical, since it is an integral element of commercial and social interactivity and a tool that could be used by the world community to forge closer interactivity between the people of the world.

The requirements of ADIZ serve well the defence of a sovereign State against attacks and accord with neo post modernist views that aviation should first serve the safety and security of a society and that any damage posed by the misuse of aviation should be effectively precluded. The real significance of the Convention, particularly as a tool for ensuring political will of individual States, lies in the fundamental philosophy contained in its Preamble. In its Preamble, the Convention enunciates a message of peace through aviation. It makes mention of the future development of international civil aviation being able to help preserve friendship and understanding among the nations of the world, while its abuse (i.e. abuse of future development of international civil aviation) can become a threat to "the general security". By "general security" the Chicago Conference meant the prevention of threats to peace. As already discussed, these words have been interpreted in the widest possible sense by the Assembly of ICAO at its various sessions to cover instances of social injustice such as racial discrimination as well as threats to commercial expediency achieved through civil aviation.

References

Baldwin SE (1910) The law of the airship. Am J Int Law 4(95):97

Bouve (1930) The development of international rules in air navigation. Air Law Rev 1(1):6

Brownlie I (1990) Principles of public international law, 4th edn. Clarendon, Oxford

Cardozo BN (1924) The growth of the law. Yale University Press, New Haven

Cheng B (1962) The law of international air transport. London/New York, Stevens/Oceana

Cooper JC (1968) Roman law and the maxim "*Cujus est Solum*" in international law. Explor Aerosp Law

Dutton PA (2009) Caelium Liberum: air defence identification zones outside sovereign airspace. Am J Int Law 103(4):691–709

Franklin M (2007) Sovereignty and functional airspace blocks. Air Space Law 32:425

[74]Preamble to the Chicago Convention, *Id.*

Havel BF, Mulligan JQ (2012) The triumph of politics: reflections on the judgment of the court of justice of the European Union validating the inclusion of non-EU airlines in the emissions trading scheme. Air Space Law 37(1):3–33

Marr S (2003) The precautionary principle in the law of the sea, modern decision making in international law. Kluwer Law International, Alphen aan den rijn

McDougal MS et al (1963) Law and public order in outer space. Yale University Press, New Haven

Milde M (2012) International air law and ICAO. In: Benko M (ed) Essential air and space law. Eleven International Publishing, The Hague, pp 8–10

Petras CM (2010) The law of air mobility—the international legal principles behind the U.S. mobility air forces' mission. Air Force Law Rev 66:1–78

Posner EA, Sykes AO (2010) Economic foundation of the law of the sea. Am J Int Law 104(4):569

Rehnquist WH (1998) All the laws but one–civil liberties in wartime. Alfred A. Knopf, New York

Schmitt C (1988) Political theology: four chapters on the theory of sovereignty (trans: George Schwab). MIT Press, Cambridge

Starke JG (1977) An introduction to international law, 7th edn. Butterworth's, London

Williams A (2007) The interception of civil aircraft over the high seas in the global war on terror. Air Force Law Rev 73(73):95–96

Article 2
Territory

For the purposes of this Convention the territory of a State shall be deemed to be the land areas and territorial waters adjacent thereto under the sovereignty, suzerainty, protection or mandate of such State.

Contents

1 Territorial Waters

The United Nations Convention on the Law of the Sea (UNCLOS)[1] entered into force on November 16, 1994 after receiving 60 ratifications or accessions. UNCLOS in Article 76 titled "Definition of the Continental Shelf" provides that a States Party may extend its continental margin beyond the 200 nautical mile Exclusive Economic Zone (EEZ) if certain criteria are fulfilled. Also, according to UNCLOS seas are divided into zones over which States have varying degrees of rights and controls. The territorial sea, which is exclusively controlled by the State, is the first zone which extends 12 nautical miles from the coast or coastal baselines.[2] It is arguable that this 12 line demarcation is not what is meant by Article 2 of the Chicago Convention which refers to territorial waters adjacent to the land areas of the State concerned. A state usually defines its area of sovereignty in the waters surrounding it, for purposes of navigation, fishing rights and national security. The natural question therefore is whether the contiguous zone which is adjacent to the territorial zone as defined by UNCLOS, of another 12 nautical miles followed by the exclusive economic zone of 200 nautical miles from the coastal baseline would be encapsulated in the territorial waters referred to in Article 2 of the Chicago Convention. Of course it is open for a State to claim all these boundaries although UNCLOS is very clear as to what the territorial waters are, which is a breadth of up to a limit not exceeding 12 nautical miles, measured from baselines determined in accordance with UNCLOS.

It is an incontrovertible principle of international maritime law that international navigation, however founded and whatever the right of innocent passage may be, often conflicts with a State's desire to protect itself from activities that infringe

[1] United Nations Convention on the Law of the Sea (UNCLOS), with Index and Final Act of the Third United Nations Convention on the Law of the Sea, United Nations: New York, 1983.

[2] UNCLOS *Id.* Article 3.

R. Abeyratne, *Convention on International Civil Aviation*,
DOI 10.1007/978-3-319-00068-8_3, © Springer International Publishing Switzerland 2014

its sovereignty, resource rights or more importantly, internal security. In such instances jurisdiction inevitably vests with the coastal State, and, both according to the treaties of 1974 and 1976 as well as UNCLOS, jurisdiction would be allocated to the coastal State with the proximate coastline. A coastal State can therefore prohibit foreign vessels from passing near its coast for purposes of security. It is also worthy of note that Article 78 of UNCLOS provides that the rights of the coastal State over the continental shelf do not affect the legal status of the superjacent waters or of the air space above those waters.

The geographic delimitations of a coastal State could cause ambivalence and uncertainty in certain instances. This can particularly be seen in the indenting of coastlines; the juxtaposition of islands next or parallel to a coastal State or the irregular incursion of bays into the coastline. One such example was the *Anglo-Norwegian Fisheries Case* which brought to bear a disparity between the demarcation by Norwegian authorities of its territorial sea to amount to 1,000 miles off its coastline by a methodology that constructed a series of baselines creating a nexus between the outermost parts of the land constituting a fringe of islands and rocks, whereas the conventional method would have been to measure the territorial sea from the low water line. The court held with the approach of the Norwegian authorities in this instance.[3] The Court also held that this methodology had been used for numerous years and had been customarily acquiesced to by the contending British authorities.

The defining parameters of the width of a territorial sea is usually drawn from the low water mark around the coasts of a particular State. The low water line along the coast has been defined as marked on large-scale charts which are officially recognized by the coastal State concerned. This customary international law principle was enshrined in the Geneva Convention on the Territorial Sea and the Contiguous Zone of 1958 in Article 3. A later Convention of 1982 in Article 5 reiterated this principle.[4]

In view of the background to this issue, certain conclusions may be drawn: The first is that UNCLOS comes under the umbrella Convention called the Vienna Convention on the Law of Treaties, which provides that such treaties shall be binding upon the Parties and be performed by them in good faith. The Vienna Convention further states that a treaty shall be interpreted in good faith in accordance with the ordinary meaning to be given to the terms of the treaty. Furthermore, the Vienna Convention stipulates that, unless a different intention appears from the treaty or is otherwise established, a treaty is binding upon each Party in respect of its entire territory. Therefore, a part of a State, however formed as a province, cannot take a unilateral decision to contravene the provisions of a treaty which the

[3]ICJ Reports 1951 at p. 128.

[4]Eritrea/Yemen (Phase Two: Maritime Delimitation) 119 ILR 417 at 458. Also *Qatar* v. *Bahrain*, ICJ Reports 2001 at para 184.

sovereign State has entered into. The Convention goes on to state that a Party may not invoke the provisions of its internal law as justification for its failure to perform a treaty.

2 Sovereignty, Suzerainty, Protection or Mandate...

Since sovereignty has already been discussed the next concept under consideration is suzerainty. **Suzerainty** is an archaic term in modern State practice. Suzerainty is practiced in instances where a region or people is the arm of a more powerful entity which controls its foreign affairs while permitting the subservient arm or nation some limited domestic autonomy.[5] The dominant entity in the suzerainty relationship, or the more powerful entity itself, is called a suzerain. The term suzerainty was originally used to describe the relationship between the Ottoman Empire and its surrounding regions. The basic difference between sovereignty and suzerainty is that the latter enjoys some degree of self-rule whereas a nation under the sovereignty of a State has no autonomy in this regard.

Under contemporary international law, either sovereignty exists or it does not, There are no known suzerains or feudal lords in the 191 member States of ICAO. Under public international law a nation or people can agree by treaty to become a protectorate of a stronger power, contemporary international law does not recognize any way of making this relationship compulsory on the weaker nation.

With regard to **protection**, a distinction should be drawn between protected States and protectorates. A protected State retains its status as a separate State but enters into a valid treaty with another State granting to that State certain external and domestic powers. In the case of a protectorate, there is an arrangement between two States whereby one State recognizes the other as a separate legal personality but does not recognize its Statehood.[6] An example of a protected State can be seen in the Treaty of Fez of 1912 between Morocco and France where the former gave certain external and internal sovereign power to the latter, which included international relations. It was the view of the International Court of Justice in this case that despite the abdication by Morocco of its international relations functions to France, Morocco continued to be a sovereign State.[7]

[5]There were two types of covenants in ancient times. One was a parity covenant. This was where two parties who were equal bound themselves together in a contract. The other covenant was the suzerainty covenant. This was an agreement between a strong party and a weak party. For example, a strong nation might overpower a weak nation. The victors would agree to protect the weaker nation if the weaker nation would keep certain rules.

[6]In the case of sub-Saharan Africa during Colonial times, tribal entities entered into treaties with their Colonial overlords. These entities, which were not sovereign States, were commonly called "colonial protectorates" under internal colonial arrangements.

[7]*Rights of Nationals of the United States of America in Morocco*, ICJ Reports, 1952 at 176–188.

A **mandate** came into being between the two world wars where colonies of defeated powers were administered by the conquering allied powers for the benefit of the inhabitants of such colonies. This system of mandates precluded the outright annexure of colonies and provided a system of protection under the League of Nations umbrella. This created *sui generis* mandated and trust territories under the principle that the well- being of people coming under a mandate were placed under "a sacred trust of civilization". Such people came under the tutelage of their caretakers.

In this context, one could conclude that the intent of the forefathers of the Chicago Convention was to extend State sovereignty to the territory—both land and sea—to all instances of control whether it be sovereignty or other forms of protection which a State offered. Therefore, in matters of international civil aviation, overall power and authority over airspace were vested in the State which had control over a territory. The Moroccan case would be a deviation in this regard and one can infer that in such an instance, whether protectorate or not, A state would not lose its sovereignty in certain circumstances and under particular conditions.

One has to appreciate the fact that in 1944, there were two overriding trends prevalent. The first was that the world had seen tremendous and prolific air power and attacks over the airspace of certain States. Therefore, Articles 1 and 2 of the Chicago Convention were seemingly to make sure that in peacetime, no one transgressed national airspace and threatened the security of a State. The second was that Colonial powers, who exercised suzeraint y, protection and mandates over colonies wanted to zealously guard their geographic assets.

As the earlier discussion on Sovereignty reflected, both these articles have to be perceived and indeed interpreted in their modern contexts relating to international civil aviation. They have to be seen through the system of governance prevalent in aviation and the overall principle that in the overall analysis, we are talking about the air transport product and connecting people around the globe through air transport as to whether governance in aviation has been for the past 68 years on the right track and if it has not, how we can get it back on track. We have to consider performance governance, otherwise known as business governance, which speaks to performance and value creation through resource utilization.

What do we stand for as a global aviation community? How do we strike the balance between growth and development? Where does aviation and its governance fit into a world transformed by the winds of globalization and change through technology?

Governance is a set of responsibilities and practices that are aimed at achieving strategic direction and ensuring that objectives are achieved. The question then is: "have we a strategic direction and have we achieved our goals in aviation?" In other words, do we rigidly enforce archaic perceptions on sovereignty and territoriality to air transport in the twenty-first century?

Let us start at the beginning. What is aviation's strategic direction? As far back as 1944 The Chicago Convention, intent on "creating and preserving friendship and understanding among the nations and people of the world" declared in its Article 44 that we should strive, through ICAO, to meet the needs of the peoples of the world for safe regular, efficient and economical air transport. One would not imagine, for a moment, that there is written anywhere in a global document that aviation's strategic direction is to make as much money as possible to the exclusion of others or to give priority to the interests of States or the air transport industry or any other business. Creating and preserving friendship and understanding among people can only be achieved through optimum connectivity.

Having given this direction, the same multilateral treaty goes on to say that no scheduled international air service may be operated over or into the territory of a Contracting State except with the special permission or other authorization of that State. It is a curious fact that in 1609, Hugo Grotius wrote in his magnum opus *Maré Liberum* (free seas) that the oceans should be open to sea faring by anyone. Yet more than three centuries later, precisely the opposite principle was adopted in 1944 by the aviation powers that assembled in Chicago.

The trouble with air transport is that, while on the one hand it is a product, on the other hand regulations pertaining to this product may constrain its availability to the consumer by depriving him of the various choices of air travel he might have under a liberalized system. In other words, State policy and the protection of national interests take precedence over the interest of the user of air transport. The aviation industry offers only one product to the ultimate consumer and that is the air transport product.

The air transport industry is cyclical and is profoundly affected by the world's economic health. I need not elaborate the economic vicissitudes of Europe and the significant growth elsewhere in Asia.

Aviation is a global industry and the need for air transportation continues to grow, as major cities continue to grow in population and prosper. A recent forecast has revealed that while in the 1970s there were just four major agglomerations of over ten million people there are 26 today and there will be more than 30 by 2015. As economic prosperity grows more and more people will demand access to air transport and traffic growth is expected to double in 15 years.

"Connectivity" which is the most compelling need in aviation, and embodied in the Chicago Convention as inter alia "meeting the needs of the people of the world for efficient and economical air transport" is stultified by interests of commercial and national policy.

When Emirates commenced its operations to Australia in 1997, the airline was viewed with trepidation and concern by QANTAS, as a threat to its market share. This concern was shared by the Australian authorities. However, attitudes quickly changed, and this concern was obviated when they realized the added economic benefit quickly enjoyed by the places Emirates flew to. Currently, Emirates operates 49 flights a week to Australian cities and hopes to expand this number to 80. QANTAS and Emirates are now partners and that...is the way to go.

The air services agreement between the UAE and the United States allows Emirates to operate to any point in the States, how often they wish with no capacity restriction, and with rights to carry traffic from intermediate points.

In such an environment, how would one view the traditionalist and restrictive interpretations some attribute to Articles 1 and 2 of the Chicago Convention?

Article 3
Civil and State Aircraft

(a) This Convention shall be applicable only to civil aircraft, and shall not be applicable to state aircraft.
(b) Aircraft used in military, customs and police services shall be deemed to be state aircraft.
(c) No state aircraft of a contracting State shall fly over the territory of another State or land thereon without authorization by special agreement or otherwise, and in accordance with the terms thereof.
(d) The contracting States undertake, when issuing regulations for their state aircraft, that they will have due regard for the safety of navigation of civil aircraft.

Contents

1 Identity and Use of Aircraft

Annex 6 to the Chicago Convention defines an aircraft as any machine that can derive its support in the atmosphere from the reactions of the air other than reactions of air on the earth's surface.

Article 3 is indeed an interesting provision which in (b) identifies three types of aircraft as military aircraft but is not clear as to whether the definition is comprehensive or merely inclusive and is open to other types of aircraft to be identified within its umbrella. For example, are aircraft used by a government to douse forest fires or aerial spraying included in this provision? The more plausible approach has been to consider the purpose for which aircraft are used rather than the label they carry. One commentator states:

> The status of military aircraft is not clearly determined by positive rules of international law and is not particularly transparent or unequivocal. The issue is not addressed in international law with any specificity, could not be located in any one single international instrument and only some fragmentary aspects can be deduced directly or indirectly from different sources of international law (international treaties). The identifiable rules are mostly "negative"- stating what does not apply to military aircraft or what such aircraft

R. Abeyratne, *Convention on International Civil Aviation*,
DOI 10.1007/978-3-319-00068-8_4, © Springer International Publishing Switzerland 2014

are not permitted to do. The practice of States that could form a basis for the development of customary law is also not transparent or uniform and is often shrouded in secrecy.[1]

The predecessor of the Chicago Convention—The Paris Convention of 1919 is much clearer when it provides that State aircraft are military aircraft and aircraft exclusively used in State service such as posts, customs and police sand that every other aircraft shall be deemed to be private aircraft. The Paris Convention goes on to say

All State aircraft other than military, customs and police aircraft shall be treated as private aircraft and as such shall be subject to all the provisions of the present Convention.

The clarity of the Paris Convention, which presumably was considered by the forefathers of the Chicago Convention, provides further argument that the Article 3 definition of State Aircraft is an open ended and inclusive one.

Another drawback of the Chicago Convention is that, while on the one hand it explicitly mentions that the treaty will not apply top State (i.e. military aircraft) on the other hand it provides in Article 3 (c) that

No state aircraft of a Contracting State shall fly over the territory of another State or land thereon without authorization by special agreement or otherwise, and in accordance with the terms thereof,

making the Convention applicable to military aircraft to the extent of Article 3 (c).

The ICAO Assembly, at its 14th Session (Rome, 21 August–15 September 1962) adopted Resolution A14-15 (Coordination of Civil and Military Air Traffic), where the Assembly directed the Council to develop guidance material for the joint civil and military use of airspace, taking into account the various policies, practices and means already employed by States to promote the satisfactory practices and means already employed by States to promote the satisfactory coordination or integration of their civil and military air traffic. Following this measure, the Assembly, at its 21st Session (Montreal, 24 September–15 October 1974) adopted Resolution A 21-21 (Consolidated Statement of Continuing Policies and Associated Practices Related Specifically to Air Navigation) whereby in Appendix O, The Assembly, while recognizing that airspace as well as many facilities and services should be used in common by civil aviation and military aviation, resolved that the common use of civil and military aviation of airspace and of certain facilities and services shall be arranged so as to ensure safety, regularity and efficiency of international civil air traffic, particularly in the context of air navigation over the high seas where the internal regulations of States should not compromise or adversely affect the regularity and efficiency of international air traffic.

At its 37th Session held in Montreal from 28 September to 8 October 2010, the ICAO Assembly adopted Resolution A37-15 (Consolidated statement of continuing ICAO policies and associated practices related specifically to air navigation),

[1]Milde (2012) at 63.

Appendix O of which requires States to take appropriate action to coordinate with military authorities to implement a flexible and cooperative approach to airspace organization and management. It also reiterated the statement in Resolution A 21-21 that regulations and procedures established by States to govern the operation of their State aircraft over the high seas shall ensure that these operations did not compromise the safety, regularity and efficiency of international civil air traffic and that to the extent possible such operations should comply with the Rules of the Air contained in Annex 2 to the Chicago Convention.

It must be noted that the laws of air navigation as embodied in the Chicago Convention and derive their genesis therefrom are applicable only to civil aircraft and not to State aircraft. Article 3 of the Convention simply states that aircraft used in military customs and police services shall be deemed to be State aircraft. There are no clear international rules, generally accepted, whether conventional or customary, as to what constitutes state aircraft and what constitutes civil aircraft. Military aircraft, more than any other kind of aircraft including customs and police aircraft, personifies the public or sovereign power of a State, and several attempts have been made to arrive at an internationally acceptable definition thereof. A simplistic but apt definition of civil aviation is "aviation activities carried out by civil aircraft". A civil aircraft has been defined as any aircraft, excluding government and military aircraft, used for the carriage of passengers, baggage, cargo and mail. However, civil aviation comprises in general all aviation activities other than government and military air services which can be divided into three main categories: commercial air transport provided to the public by scheduled or non-scheduled carriers; private flying for business or pleasure; and a wide range of specialized services commonly called aerial work, such as agriculture, construction, photography, surveying, observation and patrol, search and rescue, aerial advertisement et al. By the same token, military aviation must be aviation activities carried out by military aircraft. Military aircraft have been defined as aircraft that are designed or modified for highly specialized use by the armed services of a nation.

Military aviation therefore can be identified as the use of aircraft and other flying machines for the purposes of conducting or enabling warfare, which could include the carriage of military personnel and cargo used in military activities such as the logistical supply to forces stationed along a front. Usually these aircraft include bombers, fighters, fighter bombers and reconnaissance and unmanned attack aircraft such as drones. These varied types of aircraft allow for the completion of a wide variety of objectives.

Arguably, the most fundamental difference between the operation of civil and military aircraft lies in the fact that, although they are expected to share the same skies, the procedures by which they do this vary greatly. Civil aircraft depend entirely on predetermined flight paths and code of commercial conduct which vary depending on aircraft type and types of traffic carried, whereas military aircraft operate in line with the exigency of a situation and are not necessarily always guided by predetermined flight paths. This dichotomy led to the adoption, at the 10th Session of the ICAO Assembly (Caracas, 19 June–16 July 1956) of Resolution

A10-19 which, while recognizing that the skies (airspace) as well as many other facilities and services were commonly shared between civil and military aviation, focused on ICAO's mandate to promote the safety of flight.

The distinction between civil and military aviation cannot be made without addressing the purpose for which an aircraft is employed. This is particularly significant in instances where civil aircraft are used for military purposes. The fact that military strategists have come to expect support services from civil aviation is becoming more evident with the increasing need for military operations both in war situations and in instances of human tragedy brought about by civil conflict or natural disasters. There have been many such instances, ranging from the use by British military of chartered commercial cargo aircraft in the Falklands in 1982 to earlier practices of India and Pakistan in 1971 when both countries used civilian passenger aircraft for the transportation of their troops during the Indo-Pakistan war.

The use of civil aircraft for military purposes and vice versa intrinsically brings to bear the issue of sharing of airspace between civil and military aircraft. Military aviation and civil aviation are intrinsically different from each other in their nature and functions. However, both operate in the same air traffic management environment and therefore use common airspace which needs to be stringently managed, not only for safety reasons but also for reasons of efficiency. While military aviation is essential for national security and defence and therefore is a legitimate and indispensable activity, civil air transport is not only necessary for global interaction between nations but it also makes a significant contribution to the global economy. These two equally important activities call for uncompromising cooperation between one another in the shared use of airspace and an enduring understanding of each other's needs. Military aviation not only includes the operation of conventional aircraft for military purposes but also involves the use of unmanned aircraft systems (RPAS) and missile testing, all of which call for a close look at the use of airspace in the modern context.

ICAO has issued guidelines on the coordination between military authorities and air traffic services (ATS) authorities which recognize *in limine* that coordination between the responsible military authorities and appropriate ATS authorities is essential to the safety of civil aircraft operations whenever activities potentially hazardous to such operations are planned and conducted by any military units. These guidelines go on to state that in the event that a sudden outbreak of armed hostilities or any other factors preclude this normal coordination process, appropriate State and ATS authorities, civil aircraft operators and pilots-in-command of aircraft must assess the situation based on the information available and plan their actions so as not to jeopardize safety.

The Guidelines recommend that, in the event that a military unit observes that a civil aircraft is entering, or is about to enter, a designated prohibited, restricted or danger area or any other area of activity which constitutes potential hazards, a warning to the aircraft should be issued through the responsible ATS unit. The warning should include advice on the change of heading required to leave, or circumvent, the area. If the military unit is unable to contact the responsible ATS unit immediately and the situation is deemed to be a genuine emergency, an appropriate warning to the aircraft may be transmitted on the Very High Frequency (VHF) emergency channel 121.5 MHz. If the identity of the aircraft is not known, it is important that the warning include the Special Service Request (SSR) code, if observed, and describe the position of the aircraft in a form meaningful to the pilot, e.g. by reference to an ATS route and/or the direction and distance from an airport or an aeronautical radio navigation aid, an established waypoint or reporting point. In the case where an unauthorized aircraft is observed visually to be flying in, or about to enter a prohibited, restricted or danger area, the following visual signal is prescribed by the International Standards in Annex 2 to the Chicago Convention—Rules of the Air, Appendix 1 to indicate that the aircraft is to take such remedial action as is necessary. The Guidelines caution that the importance of co-ordinating with the responsible ATS unit(s), whenever possible, the issuance of any warnings and advice to civil aircraft regarding changes of flight path should be emphasized in any briefings or instructions given by military authorities to their units, since uncoordinated warnings and associated navigational advice, when followed, may result in a potential risk of collision with other aircraft in the area.

The objective of the co-ordination between the military authorities planning activities potentially hazardous to civil aircraft and the responsible ATS authorities is to reach agreement on the best arrangements which will avoid hazards to civil aircraft and minimize interference with the normal operations of civil aircraft. Ideally, this means the selection of locations outside promulgated ATS routes and controlled airspace for the conduct of the potentially hazardous activities. If the selection of such locations is not possible due to the nature and scope of the planned activities, temporary restrictions imposed on civil air traffic should be kept to a minimum through close co-ordination between the military units and the ATS unit.

In recent times, both statesmen and members of the aviation community have been consciously aware of the dual role played by civil and military aviation, while sharing the same sky. This has called for delicate diplomacy and political compromise. When dealing with issues of aviation to which politics is applied, it is important to remember that from the distant past, it has been recognized that a nation's air power is the sum total of all its civil and military aviation resources.[2] Furthermore, the importance of aviation toward maintaining peace has been accepted since World War 2 and is aptly reflected in the Statement of the British at that time, that civil aviation holds the key to power and importance of a nation and therefore it must be regulated or controlled by international authority.[3] Lord Beaverbrook for the British Government of that time stated in Parliament:

---

[2]van Zandt (1944) at pp. 28, 93.

[3]Wings for Peace—Labour's Post War Policy for Civil Flying, published by the Labour Party of England, April 1944, cited in van Zandt, *Id.* at 1.

> Our first concern will be to gain general acceptance of certain broad principles whereby civil aviation can be made into a benign influence for welding the nations of the world together into a closer cooperation. . .it will be our aim to make civil aviation a guarantee of international solidarity, a mainstay of world peace.[4]

The intensely political overtones that moulded the incipient civil aviation system of the world immediately after the War, thereby incontrovertibly establishing the relevance of diplomacy, international politics and international relations in civil aviation, is borne out by the statement of the first President of the ICAO Council when he said:

It is well that we should be reminded. . .if the extent of the part which diplomatic and military considerations have played in international air transport, even in periods of undisturbed peace. We shall have a false idea of air transport's history, and a very false view of the problems of planning its future, if we think of it purely as a commercial enterprise, or neglect the extent to which political considerations have been controlling in shaping its course.[5]

In retrospect, it must be noted that this statement is a true reflection of what civil aviation stood for at that time, and, more importantly, that the statement has weathered the passage of time and is true even in the present context. A more recent commentator correctly observes that over the past decades, civil aviation has had to serve the political and economic interests of States and that, in this regard, ICAO has alternated between two positions, in its unobtrusive diplomatic role and its more pronounced regulatory role.[6]

An inherent characteristic of aviation is its ability to forge inroads into human affairs and promote international discourse. It also promotes international goodwill and develops "a feeling of brotherhood among the peoples of the world".[7] Therefore, it has been claimed that problems of international civil aviation constitute an integral part of the universal political problems of world organization and therefore aviation problems cannot be solved without involving the world political and diplomatic machinery.[8] It is at these crossroads that one encounters the profound involvement of the United Nations mechanism in general and ICAO in particular.

Military aviation and civil aviation are intrinsically different from each other in their nature and functions. However, both operate in the same air traffic management environment and therefore use common airspace which needs to be stringently managed, not only for safety reasons but also for reasons of efficiency. While military aviation is essential for national security and defence and therefore is a legitimate and indispensable activity, civil air transport is not only necessary for global interaction between nations but it also makes a significant contribution to the

---

[4] *Flight*, Vol. XLV No. 1331, January 27, 1944, at pp. 97–98.

[5] Warner (1942), p. V.

[6] Sochor (1991) at xvi.

[7] Schenkman (1955), at p. 6.

[8] *Id.* vi.

global economy.[9] These two equally important activities call for uncompromising cooperation between one another in the shared use of airspace and an enduring understanding of each other's needs. Military aviation not only includes the operation of conventional aircraft for military purposes but also involves the use of unmanned aerial systems (RPAS)[10] and missile testing, all of which call for a close look at the use of airspace in the modern context.

The above considerations of safety notwithstanding, it is incontrovertible that cooperation in the activities of military and civil aviation is not only about sharing airspace. It is also about the efficient allocation of airspace to both categories of activity in separating such flights, particularly in the context of military flights which operate in special use airspace and those proceeding to special use airspace across civilian air routes. This brings to bear the inevitable conclusion that there must essentially be coordination between military authorities and air navigation service authorities.

At the Global Air Traffic Management Forum on Civil and Military Cooperation,[11] convened by ICAO on 19 October 2009, the International Air Transport Association (IATA)[12] noted that, given the equal importance of civil and military

---

[9]Abeyratne (2007) at 25–47.

[10]The potential explosion of unmanned aircraft Systems (commonly called RPASs) in airspace also brings to bear the need to have a closer look at the civil–military aviation airspace demarcation. RPASs are commonly associated with military operations in many parts of the world. The question that would arise in this context is how would a State feel about sharing airspace over contiguous States with a swarm of RPASs operated by a mix of military/law enforcement and commercial enterprises? For more information see Abeyratne (2009).

[11]The theme of the Forum was "*Time to take it global: Meeting each other's needs without compromising the Mission.*" The event was held as a follow up to recommendations of the Eleventh Air Navigation Conference (Doc 9828, Rec. 1/2) concerning coordination with military authorities with a view to achieving enhanced airspace organization and management and as an integral supporting mechanism of the successful series of civil/military air traffic management summits instituted by the Air Traffic Control Association (ATCA). It was also a follow up to ICAO Assembly Resolution A36-13, Appendix O, *Coordination of civil and military air traffic* wherein States are asked to take appropriate action to coordinate with military authorities to implement a flexible and cooperative approach to airspace organization and management. The Forum was intended to create awareness among civil and military policy makers and regulators, civil and military air navigation service providers (ANSPs) and civil and military airspace users, on the need to improve civil/military cooperation and coordination in support of an optimum use of airspace by all users.

[12]The International Air Transport Association, an association of air carriers, was formed in 1919 as the International Air Traffic Association. Encapsulated in IATA's overall mission are seven core objectives: to promote safe, reliable and secure air services; to achieve recognition of the importance of a healthy air transport industry to worldwide social and economic development; to assist the air transport industry in achieving adequate levels of profitability; to provide high quality, value for money, industry-required products and services that meet the needs of the customer; to develop cost effective, environmentally-friendly standards and procedures to facilitate the operation of international air transport; to identify and articulate common industry positions and support the resolution of key industry issues; and to provide a working environment which attracts, retains and develops committed employees.

aviation, it was imperative that airspace, which is an international and national resource, be managed as a whole, as a continuum and one common source and not a collection of segregated areas. This called for minimal restrictions on the use of airspace by both users, which in turn called for a structured and systematic management of the scope and duration of the use of airspace.

At the ICAO Forum, the Civil Air Navigation Services Organization (CANSO)[13] underscored the fact that increasing growth in civil air transport and traffic was putting pressure on limited airspace resources and that civil–military cooperation was becoming imperative. CANSO, while calling for a global platform of cooperation, emphasized that the key to successful cooperation is the establishment of trust, respect, transparency and flexibility on all key players and that States could play a lead role in developing a framework of cooperation. It also stated that a regional approach (as against a national approach) was essential, citing EUROCONTROL[14] as a true civil military agency which involved both civil and military offices at policy making level. In summing up, CANSO called for a fully integrated Civil–Military ATM, leading to the complete union of Civil–Military partners at national, regional and global level.[15]

A good example of the management system called for by IATA, and balanced cooperation as referred to by CANSO is the establishment of a Single European Sky (SES) legislation that is aimed at ensuring a harmonized regulatory framework for air traffic management and which uniformly and harmoniously applies in all 27 member States of the EU and 28 other associated States surrounding the Union. This legislation is accompanied by a technology programme called Single European Sky Air Traffic Management Research (SESAR) which modernizes and helps run the European air traffic control infrastructure modernization programme making SES and SESAR the essential components of the full air transport policy of Europe.

The outcome of this merger between policy and infrastructure technology has resulted in a robust civil–military aviation cooperation enabling all EU member States to be represented by a civilian and a military officer in the EU Single Sky Committee (which, *inter alia*, develops legislation) and military officers to be included in other bodies working on SES and SESAR.

---

[13]CANSO is the global voice of the air traffic management profession. Its members comprise over 50 air navigation service providers who control more than 85 % of global air traffic movements. CANSO seeks to promote best practices within the industry.

[14]EUROCONTROL, the European Organisation for the Safety of Air Navigation, is an intergovernmental organisation made up of 38 Member States and the European Community. Its primary objective is the development of a seamless, pan-European air traffic management (ATM) system. EUROCONTROL contributes to **making European aviation safer, performance-driven and environmentally sustainable.** It was originally founded in 1960 as a **civil–military organisation** to deal with air traffic control for civil and military users in the upper airspace of its six founding European Member States. EUROCONTROL has developed into **a vital European repository of ATM excellence**, both leading and supporting ATM improvements across Europe.

[15]See Civil–Military Cooperation—The CANSO Perspective, October 2009.

The counterpart of SESAR in the United States is the Next Generation Air Transport System (NextGen). Next Gen, which is scheduled to be effective from 2012 to 2025, calls for a shift in airspace management to a trajectory-based system. It will have the following five attributes: Automatic dependent surveillance broadcast (ADS-B) which will use the Global Positioning System (GPS) satellite signals to provide air traffic controllers and pilots with much more accurate information that will help to keep aircraft safely separated both in the air and op runways; System-wide Information Management System (SWIM) which will provide a single infrastructure and information management system to deliver high quality, timely data to many users and applications; Next Generation Data Communications which will provide an additional means of two-way communication for air traffic control clearances, instructions, advisories, flight crew requests and reports; Next Generation Network Enabled Weather (NNEW) which will cut weather-related delays at least in half; and NAS Voice Switch which will replace existing voice systems with a single air/ground and ground/ground voice communications system.

Both SESAR and NextGen, which are targeted for post 2020, would improve the performance of the air traffic management system by combining increased automation with new procedures that improve and achieve benefits related to safety, economic efficiency, capacity and environmental protection.

2 Distinction Between Civil and Military Aviation

A simplistic but apt definition of civil aviation is "aviation activities carried out by civil aircraft". A civil aircraft has been defined as any aircraft, excluding government and military aircraft, used for the carriage of passengers, baggage, cargo and mail.[16] However, civil aviation comprises in general all aviation activities other than government and military air services which can be divided into three main categories: commercial air transport provided to the public by scheduled or non scheduled carriers; private flying for business or pleasure; and a wide range of specialized services commonly called aerial work, such as agriculture, construction, photography, surveying, observation and patrol, search and rescue, aerial advertisement et al.[17] By the same token, military aviation must be aviation activities carried out by military aircraft. Military aircraft have been defined as aircraft that are designed or modified for highly specialized use by the armed services of a nation.[18]

Military aviation therefore can be identified as the use of aircraft and other flying machines for the purposes of conducting or enabling warfare, which could include the carriage of military personnel and cargo used in military activities such as the

[16]Groenewege (1999), at 437. It must also be noted that an aircraft has been defined in Annexes 6, 7 and 8 to the Chicago Convention as any machine which can derive support in the atmosphere from the reactions of air other than the reactions of air on the Earth's surface.

[17]*Ibid.*

[18]http://www.answers.com/topic/military-aircraft.

logistical supply to forces stationed along a front. Usually these aircraft include bombers, fighters, fighter bombers and reconnaissance and unmanned attack aircraft such as drones.[19] These varied types of aircraft allow for the completion of a wide variety of objectives.

Arguably, the most fundamental difference between the operation of civil and military aircraft lies in the fact that, although they are expected to share the same skies, the procedures by which they do this vary greatly. Civil aircraft depend entirely on predetermined flight paths and code of commercial conduct which vary depending on aircraft type and types of traffic carried, whereas military aircraft operate in line with the exigency of a situation and are not necessarily always guided by predetermined flight paths. This dichotomy led to the adoption, at the 10th Session of the ICAO Assembly (Caracas, 19 June–16 July 1956) of Resolution A10-19 which, while recognizing that the skies (airspace) as well as many other facilities and services were commonly shared between civil and military aviation, focused on ICAO's mandate to promote the safety of flight.[20]

The preponderance of weight in prioritizing civil and military aviation seems therefore to be in favour of civil aviation, particularly when taking into consideration this Resolution and the earlier discussion on Annex 11 to the Chicago Convention, thus attenuating the principle that military aviation should, of necessity, consider the compelling need to protect civil aviation from the spontaneous risks that the former may carry with it.

The above notwithstanding, a glaring example of conflict in the civil and military aviation environment can be seen in the ongoing conflict between Greece and Turkey (Aegean crisis). According to reports,[21] the core of the conflict is the persistent abuse of "Flight Information Region" (FIR)[22] responsibility by Greece. FIR responsibility over the Aegean international airspace was assumed in 1952 by Greece. The report goes on to say that Greece considers the FIR as a national boundary line and a defence perimeter (i.e. Western boundary of Turkey and Eastern boundary of Greece, embracing all international airspace in the Aegean beyond Turkish territorial sea within Greek sovereignty area) and that consequently,

---

[19]In a report released on 21 December 2009, Venezuelan President Hugo Chavez is reported to have announced that, on Sunday 20 December, military drones had penetrated Venezuelan airspace along the North-western border with Colombia He had warned that Venezuela was prepared to defend itself if any State were to violate its sovereignty. See http://www.venezuelanalysis.com/news/5022. On 4 January 2010, it was reported that a US drone had fired two missiles in Pakistan, flattening an extremist hideout in Pakistan's lawless tribal belt on Sunday 3 January 2010, killing five militants in a recent spike in drone attacks. See http://www.channelnewsasia.com/stories/afp_asiapacific/view/1028351/1/.html.

[20]As per Article 44 of the Chicago Convention.

[21]http://www.aegeancrisis.org/category/air-space/.

[22]FIRs were devised by ICAO in the 1950s to provide facilities and services to the civilian aircraft in the international airspace. FIR arrangements solely entail technical responsibility. It does not change the free status of the airspace over the high seas under international law.

Greece maintains the view that military aircraft entering into Athens FIR and flying in international airspace should submit flight plans and come under control of Greek air traffic control authorities.

Against Greece's alleged claim that non-submission of flight plans by Turkish military aircraft constitutes a violation of the Greek FIR, some have argued that there is no need for Turkish military aircraft to file flight plans under the Chicago Convention as the Convention explicitly states in Article 3 that it would not apply to State aircraft (which includes military aircraft) and that there is no possibility of one violating an FIR which has nothing to do with the territorial sovereignty of a State, thus leading to the conclusion that Greek abuse of FIR responsibility is yet another manifestation of her claim of "de facto sovereignty" over the whole Aegean airspace.

3 The Use of Civil Aircraft for Military Purposes

The distinction between civil and military aviation cannot be made without addressing the purpose for which an aircraft is employed. This is particularly significant in instances where civil aircraft are used for military purposes. The fact that military strategists have come to expect support services from civil aviation is becoming more evident with the increasing need for military operations both in war situations and in instances of human tragedy brought about by civil conflict or natural disasters. There have been many such instances, ranging from the use by British military of chartered commercial cargo aircraft in the Falklands in 1982 to earlier practices of India and Pakistan in 1971 when both countries used civilian passenger aircraft for the transportation of their troops during the Indo-Pakistan war.

The use of civil aircraft for military purposes brings to bear issues of identification of aircraft and the status of aircraft under article 3 of the Chicago Convention.[23] The question as to whether civil aviation and military aviation have demarcated operational regimes or whether they can still function in symbiosis has become an argumentative one in view of developments in the air transport industry which have occurred over the years. There are some determinants in this regard. Firstly, the nature of the cargo carried. Are they supplies or equipment for the military, customs or police services of a State? Article 35 of the Chicago Convention recognizes that the mere carriage "of munitions or implements of war" does not by itself make an aircraft a state aircraft. Then there is the question of ownership of the aircraft. Is it owned privately or by the State? The degree of control and supervision of the operation of the aircraft by the specified services are also factors to be considered in this equation. The nature of the passengers or personnel carried is also a consideration. Are they military, customs or police officials, or members of the

[23] Article 3 of the Chicago Convention states that the Convention applies only to civil aircraft and not to State aircraft, and goes on to explain that aircraft used in military, customs and police services shall be deemed to be State aircraft. Article 3 (c) prohibits State aircraft of one State from flying over the territory of another State or landing thereon without special agreement or otherwise.

public at large? Is the particular flight open for use by members of the public? Do aircraft registration and nationality markings become relevant? Will a usual civil (ICAO) flight plan be submitted and the usual air traffic clearances obtained? What is the nature of crew? Are the crew civilian, or are they military, customs or police personnel, or employed by these services? Who is the operator? Is the operator a military, customs or police agent? What sort of documentation is carried in the aircraft? Are the documents required by the Chicago Convention and its Annexes to be carried on civil aircraft in fact being carried (e.g. certificate of registration, certificate of airworthiness, licences for the crew, journey log book, etc.)? What would the area of operations be? Will the aircraft fly to, or over, areas in a situation of on-going or imminent armed conflict? What about customs clearances? Will the normal clearances be obtained?

The broad answer to all these questions would lie in the fact that, in the ultimate analysis, the responsibility of using civil aircraft and crew for military purposes rests with the State concerned. The fundamental legal premise which applies in such situations is that, in international relations, the erosion of one's legal interests by another brings to bear the latter's responsibility. State responsibility is a recognized principle of international law in the current context. The law of international responsibility involves the incidence and consequence of acts which are irregular at international law, leading to the payment of compensation for the loss caused. It might therefore just be worthwhile to inquire as to whether Article 89[24] of the Chicago Convention should be reviewed so that the international community and ICAO could be given more flexibility in the determination of propriety in the use of civil aircraft for military purposes.

4 Some Recent Developments

At the Global Air Traffic Management Forum on Civil and Military Cooperation[25] ICAO subsumed its position by stating that airspace is a natural resource with finite capacity for which demand from all users is constantly expanding and that there has been an increased requirement on airspace use to meet a fast-growing aviation demand. States elected to be parties to the Chicago Convention in order that international civil aviation may be developed in a safe and orderly manner and that international air transport services may be established on the basis of equality of opportunity and operated soundly and economically. To achieve these objectives in recent years and to take due account of current and future needs in aviation, ICAO developed its vision of a seamless air traffic management (ATM) system.[26]

[24]See McDougal et al. (1963) at 306–311 where the author suggests that if for security reasons States have certain claims on those who enter their sovereign territories, such claims may not be inconsistent with the principles of international law.

[25]See Dutton (2009) at 691.

[26]Global Air Traffic Management Operational Concept, *ICAO Doc 9854*, AN/458, First Edition-2005, Chapter 1, Para 1.1.

ICAO further advised that, although the Chicago Convention governs international civil aviation and is not applicable to State aircraft (aircraft used in military, customs and police services)[27] State aircraft as well as military CNS/ATM systems and services are an integral part of the aviation community. A much closer cooperation between civil and military organizations will contribute to the vision encapsulated in the preamble to the Chicago Convention, leading to the optimum use of the airspace and balancing State requirements for both civil and military aviation.

ICAO drew the attention of the Forum to Assembly Resolution A 36-13[28] adopted at the 36th ICAO Assembly (Montreal, 18–28 September 2007), Appendix O of which recognizes that the airspace as well as many facilities and services should be used in common by civil aviation and military aviation and that the ICAO Global ATB Operational Concept[29] states that all airspace should be a usable resource and that therefore any restriction on the use of any particular volume of airspace should be considered transitory, and all airspace should be managed flexibly. It was noted by the Forum that, through A36-13, the Assembly resolved that the common use by civil and military aviation of airspace and of certain facilities and services shall be arranged so as to ensure safety, regularity and efficiency of international civil air traffic and that the regulations established by ICAO member States to govern the operation of their State aircraft over the high seas shall ensure that these operations did not compromise the safety, regularity and efficiency of international civil air traffic and to the extent possible such operations conformed to the Rules of the Air contained in Annex 2 to the Chicago Convention. The resolution also requested the Council of ICAO to provide guidance and advice to States that wished to establish civil/military agreements.

Against this backdrop, ICAO advised the Forum of the need for a strengthened civil/military cooperation and coordination which called upon ICAO Member States to initiate as necessary or improve the coordination between their civil and military air traffic services. It was important that States, in view of the increasing need to cooperate with multiple airspace users, developed an integrated and cohesive civil–military coordination strategy with a roadmap indicating short, mid and long term objectives. ICAO further advised that the benefits of enhancing civil–military cooperation should be considered at the global level with a view to identifying best practices through dialogue and exchange of information. Effective civil/military cooperation and coordination is required not only to meet future civil and military air traffic requirements for increased safety, security, capacity,

[27]Chicago Convention, Preamble (*supra* note 1), Article 3.

[28]Consolidated statement of continuing ICAO policies and associated practices related specifically to air navigation, Assembly Resolutions in Force (as of 28 September 2007), *Doc 9902*, II-2.

[29]*Hamdi* v. *Rumsfeld*, 542 US 5047 (2004).

efficiency, environmental sustainability but also to achieve interoperability, seamlessness and harmonisation through sound policy, a structured framework, effective liaisons and management at working level.[30]

With regard to Conventions other than the Chicago Convention, one can see some provisions which are relevant to the discussion on the distinction between civil and military aircraft, the latter of which, by implication, includes RPASs. The Convention on the International Recognition of Rights in Aircraft (Geneva. 1948), the Convention on Offences and Certain Other Acts Committed on Board Aircraft (Tokyo, 1963), the Convention for the Suppression of Unlawful Seizure of Aircraft (The Hague, 1970) and the Convention for the Suppression of Unlawful Acts Against the Safety of Civil Aviation (Montreal, 1971), all contain a provision that "this Convention shall not apply to aircraft used in military, customs or police services." This appears to be a more simple way to indicate the scope of applicability of these Conventions than the provisions of Article 3 (a) and (b) of the Chicago Convention, although the end result seems to be the same. Furthermore the clear implication is that all aircraft not so used would be subject to the provisions of the respective Conventions (paragraph 1.4 above refers).

The Convention on Damage Caused by Foreign Aircraft to Third Parties on the Surface (Rome, 1952) states in Article 26 that, "this Convention shall not apply to damage caused by military, customs or police aircraft." It should be noted that a "military, customs or police aircraft" is not necessarily the same thing as an "aircraft used in military, customs and police services" although again the expression "military, customs or police aircraft" was left undefined. Similarly, other "state" aircraft fall within the scope of the Convention. However, the 1978 Protocol to amend this Convention reverts to more familiar language; it would amend Article 26 by replacing it with, "this Convention shall not apply to damage caused by aircraft used in military, customs and police services."

The Convention for the Unification of Certain Rules Relating to the Precautionary Attachment of Aircraft (Rome, 1933) provides that certain categories of aircraft are exempt from precautionary attachment, including aircraft assigned exclusively to a government service, including postal services, but not commercial aircraft. On the other hand, the Convention for the Unification of Certain Rules Relating to Assistance and Salvage of Aircraft or by Aircraft at Sea (Brussels), 1938 "apply to government vessels and aircraft, with the exception of military, customs and police vessels or aircraft . . ."

The Convention for the Unification of Certain Rules Relating to International Carriage By Air (Warsaw, 1929) applies, *inter alia* to all international carriage of persons, luggage or goods performed by aircraft for reward, regardless of the classification of the aircraft. Article 2 specifically provides that the Convention

[30]In its briefing, ICAO emphasized that cooperation between civil and military authorities should be aimed at achieving optimal use of the airspace resulting in increased airspace capacity, operational flexibility, and savings in flying time, fuel and CO_2 emissions. The Forum noted that safety, economical impact, efficiency and interoperability are objectives shared by both civil and military aviation communities.

applies to carriage performed by the State or by legally constituted public bodies, but by virtue of the Additional Protocol, Parties may make a declaration at the time of ratification or accession that Article 2 (1) shall not apply to international carriage performed directly by the State. The Hague Protocol of 1955 to amend this Convention, in Article XXVI allows a State to declare that the Convention as amended by the Protocol shall not apply to the carriage of persons, cargo and baggage for its military authorities on aircraft, registered in that State, the whole capacity of which has been reserved by or on behalf of such authorities. Identical provisions are contained, *mutatis mutandis,* in the Guatemala City Protocol of 1971 (Article XXIII) the *1975* Additional Protocol No. 2 (Montreal), the *1975* Additional Protocol No. 3 (Montreal) and in Montreal Protocol No. 4 of 1975. It is submitted that Article 3 (b) of the Chicago Convention has no bearing on the applicability of these instruments of the "Warsaw System" which specify their own scope of applicability.

The Montreal Convention of 1999[31] which replaced the Warsaw Convention of 1929 also stipulates in its Article 1 that the Convention applies to all international carriage of persons, baggage or cargo performed by aircraft for reward. Like its predecessor, the Montreal Convention does not distinguish between civil and military or other State aircraft.

This analysis of some international air law instruments illustrates that many post-Chicago air law instruments (Geneva 1948, Tokyo 1963, The Hague 1970, Montreal 1971 and Rome 1952 and as amended in 1978) all have broadly similar provisions to Article 3 (a) and (b) of the Chicago Convention. The private air law instruments of the Warsaw System on the other hand, because of their nature, have adopted different formulae.

The provisions of the Chicago Convention and Annexes would not apply in a case where a state aircraft is (mistakenly or otherwise) operated on the basis that it is a civil aircraft. Similarly, the Geneva Convention of 1948, the Tokyo Convention of 1963, The Hague Convention of 1970, the Montreal Convention of 1971 and the Rome Convention (*1952*) as amended in 1978, will also not be applicable where it is determined that the aircraft was "used in military, customs or police services". The converse, of a civil aircraft being operated on the basis that it is a state aircraft, would theoretically raise the same problems (i.e. legal regimes thought to be inapplicable are in fact applicable). Concern is not often expressed in this regard.

Another frequently mentioned difficulty is claimed to be the loss of insurance coverage in respect of the aircraft (hull), operator, crew and passengers or other parties where the aircraft is in fact state aircraft. The question whether a particular insurance coverage is rendered invalid in such situations is primarily a private law matter of the construction and interpretation of the insurance contract. Unless the contract has an exclusion clause which specifically makes reference to the classification in Article 3 of the Chicago Convention (e.g. loss of coverage where the

[31] *Convention for the Unification of Certain Rules for International Carriage by Air*, signed at Montreal on 28 May 1999. ICAO Doc 9740.

operation is of a state (or civil) aircraft as defined in the Chicago Convention), then the Convention will have no bearing on the contract, and this issue of the loss of insurance coverage is not germane to this study. Frequently, the policy will exclude usage of the aircraft "for any purpose other than those stated" in a Schedule; among the exclusions would be any use involving abnormal hazards. Nearly every aviation hull and liability policy now excludes losses due to war, invasion, hostilities, rebellion. etc., although insurance to cover such losses can usually be obtained by the payment of a higher premium. However, the instances mentioned do not require a determination of whether the aircraft is considered to be state or civil under the Chicago Convention.

A question sometimes asked is whether national civil laws and regulations would apply to civilian flight crews operating what is a state aircraft under the Chicago Convention. Would civil or military investigative and judicial processes be applied, for example in the case of an accident? The answer would depend largely on the domestic laws of the State concerned. The fundamental principle is stated in Article 1 of the Convention: every State has complete and exclusive sovereignty over the airspace above its territory. Furthermore, subject to the provisions of the Convention, the laws and regulations of a Contracting State relating to the admission to or departure from its territory of aircraft engaged in international air navigation, or the operation and navigation of such aircraft within its territory, shall be complied with by (civil) aircraft of other Contracting States, upon entering or departing from or while in the territory of that State. *A fortiori,* state aircraft are also subject to the laws of the subjacent State.

In the case of an accident involving state aircraft, States are not bound by Article 26 of the Chicago Convention and Annex 13. However, they can voluntarily (through their legislation) apply' these provisions. Sometimes, the legislation specifies a different procedure in relation to military aircraft only; all other aircraft, including those used in customs or police services, are treated as civilian in this regard. In the case of other incidents, where for example the requisite over-flight permission has not been obtained by a state aircraft, which is then forced to land and charges brought against the crew, again the answer would depend on the domestic laws of the over-flown State and the factual circumstances. The classification of an aircraft as "state" aircraft under the Convention does not necessarily mean that military laws and procedures of a State would apply to that aircraft or its crew. The current or any different classification of aircraft under the Convention would not be determinative whether a particular State, in the exercise of its sovereignty, would make that aircraft and/or its crew subject to civil or military laws and regulations. As a matter of practice States usually apply military rules and processes to military aircraft and personnel only. At the international level, attempts to arrive at a common, acceptable definition of military aircraft have met with a singular lack of success.

Even though there are no international regulations applicable to RPASs, it is clear that there are certain rules that States are required to adhere to in order to ensure that RPASs operated under their control do not adversely affect civil air transport. The various provisions of the Chicago Convention and its Annexes cited in this article as well as the numerous ICAO Assembly resolutions quoted leave no room for doubt that there is an existing regime that addresses the safety of de-segregated air space when it comes to the operation of civil and State aircraft. This regime derives its legal legitimacy from the principles of State responsibility which are now accepted as binding on States. Article 1 of the *Articles of Responsibility* of the International Law Commission (ILC) expressly stipulates that every internationally wrongful act entails the international responsibility of a State.[32]

Paul Stephen Dempsey[33] sums it up well, when he says that the issue of air traffic management has two critical considerations, one relating to legal issues and the other impacting public policy. Dempsey states correctly that the skies belong to the public and the sovereign is but the trustee in this regard. Therefore, inasmuch as States cannot abdicate or pass on their responsibility and accountability of their traditional function and fiduciary responsibility, ICAO too has responsibility under Chapter XV of the Chicago Convention to assist States needing help with regard to the provision of air navigation services.

These obligations are *erga omnes* affecting all States and thus cannot be made inapplicable to a State or group of States by an exclusive clause in a treaty or other document reflecting legal obligations without the consent of the international community as a whole. Besides, holding governments responsible will ensure proper quality control in the provision of air navigation services.

Article 3 *bis*

(a) The contracting States recognize that every State must (a) refrain from resorting to the use of weapons against civil aircraft in flight and that, in case of interception, the lives of persons on board and the safety of aircraft must not be endangered. This provision shall not be interpreted as modifying in any way the rights and obligations of States set forth in the Charter of the United Nations.

(b) The contracting States recognize that every State, in the exercise of its sovereignty, is entitled to require the landing at some designated airport of a civil aircraft flying above its territory without authority or if there are reasonable grounds to conclude that it is being used for any purpose inconsistent with the aims of this Convention; it may also give such aircraft any other instructions to put an end to such violations. For this purpose, the contracting States may resort to any appropriate means consistent with relevant rules of international law, including the relevant provisions of this Convention, specifically paragraph (a) of this Article. Each contracting State agrees to publish its regulations in force regarding the interception of civil aircraft.

32. See Crawford (2002), p. 77.

33. See Dempsey (2003) at 118–119.

(c) Every civil aircraft shall comply with an order given inconformity with paragraph (b) of this Article. To this end each contracting State shall establish all necessary provisions in its national laws or regulations to make such compliance mandatory for any civil aircraft registered in that State or operated by an operator who has his principal place of business or permanent residence in that State. Each contracting State shall make any violation of such applicable laws or regulations punishable by severe penalties and shall submit the case to its competent authorities in accordance with its laws or regulations.

(d) Each contracting State shall take appropriate measures to prohibit the deliberate use of any civil aircraft registered in that State or operated by an operator who has his principal place of business or permanent residence in that State for any purpose inconsistent with the aims of this Convention. This provision shall not affect paragraph (a) or derogate from paragraph (b) and (c) of this Article.

Conduct of and Action Against Aircraft

Article 3 *bis* to the Chicago Convention was adopted by consensus on 10 May 1984 by the 25th Session (Extraordinary) of the Assembly. The Protocol, which entered into force on 1 October 1998, and currently has 143 parties, is commonly regarded as codifying existing rules of customary international law. The term "purpose[s] inconsistent with the aims of [the] [this] Convention" is used twice in the Preambular clauses, as well as in substantive paragraphs (b) and (d). However, C-WP/8217 points out that the drafting history indicates that the scope of the phrase is different in Article 3 *bis* than in Article 4 which will be discussed next.

At first glance, one would conclude correctly that this is indeed a watered down provision that reflects the sensitive nuances of the wording of the Chicago Convention as discussed in the Preface of this book. For one, the provision states that

> *Contracting States recognize that every State must a) refrain from resorting to the use of weapons against civil aircraft in flight...*

The two operative words are "recognize" and "refrain". This provision came into being as a result of ICAO Assembly deliberations on the shooting down of Korean Air Flight 007 over Soviet airspace. It becomes obvious that this provision was cobbled together amidst highly charged and polarized discussions at the Assembly, and the consensus reached—which most participating States could "live with" (a popular phrase within ICAO Assemblies and Council sessions) were these two words. There were of course States which insisted on more peremptory words to deal with this extremely important issue of aviation safety. Had these States prevailed Article 3 *bis* would have started with words to the effect that:

> Contracting States shall *ensure* that every State must a) *not* resort to to the use of weapons against civil aircraft in flight...

A grave concern confronting the civil aviation community is that, with the proliferation of military activity will inevitably come the issue of endangerment of air routes.

The consequences of the nuclear missile firings of 5 July 2006 by the Democratic Peoples' Republic of Korea (DPRK) brought to bear the hazards and grave dangers such activities pose to civil aviation. In this instance, missiles launched by DPRK crossed several international air routes over the high seas. It was revealed that, when extrapolating the projected paths of some of the missiles, it appeared that they could have interfered with many more air routes, both over Japan and the air space of the North Pacific Ocean.

This is not the first instance of its kind. A similar incident took place on 31 August 1998 in the same vicinity in which the North Korean missiles were fired in July 2006. An object propelled by rockets was launched by North Korea and a part of the object hit the sea in the Pacific Ocean off the coast of Sanriku in north-eastern Japan. The impact area of the object was in the vicinity of the international airway A590 which is known as composing NOPAC Composite Route System, a trunk route connecting Asia and North America where some 180 flights of various countries fly every day.

The member States of ICAO at the 32nd Session of the Assembly (Montreal 22 September–2 October 1998) adopted Resolution A32-6 (Safety of Navigation) which considered that, on August 31, 1998, an object propelled by rockets was launched by a certain Contracting State and a part of the object hit the sea in the Pacific Ocean off the coast of Sanriku in North-eastern Japan and that the impact area of the object was in the vicinity of the international airway A590 which was known as composing NOPAC Composite Route System, a trunk route connecting Asia and North America where some 180 flights of various countries fly every day and concluded that the launching of such an object vehicle was done in a way not compatible with the fundamental principles, standards and recommended practices of the Chicago Convention and noted that it was necessary that international aviation should be developed in a safe and orderly manner, and that the Member States of ICAO will take appropriate measures to enhance further the safety of international civil aviation.

The Assembly urged all Member States to reaffirm that air traffic safety is of paramount importance for the sound development of international civil aviation and to strictly comply with the provisions of the Chicago Convention its Annexes and its related procedures, in order to prevent a recurrence of such potentially hazardous activities, while instructing the Secretary General; to immediately draw the attention of all Contracting States to the Resolution.

Applicable Aeronautical Principles

From an aeronautical perspective, Annex 11 to the Chicago Convention, which deals with the subject of air traffic services, lays down requirements for coordination of activities that are potentially hazardous to civil aircraft. The International

Standards and Recommended Practices in the Annex, Chapter 2 (2.17 and 2.18 in particular) contain provisions for co-ordination between military authorities and air traffic services and co-ordination of activities potentially hazardous to civil aircraft. These provisions specify that air traffic services authorities shall establish and maintain close co-operation with military authorities responsible for activities that may affect flights of civil aircraft. The provisions also prescribe that the arrangements for activities potentially hazardous to civil aircraft shall be co-ordinated with the appropriate air traffic services authorities and that the objective of this co-ordination shall be to achieve the best arrangements which will avoid hazards to civil aircraft and minimize interference with the normal operations of such aircraft. Standard 2.17.1 stipulates that arrangements for activities potentially hazardous to civil aircraft, whether over the territory of a State or over the high seas, shall be coordinated with the appropriate air traffic services authorities, such coordination to be effected early enough to permit timely promulgation of information regarding the activities in accordance with the provisions of Annex 15 to the Chicago Convention. Standard 2.17.2 of Annex 11 explains that the objective of the coordination referred to in the earlier provision shall be to achieve the best arrangements that are calculated to avoid hazards to civil aircraft and minimize interference with the normal operations of aircraft. One must of course hasten to add that Article 89 of the Convention stipulates that in case of war, the provisions of the Convention (and, by implication its Annexes) shall not affect the freedom of action of any of ICAO's member States affected, whether as belligerents or as neutrals. The same principle would apply in the case of any member State which declares a state of national emergency and notifies the fact to the ICAO Council.

Article 3 *bis* is the result of attacks on aerial intrusions and what are deemed to be incursions into sovereign airspace. This was the argument in KL 007 as with other instances of aerial intrusions.

One commentator said in 1953:

> Aerial intrusions may occur for a variety of reasons and in a variety of circumstances. They may be deliberate and with hostile and illicit intentions such as attack, reconnaissance, aid to subversive activities, smuggling, or calculated defiance of the territorial sovereign. They may be deliberate but with essentially harmless intentions such as shortening a flight or avoiding bad weather. They may be necessitated by distress or caused by mistakes. They may occur in peacetime or wartime.[34]

Article 3 *bis* is intended to effectively preclude a State from using its unfettered discretion to use weapons against an intruding aircraft and to ensure that people on board are not harmed. In this sense, and in the context of the aforementioned instances where an aircraft may unknowingly or innocently transgress airspace, the wording in Article 3 *bis* is woefully inadequate.

[34]Lissitzyn (1953).

At the Council session, held on 25 and 16 September 1983 which discussed the shooting down of KAL 007 the President of the Council succinctly summarized ICAO's role in the investigation of the KAL 007 incident:

> It falls clearly to ICAO ... to focus its attention on gaining a full and complete technical understanding of how this tragic event occurred and to examine every element in ICAO's existing technical provisions for promoting the safety of air navigation ...[35]

At its 138th Session, the Council examined the interim report of the ICAO investigative team into the KAL 007 incident and progress made in collecting facts regarding the shooting down of the aircraft. The Council noted the excellent cooperation provided to the ICAO investigative team by the Contracting States concerned and noted that a final report on the ICAO investigation would be placed by the Secretary General before the Council at its 139th Session.

The completed report of the Secretary General was presented to the Council during its 139th Session[36] and the Council closed the matter of KAL 007 on 14 June 1993. From a diplomatic perspective, and irrespective of the findings of the Report which are not relevant to this work, it must be noted that the outcome of the Report and discussions that ensued in the Council endorsed the usefulness of the Council. As reflected in the Statement issued in Council by the Republic of Korea:

> The Council must once again make it clear to the world that, while reaffirming the principle of prohibition of the use of arms against civil aircraft, it unreservedly condemns the destruction of a civilian aircraft simply[37] because it strayed into the airspace of another country.

The role of the ICAO Council was aptly brought to bear by the United Kingdom, supported by several other States, that the Council should not seek to endorse the conclusions and recommendations in the Report since it was not a tribunal seeking to reach a judgment on the facts.[38] The significance of the Council's role as a diplomatic tool in international civil aviation is borne out by the Summary of the President of the Council which formed the substance of the Council Resolution which followed and which, *inter alia*, expressed appreciation for the full cooperation extended to the fact finding mission by the authorities of all the States concerned. The President appealed to all Contracting States to ratify Article 3 *bis* to the Chicago Convention which approved the fundamental principle of general international law that States must refrain from resorting to the use of weapons against civil aircraft.

The KAL 007 investigation and the ICAO approach to the issue of dispute resolution was clearly a reiteration of the position taken by the Council in its earlier determination of the Libya–Israel dispute in 1973. The incident concerned the

---

[35] ICAO Doc 9416, Chapter II, note 12 at 4.

[36] See C-WP/9781 Appendix for the Secretary General's Report.

[37] ICAO Doc 9615-C/1110, C-MIN 139/1-17: Council—139th Session, Summary Minutes with Subject Index at 69.

[38] *Id.* at 72.

shooting down of a Libyan Airlines Boeing 727 aircraft by Israeli fighter aircraft on 21 February 1973 over Israeli occupied Sinai territory. One hundred and ten persons were killed in the incident and the Boeing 727 aircraft involved was completely destroyed. As an immediate response, the ICAO Council convened the 19th Session (Extraordinary) of the Assembly, at which speakers generally condemned the act of destruction. An investigation was called for and the Assembly proceeded to adopt Resolution A19-1 which stated that the Assembly, having considered the item concerning the Libyan civil aircraft shot down on 21 February 1973 by Israeli fighters over the occupied Egyptian territory of Sinai, condemned the Israeli action which resulted in the loss of innocent lives. Convinced that such an action adversely affects and jeopardizes the safety of international civil aviation and therefore, emphasizing the urgency of undertaking an immediate investigation, the Assembly directed the Council to instruct the Secretary General to institute an investigation in order to undertake fact findings and report to the Council. The Assembly also called upon all parties involved to cooperate fully in the investigation.[39]

Consequently, the Secretary General of ICAO presented his report,[40] which was in effect a report of the Secretariat investigative team containing, *inter alia*, a draft resolution[41] developed by numerous ICAO Contracting States. Pursuant to sustained discussion in Council, the Representatives on Council agreed upon a Resolution which was adopted by the Council. The Resolution, while recalling United Nations Security Council Resolution 262 of 1969, which condemned Israel for premeditated action against Beirut International Airport resulting in the destruction of 13 commercial and civil aircraft, expressed its deep conviction and belief that such acts constitute a serious danger against the safety of international civil aviation, and recognized that such an attitude is a flagrant violation of the principles enshrined in the Chicago Convention.

The above statement of the ICAO Council truly typifies the quintessentially diplomatic approach taken by ICAO on contentious issues between ICAO Contracting States. If one analyses the first part of the Council Resolution as given above, it is difficult not to note that the Council has skilfully restated an already adopted resolution of the United Nations, ensuring that, while avoiding being judgmental, it nonetheless conveys to the international aviation community its position on the issue at hand.

In the second part of the Resolution, the Council proved to be even more dexterous, in courageously taking a stand by strongly condemning the Israeli action which resulted in the destruction of the Libyan civil aircraft and the loss of 108 innocent lives, and urging Israel to comply with the aims and objectives of the Chicago Convention. The mastery of the Council, in encompassing into a single resolution compelling precedent established by a United Nations resolution together

[39]ICAO Doc 9061, Chapter II, note 60.

[40]C-WP/5764, Attachment.

[41]C-WP/5792 at p. 33.

with its own resolute position, is diplomacy at its most astute. The dexterity of the Council in this instance must not be mistaken for tendentiousness nor deviousness as the Council Resolution is clearly forthright and direct.

It must also be mentioned that the Chicago Convention by virtue of its Article 89 which stipulates:

> In case of war, the provisions of this Convention shall not affect the freedom of action of any of the Contracting States affected, whether as belligerents or as neutrals. The same principle shall apply in the case of any Contracting State which declares a state of national emergency and notifies the fact to the Council.

In this context, States may prescribe aerial conduct as was seen in, the United Nations Security Council adopted Resolution 1973 which *inter alia* decided to establish a ban on all flights in the airspace of the Libyan Arab Jamahiriya in order to help protect civilians. The Resolution also authorized Member States to take all necessary measures, to protect civilians and civilian populated areas under threat of attack in the Libyan Arab Jamahiriya, including Benghazi. This resulted in concerted air attacks by NATO forces against Libya in 2011.

References

Abeyratne R (2007) Air law and policy. PublishAmerica, Baltimore

Abeyratne R (2009) Regulating unmanned aerial vehicles - issues and challenges. Eur Transport Law XLIV(5):503–520

Crawford J (2002) The international law commission's articles on state responsibility: introduction, text and commentaries. Butterworth, Oxford

Dempsey PS (2003) Privatization of the air: governmental liability for privatized air traffic services. Ann Air Space Law XXVII(95):118–119

Dutton PA (2009) *Caelium liberum*: air defence identification zones outside sovereign airspace. Am J Int Law 103(4):691–709

Groenewege AD (1999) Compendium of international civil aviation, 2nd edn. International Aviation Development Corporation, Canada

Lissitzyn OJ (1953) The treatment of aerial intruders in recent practice and international law. Am J Int L 47:559

McDougal MS et al (1963) Law and public order in outer space. Yale University Press, New Haven/London

Milde M (2012) In: Marietta B (ed) International air law and ICAO, essential air and space law, 2nd edn. Eleven International Publishing, The Hague

Schenkman J (1955) International civil aviation organization. Librairie E. Dross, Geneve

Sochor E (1991) The politics of international aviation. Macmillan, London

van Zandt P (1944) Civil aviation and peace. Brookings Institution, Washington

Warner E (1942) Foreword to international air transport and national policy by Lissitzyn O.J., Oceana Publishers, New York

Article 4
Misuse of Civil Aviation

Each Contracting State agrees not to use civil aviation for any purpose inconsistent with the aims of this Convention.

Contents

1 Peaceful Uses of Civil Aviation

During the 25th (Extraordinary) Session of the Assembly (Montreal, 24 April–10 May 1984), a distinction was drawn between aerial violations of sovereignty, and other usage for purposes inconsistent with the aims of the Convention. Cuba defined "acts inconsistent with the aims of the Convention" as

> Acts of aggression, infiltration or espionage involving discharge of harmful substances or pathogenic agents; transport of contraband or prohibited traffic using the airspace of another State, even with destination to a third State or with any other purpose inconsistent with the aims of the Convention[1]

References were made by other States to civil aircraft being used for military reconnaissance; to violate mandatory laws and requirements for State security; to engage in provocative activities, acts of espionage and aggression; for the transportation of mercenaries, narcotics, guns, or ammunition; or otherwise engaged in criminal activities. It must be concluded that the intention was to cover activities of foreign civil aircraft (as opposed to civil aviation) not only contrary to the "aims" of the Convention, but also contrary to the law and public order of the over-flown State.

[1] A25-Min. EX/6.

R. Abeyratne, *Convention on International Civil Aviation*,
DOI 10.1007/978-3-319-00068-8_5,

2 The Convention

> does not contain any provision which would foresee the specific situations when an aircraft is used for or involved in criminal activities or other activities violating the law and public order of the State,

but that there were numerous provisions offering

> effective safeguards to States that their applicable laws and public order are observed by foreign aircraft.

Particular reference was made to Articles 3, 5, 9–13, 16 and 35. Article 1 recognizes that every State has complete and exclusive sovereignty over the airspace above its territory. A State may consequently, in accordance with international law, enforce its domestic laws and regulations insofar as such laws and regulations and method of enforcement are not contrary to an express agreement with other States (e.g. the Chicago Convention) or other norms of international law; this power and authority is an essential attribute of sovereignty.

All States possess within the existing framework of the Chicago Convention full jurisdiction in the application of their respective laws to prevent or prohibit "the use of civil aircraft for unlawful purposes". In accordance with Article 3, the Chicago Convention is applicable only to civil aircraft and not to State aircraft. Aircraft "used in military, customs and police services shall be deemed to be state aircraft." Consequently, the Convention recognizes two categories of aircraft only: civil and state.

It has been argued that no aircraft used in military, customs or police services should be considered to be civil aircraft under the Convention[2]; conversely all aircraft used in other than the government services specified in Article 3 should be regarded as civil aircraft. It was stated that the usage of the aircraft in question is the determining criterion, and not, by themselves, other factors such as aircraft registration and markings, call signed used, ownership (public or private), type of operator (private/state), except insofar as these criteria go towards showing the type of usage. Nevertheless, it is submitted that in the vast majority of cases, the predominant and primary criterion and the strongest evidence is the aircraft registration. In cases where the usage of the aircraft manifestly differs from its type of registration, a determination of whether it is a State or civil aircraft under the Chicago Convention should be made by taking into account all the circumstances surrounding the flight to determine usage, such as: nature of cargo or personnel carried; ownership; control and supervision by the specified services; and nationality markings.

There have been several instances of misuse of civil aviation, whether it be by attacking aircraft in flight or attacking ground facilities.

---

[2]C-WP/9835, 22/9/93, ICAO: Montreal at 5.

3 Destruction of Gaza International Airport

A recent manifestation of the link between aviation and global efforts at striving for peace is reflected in the consideration by the ICAO Council of the destruction of Gaza International Airport. At the High-level, Ministerial Conference on Aviation Security (Montreal, 19–20 February 2002), an information paper[3] was presented by Arab States Members of the Arab Civil Aviation Commission. Consequent upon the Conference referring this matter to the Council,[4] the Council, at the sixth meeting of its 165th Session on 4 March 2002, was advised[5] that, on 4 December 2001, Israeli military forces attacked Gaza International airport, destroyed air navigation facilities and bombarded runways and taxiways until the airport became unserviceable. It was reported that, when the Palestinian Authority attempted a repair on 11 January 2002, the Israeli military forces bombarded once again the airport and its facilities by aircraft, artillery and tanks, thereby destroying the runway, the taxiways and all facilities.

The Palestinian Authority claimed that the destroyed airport and air navigation facilities were used for the transportation of civilian passengers, search and rescue operations in case of emergencies, transportation of rescue material, including medical equipment, medicines and survival kits for safeguarding human lives.

It was noted by the Council that the airport was developed with voluntary contributions from a number of European countries, which recognized beyond doubt the urgent need for the airport. Nevertheless, the airport was destroyed without paying attention to any humanitarian consideration. This led the European Union to condemn the Israeli actions and reserve the right to demand compensation for the damages. The Council was further advised that the destruction of the civil airport in Gaza was an act deliberately perpetrated by a Contracting State. It was claimed that such destruction took place under the watchful eyes of the international community and was widely covered by local and international media reports and the Council was requested to consider the ramifications of the act, i.e. contempt of respect for human life, the disrespect of international laws, including the relevant conventions on civil aviation security.

Among the considerations of the Council were relevant provisions of the Chicago Convention, the first being Article 4 which stipulates that

> each Contracting state agrees not to use civil aviation for any purpose inconsistent with the aims of this Convention. Also considered was Article 44 which lays down the objectives of ICAO, particularly to meet the needs of the peoples of the world for safe, regular, efficient and economical air transport.[6]

---

[3]AVSEC-Conf/02-IP/29.

[4]In accordance with Article 54 (n) of the *Convention on International Civil Aviation*, the Council is empowered to consider any matter relating to the Convention which any Contracting State refers to it. See *ICAO Doc 7300/8*: Eighth Edition, Montreal: 2000.

[5]C-WP/11790 (Destruction of Gaza International Airport).

[6]Chicago Convention, Preamble (*supra* note 1), Article 44 (d).

Another Convention, the Montreal Convention (1971)[7] was also considered by the Council, particularly the views of the State parties to the Convention, to the effect that:

> "Unlawful acts against the safety of civil aviation jeopardize the safety of persons and property, seriously affect the operation of air services, and undermine the confidence of the peoples of the world in the safety of civil aviation",
>
> that "the occurrence of such acts is a matter of grave concern" and that "for the purpose of deterring such acts, there is an urgent need to provide appropriate measures for punishment of offenders".[8]

The Council was reminded of Resolution A20-1, adopted at the 20th Session of the Assembly (Rome, 28 August, 21 September 1973), in particular Resolving Clause (3), where the Assembly solemnly warned Israel that if it continued committing such acts the Assembly will take further measures against Israel to protect international civil aviation. Also recalled was Resolution A33-2, adopted by the 33rd Session of the Assembly (Montreal, 25 September–5 October 2001) where the Assembly stated that *Whereas* acts of unlawful interference against civil aviation have become the main threat to its safe and orderly development; and *Recognizing* that all acts of unlawful interference against international civil aviation constitute a grave offence in violation of International law; the Assembly *Strongly condemned* all acts of unlawful interference against civil aviation wherever and by whomever and for whatever reason they are perpetrated. It was noted that the challenge facing this High-Level, Ministerial Conference is to take effective measures in order to help States in responding to unlawful interference against civil aviation security and to reject and condemn the use of civil aviation as weapon of destruction against human lives and properties.

Based on its considerations of the issue, the Council, on 13 March 2002, adopted a resolution strongly condemning the destruction of Gaza International Airport and its air navigation facilities. In its Resolution, the Council, strongly condemned all acts of unlawful interference against civil aviation, wherever, by whomsoever and for whatever reasons they are perpetrated. It also strongly condemned the destruction of Gaza International Airport and its air navigation facilities while reaffirming the important role of ICAO in facilitating the resolution of questions which may arise between Contracting States in relation to matters affecting the safe and orderly operation of international civil aviation throughout the world. The Council urged Israel to fully comply with the aims and objectives of the Chicago Convention, while strongly urging Israel to take the necessary measures to restore Gaza International Airport so as to allow its reopening as soon as possible. Additionally, the Council requested the President of the Council to attend to the implementation of this Resolution, and to secure the full cooperation of the parties with respect to the application of the Chicago Convention and of the above-mentioned principles.

---

[7] *Convention for the Suppression of Unlawful Acts Against Civil Aviation*, Signed at Montreal on 23 September 1971. ICAO Doc. 8966.

[8] *Id.* Preamble.

Finally, the Council requested the Secretary General to inform all Contracting States of the Resolution.[9]

4 Korean Airlines (South Korea–USSR, 1983)

On 1 September 1983, the President of the Council of ICAO received a communique from the Minister of Foreign Affairs of the Republic of Korea that Flight KE 007 which was being carried out by a Korean Airlines Boeing 747 passenger airliner had disappeared off the radar screens after it took off from Anchorage, Alaska on 31 August 1983 bound for Seoul. The Minister requested ICAO's assistance with regard to ensuring the safety of the passengers, crew and aircraft.[10] The diplomatic response of the President was instantaneous and immediate, containing a message to the Minister of Civil Aviation of the USSR. It stated that information had been received by ICAO that an aircraft may have possibly landed in Soviet territory and that ICAO was confident that the Soviet authorities were rendering all assistance to persons and property concerned.[11]

As an initial response to the incident, the ICAO Council met in extraordinary session on 15 and 16 September 1983 at the request of the Government of the Republic of Korea and the Government of Canada, and adopted a resolution which averred to the fact that a Korean Air Lines civil aircraft was destroyed on September 1, 1983 by Soviet military aircraft. The Council, by Resolution, expressed its deepest sympathy to the families bereaved in this tragic incident; and reaffirmed the principle that States, when intercepting civil aircraft, should not use weapons against them. *Inter alia*, the Resolution also deplored the destruction of an aircraft in commercial international service resulting in the loss of 269 innocent lives and recognized that such use of armed force against international civil aviation is incompatible with the norms governing international behavior and elementary considerations of humanity and with the rules, Standards and Recommended Practices enshrined in the Chicago Convention and its Annexes. The Council directed the Secretary General to institute an investigation to determine the facts and technical aspects relating to the flight and destruction of the aircraft and to provide an interim report to the Council within 30 days of the adoption of this Resolution and a complete report during the 110th Session of the Council. All parties were requested to cooperate fully in the investigation.

The issue was further discussed under the auspices of ICAO at the 24th (Extraordinary) Session of the ICAO Assembly which met at Montreal from

[9]By State Letter E 5/61-02/42 dated 26 April 2002, the Secretary General of ICAO advised all ICAO Contracting States of the Council Resolution. See also, ICAO official press release PIO 03/02 (ICAO Council Adopts Resolution Strongly Condemning The Destruction of Gaza International Airport.

[10]Memorandum dated 2 September 1983 from President of the Council to the Representatives on the Council, Attachment 1.

[11]*Id.* Attachment 2.

20 September to 7 October 1983 with the participation of 131 Contracting States. In the general discussion, much attention focused on the tragedy of the Korean Airlines flight 007 and on the resolutions of the Extraordinary Session of the Council. The Assembly adopted Resolution A24-5 which, while endorsing Council action taken so far, urged all Member States to cooperate fully in their implementation.

During the Assembly, the Delegation of Canada presented a proposal for a new Convention on the Interception of Civil Aircraft[12] and the Assembly referred the proposal to the Council of ICAO for further study on the understanding that the Council was empowered to consider the inclusion of this item into the General Work Programme of the Legal Committee.

Pursuant to requests[13] from the Governments of Japan, the Republic of Korea, the Russian Federation and the United States, where all but the Russian Federation had made direct reference to Article 54 (n) of the Chicago Convention,[14] the President quoted Rules 27 d) and 25 b) of the Rules of Procedure of the Council—the former of which provides for an item to be included on the Agenda of a Council meeting where the President, Secretary General or a Contracting State requests a new subject to be included—and the latter of which provides that any additional subject which fulfils the conditions in Rule 27 (d) should be included in the work programme of the Council. Accordingly, the Council decided to include the Korean Air incident in the work programme of the 137th Session of the Council. The subject was documented accordingly[15] and subjected to sustained discussion by the Council with attention to detail and with views being expressed by many representatives.[16] These discussions resulted in the Council, *inter alia*, deciding to complete a fact finding investigation which ICAO initiated in 1983 and instructing the Secretary General to request all parties involved in the fact finding investigation relating to Korean Airlines Flight KAL 007 to cooperate fully with ICAO in turning over to the Organization, as soon as possible, all relevant materials.[17]

The intervention of the ICAO Council with regard to the Korean Air incident and its instructions to the Secretary General are good examples of the ICAO diplomatic machinery in action. The almost instantaneous galvanizing into action of the ICAO Council, through which the diplomatic voice of ICAO as the global regulatory agency of the United Nations, is heard, and the thoroughness and meticulous attention to detail (particularly regarding procedure) reflects a good example of the legal maxim *omnia preasumuntur rite esse acta* (everything is presumed to be done the proper way).

---

[12]A24-WP/85.

[13]The requests of the four States were circulated to Representatives on the Council by the President through his memo PRES AK/333 dated 14 December 1992.

[14]Article 54 (n) of the Chicago Convention makes it a mandatory function of the ICAO Council to consider any matter of relevance to the Convention which a Contracting State refers to.

[15]See C-WP/9684, 14/12/92, ICAO C-WP/9616–9686: 1992.

[16]C-MIN 137/15 contained in ICAO Doc C-MIN 137th Session, 1992 at p. 131.

[17]*Ibid.*

At the Council session, held on 25 and 16 September 1983, the President of the Council succinctly summarized ICAO's role in the investigation of the KAL 007 incident:

> It falls clearly to ICAO ... to focus its attention on gaining a full and complete technical understanding of how this tragic event occurred and to examine every element in ICAO's existing technical provisions for promoting the safety of air navigation ...[18]

At its 138th Session, the Council examined the interim report of the ICAO investigative team into the KAL 007 incident and progress made in collecting facts regarding the shooting down of the aircraft. The Council noted the excellent cooperation provided to the ICAO investigative team by the Contracting States concerned and noted that a final report on the ICAO investigation would be placed by the Secretary General before the Council at its 139th Session.

The completed report of the Secretary General was presented to the Council during its 139th Session[19] and the Council closed the matter of KAL 007 on 14 June 1993. From a diplomatic perspective, and irrespective of the findings of the Report—which are not relevant to this article—it must be noted that the outcome of the Report and discussions that ensued in the Council endorsed the usefulness of the Council. As reflected in the Statement issued in Council by the Republic of Korea:

> The Council must once again make it clear to the world that, while reaffirming the principle of prohibition of the use of arms against civil aircraft, it unreservedly condemns the destruction of a civilian aircraft simply because it strayed into the airspace of another country.[20]

The role of the ICAO Council was aptly brought to bear by the United Kingdom during the Council's deliberations on KAL 007, which was supported by several other States, that the Council should not seek to endorse the conclusions and recommendations in the Report since it was not a tribunal seeking to reach a judgment on the facts.[21] The significance of the Council's role as a diplomatic tool in international civil aviation is borne out by the Summary of the President of the Council which formed the substance of the Council Resolution which followed and which, *inter alia*, expressed appreciation for the full cooperation extended to the fact finding mission by the authorities of all the States concerned. The President appealed to all Contracting States to ratify Article 3 *bis* to the Chicago Convention which approved the fundamental principle of general international law that States must refrain from resorting to the use of weapons against civil aircraft.

---

[18]ICAO Doc 9416, Chapter II, note 12 at 4.

[19]See C-WP/9781 Appendix for the Secretary General's Report.

[20]ICAO Doc 9615-C/1110, C-MIN 139/1-17: Council—139th Session, Summary Minutes with Subject Index at 69.

[21]*Id.* at 72.

5 Libyan Airlines (Libyan Arab Jamahiriya, United States, 1973)

The KAL 007 investigation and the ICAO approach to the issue of dispute resolution was clearly a reiteration of the position taken by the Council in its earlier determination of the Libya–Israel dispute in 1973. The incident concerned the shooting down of a Libyan Airlines Boeing 727 aircraft by Israeli fighter aircraft on 21 February 1973 over Israeli occupied Sinai territory. One hundred and ten persons were killed in the incident and the Boeing 727 aircraft involved was completely destroyed. As an immediate response, the ICAO Council convened the 19th Session (Extraordinary) of the Assembly, at which speakers generally condemned the act of destruction. An investigation was called for and the Assembly proceeded to adopt Resolution A19-1 which stated that the Assembly, having considered the item concerning the Libyan civil aircraft shot down on 21 February 1973 by Israeli fighters over the occupied Egyptian territory of Sinai, condemned the Israeli action which resulted in the loss of innocent lives. Convinced that such an action adversely affects and jeopardizes the safety of international civil aviation and therefore, emphasizing the urgency of undertaking an immediate investigation, the Assembly directed the Council to instruct the Secretary General to institute an investigation in order to undertake fact findings and report to the Council. The Assembly also called upon all parties involved to cooperate fully in the investigation.[22]

Consequently, the Secretary General of ICAO presented his report,[23] which was in effect a report of the Secretariat investigative team containing, *inter alia*, a draft resolution[24] developed by numerous ICAO Contracting States. Pursuant to sustained discussion in Council, the Representatives on Council agreed upon a Resolution which was adopted by the Council. The Resolution, while recalling United Nations Security Council Resolution 262 of 1969, which condemned Israel for premeditated action against Beirut International Airport resulting in the destruction of 13 commercial and civil aircraft, expressed its deep conviction and belief that such acts constitute a serious danger against the safety of international civil aviation, and recognized that such an attitude is a flagrant violation of the principles enshrined in the Chicago Convention.

The above statement of the ICAO Council truly typifies the quintessentially diplomatic approach taken by ICAO on contentious issues between ICAO Contracting States. If one analyses the first part of the Council Resolution as given above, it is difficult not to note that the Council has skilfully restated an already adopted resolution of the United Nations, ensuring that, while avoiding being judgmental, it nonetheless conveys to the international aviation community its position on the issue at hand.

---

[22] ICAO Doc 9061, Chapter II, note 60.

[23] C-WP/5764, Attachment.

[24] C-WP/5792 at p. 33.

In the second part of the Resolution, the Council proved to be even more dexterous, in courageously taking a stand by strongly condemning the Israeli action which resulted in the destruction of the Libyan civil aircraft and the loss of 108 innocent lives, and urging Israel to comply with the aims and objectives of the Chicago Convention. The mastery of the Council, in encompassing into a single resolution compelling precedent established by a United Nations resolution together with its own resolute position, is diplomacy at its most astute. The dexterity of the Council in this instance must not be mistaken for tendentiousness nor deviousness as the Council Resolution is clearly forthright and direct.

6 USA–Cuba, 1996

On 24 February 1996, two United States registered private (general aviation) civil aircraft were shot down by Cuban military aircraft, which resulted in the loss of four lives. Consequent upon information received from the United States authorities of the incident, the President of the ICAO Council, on 26 February 1996, wrote to the Government of Cuba expressing his deep concern and requesting authentic and authoritative information pertaining to the incidents.[25] Further developments ensued on 27 February 1996 when the United States formally requested that the Council of ICAO consider the matter under Article 54 (n) of the Chicago Convention, and, on the same day, the United Nations Security Council issued a statement through its President deploring the shooting down, by Cuban military aircraft, of the two United States registered aircraft. The Security Council also alluded to Article 3 *bis* of the Chicago Convention and the Montreal Protocol of 1984 which provide that States must refrain from the use of weapons against civil aircraft in flight and must not endanger the lives of persons on board and the safety of aircraft. The Security Council requested the ICAO Council to look into the matter and report to it as soon as possible.[26] For its part, Cuba, in its communications to the President of the Council, chronicled a series of chronological violations by United States registered aircraft. This was followed by a further communication on 28 February 1996 from the Cuban Ministry of Foreign Affairs addressed to the Secretary General of ICAO alluding to a series of violations, which had allegedly increased in number over a 20 month period, of Cuban airspace by civil aircraft registered and based in the United States. The Government of Cuba urged ICAO to carry out an extensive investigation into the violations, repeated over the years, of Cuban airspace by aircraft coming from the United States, including the incidents of 24 February 1996.

[25]Memorandum PRES AK/97 dated 26 February 1996 from President of the Council to Representatives on the Council, Attachment.

[26]S/PRST/1996/9, 27 February 1996 at 35 I.L.M. 493 (1996).

The communications received by ICAO with regard to the incidents of 24 February 1996 clearly required the Organization, under Article 54 (n) of the Chicago Convention, to investigate two issues:

1. The incidents of 24 February 1996, an investigation into which was requested both by the United States and Cuba; and
2. Repeated violations of Cuban airspace by aircraft registered and based in and coming from the United States, alleged by Cuba which requested an investigation.

When the abovementioned issues were addressed by the ICAO Council on 6 March 1996, the position taken by the United States was primarily based on Article 3 *bis* of the Chicago Convention, whereby the US claimed that there was a duty incumbent upon every State to refrain from resorting to the use of weapons against civil aircraft in flight. Accordingly, the United States claimed that the Cuban action was a blatant violation of international law and that firing on unarmed, known civil aircraft could never be justified. The United States claimed that, consequently, as required at international law, the Cuban Government should pay appropriate compensation to the families of those whose lives were lost.[27]

In response, the Cuban Delegation claimed that Cuba had been a victim of violations of its sovereignty and territorial integrity for many years which involved aircraft coming from the territory of the United States and that, over the past 20 months, as many as 25 such incursions and violations had been detected by Cuba. Cuba also counterclaimed that, in response to the reference by the United States of Article 3 *bis*, there was a stipulation in the Article obliging every civil aircraft to comply with orders of the subjacent State making the State of origin of the aircraft obligated to ensure compliance with such orders. Another argument adduced by Cuba was that paragraph (d) of Article 3 *bis*, that each Contracting State was required to take appropriate measures to prohibit the deliberate use of any civil aircraft registered in that State, *inter alia*, for any purpose inconsistent with the Chicago Convention, was applicable to the instances concerned.

The overall trend in the Council, when the US–Cuba dispute was taken up, was indicative of a consensus that action taken by Cuba was deplorable[28] and, in the words of the United Kingdom which seemingly echoed the general view: "the principle is simple. Weapons must not be used against civil aircraft in international and civil aviation".[29] On the issue of violation of airspace, which was brought up by Cuba, many States voiced the view that there was indeed an obligation on the part of all States to refrain from violating the sovereignty of States, while some States focussed their attention on Article 4 of the Convention which requires that civil aviation must not be used for any purpose inconsistent with the aims of the Convention.

---

[27]ICAO Doc 9676-C/1118, C-MIN 147/1-16: Council—147th Session, Summary Minutes with Subject Index at 68–71.

[28]*Id.* at 79–92.

[29]*Id.* at 88.

Due to its inherent complexities, this was clearly one issue which demanded that ICAO's diplomatic fabric be tested to its limits. The wisdom and diplomacy of the President of the Council proved invaluable when he advised the Council of the three alternatives available to Council in its pronouncement: resolution; decision; or conclusion. The President further advised the Council that whether the Council pronounced by resolution, by decision or by conclusion, any one of these would be binding in terms of implementation. Consequently, the President of the Council presented a revised version of the draft Resolutions presented by both the United States and by Cuba, for consideration of the Council. The draft Resolution suggested by the President, while recognizing that the use of weapons against civil aircraft in flight is incompatible with elementary considerations of humanity and the norms governing international behaviour, reaffirmed that States must refrain from the use of weapons against civil aircraft in flight and that, when intercepting aircraft, the lives of persons on board and the safety of the aircraft must not be endangered. For action, the draft Resolution required that the Secretary General initiate an immediate investigation into the shooting down of the aircraft, in particular with reference to the request of the United Nations Security Council Resolution, and that the Report of such investigation should be made available to the Council within 60 days in order to be transmitted to the United Nations Security Council.[30]

As to the relevance of including a reference to Article 3 *bis* in the Resolution, the President of the Council advised that Article 3 *bis* merely recognized a principle of customary international law and there was an addition to the principles embodied in the Convention. As such, it was the President's view that there was no need for the resolution to reaffirm an Article which in effect was an affirmation of the humanitarian principles already incorporated in the text.[31] It is noted that, by effectively precluding the express mention of a principle of customary international law as incorporated into the Chicago Convention, the Council played its ultimate role in diplomacy and political rectitude, by staying within the parameters of its own jurisdiction and avoiding incursions into judgment prior to facts being properly ascertained.

The final Resolution of the ICAO Council, adopted on 27 June 1996 following the Report of the Secretary General, embodies two critical principles. These were that the Council recalled and recognized the principle that every State has complete and exclusive sovereignty over the airspace above its territory and that the territory of a State shall be deemed to be the land areas and territorial waters adjacent thereto; and that States must refrain from the use of weapons against civil aircraft in flight and that, when intercepting civil aircraft, the lives of persons on board and the safety of the aircraft must not be endangered. Integral to the Resolution was also the principle that each Contracting State should ensure that appropriate measures are taken to prohibit the deliberate use of any civil aircraft registered in that State or

[30] *Id.* at 102–103.

[31] *Ibid.* at 103.

operated by an operator who has his principal place of business or permanent residence in that State for any purpose inconsistent with the Chicago Convention. The Council's condemnation of the use of weapons against civil aircraft involved the explicit mention of Article 3 *bis* at this advanced stage of the resolution making process, which, when examined from a diplomatic perspective, is seemingly appropriate and purposeful.

The Council Resolution was an example of the comprehensive manner in which the Council addresses issues referred to it under Article 54 (n). Additionally, the Resolution masterfully indicates the views of the Council by recognizing that, while on the one hand it should be recognized that all states have complete and exclusive sovereignty over the air space above their territories and that such sovereignty should not be encroached upon, on the other hand States do not have the right to use weapons against aircraft endangering the lives of those on board, no matter what the circumstances.

In the consideration of ICAO's role as a specialized agency of the United Nations, which is from time to time called upon to address contentious issues at the request of its Contracting States, it is inevitable that some determination must be made on whether ICAO should refrain from transgressing the parameters of international politics within its diplomatic efforts. The US–Cuba issue was clearly one where the ICAO Council traversed the diplomatic rope with a balanced sense of purpose and dedication to its role. The duality of sovereignty and protection of its territory by a State balanced well with the somewhat peremptory admonition that whatever the rights of a State may be, the use of weaponry could not be condoned under any circumstance.

7 The Iranair Incident: IR 655 (Iran, United States 1998)

The extent to which ICAO will be exposed politically in issues addressed by the Council is perhaps best illustrated by the consideration of the Council, in 1988 of the Iran Air incident. This concerned the shooting down of an Iran Air Airbus A300 (IR655) carrying commercial passengers on a scheduled flight from Bandar-Abbas (Iran) to Dubai. The aircraft was brought down by the *U.S.S. Vincennes* over the Persian Gulf, resulting in the death of all 290 persons on board the aircraft. The incident, which occurred on 3 July 1988, was considered by the Council at several of its meetings, notably on 7 December 1988 when the Council adopted its decision. The Council decision, while recalling the event of 3 July 1988, acknowledged the fact finding investigation report of the Secretary General of ICAO, and urged all States to take all necessary action for the safety of navigation of civil aircraft, particularly by assuring effective coordination of civil and military activities. The Resolution went on to refer to the fundamental principle of general international law—that States must refrain from resorting to the use of weapons against civil aircraft—and urged States to ratify Article 3 *bis* as soon as possible if they had already not done so.

One of the emergent features of the ICAO Council which became clear at its deliberations was the Council's resolve to address its deliberations to purely technical issues pertaining to the incident, while stringently avoiding political issues and diplomatic pitfalls. This is certainly true of all incidents discussed above, where the Council restricted its scope to technical issues as applicable to the principles embodied in the Chicago Convention.

Although ICAO has so far successfully avoided underlying political contentions brought to bear by the issues it addressed, the question has been asked as to whether ICAO could continue to divorce aeronautical or technical issues from underlying political nuances. The answer would seem to lie in the environment within which ICAO functions and the principles upon which, under the Chicago Convention, ICAO could work. Primarily, ICAO's objective is to develop principles and techniques of air navigation and to foster the planning and development of international air transport so as to insure the sage and orderly growth of air navigation.[32] When this fundamental postulate is applied to the Preamble of the Chicago Convention, which provides that the abuse of international civil aviation can become a threat to the general security, ICAO's mandate becomes clear. Taken together, those two principles bring to bear the fundamental truth about ICAO—that the Organization has to ensure safety and orderly (economic) growth and ensure, and at the same time, that civil aviation not be abused to the extent of becoming a threat to general global security. What this generally means is that ICAO has to ensure adherence by States to the principles of aviation as adopted within the ICAO regulatory umbrella.

In this context, the principles of the Chicago Convention and its Annexes become relevant, as pointed out by member States during discussions in Council on the issues addressed above. However, the responsibility is not merely one-sided. ICAO cannot, and will not turn a blind eye on the non-aviation practices of a State if it would endanger the objectives of civil aviation. For example, and as mentioned earlier, at its 15th Session in June/July 1965, the Assembly adopted Resolution A15-7 (now no longer applicable)—*Condemnation of the Policies of Apartheid and Racial Discrimination of South Africa*—which recognized that the then Apartheid policies of South Africa constituted a permanent source of conflict between the peoples and the nations of the world and that the policies of Apartheid and racial discrimination are a flagrant violation of the principles enshrined in the Preamble of the Chicago Convention. The Resolution urged South Africa to comply with the aims and objectives of the Chicago Convention. A similar initiative was seen later when, at its 17th Session in June 1970, the Assembly adopted Resolution A17-1—*Declaration by the Assembly*—which recognized that international civil air transport helped to create and preserve friendship and understanding among the people of the world and to promote commerce between nations and requested Contracting States to take concerted action towards suppressing all acts which jeopardized the sage and orderly development of international air transport. In this context, the most

[32]Chicago Convention, Preamble (*supra* note 1), Article 44 (a).

forceful example of ICAO's role can be seen in Resolution A20-2—*Acts of Unlawful Interference with Civil Aviation*, adopted in March 1973 by the Assembly, which reaffirmed ICAO's role as facilitating the resolution of questions that may arise between Contracting States in matters affecting the safe and orderly operation of civil aviation throughout the world.[33]

It may be noted that an inevitable corollary to the establishment of ICAO by the world community, as a "club" of States, is that most problems which are directed at the ICAO Council could involve or be generated by intractable political disagreements or conflicts between States. As such, it would be naive for ICAO not to be aware of the nature of conflicts before its Council. However, ICAO remains a specialized agency of the United Nations with a specific agenda as embodied in the Chicago Convention. In this regard one must bear in mind the observation of a former Secretary General of the United Nations, Javier Pérez de Cuellar, when he said that the world must be cautious not to blur, mix or separate specific functions of the main organs and specialized agencies by treating them as interchangeable platforms for pursuing the same political aims.[34] States bear an enormous responsibility in not letting this happen.

By letter dated 4 September 9 1996 to the President of the Council, the Government of Cuba requested to have included in the Work Programme of the Council a study on the "Premeditated misuse of civil aviation by aircraft registered as civil aircraft."

Attached to this letter was a document expanding further on the request. It stated that

> more and more frequently, situations arise involving aircraft registered as civil aircraft which are deliberately used for flights, the objectives of which are clearly inconsistent with the Convention on International Civil Aviation and which consequently cannot enjoy ICAO's guarantees and facilities.

Further, the document expressed the view that

> every Contracting State, when registering an aircraft as a civil aircraft, assumes the full responsibilities of Article 4 of the Convention, that aircraft are used in a manner consistent with the Convention, as well as the obligations of Article 12 with regard to compliance with the Rules of the Air, by having to take preventive and punitive action against all those aircraft which violate the rules of international civil aviation and which are used in manner inconsistent with the friendship and understanding among nations and general security, as called for by the Chicago Convention.

The Government of Cuba provided further clarifications in a letter dated 21 October 1996. It stated that the aim of the request was that the Council study the matter of the premeditated misuse of civil aviation for any purpose inconsistent with the Chicago Convention. Reference was made to the deliberate misuse of aircraft registered as civil aircraft and in particular, of

[33]For the resolutions quoted above, see Doc 8900/2 *Repertory—Guide to the Convention on International Civil Aviation*, Second Edition, 1977 at pp. 1–3.

[34]A26/MIN EX/10 pp. 148–149.

general aviation light aircraft being used in the illicit trafficking of narcotics, smuggling of people and goods, terrorism, aggression, piracy, acts against the environment and other acts considered as international offences or actions repudiated and condemned by the international community. The line of argument must be understood as referring to unlawful acts generally.

The aim of the request was "for the Council to consider and take action in resolving these unlawful practices".

The problem of "misuse of civil aviation" was raised at the 25th Session (Extraordinary) of the Assembly, held in Montreal from 24 April to 10 May 1984 and was mentioned by several Representatives on the Council in the context of discussions on Amendment 27 to Annex 2 to the Chicago Convention with respect to interception of civil aircraft; in fact it was during this discussion that the Secretary General had been requested to prepare a paper on the "*Misuse of Civil Aviation*". Several Representatives had expressed

concern that necessary procedures must be foreseen to prevent the use of civil aviation for unlawful purposes.

It was necessary to find out whether the existing framework of the Chicago Convention offered any solution to the problem of treatment of civil aircraft engaged in an activity violating the laws of the State being over-flown.

Article 4 is the only provision in the Convention explicitly using the words "misuse of civil aviation" and the term "misuse" is contained only in the heading. Article 4 of the Chicago Convention provides that each Contracting State agrees not to use civil aviation for any purpose inconsistent with the aims of the Convention. There have been several instances where the principle of this article has been imputed to disputes among States where one State claimed that the other had misused aviation. These instances will be discussed later in this book under the subject of ICAO's role in diplomacy and dispute settlement. One of the issues which exemplify Article 4 and its significance is reflected in contentions on the issue of misuse of civil aviation, as claimed by Cuba against the United States, which culminated in the US–Cuba incident of 1997[35] The Government of Cuba, by letter dated 4 September 9 1996 addressed to the President of the Council, requested to have included in the Work Programme of the Council a study on the "Premeditated misuse of civil aviation by aircraft registered as civil aircraft."

Attached to this letter was a document expanding further on the request. It stated that

more and more frequently, situations arise involving aircraft registered as civil aircraft which are deliberately used for flights, the objectives of which are clearly inconsistent with the Convention on International Civil Aviation and which consequently cannot enjoy ICAO's guarantees and facilities.

[35]The following text is taken from C-WP/10588 (25/2/97) Misuse of Civil Aviation (Request From Cuba).

Further, the document expressed the view that

> every Contracting State, when registering an aircraft as a civil aircraft, assumes the full responsibilities of Article 4 of the Convention, that aircraft are used in a manner consistent with the Convention, as well as the obligations of Article 12 with regard to compliance with the Rules of the Air, by having to take preventive and punitive action against all those aircraft which violate the rules of international civil aviation and which are used in manner inconsistent with the friendship and understanding among nations and general security, as called for by the Chicago Convention.

The Government of Cuba provided further clarifications in a letter dated 21 October 1996. It stated that the aim of the request was that the Council study the matter of the premeditated misuse of civil aviation for any purpose inconsistent with the Chicago Convention. Reference was made to the deliberate misuse of aircraft registered as civil aircraft and in particular, of

> general aviation light aircraft being used in the illicit trafficking of narcotics, smuggling of people and goods, terrorism, aggression, piracy, acts against the environment and other acts considered as international offences or actions repudiated and condemned by the international community. The line of argument must be understood as referring to unlawful acts generally.

The aim of the request was "for the Council to consider and take action in resolving these unlawful practices".

During its 117th and 118th Sessions, the Council considered a Secretariat Study entitled, *"Misuse" of Civil Aviation* (C-WP/8217). The Secretariat Study recalled that the problem of "misuse of civil aviation" was raised at the 25th Session (Extraordinary) of the Assembly and was mentioned by several Representatives on the Council in the context of discussions on Amendment 27 to Annex 2 to the Chicago Convention with respect to interception of civil aircraft; in fact it was during this discussion that the Secretary General had been requested to prepare a paper on the "*Misuse of Civil Aviation*". Several Representatives had expressed "concern that necessary procedures must be foreseen to prevent the use of civil aviation for unlawful purposes." C-WP/8217 stated that it was necessary to find out whether the existing framework of the Chicago Convention offered any solution to the problem of treatment of civil aircraft engaged in an activity violating the laws of the State being overflown.

On 9 June 1986, the Council noted the study and the fact that it would be taken into account to the extent required in further work on all aspects of ICAO's role in efforts to combat illicit transport of narcotic drugs and psychotropic substances by air. It was understood that the legal analysis in C-WP/8217 would be used in future work on any other study related to misuse of civil aviation. Article 4 is the only provision in the Convention explicitly using the words "misuse of civil aviation" and the term "misuse" is contained only in the heading.

It may also be useful to refer to Article 44 which sets out the "Objectives" of the Organization (as opposed to the aims of the Convention). In a case like this where it is difficult to attach a clear and unambiguous interpretation to the phrase "purpose inconsistent with the aims of this Convention", the Vienna Convention allows recourse to supplementary means of interpretation, including the preparatory work of the treaty and the circumstances of its conclusion.

The drafting history of Article 4 indicates that the underlying intent was to prevent the use of civil aviation for purposes which might create a threat to the security of other nations. Article 4 originated in the "Canadian Revised Preliminary Draft of an International Air Convention" which provided that the proposed International Air Authority shall plan and foster the organization of international air services so as, *inter alia*,

> to avert the possibility of the misuse of civil aviation creating a threat to the security of nations, and to make the most effective contribution to the establishment and maintenance of a permanent system of general security.

In the "Tripartite Proposal" of the United States, United Kingdom and Canada, this wording was changed in Article II to "Each member state rejects the use of civil air transport as an instrument of national policy in international relations." This wording practically repeats the text of the 1928 Treaty for the Renunciation of War, in which the parties agreed to renounce war "as an instrument of national policy in their relations with one another". The Tripartite Proposal was referred to a drafting committee "to resolve questions of language only". In particular, the above provision of Article II was referred to the committee "to find more suitable language for the desire of all to prevent the use of civil air transport for purposes of aggression." The words "purposes inconsistent with the aims of this Convention" in Article 4 essentially mean "threats to general security" and do not offer any solution to the problem where an aircraft is used for criminal purposes or other unlawful purposes, not associated with threats to the general security. The undertaking of States not to threaten the general security of other States through the usage of civil aviation under Article 4 should be seen as an overriding principle and a prerequisite to the attainment of the other objectives of the Convention.

This interpretation of the intent of Article 4 is logical when one considers the circumstances prevailing in the world at the time the Convention was drafted. Therefore, under Article 4 States agree in particular not to use civil aviation as a means to threaten the security of other States. The term "civil aviation" must also be read broadly to encompass not just civil aircraft, but also "civil" aeronautical facilities and services. This distinction was illustrated by one Delegation (United States) during the 25th Session (Extraordinary) of the Assembly which adopted Article 3 *bis*. In presenting its proposal for a draft amendment, that Delegation stated that the term "civil aviation" was used to provide protection to civil aircraft as well as other elements of civil aviation.

It may also be useful to refer to Article 44 which sets out the "Objectives" of the Organization (as opposed to the aims of the Convention). In a case like this where it is difficult to attach a clear and unambiguous interpretation to the phrase in Article 4 that States agree not to use civil aviation for any "purpose inconsistent with the aims of this Convention".

The drafting history of Article 4 indicates that the underlying intent was to prevent the use of civil aviation for purposes which might create a threat to the security of other nations. Article 4 originated in the "Canadian Revised Preliminary Draft of an International Air Convention" which provided that the proposed International Air Authority shall plan and foster the organization of international air services so as, *inter alia*,

> to avert the possibility of the misuse of civil aviation creating a threat to the security of nations, and to make the most effective contribution to the establishment and maintenance of a permanent system of general security.

In the "Tripartite Proposal" of the United States, United Kingdom and Canada, this wording was changed in Article II to "Each member state rejects the use of civil air transport as an instrument of national policy in international relations." This wording practically repeats the text of the 1928 Treaty for the Renunciation of War, in which the parties agreed to renounce war "as an instrument of national policy in their relations with one another". The Tripartite Proposal was referred to a drafting committee "to resolve questions of language only." In particular, the above provision of Article II was referred to the committee "to find more suitable language for the desire of all to prevent the use of civil air transport for purposes of aggression." The words "purposes inconsistent with the aims of this Convention" in Article 4 essentially mean "threats to general security" and do not offer any solution to the problem where an aircraft is used for criminal purposes or other unlawful purposes, not associated with threats to the general security. The undertaking of States not to threaten the general security of other States through the usage of civil aviation under Article 4 should be seen as an overriding principle and a prerequisite to the attainment of the other objectives of the Convention.

This interpretation of the intent of Article 4 is logical when one considers the circumstances prevailing in the world at the time the Convention was drafted.

Therefore, under Article 4 States agree in particular not to use civil aviation as a means to threaten the security of other States. The term "civil aviation" must also be read broadly to encompass not just civil aircraft, but also "civil" aeronautical facilities and services. This distinction was illustrated by one Delegation (United States) during the 25th Session (Extraordinary) of the Assembly which adopted Article 3 *bis*. In presenting its proposal for a draft amendment, that Delegation stated that the term "civil aviation" was used to provide protection to civil aircraft as well as other elements of civil aviation.

During the 25th Session of the Assembly, a distinction was often drawn between aerial violations of sovereignty, and other usage for purposes inconsistent with the aims of the Convention. Cuba defined "acts inconsistent with the aims of the Convention" as

> Acts of aggression, infiltration or espionage involving discharge of harmful substances or pathogenic agents; transport of contraband or prohibited traffic using the airspace of another State, even with destination to a third State or with any other purpose inconsistent with the aims of the Convention.[36]

References were made by other States to civil aircraft being used for military reconnaissance; to violate mandatory laws and requirements for State security; to engage in provocative activities, acts of espionage and aggression; for the transportation of mercenaries, narcotics, guns, or ammunition; or otherwise engaged in criminal activities. It must be concluded that the intention was to cover activities of foreign civil aircraft (as opposed to civil aviation) not only contrary to the "aims" of the Convention, but also contrary to the law and public order of the over-flown State.

8 The Convention

Does not contain any provision which would foresee the specific situations when an aircraft is used for or involved in criminal activities or other activities violating the law and public order of the State, but that were numerous provisions offering "effective safeguards to States that their applicable laws and public order are observed by foreign aircraft". Particular reference was made to Articles 3, 5, 9, 10, 11, 12, 13, 16 and 35. Article 1 recognizes that every State has complete and exclusive sovereignty over the airspace above its territory. A State may consequently, in accordance with international law, enforce its domestic laws and regulations insofar as such laws and regulations and method of enforcement are not contrary to an express agreement with other States (e.g. the Chicago Convention) or other norms of international law; this power and authority is an essential attribute of sovereignty.

The following conclusions may be drawn from the above:

- Article 4 of the Chicago Convention, and the words "purpose inconsistent with the aims of this Convention" essentially mean "threats to general security".
- The words "purpose inconsistent with the aims of this Convention" in Article 3 *bis* have a wider meaning and covers breaches of the law and public order of the overflown State by foreign civil aircraft.
- The Chicago Convention, and rules of customary international law as codified in Article 3 *bis*, contain effective provisions safeguarding the full jurisdiction of States to prohibit or prevent the use of foreign aircraft for unlawful purposes in their territory.

At the 33rd Session of the Assembly, ICAO adopted Resolution 33/1 entitled Declaration on misuse of civil aircraft as weapons of destruction and other terrorist acts involving civil aviation. This Resolution, while singling out for consideration the terrorist acts which occurred in the United States on 11 September 2001, and, inter alia, recognizing that the new type of threat posed by terrorist organizations

[36]A25-Min. EX/6.

requires new concerted efforts and policies of cooperation on the part of States, urges all Contracting States to intensify their efforts in order to achieve the full implementation and enforcement of the multilateral conventions on aviation security, as well as of the ICAO Standards and Recommended Practices and Procedures (SARPs) relating to aviation security, to monitor such implementation, and to take within their territories appropriate additional security measures commensurate to the level of threat in order to prevent and eradicate terrorist acts involving civil aviation. The Resolution also urges all Contracting States to make contributions in the form of financial or human resources to ICAO's aviation security mechanism to support and strengthen the combat against terrorism and unlawful interference in civil aviation; calls on Contracting States to agree on special funding for urgent action by ICAO in the field of aviation security referred to in paragraph 7 below; and directs the Council to develop proposals and take appropriate decisions for a more stable funding of ICAO action in the field of aviation security, including appropriate remedial action.

Resolution 33/1 directed the Council to convene, at the earliest date, if possible in the year 2001, an international high level, ministerial conference on aviation security in Montreal with the objectives of preventing, combatting and eradicating acts of terrorism involving civil aviation; of strengthening ICAO's role in the adoption of SARPs in the field of security and the audit of their implementation; and of ensuring the necessary financial means to strengthen ICAO's AVSEC Mechanism, while providing special funding for urgent action by ICAO in the field of aviation security.

The effects of this resolution can be seen in certain concerted efforts made both in the United States and Europe to take immediate measures to strengthen aviation security. The European Transport and Telecommunications Council, at its meeting in Luxemburg on 16 October 2001, welcomed the proposal by the Commission for a Regulation establishing common rules in the field of civil aviation security. The Council invited Member States and the European Commission to contribute to the preparation for the ICAO High Level/Ministerial Conference as referred to in Resolution 13/1.

In the United States, Ms. Jane Garvey, FAA Administrator, stated on 17 October 2001 at a meeting in Washington, that the United States will start using new technology called the Computer Assisted Passenger Pre-screening System which would introduce new technologies to detect plastic weapon, and greater use of explosive detection equipment. Ms. Garvey further added that Transportation Secretary Mineta had created a $20 million dollar fund to explore new technologies to improve aircraft security. These grants could be used to test any new technology that leads to safer, more secure aircraft.

Article 5
Right of Non-scheduled Flight

Each contracting State agrees that all aircraft of the other contracting States, being aircraft not engaged in scheduled international air services shall have the right, subject to the observance of the terms of this Convention, to make flights into or in transit non-stop across its territory and to make stops for non-traffic purposes without the necessity of obtaining prior permission, and subject to the right of the State flown over to require landing. Each contracting State nevertheless reserves the right, for reasons of safety of flight, to require aircraft desiring to proceed over regions which are inaccessible or without adequate air navigation facilities to follow prescribed routes, or to obtain special permission for such flights.

Such aircraft, if engaged in the carriage of passengers, cargo, or mail for remuneration or hire on other than scheduled international air services, shall also, subject to the provisions of Article 7, have the privilege of taking on or discharging passengers, cargo, or mail, subject to the right of any State where such embarkation or discharge takes place to impose such regulations, conditions or limitations as it may consider desirable

Contents

1 Commercial Aspects

The ICAO Council in 1952 developed a definition of "scheduled international air service" for use by States with respect of Articles (on unscheduled air transport) and Article 6 (on scheduled international air transport).[1] On this occasion, member States of ICAO were requested by the Assembly to file copies of their regulations in this field with ICAO. At the Seventh Session of the Assembly (Brighton, 16 June–6 July 1953) the ICAO Secretariat submitted to the Assembly an analysis[2] made in 1952 of the regulations sent in by States. The first ICAO Assembly (Montreal, 6–27 May 1947) adopted Resolution A1-39 (Distinction Between Scheduled and Non-Scheduled Operations in International Air Transport) which stated *inter alia*:

[1]ICAO Doc 7278-C/841, May 1952.

[2]A7-WP/10.

DOI 10.1007/978-3-319-00068-8_6,

Whereas Article 5 of the said Convention provides, in effect that aircraft not engaged in scheduled international air services:

- May make flights into or in transit non-stop across the territory of a Contracting State and make stops for non-traffic purposes subject to certain safety requirements;
- If engaged in the carriage of passengers, cargo or mail for remuneration or reward, shall also have the privilege of taking on or discharging passengers, cargo or mail, subject to the right of any State where such embarkation or discharge takes place to impose such regulations, conditions or limitations as it may consider desirable.

The Assembly resolved that ICAO continue studies with the object of devising for international adoption a definition which clearly distinguishes for the purposes of the Convention between scheduled and non-scheduled operations. In the Assembly that followed (Geneva, 1–21 June 1948) Resolution A2-18 (Definition of Scheduled International Air Services) was adopted whereby the Assembly resolved that the Council should adopt, subject to periodic review, a definition of the term " Scheduled International Air Service" for the guidance of ICAO member States in the in the interpretation or application of Articles 5 and 6 of the Convention relating to the carriage of passengers, mail and cargo, or mail for remuneration or hire. The Assembly suggested the following definition:

> For the purposes of the Convention, a "scheduled international air service" is an international air service consisting of a recognizable systematic series of flights a) which are operated between two or more points or two or more traffic areas that, considering relevant characteristics of the service such as the distance covered and the type of aircraft used, do not materially vary; which are operated for valuable consideration; and which are open to use by members of the public(acceptable to the carrier) who, from time to time seek to take advantage of them.

The member States of ICAO at ICAO's 7th Assembly (Brighton, 16 June–6 July 1953) adopted Resolution A7-16 (Prospects of and Methods for Further International Agreement on Commercial Rights in International Air Transport–Non Scheduled Air Transport Operations) resolved that ICAO member States be requested to inform the Council whether they have accepted the definition of "a scheduled international air service" adopted by the Council on 25 March 1952 or what other definition or method they have adopted for distinguishing between scheduled international air services and non-scheduled international air services. The Assembly also requested that the Council give careful consideration to any views expressed by member States which might lead to an improvement of the definition a means of distinguishing between the two types of air services.

Consequently, the Council adopted the following definition of a scheduled international air service:

A scheduled international air service is a series of flights that possesses all the following characteristics:

- It passes through the air space over the territory of more than one State;
- It is performed by aircraft for the transport of passengers, mail or cargo for remuneration, in such manner that each flight is open to use by members of the public.

There have been some Resolutions both of ICAO and the United Nations Security Council which have some relevance to Articles 1, 5 and 6 of the Chicago Convention. Assembly Resolution A21-7 adopted by the 21st Session of the ICAO Assembly (Montreal, 24 September–15 October 1974) entitled "The Airport of Jerusalem" recalled that Jerusalem lay in the occupied Arab territories and was registered under the jurisdiction of Jordan and therefore resolved that all ICAO member States should, in implementing Articles 1, 5 and 6 of the Chicago Convention, take all necessary measures to refrain from operating, or giving permission to any airline to operate any air service whether scheduled or non-scheduled to or from Jerusalem airport, unless prior permission is granted pursuant to the provisions of the said three provisions. In 2011 the United Nations Security Council adopted Resolution 1970 whereby the Security Council demanded an immediate ceasefire in Libya, including an end to the current attacks against civilians, which it said might constitute "crimes against humanity", the Security Council imposed a ban on all flights in the country's airspace—a no-fly zone—and tightened sanctions on the Qadhafi regime and its supporters. This was followed by Resolution 1973 whereby the Security Council resolved that a ban be imposed on all flights in the airspace of the Libyan Arab Jamahiriya in order to help protect civilians. The Resolution required all Member States that have notified the Secretary-General and the Secretary-General of the League of Arab States, acting nationally or through regional organizations or arrangements, to take all necessary measures to enforce compliance with the ban on flights imposed as necessary, and requested the States concerned in cooperation with the League of Arab States to coordinate closely with the Secretary General on the measures they are taking to implement the ban. Earlier the Security Council had adopted.

The Chicago Convention distinguishes between the rights to be accorded by Contracting States to international non-scheduled flights (Article 5) and to scheduled international air services (Article 6). It refers to non-scheduled flights as the flights of all aircraft "not engaged in scheduled international air services". The first paragraph of Article 5 requires that each Contracting State grant the rights of transit and non-traffic stops to all international non-scheduled flights by aircraft of other Contracting States "without the necessity of obtaining prior permission". The second paragraph of this Article states that commercial non-scheduled flights shall also

> have the privilege of taking on or discharging passengers, cargo or mail, subject to the right of any State where such embarkation or discharge takes place to impose such regulations, conditions or limitations as it may consider desirable.

As a practical consequence of Article 5, the regulation of inter-national non-scheduled services has generally been governed by rules laid down by individual States, with only a few bilateral and multilateral agreements existing to create joint regulation.

For the guidance of States in their interpretation or application of Articles 5 and 6 of the Chicago Convention, the Council of ICAO has developed a definition for the term "scheduled international air service" which is accompanied by "Notes on the Application of the Definition and an Analysis of the Rights Conferred by Article 5 of the Convention".[3] The Council recognized, when developing the definition, that the right of Contracting States to impose regulations, conditions and limitations on the taking on or discharging of passengers, cargo or mail by commercial non-scheduled air transport is unqualified. Ithas expressed the opinion, however, that it should be under-stood that the right would not be exercised in such a manner as to render the operation of this important form of air transport impossible or non-effective.

Under a unilateral framework where international non-scheduled/charter operations continue to be regulated, States of origin and destination regulate independently of each other such services between their territories. In this situation the charterer and carrier must follow the rules of both States for the operation to be *charterworthy*, i.e. being a valid charter under the relevant regulation. These rules generally appear in national laws, government regulations, policy statements dealing with air transport and authorizing the regulation of such operations, or in the licence/permit authorizing the non-scheduled flight or flights. In some cases, ad hoc decisions are made by regulatory authorities.

National policies with respect to international non-scheduled commercial operations have taken a variety of forms, ranging from severe limitation to complete freedom. Most State policies lie between these approaches. In developing policies and regulations concerning non-scheduled air services, individual States usually take into account the role of such services in the satisfaction of the demand of the public for low-price air transport; their place in the overall air transport system; and their contribution in meeting some general national priorities and interests (e.g. promotion of tourism, exPANsion of airport utilization, job creation and community development).

A carrier must be licensed by its home State to engage in international non-scheduled air transport. Some States require evidence of this from foreign carriers. Air transport authorities may authorize international non-scheduled operations by a national or foreign carrier by issuing a licence or permit (i.e. general authorization or permission given on a relatively long-term, continuing basis, for example, for a year or a season), or an ad hoc authorization for a flight or flights.

States may also adopt procedures to:

- Require advance approval of charter programmes or individual flights; or
- Not require pre-flight approvals; and
- Require pre-flight notification and/or post-flight reporting.

Some States may continue to use the procedure of advance placement of carriers on a list of those eligible to perform charter flights. Various States use combinations of the above and take into consideration reciprocity, the origin of the traffic,

[3]See Doc 9587.

the region involved, the nationality of the carrier (national versus foreign), the type of carrier or size of aircraft, the kind of charter or other determinants. It is general practice for States to require the filing of a flight plan or some form of prior notification (usually 24 h in advance) for air traffic control, customs, immigration and public health purposes (note that Annex 9 to the Chicago Convention also contains provisions requiring Contracting States to minimize such procedures to facilitate non-scheduled operations).

Some States have concluded bilateral non-scheduled air services agreements or bilateral air transport agreements covering both scheduled and charter services to allow operation of non-scheduled services under mutually agreed terms. These agreements normally include provisions regarding charter worthiness rules (for example, acceptance of country of origin rules or harmonization of rules), points which may be served, fair and equal opportunity, pricing, traffic freedoms covered, designation and licensing of carriers; provisions similar to those found in other agreements covering technical subjects, such as customs exemption and consultation and arbitration.

Only a few multilateral agreements have been concluded on non-scheduled air services, all on a regional basis (such as the 1956 Multilateral Agreement on Commercial Rights of Non-scheduled Air Services in Europe between ECAC States, and the 1971 Multilateral Agreement on Commercial Rights of Non-scheduled Air Services Among the Association of South-East Asian Nations (ASEAN)). These agreements generally provided for a more liberalized regime in authorizing non-scheduled operations between the signatory States, for example, by permitting free admission of certain types of non-scheduled flights (e.g. Humanitarian, emergency charters, single entity charters or charter flights serving routes not being directly served by scheduled services), subject only to prior notification.

2 Inherent Practical Problems

A basic problem experienced by many States in respect of regulating international non-scheduled services is how to strike a balance between the commercial interests of the scheduled services operators and those of the charter operators in the same markets, while taking into account the overall economic interests of the country concerned. States that maintain significant regulation generally impose various restrictions or controls to ensure that non-scheduled air services do not impair the profitability and efficiency of their scheduled air services and/or to satisfy a need for some balance in the charter benefits received by carriers of each involved State. The controls which such States may use on commercial non-scheduled operations include:

- Marketing restrictions through charter definitions and rules (for example, by not permitting certain types of charters);
- Geographical and route restrictions (for example, by allowing the operation of certain types of charters only to defined areas or only on specified routes);

- Capacity control (involving, for example, a specific numerical limit or one related to a specific percentage of scheduled flights);
- And price control.

Another problem encountered by various States in regulating international non-scheduled services stems from the absence of an agreed clear definition of a non-scheduled service (it is defined only as other than a scheduled air service). This is especially true in situations when the distinction between the two types of air services has become blurred. For example, as charter services became more readily open to use by members of the public they came to be called "schedulized charters" or " programmed charters", i.e. charter flights open to the public that are so regular or frequent that they constitute a recognizable systematic series.

Most scheduled carriers now offer reduced fares and conditions which were once more common to charter services. As the air transport industry has evolved and as more States have adopted a liberal policy towards inter-national air transport regulation, the usefulness of making such distinctions for charters has been questioned. In the case of the European Union, the "third package" of air transport liberalization has effectively eliminated the regulatory distinction between the two (by allowing non-scheduled/charter carriers to operate scheduled flights and sell flights directly to the public), although the distinction tends to be retained by the industry in terms of how non-scheduled services are marketed and operated.

In 2009, non-scheduled or" charter airlines" carried fifty six million passengers at 99.8 load factors.[4] They have been described as carriers which:

> Usually offer flights as part of a holiday package that also includes transfers and hotel accommodation, most of which are provided by vertically integrated tour operators, which is a key reason why this type of airline business model has such high load factors, but low yields.[5]

The flexibility offered by Article 5 of the Chicago Convention, that non-scheduled flights do not have to obtain landing or take off permission from the grantor State to make flights into or in transit non-stop across its territory and to make stops for non-traffic purposes, augurs well for the Convention's theme of facilitating air transport to accord with the needs of the people of the world. A bureaucratic mess that is created by Article 6, a discussion on which is to follow, would have made matters worse for the travelling public.

References

O'Connell JF (2007) Airlines, an inherently turbulent industry. In: O'Connell JF, Williams G (eds) Air transport in the 21st century – key strategic developments. Ashgate, Surrey

---

[4]O'Connell (2007) at 59 at 63.

[5]*Id*. 67.

Article 6
Scheduled Air Services

No scheduled international air service may be operated over or into the territory of a contracting State, except with the special permission or other authorization of that State, and in accordance with the terms of such permission or authorization.

Contents

1 Commercial Restraints

Article 6 came into being due to a failure by the participating 52 States at the Chicago Conference to agree upon a multilateral agreement on air traffic rights. As can be seen later in this discussion, many delegates advocated multilateralism in commercial air traffic rights at the Chicago Convention, but, due to polarized protectionism of airspace by certain blocs, this trend never prevailed, resulting in Article 6. The first interim Assembly of ICAO (Resolution IV in June 1946) affirmed the opinion that a multilateral agreement on commercial rights constituted "the only solution compatible with the character of the International Civil Aviation Organization created at Chicago". However, in November 1947, consequent upon exhaustive deliberations and discussion, the Commission on Multilateral Agreement on Commercial Rights in International Civil Air Transport which was convened prior to the adoption of Resolution A1-38 decided that the divergence of views on certain important issues rendered impossible any agreement that would find wide acceptance.[1] This notwithstanding, the Commission produced Annex III to its report which contained provisions that would form part of a future multilateral agreement.

The member States of ICAO at ICAO's 7th Assembly (Brighton 16 June–6 July 1953) adopted Resolution A7-15 (Prospects and Methods of Future International Agreement on Commercial Rights in International Air Transport—Scheduled International Air Services) whereby the Assembly gave its opinion that there was no prospect at that time of achieving a universal multilateral agreement, although multilateralism in commercial rights to the greatest possible extent continues to be the objective of the Organization. Through Resolution A7-15, the Assembly

[1] Doc 5230 A2-EC/10.

R. Abeyratne, *Convention on International Civil Aviation*,
DOI 10.1007/978-3-319-00068-8_7, © Springer International Publishing Switzerland 2014

requested favourable consideration of the ICAO Council towards convening regional conference in Europe (upon a request of the Council of Europe at that time)

Article 6 is arguably the most contentious in the Chicago Convention. In brief, it contends the following:

- All scheduled international commercial air transport services are prohibited except to the extent they are permitted;
- All bilateral and open skies agreements are reciprocal and subject to the nuances of aeropolitics and protectionism and arbitrary demarcations of market share;
- Therefore airlines do not have freedom to access of markets;
- There are rigid and archaic ownership and control regulations governing so called "national carriers";
- In many instances, this effectively precludes direct foreign investment in airlines;
- All the above unduly prevent connectivity, which is the meaning and purpose of meeting the needs of the people of the world for regular, efficient and economical air transport.

It is worthy of note that ICAO's remit in this area of air transport is restricted by its aim and objective, as stipulated in Article 44 of the Chicago Convention that the Organization should "foster the development of air transport" which devolves upon ICAO the somewhat watered down responsibility of developing guidance on the regulation of air transport. This responsibility was reiterated in Assembly Resolution A37-20 Appendix A which stated: The Assembly reaffirms the primary role of ICAO in developing guidance on the regulation of international air transport and in assisting and facilitating liberalization as necessary.

With all this ambivalent rhetoric, Article 6 makes us ask: What do we stand for as a global aviation community? How do we strike the balance between growth and development? Where does aviation and its governance fit into a world transformed by the winds of globalization and change through technology? Have we a strategic direction and have we achieved our goals in aviation?

The reason for these questions is that, as already mentioned, as far back as 1944, the world, represented then by 52 signatory States to a multilateral treaty called the Chicago Convention—which now has 191 States acceding to it—called aviation's strategic direction "creating and preserving friendship and understanding among the nations and people of the world". One would not imagine, for a moment, that there is written anywhere in a global document that aviation's strategic direction is to make as much money as possible to the exclusion of others or to give priority to the interests of States or the air transport industry. Creating and preserving friendship and understanding among people can only be achieved through optimum connectivity. The original aim was to connect the world and cater to the needs of the people of the world for safer, regular, efficient and economical air transport.[2]

[2]Chicago Convention, Preamble (*supra* note 1), ICAO Doc. 7300/8: 2006, Article 44 (d).

Having given this direction, the same multilateral treaty goes on to say that no scheduled international air service may be operated over or into the territory of a Contracting State except with the special permission or other authorization of that State. It is a curious fact that in 1609, Hugo Grotius wrote in his magnum opus *Maré Liberum* (free seas) that the oceans should be open to sea faring by anyone. Yet more than three centuries later, precisely the opposite principle was adopted in 1944 by the aviation powers that assembled in Chicago.

The trouble with air transport is that, while on the one hand it is a product, on the other hand regulations pertaining to this product may constrain its availability to the consumer by depriving him of the various choices of air travel he might have under a liberalized system. In other words, State policy and the protection of national interests take precedence over the interest of the user of air transport. The aviation industry offers only one product to the ultimate consumer and that is the air transport product.

The air transport industry is cyclical and is profoundly affected by the world's economic health. I need not elaborate the economic vicissitudes of Europe and the significant growth elsewhere in Asia.

Aviation is a global industry and the need for air transportation continues to grow, as major cities continue to grow in population and prosper. A recent forecast has revealed that while in the 1970s there were just four major agglomerations of over 10 million people there are 26 today and there will be more than 30 by 2015. As economic prosperity grows more and more people will demand access to air transport and traffic growth is expected to double in 15 years.

"Connectivity" which is the most compelling need in aviation, and embodied in the Chicago Convention as inter alia "meeting the needs of the people of the world for efficient and economical air transport" is stultified by interests of commercial and national policy.

Because of Article 6, the operation of commercial international air services by member States of ICAO has been governed by bilateral agreements on the subject. This creates a stumbling block for air carriers to operate with fair and equal opportunity as enunciated by the Chicago Convention. This was not the trend that existed during the Chicago Conference that led to the adoption of the Convention.

2 The Enexplicable Turnaround from Initial Policy

At the Chicago Conference in 1944, the United States proposed a multilateral agreement calculated to guarantee commercial landing rights everywhere in the globe to all the world's airlines without restriction. The United States took the position that the use of the air and the use of the sea were both common in that they were highways given by nature to all men. They were different in that man's use of the air is subject to the sovereignty of nations over which such use is made. The United States was therefore of the opinion that nations ought to arrange among themselves for its use in such manner as would be of the greatest benefit to all humanity, wherever situated.

The United Kingdom contended:

> While recognizing national interests we want to encourage enterprise and efficiency which are indeed themselves a national as well as an international interest. And we want therefore to encourage the efficient and to stimulate the less efficient. . .only by common action on some such lines as indicated can we reduce and gradually eliminate subsidies, thereby putting civil aviation on an economic footing and incidentally very considerably relieving the tax payer. Unrestricted competition is their most fruitful soil.[3]

The United Kingdom seems to have adopted a balanced approach that supported the establishment of air services to serve the needs of the travelling public, while not unduly affecting the rights of States to have a fair share of traffic for themselves.

India, while believing that it was essential for air services to develop rationally with a certain degree of freedom of the air being the inherent right of every State, went on to say:

> We believe that the grant of commercial rights - that is to say, the right to carry traffic to and from another country, - is best negotiated and agreed to on a universal reciprocal basis, rather than by bilateral agreements. We think that only such an arrangement will secure to all countries the reciprocal rights which their interests require. But the grant of any such freedoms and rights must, in our opinion, necessarily be associated with the constitution of an authority which will regulate the use of such freedoms. It will be the function of such authority. . .to ensure that the interests of the people, both of the most powerful and of the smaller countries, are secured.[4]

India's position therefore has been to recommend a liberal approach of universal reciprocity within the parameters of control by an authority which could ensure that the smaller nations were protected from being swamped by larger States.

It is important to note that the economic significance of the Chicago Convention lies entirely in its main theme—of meeting the needs of the peoples of the world for economical air transport, whilst preventing waste through unfair competition and providing for a fair opportunity for all States concerned to operate air services. In order to accomplish this goal, the Convention, through ICAO, has to consider all aspects of economic implications that the operation of international air services by commercial air transport enterprises of the world, particularly those of the member States of ICAO, pose.

In August 1945, at the first meeting of the Opening Session of the Interim Council of the Provisional International Civil Aviation Organization (PICAO), the Hon. C.D. Howe, Minister of Reconstruction, Canada said:

> We (Canada) believe that there must be greater freedom for development of international air transport and that this freedom may best be obtained within a framework which provides equality of opportunity and rewards for efficiency.[5]

---

[3]See *Proceedings of the International Civil Aviation Conference*, Chicago, Illinois, November 1–December 7, 1944 United States Government Printing Office: Washington, 1948 at 65.

[4]*Id.* 76.

[5]*PICAO Documents*, Montreal, 1945, Volume 1, Doc 1, at 3.

Dr. Edward Warner, Representative of the United States of America (later the first President of the ICAO Council) said at the same meeting:

> Our first purpose will be to smooth the paths for civil flying wherever we are able. We shall seek to make it physically easier, safer, more reliable, more pleasant; but I believe it will be agreed also that we should maintain the constant goal that civil aviation should contribute to international harmony. The civil use of aircraft must so develop as to bring the peoples closer together, letting nation speak more understandingly unto nation.[6]

Dr. Warner had notably stressed on the purpose of civil aviation to be the promotion of international harmony and dialogue between nations. He had also made it clear that the seminal task of civil aviation is to bring the people of the world together through understanding and interaction. It is clear that at this stage at least, civil aviation was recognised more as a social necessity rather than a mere economic factor. In addition, through the statements of Minister Howe and Dr. Warner, one can glean the attitude of the international community towards aviation at that time:

- That civil aviation was based on equality of opportunity: and,
- That it was a social need rather than a fiscal tool.

The above notwithstanding, the American approach at the Conference to market access, particularly in terms of air traffic rights, is embodied in the statement of Adolf Berle, the Assistant Secretary at that time in the State Department when he said:

> I feel that aviation will have a great influence on American foreign interests and American foreign policy than any other non-political consideration. . .it may well be determinative in certain territorial matters which have to do with American defence, as well as with transportation matters affecting American commerce, in a degree comparable to that which sea power has had on our interests and policy.[7]

This certainly goes above and beyond using air transport as a social need on the basis of equality of opportunity.

The First Interim Assembly of the Provisional International Civil Aviation Organization (PICAO) was held in May 1946. This Session set the scene for identifying issues that had culminated in the provisions of the Chicago Convention. In the period that followed the First Interim Assembly Session, PICAO commissioned a group of experts called Commission 3 to draft a multilateral agreement on commercial rights for aircraft, which culminated in a Draft Multilateral Agreement on Commercial Rights. The Draft Agreement contained three basic elements:

- A grant of the right to operate commercially to a reasonable number of traffic centres serving as conveniently as is practicable each State's international traffic;

---

[6] *Id.* Doc 2, at 2.

[7] Mackenzie (2010) at 3.

- A basic regulatory provision dealing with the amount of capacity to be provided, with subsidiary provisions designed to prevent abuses; and,
- A provision for the settlement of differences between Contracting States through arbitral tribunals with power to render binding decisions.[8]

The only provisions of the draft on which unanimous agreement was not reached were those concerning routes and airports and capacity. Commission 3 also inquired into the distinction between scheduled and non-scheduled services as they appeared in Articles 5 and 6 of the Chicago Convention.

As a result of the study of Commission 3 on scheduled and non-scheduled air transport, the Air Transport Committee, at the 17th Session of the ICAO Council, examined in 1952, a Secretariat study on regulations in international non-scheduled aviation. The study found that at the time, national policies with respect to the taking on or discharging of traffic in their territories by foreign non-scheduled aircraft had taken a variety of forms. There were 13 States which required prior permission for each individual flight or series of flights where the granting of permission was based on the circumstances of each case. Ten States required that permission for non-scheduled flights should be granted for each flight or series of flights subject to prescribed regulations. Some States required specific bilateral agreements, while others demanded reciprocal treatment for their carriers.[9] Five European States were known to have made arrangements by means of formal bilateral arrangements for the regulation of non-scheduled commercial flights between their territories.[10]

The Committee also noted that the Council had expressed the view that a "stop for non-traffic purposes" as referred to in Article 5 of the Convention should be taken to include the freedom to load and unload passengers or goods not carried for remuneration or hire. The Council had also considered "remuneration or hire" to mean something received for the act of transportation from someone other than the operator. This interpretation would mean that flights carried out on the business of the operator would receive the freedom granted by the first paragraph of Article 5.[11] The Council's analysis of Article 5 also indicated that the State flown over must not consider its right to require landing as a matter of course and that this right, as granted in the provision, must not be exercised too restrictively. Consideration was also given to the fact that although the Chicago Convention, by Article 3, precludes its application to State aircraft, most States may be prepared to agree that civilian State aircraft should be given the type off free passage described in the first

[8] *Views of Commission No 3, Doc 4023*, A-1 - P/3, 1/4/47. See also *C-WP/369*, 22/6/49 for a detailed discussion on the Commission's work on the Agreement.

[9] *AT-WP/295, 15 Dec 52* at 5.

[10] *Ibid.*

[11] See *AT-WP/296*, 15/12/52, at 9.

paragraph of Article 5.[12] The same right may be given to emergency operations, taxi type flights and all inclusive charter tours.[13]

An analysis containing the above views of the ICAO Council, together with a definitive report by Council to Contracting States of scheduled international air services[14] as referred to in Article 6 of the Chicago Convention was adopted by Council at its Fifteenth Session on 28 March 1952. This Report contained the fact that a scheduled international air service must in the first instance consist of a series of flights. A single flight by itself could thus not constitute a scheduled international air service. Article 6 therefore requires that a series of flights must be performed through the air space over the territory of more than one State and performed by aircraft for the transport of passengers, cargo or mail for remuneration in order to constitute a scheduled national air service. The service must be performed so as to serve traffic between the same two or more points, either according to a published time table, or with flights so regular or frequent that they constitute a recognizably systematic series.[15] The word "remuneration" in the provision has the same application and meaning as in Article 5.

3 The Inevitable Compromise

In the meanwhile, in 1946, the United States and the United Kingdom, as a means of compromise between the "free market" approach of the former and the somewhat more cautious and conservative approach by the latter entered into a bilateral agreement for air services between their two territories. Called "Bermuda 1" this agreement represented a compromise between the philosophies of the two States that had been so divergent during the Chicago Conference. The Bermuda 1 agreement was typified by is restrictive pricing regime and liberal capacity arrangements and route descriptions. In the Agreement, while the United States compromised by withdrawing its opposition to the international regulation of fares and agreed that primary fare-setting functions should devolve upon the International Air Transport Association (IATA), the United Kingdom agreed to retract its earlier position that capacity should be regulated and recognized that airlines should be allowed to regulate capacity by determining their frequency on a given route provided that Governments were the ultimate arbiters of the control of capacity on the routes

[12] *Ibid.*

[13] AT-WP/296, 15/12/52 at 10.

[14] The ICAO Assembly, at its Second Session held in Geneva in June 1948, adopted Resolution A2-18 which called for the adoption by Council of a definition of the term "scheduled international air service." See Doc 7670, United Nations Convention on the Law of the Sea (UNCLOS), with Index and Final Act of the Third United Nations Convention on the Law of the Sea, United Nations: New York, 1983 at 79–80.

[15] See, *Report By the Council to Contracting States on the Definition of a Scheduled International Air Service and the Analysis of the Rights Conferred by Article 5 of the Convention. Doc 7278*, C/841, 10/5/52.

that were relevant to their territories. Accordingly, the Bermuda 1 Agreement determined that capacity should bear a strong and close relationship to the requirements of the public for air transport.

Many other States followed the Bermuda model in their air services agreements for nearly 30 years following its conclusion. One of the advantages of the Bermuda model was recognized as the IATA tariff setting clause which achieved a certain multilateralism through bilateralism, while one of its main disadvantages has been known to be that it gave governments a basis to formulate their civil aviation policies and sometimes adopt an unduly restrictive stance on their sovereignty in airspace, leading to air traffic rights that were being enjoyed by airlines being frequently withdrawn by States. Due to these shortcomings, Bermuda 1 collapsed predictably after 30 years.[16]

The first ICAO Assembly in 1947 followed up on the development of a Multilateral Agreement on Commercial Rights in International Civil Air Transport that was commenced by PICAO. At this Assembly, The United Kingdom felt that certain general principles should govern route agreements.[17] The concern of the US Government was that in matters of frequencies, capacity route exchanges and fifth freedom traffic rights, there would be disorder in operating on a general multilateral basis.[18] At this meeting, the Delegate of Canada analyzed the reason for seeking multilateralism in air services by stating:

> So we looked at the matter basically and said, "Why do we want Multilateralism?" and the feeling that I had, speaking for Canada, was not that we wanted uniformity, although that is desirable, in as much as I see no end result in uniformity for its own sake. We had a much loftier purpose in mind, and that was the idea of creating a set of conditions that all nations who wanted to fly could use so that they would know in advance what their opportunities were, what the conditions were that they would be up against, so that it would not be possible for nation to discriminate against another, and grant to another nation privileges that they would not be willing to grant to others equally entitled to them, so that these things would not lead to friction between nations and quarrels and eventually be the seed from which might spring a war. For this reason, it was said we wanted multilateralism, not merely uniform clauses.[19]

The views of the developing world were placed before the Assembly by the delegate of Peru:

> The multilateral agreement is a high ideal for which we have already fought and must continue to fight, but a firm fighting spirit should not allow eagerness to obscure reality. The latter, as we Peruvians see it, places grave difficulties in the way of an absolute and universal multilateral agreement. Those difficulties emanate from the different stages of development in commercial aviation among various nations, from the different aeronautical

---

[16]The "Bermuda II" Agreement, which was signed in 1977, contained a system of multiple designation of airlines by one State and other liberal provisions that toned down the harshness of capacity and route designation of its predecessor.

[17]ICAO *Doc 4510, A1-EC/72*, May 1947, 12–13.

[18]*Id.* 23.

[19]*Id.* 35

potential of each country, from the variations found when considering each country in international air transport, according to its climatic or geographical conditions and lastly, what is more important, the substantial differences between the countries already in commercial aeronautics, and these countries, such as ours, which can only look to the future.[20]

The ICAO Assembly, at its Second Session held in Geneva in June 1948 adopted Resolution A2-16 which called for further action on a Multilateral Agreement on Commercial Rights and resolved that Contracting States study and consider the above elements.[21]

Pursuant to the inability of Contracting States to reach multilateral agreement on uniformity in the award of air traffic rights, two agreements emerged that attempted to group States into accepting a limited common base on commercial aviation. The first—the Transit or Two Freedoms Agreement—was signed by 32 States and admitted of aircraft of those States being able to fly across each other's territories or land in them for non-traffic purposes, without having to obtain permission from the grantor State concerned. The second—the Five Freedoms or Transport Agreement, was signed by 20 States who granted each other the Five Freedoms of the air as they are known today, which their carriers could use freely in each others territories.[22] Those States which did not sign any of these agreements were required to sign bilateral air services agreements with each other, if their aircraft were to operate commercial air services into each other's territories involving the taking on or discharging passengers, mail and cargo in each other's territories. In addition, cabotage was introduced in Article 7 of the Convention, prohibiting aircraft from a State from picking up or discharging passengers, mail and cargo destined from one point of a State to another.

Two things have to be done if the air transport industry were to be recognized as a major contributor to the world economy and trading process and assisted accordingly. The first is to treat air transport as a trading tool and not as a luxury. A liberalized trading process must be applied in the context of air transport. It is incontrovertible that liberalization of air transport is a global trend that is irreversible and has been on-going since the eighties. In the liberalization process, fluctuations of global economic factors and their effect on the role and national approaches to market access continues to be the most critical element in air services agreements

---

[20] *Id.* 45–46.

[21] *Resolutions and Recommendations of the Assembly 1-9th Sessions* (1947–1955) Part II, Doc 7670 at 78.

[22] See Shawcross and Beaumont (1977) Paras. 207–209. There are three other freedoms of the air that have been added since the Chicago Convention was signed: The Sixth Freedom provides that an airline has the right to carry traffic between two foreign States *via* its own State or registry. This freedom can also be considered a combination of third and fourth freedoms secured by the State of registry from two different States producing the same effect as the fifth freedom *vis a vis* both foreign States; The Seventh Freedom allows an airline operating air services entirely outside the territory of its State of registry, to fly into the territory of another State and there discharge, or take on, traffic coming from, or destined for, a third State or States; and, the Eighth Freedom is Cabotage, as referred to in Article 7 of the Chicago Convention. See Dempsey (1987) at 50.

between States. These factors remain integral to substantive regulatory liberalization should a State decide to radically alter its stance toward opening the skies. In considering liberalization of market access, States invariably face two basic issues: the extent of liberalization, i.e. how open the market access should be in terms of the grant of traffic rights; and the approach to liberalization, i.e. whether liberalization should be national, bilateral, regional, plurilateral, or multilateral and the pace with which liberalization should be pursued. The second measure should be to liberalize the restrictive and antiquated "national ownership and control of airlines" requirement to the extent that foreign direct investment in air transport is encouraged.

References

Dempsey PS (1987) Law and foreign policy in international aviation. Transnational Publishers Inc, Dobbs Ferry

Mackenzie D (2010) ICAO – a history of the international civil aviation organization. University of Toronto Press, Toronto

Shawcross, Beaumont (1977) Air law, vol 1, 4th edn. Butterworths, London

Article 7
Cabotage

Each contracting State shall have the right to refuse permission to the aircraft of other contracting States to take on in its territory passengers, mail and cargo carried for remuneration or hire and destined for another point within its territory. Each contracting State undertakes not to enter into any arrangements which specifically grant any such privilege on an exclusive basis to any other State or an airline of any other State, and not to obtain any such exclusive privilege from any other State.

Contents

1 Commercial Equity

The first premise that has to be recognized is that cabotage—which is the exercise of commercial air traffic rights by a carrier of one State between points in another State[1]—is not prohibited by this provision. Article 7 merely states that States have a right to refuse permission for such carriage and, if it grants permission to one carrier, such permission shall not be privileged to just one carrier.

The terms "cabotage" and cabotage traffic in air trans port usage are derived, respectively, from maritime terms for the prohibition of coastwise carriage of traffic by foreign carriers and from the traffic thus prohibited which could be equated with domestic traffic, i.e. traffic moving on a single transportation document (ticket or waybill) involving no origination, stop-over or termination outside the territory of one State are sometimes expanded to also include (and thus prohibit) certain portions of international movements such as those between two points on an international route which are located in the territory of the same State (of which the carrier is not a national), before or after a connection or stopover at one such point, with an exception sometimes made to allow online on-route connections and stopovers; are sometimes erroneously applied to traffic moving between two States in the same group of States or economic union of States, when the group or union decides to reserve such traffic for its own air carriers; and can be applied to a traffic

[1]A cabotage right or cabotage privilege is a right or privilege granted to a foreign State or foreign carrier to carry revenue traffic from one airport of a State to another in the same contiguous territory of that State. see *Manual on the Regulation of International Air Transport*, ICAO Doc 9626, at 4.1–10.

R. Abeyratne, *Convention on International Civil Aviation*,
DOI 10.1007/978-3-319-00068-8_8, © Springer International Publishing Switzerland 2014

movement that constitutes prima facie cabotage such as a movement by air or surface across a national border followed immediately by a similar movement back across the same border, even when pursuant to separate tickets or waybills.

A cabotage right or cabotage privilege is a right or privilege, granted to a foreign State or a foreign carrier, to transport otherwise prohibited cabotage traffic. *Petit cabotage* involves traffic movements between two ports on the same coast of the same country (in maritime usage) and, by extension to air transport, between two airports in the same contiguous territory of a State. *Grand cabotage* involves traffic movements beginning and ending on different coasts of the same country (in maritime usage) and, by extension to air transport, movements between a State and a noncontiguous territory of that State.

The so-called *Eighth Freedom of the Air* is the right or privilege, in respect of scheduled international air services, of transporting cabotage traffic between two points in the territory of the granting State on a service which originates or terminates in the home territory of the foreign carrier or (in connection with the so-called Seventh Freedom of the Air) outside the territory of the granting State (also known *as an Eighth Freedom Right* or *"consecutive cabotage"*).

No decision either by the Assembly or the Council has been taken on this provision. It is recorded that Sweden had made proposals both at the Sixteenth and Eighteenth Sessions of the Assembly that the second sentence of Article, which was excessively restrictive, should be deleted since in practicality no State would leave itself open to granting cabotage rights to all airlines just because they wished to grant the right to one carrier. Sweden also anchored itself on Article 1 and the sovereignty of States over their airspace which was at variance with the untrammelled flexibility in Article 7. The Swedish proposal was never adopted by the Assembly.[2]

2 Cabotage in the US and EU

Cabotage becomes relevant in open skies agreements,[3] particularly when it concerns a whole region such as Europe. In the Open Skies Agreement between the United States and the European Union. The fundamental aim of the US/EU Agreement is to do away with the existing tapestry of bilateral air services

[2]See Doc 8771 A16-EX—report of the executive Committee, p. 43 at paragraphs 39:1–39:6. Also Doc 8960 A18-EX—Report of the Executive Committee, p. 32, 33, paragraphs 37:1–37:7. A16-WP/7 EX/1, A16-Min. P/7, A18-WP/26 EX/4; A18-WP/27 EX/5 and A18-Min. P/12.

[3]An open skies agreement is defined as a type of agreement which, while not uniformly defined by its various advocates, would create a regulatory regime that relies chiefly on sustained market competition for the achievement of its air services goals and is largely or entirely devoid of *a priori* governmental management of access rights, capacity and pricing, while having safeguards appropriate to maintaining the minimum regulation necessary to achieve the goals of the agreement. Open skies agreements are believed to provide for more competition, lower prices and higher passenger volumes in markets between signatory nations. See ICAO Doc 9626, S/PRST/1996/9, 27 February 1996 at 35 I.L.M. 493 (1996) at 2.2.2.

agreements between individual European Union member States and the United States and set up one system regulating transatlantic aviation. One of the issues on the table is cabotage rights, particularly for European carriers who cannot carry revenue passengers from point to point in the United States. On offer by the United States has been the right for European carriers to fly from anywhere in the EU to any point in the US. However, the US has sought in return beyond fifth freedom rights (i.e. the right for US carriers to carry revenue passengers from a European point to points beyond Europe) and vice versa offered rights beyond the US for European carriers.

Individual European States, which up until 1987 were separately chartering their destinies and their carriers' fortunes in their operations of international air services, showed an initial inclination to work towards collective interests by partially liberalizing European pricing policy in 1987. In 1993 the European countries of the European Economic Community agreed to full liberalization of pricing and liberalization of market access to apply on an intra-European basis. The culmination of the unification of European air transport came in 1997 when the European Union agreed to accord cabotage rights to carriers of the EU member States within the Union.

From the perspective of the European carriers, they would like the right to operate between the EU and the US from any point within the EU (which would translate as a seventh freedom[4] right in operating from a country other than other than the carrier's national territory) and extend that service to points within the US (which is the eighth freedom[5] right or consecutive cabotage). Also, the EU carriers are seeking the right to own and control US carriers and therefore be able to operate air services between points in the US, which is identified in the context of air law as ninth freedom[6] or "stand alone cabotage". In order to obtain these rights, the European carriers are seeking the abolition of ownership of US carriers by US nationals so that they (the European carriers) can attract capital from international money markets and enter into merges and acquisitions of foreign carriers. If this were to be at all allowed by the US (which is seemingly an impossibility according to current US policy) the European carriers would still have to operate on the basis that they remain "Community carriers" by their European ownership as they have to be owned in the majority by EU member States or their nationals.

[4]The seventh freedom of the air is the right or privilege in respect of scheduled international air services, granted by one State to another State, of transporting traffic between the territory of the granting State and any third State with no requirement to include on such operation any point in the territory of the recipient State. i.e. the service need not connect to or be an extension of service to/from the home State of the carrier.

[5]The eighth freedom of the air is the right or privilege, in respect of scheduled international air services, of transporting cabotage traffic between two points in the territory of the granting State on a service which originates or terminates in the home territory of the foreign carrier or (in connection with the seventh freedom of the air) outside the territory of the granting State).

[6]The ninth freedom of the air is the right or privilege of transporting cabotage traffic of the granting State on a service performed entirely within the territory of the granting State.

In the US context, what is sought by the US carriers are free access to London Heathrow and seventh freedom carriage within the EU for express carriers. In broad terms, the US interests are focused on turning the North Atlantic aviation market into an open skies area giving rise to a common international air traffic market untrammelled by any conditions on market access, capacity and pricing. This would of course exclude the internal US market and any incursion into current US policy on majority ownership of US carriers by US nationals.

In reality, the US has already acquired for its carriers the rights to operate between European States through current bilateral air services agreements negotiated with individual European States and as such, any demand by the US for fifth freedom rights within the EU cannot be considered cabotage. As Wassenbergh correctly observes, in the absence of a single, unified, sovereign EU airspace, the EU cannot consider operations between sovereign States within the EU as cabotage.[7] Nonetheless, the open skies judgments of the European Court of Justice[8] of 5 November 2002 were to the effect that the eight EU members, by concluding individual bilateral agreements with the US, had breached EC law in that the individual nationality clauses in all agreements infringed the right of establishment under Article 43 of the EC Treaty as they discriminated on grounds of nationality.[9] The ECJ also held that the agreements infringed the exclusive external competence of the European Commission. The essence of the judgement was that in areas where EC legislation affects third countries, only the EU could enter into international commitments. The new framework of EU air services negotiations are enshrined in Regulation 847/2004 which allows the EC to exercise a "horizontal mandate" to negotiate comprehensive agreements with third countries. This would mean that the third country acknowledges the existence of a single European market and the concomitant fact that EU airlines can operate international flights from any member State where they are established.

---

[7]Wassenbergh (2005) at p. 55.

[8]The European Court of Justice (ECJ) is formally known as the *Court of Justice of the European Communities*', i.e. the court of the European Union (EU). It is based in Luxembourg, unlike most of the rest of the European Union institutions, which are based in Brussels and Strasbourg. The ECJ is the Supreme Court of the European Union. It adjudicates on matters of interpretation of European law, most commonly: claims by the European Commission that a member state has not implemented a European Union Directive or other legal requirement; claims by member states that the European Commission has exceeded its authority; and references from national courts in the EU member states asking the ECJ questions about the meaning or validity of a particular piece of EC law. The Union has many languages and competing political interests, and so local courts often have difficulty deciding what a particular piece of legislation means in any given context. The ECJ steps in, giving its ruling which is binding on the national court, to which, the case will be returned to be disposed of. The ECJ is only permitted to aid in interpretation of the law, and not decide the facts of the case itself. Individuals cannot bring cases to the ECJ directly. An individual who is sufficiently concerned by an act of one of the institutions of the European Union can challenge that act in a lower court, called the Court of First Instance. An appeal on points of law lies against the decisions of the Court of First Instance to the ECJ.

[9]See generally, Abeyratne (2003), pp. 485–518.

Whatever the outcome of the US/EU negotiations are in terms of cabotage for EU carriers and the US rule on ownership and control of its carriers, it seems certain that the two parties could agree on free market access between points in the US and points in the EU along with beyond fifth freedom rights respectively. As a corollary, free market access along with no limitations on pricing and capacity would certainly open up competition between US and EU carriers.

References

Wassenbergh H (2005) Open skies and a global common air traffic market. J LuchtRecht. Special Edition, Liber Amicorum in Honour of Prof. Dr. I.H.Ph. Diederiks-Verschoor, Nr. 9/10.
Abeyratne RIR (2003) The decision of the European court of justice on open skies - how can we take liberalization to the next level? J Air Law Commerce 68(3):485–518

Article 8
Pilotless Aircraft

No aircraft capable of being flown without a pilot shall be flown without a pilot over the territory of a contracting State without special authorization by that State and in accordance with the terms of such authorization. Each contracting State undertakes to insure that the flight of such aircraft without a pilot in regions open to civil aircraft shall be so controlled as to obviate danger to civil aircraft.

Contents

1 Remote Control of Civilian Aircraft

On 13 June 1960, the ICAO Council, at the sixth meeting of its fortieth session adoption a Resolution whereby the Council declared that the flight of uncontrolled balloons[1] not released under appropriate safeguards and conditions may constitute a definite hazard to the safety of air navigation.[2] The Council, while drawing attention to ICAO member States to Article 8, urged States to take whatever action they deemed appropriate or necessary to ensure the safety of flight.

In modern parlance the most contentious pilotless aircraft is the drone, more technically referred to as Remotely piloted aircraft system (RPAS). Originally called An Unmanned Aerial Vehicle (UAV)[3] it is a self-piloted or remotely piloted aircraft[4] that can carry cameras, sensors, communications equipment or other payloads. They have been used to conduct reconnaissance and intelligence-gathering for nearly sixty years (since the 1950s). The future role of the RPAS is a more

---

[1]A balloon falls within the definition of an aircraft as defined in Annex 7 to the Chicago Convention. See CANSO is the global voice of the air traffic management profession. Its members comprise over 50 air navigation service providers who control more than 85% of global air traffic movements. CANSO seeks to promote best practices within the industry.

[2]Doc 8097-C/926 at p.9.

[3]For more details of UAV operations and their nature visit www.uvs-info.com.

[4]An aircraft is defined as "any machine that can derive support in the atmosphere from the reactions of the air other than the reactions of the air against the earth's surface." This definition appears in Annexes 1, 2, 3, 7, 8, 11, 13, 16 and 17 to the Convention on International Civil Aviation, signed at Chicago on 7 December 1944. See ICAO Doc 7300/8 Eighth Edition, 2000.

R. Abeyratne, *Convention on International Civil Aviation*,
DOI 10.1007/978-3-319-00068-8_9,

challenging one which, in addition to its current uses will include involvement in combat missions.[5] The issues and challenges that RPASs bring to civil aviation can be bifurcated into two main areas. The first concerns airworthiness regulations which are required to ensure that a RPAS is built, maintained and operated at high standards that ensure the safety of all involved including crew and passengers of manned civilian and military aircraft with which RPASs will share de-segregated airspace as well as persons and property on the ground.[6] There is currently no international Standards and Recommended Practices (SARPs) adopted under the auspices of the International Civil Aviation Organization[7] applicable to the RPAS and the Unmanned Aircraft System (RPAS)[8] although RPASs are increasingly requiring access to all categories of airspace including non segregated airspace.

The second challenge is more far reaching and concerns the possibility of the RPAS encroaching on air traffic control (ATC) functions in non segregated airspace. In doing so, RPASs should not place an added burden and demands on airspace management and the flow of general air traffic within the en-route air space structure which must not be impeded by the presence of RPASs. In this context, the priority would lie in collision avoidance, primarily through effective separation of aircraft by which aircraft could be kept apart by the application of appropriate separation minima. The two key players in this exercise would be the pilot of the manned aircraft involved and the air navigation service provider who would be jointly or severally liable if a separation minima were compromised.

Although there are international regulations in place that address the operation of RPASs in non segregated airspace, there is provision under ICAO regulations for the appropriate procedure to be followed. Annex 11 to the Chicago Convention, which deals with the subject of air traffic services, lays down requirements for coordination of activities that are potentially hazardous to civil aircraft. Standard 2.17.1 stipulates that arrangements for activities potentially hazardous to civil

---

[5]Since 1964 the US Defense Department has developed 11 different UAVs, though due to acquisition and development problems only 3 entered production. The US Navy has studied the feasibility of operating Vertical Take off and Landing (VTOL) UAVs since the early 1960s, the QH-50 Gyrodyne torpedo-delivery drone being an early example. However, high cost and technological immaturity have precluded acquiring and fielding operational VTOL UAV systems.

[6]The main concern of the International Civil Aviation Organization in its role as regulator in this context is with international civil RPAS operations and those standards that affect such operations. ICAO should therefore, not be expected to take on a leading role in the development of aircraft performance specifications.

[7]An ICAO Exploratory Meeting on Unmanned Aerial Vehicles (UAVs) was held at ICAO Headquarters in Montreal from 23 to 24 May 2006. The primary objective of the meeting was to explore the current state of affairs with respect to development of regulatory material related to UAVs and to discuss the possible role of ICAO in the regulatory process. The meeting was informed that the ICAO Secretariat would use the results of the meeting as the basis for developing a report to the ICAO Air Navigation Commission (ANC) along with recommendations on an ICAO work programme.

[8]At least four States: Australia; France; South Africa; and the United States are known to have commenced a programme developing standards for RPAS operations. See ter Kuille, p. 24 at 25.

aircraft, whether over the territory of a State or over the high seas, shall be coordinated with the appropriate air traffic services authorities, such coordination to be effected early enough to permit timely promulgation of information regarding the activities in accordance with the provisions of Annex 15 to the Chicago Convention.[9] Standard 2.17.2 of Annex 11 explains that the objective of the coordination referred to in the earlier provision shall be to achieve the best arrangements that are calculated to avoid hazards to civil aircraft and minimize interference with the normal operations of aircraft.

The conduct of operations of RPASs are essentially State based and therefore becomes an issue of State Responsibility. State responsibility in turn is founded on the basic legal principle of sovereignty and the rights and liabilities of States. The principle of State sovereignty in airspace is embodied in Article 1 of the Chicago Convention which recognizes that every State has sovereignty over the air space above its territory, the latter being defined in Article 2 as land situated within and water adjacent to the State concerned. As for rights over airspace over the high seas, Article 87 of the *United Nations Convention on the Law of the Sea* of 1982[10] awards freedom for the aircraft of all States to fly over the high seas. An important consideration in delineating territorial sovereignty lies in the exPANsion of Flight Information Regions (FIR) and the provision of air traffic management services by States particularly when such measures are influenced by the revenue generating capabilities that are inherent in such an exPANsion of scope. The Chicago Convention, in its vision and wisdom, incorporates various provisions regarding the provision of air navigation services by States to aircraft flying over their territories. Firstly, the Convention guarantees, through provisions included in Chapter XV, that States which are unable to provide air navigation services to aircraft will be assisted. Secondly, Article 15 of the Convention assures airlines that every airport in a Contracting State that is open to public use by its national aircraft shall also be open under uniform conditions to the aircraft of all the other Contracting States. The conditions are deemed to apply to the use, by aircraft, of every Contracting State of all air navigation facilities, including radio and meteorological services, which may be provided for public use for the safety and expedition of air navigation. Charges levied for such services are deemed by Article 15 to be anti-discriminatory whereby aircraft are not to be charged for airports and air navigation services provided to them at a rate higher than those levied on the national carrier of the State which provides the service. To this end, Article 28 of the Convention obligates Contracting States to provide, as far as practicable in their territories, airports, radio services, meteorological services and other air navigation facilities to facilitate international air navigation according to Standards established pursuant to the Convention.

[9]Annex 15 contains Standards and Recommended Practices relating to Aeronautical Information Services.

[10]The Law of the Sea, Original Text of the *United Nations Convention on the Law of the Sea*, all Annexes and Index, United Nations: New York, 1983.

The tightly-set legal parameters of the Chicago Convention, particularly the assurance of air navigation services on an equal and non-discriminatory basis, are relevant in the twenty-first Century, where service providers and airline operators have to collaborate in ensuring a seamless global air navigation system. Modern technology offers sophisticated air-ground data communications by VHF (very high frequency) and satellite, assisted by precise navigation by inertial/GNSS and computing in air traffic services. These will be used in the negotiation of dynamic user preferred routes offering various alternatives to airline operators which provide fuel and time savings. However, such preferences for flight profiles and uses thereof will be subject to growing air traffic demands which have to be cautiously assessed. This imposes an added burden on both the service provider and airline operator. Judgment and interpretation will be critical factors in this process, an inevitable corollary of which will be the need to examine legal aspects of the modern seamless air traffic management system.

As already stated, responsibility of States for the provision of air navigation services in their territories is founded in principles contained in Article 28 of the Chicago Convention of 1944. It must be noted that this is not an absolute obligation as the State is called upon to provide such services only in so far as it finds practicable to do so. In order to cover an eventuality of a State not being able to provide adequate air navigation services, the Convention imposes an overall obligation on the Council of ICAO in Article 69 to the effect that the Council shall consult with a State which is not in a position to provide reasonably adequate air navigation services for the safe, regular, efficient and economical operations of aircraft. Such consultations will be with a view to finding means by which the situation may be remedied. Article 70 of the Chicago Convention even allows for a State to conclude an arrangement with the Council regarding the financing of air navigation facilities and the Council is given the option in Article 71 of agreeing to provide, man, maintain and administer such services at the request of a State.

The provision of air navigation services are mainly regulated by three Annexes to the Chicago Convention, namely Annex 2 (Rules of the Air), Annex 3 (Meteorological Service for International Air Navigation) and Annex 11 (Air Traffic Services).[11]

[11]Article 54 (l) of the Chicago Convention stipulates as a mandatory function of the Council the act of adopting, in accordance with Chapter VI of the Convention, international standards and recommended practices (SARPs) and for convenience designate them as Annexes to the Convention. Article 37 of the Convention reflects the areas in which SARPs should be developed and Annexes formed. Article 38 obliges contracting States to notify ICAO of any differences between their own regulations and practices and those established by international standards or procedures. The notification of differences however, does not absolve States from their continuing obligation under Article 37 to collaborate in securing the highest practicable degree of uniformity in international regulations, standards, and procedures.

Of these, compliance with Annex 2 is mandatory[12] and does not give the States the flexibility provided in Article 38 of the Chicago Convention to register differences from any provisions of the Annex.

Before UAVs became RPASs pilotless aircraft were known as an *Unmanned Aircraft System* (RPAS) is an aircraft[13] and its associated elements which are operated with no pilot on board. RPAS was an overarching term for the entire system comprising an *Unmanned Aerial Vehicle* (UAV)[14] which is applied to describe a self piloted or remotely piloted aircraft that can carry cameras, sensors, communications equipment or other payloads,[15] as well those which support unmanned flights such as air traffic management and remote controllers of such aircraft. The United States Department of Defence defines a UAV as

> a powered aerial vehicle that does not carry a human operator, uses aerodynamic forces to provide vehicle lift, can fly autonomously or be piloted remotely, can be expendable or recoverable, and carry a lethal or non-lethal payload.[16]

---

[12]In October 1945, the Rules of the Air and Air Traffic Control (RAC) Division at its first session made recommendations for Standards, Practices and Procedures for the Rules of the Air. These were reviewed by the then Air Navigation Committee and approved by the Council on 25 February 1946. They were published as *Recommendations for Standards, Practices and Procedures—Rules of the Air* in the first part of Doc 2010, published in February 1946. The RAC Division, at its second session in December 1946–January 1947, reviewed Doc 2010 and proposed Standards and Recommended Practices for the Rules of the Air. These were adopted by the Council as Standards and Recommended Practices relating to Rules of the Air on 15 April 1948, pursuant to Article 37 of the Convention on International Civil Aviation (Chicago, 1944) and designated as Annex 2 to the Convention with the title *International Standards and Recommended Practices—Rules of the Air*. They became effective on 15 September 1948. On 27 November 1951, the Council adopted a complete new text of the Annex, which no longer contained Recommended Practices. The Standards of the amended Annex 2 (Amendment 1) became effective on 1 April 1952 and applicable on 1 September 1952.

[13]An aircraft is defined as "any machine that can derive support in the atmosphere from the reactions of the air other than the reactions of the air against the earth's surface." This definition appears in Annexes 1,2,3,7,8,11,13,16 and 17 to the Convention on International Civil Aviation, signed at Chicago on 7 December 1944. See ICAO Doc 7300/9 Ninth Edition, 2006.

[14]For more details of RPAS operations and their nature visit www.uvs-info.com.

[15]In January 2007, the Air Navigation Commission (ANC) of the International Civil Aviation Organization (ICAO) consulted with States and appropriate international organizations on convening an Accident Investigation and Prevention Group (AIG) meeting in 2008 to discuss subjects in the field of accident investigation. One of the proposed subjects for discussion is the amendment of the definition of accident in Chapter 1 of Annex 13 to include events involving unmanned aerial vehicles (UAV). See Addressing Unmanned Aircraft System Accident Investigation and Prevention (Paper presented by the United States at the 36th ICAO Assembly) A36-WP/217 TE/70 18/09/07.

[16]*Unmanned Aerial Vehicles: Background and Issues for Congress*, Report for Congress written by Elizabeth Bone and Christopher Bolkcom, Congressional Research Service: The Library of Congress, April 25 2003 CRS 1.

Ballistic or semi-ballistic vehicles, cruise missiles and artillery projectiles were not considered UAVs by this definition.[17] The Federal Aviation Administration of the United States defines a RPAS as

> a device that is used or intended to be used for flight in the air that has no onboard pilot. This includes all classes of airplanes, helicopters, airships, and translational lift aircraft that have no onboard pilot.[18]

All references to RPAS that follow in this article therefore necessarily include UAVs.

RPAS are used to serve different purposes and therefore come in a variety of models, shapes and sizes. Their sizes may differ from having as wide a wing span as a Boeing 737 aircraft to that of a radio-controlled model airplane. A RPAS has of necessity to be guided and operated by a pilot on the ground. A strategic use of a RPAS is military reconnaissance and attack where they are commonly called drones. However, they now also serve to increase efficiency, and be cost effective, enhance safety and even save lives. They could also be used in aerial photography, surveying land and crops, monitoring forest fires and environmental conditions, and protecting borders and ports against intruders.[19]

RPAS have been used to conduct reconnaissance and intelligence-gathering for nearly sixty years (since the 1950s). The future role of the RPAS is a more challenging one which, in addition to its current uses will include involvement in combat missions.[20] The issues and challenges that RPAS bring to civil aviation can be bifurcated into two main areas. The first concerns airworthiness regulations which are required to ensure that a RPAS is built, maintained and operated at high standards that ensure the safety of all involved including crew and passengers of manned civilian and military aircraft with which RPAS will share de-segregated airspace as well as persons and property on the ground.[21] The International Civil Aviation Organization[22] began addressing issues concerned with the operation of

---

[17]*Ibid.*

[18]Unmanned Aircraft System Regulation Review, September 2009, Final Report, DOT/FAA/AR-09/7, 14.

[19]See FAA Fact Sheet – Unmanned Aircraft Systems (RPAS), December 1, 2010 at http://www.faa.gov/news/fact_sheets/news_story.cfm?newsId=6287.

[20]Since 1964 the US Defense Department has developed 11 different RPAS, though due to acquisition and development problems only 3 entered production. The US Navy has studied the feasibility of operating Vertical Take off and Landing (VTOL) RPAS since the early 1960s, the QH-50 Gyrodyne torpedo-delivery drone being an early example. However, high cost and technological immaturity have precluded acquiring and fielding operational VTOL RPAS systems.

[21]The main concern of the International Civil Aviation Organization in its role as regulator in this context is with international civil RPAS operations and those standards that affect such operations.

[22]An ICAO Exploratory Meeting on Unmanned Aerial Vehicles (RPAS) was held at ICAO Headquarters in Montreal from 23 to 24 May 2006. The primary objective of the meeting was to explore the current state of affairs with respect to development of regulatory material related to RPAS and to discuss the possible role of ICAO in the regulatory process. The meeting was informed that the ICAO Secretariat would use the results of the meeting as the basis for developing a report to the ICAO Air Navigation Commission (ANC) along with recommendations on an ICAO work programme.

RPAS and principles applicable thereto[23] since RPAS were increasingly requiring access to all categories of airspace including non segregated airspace.

The second challenge is more far reaching and concerns the possibility of the RPAS encroaching on air traffic control (ATC) functions in non segregated airspace. In doing so, RPAS should not place an added burden and demands on airspace management and the flow of general air traffic within the en-route air space structure which must not be impeded by the presence of RPAS. In this context, the priority would lie in collision avoidance, primarily through effective separation of aircraft by which aircraft could be kept apart by the application of appropriate separation minima. The two key players in this exercise would be the pilot of the manned aircraft involved and the air navigation service provider who would be jointly or severally liable if separation minima were compromised.

Although there are international regulations in place that address the operation of RPAS in non segregated airspace, there is provision under ICAO regulations for the appropriate procedure to be followed. Annex 11 to the Chicago Convention, which deals with the subject of air traffic services, lays down requirements for coordination of activities that are potentially hazardous to civil aircraft. Standard 2.17.1 stipulates that arrangements for activities potentially hazardous to civil aircraft, whether over the territory of a State or over the high seas, shall be coordinated with the appropriate air traffic services authorities, such coordination to be effected early enough to permit timely promulgation of information regarding the activities in accordance with the provisions of Annex 15 to the Chicago Convention.[24] Standard 2.17.2 of Annex 11 explains that the objective of the coordination referred to in the earlier provision shall be to achieve the best arrangements that are calculated to avoid hazards to civil aircraft and minimize interference with the normal operations of aircraft.

The Chicago Convention is focused on civil aviation, and applies to civil aircraft. The Convention does not apply to State aircraft, which are identified as aircraft engaged in police, military an customs services.[25] Therefore, principles of the Convention will apply only to RPAS not engaged in such activities as are excluded. One of the provisions which may have a bearing on RPAS in the Convention is Article 8 which stipulates that no aircraft capable of being flown without a pilot shall be flown without a pilot over the territory of a Contracting State without special authorization by that State. Furthermore states allowing the operation of aircraft that do not have a pilot in air space open to civil aircraft are required to ensure that they are so controlled as to obviate danger to civil aircraft. One of the common usages of RPAS—aerial photography—is affected by Article 36 of the Chicago Convention which empowers Contracting states to prohibit or regulate

---

[23]At least four States: Australia; France; South Africa; and the United States are known to have commenced a programme developing standards for RPAS operations. See ter Kuille, p. 24 at 25.

[24]Annex 15 contains Standards and Recommended Practices relating to Aeronautical Information Services.

[25]Chicago Convention, Preamble (*supra* note 1), at Article 3.

the use of photographic apparatus in aircraft over its territory. Presumably this provision can be tagged on to Article 1 of the Convention whereby every State has complete and exclusive sovereignty over the airspace above its territory. Another important consideration could lie in Finally Annex 17[26] to the Chicago Convention, on the subject of aviation security where. Article 2.1.2 of the Annex states that each Contracting State shall establish an organization and develop and implement regulations, practices and procedures to safeguard civil aviation against acts of unlawful interference taking into account the safety, regularity and efficiency of flights. This could impel States to develop regulations and practices addressing the interference of control signals or even the hostile takeover of the command of an RPAS which is a very common hazard to the operation of RPAS.

Another challenge in the operation of RPAS is licensing of personnel in charge of the operation of the vehicle and certification of the RPAS. Article 31 of the Convention provides that every aircraft engaged in international navigation shall be provided with a certificate of airworthiness issued or rendered valid by the State in which it is registered. The Standards and Recommended Practices (SARP) for the issuance of an airworthiness certificate are laid down in Annex 8[27] to the Chicago Convention. Annex 8 (in its 9th Edition) only addresses aeroplanes[28] over 5700 kg certificated take-off mass and helicopters[29] without a limitation on the mass of an aircraft which is intended for the carriage of passengers or cargo or mail in international air navigation[30] This might provoke the argument that Annex 8 would not usually apply to RPAS since only large RPAS exceed the weight of 5700 kg. The lack of internationally recommended and accepted standards and practices for smaller aeroplanes is a challenge for the operation of RPAS as well as for aeroplanes with a pilot on board. This point is covered in the 10th edition of Annex 8 which, in addition to the provisions in part VI on helicopters has been

---

[26] Annex 2 to the Convention on International Civil Aviation, "Safety – Safeguarding International Civil Aviation Against Acts of Unlawful Interference", 8th edition, April 2006.

[27] Annex 8 to the Convention on International Civil Aviation, "Airworthiness of Aircraft", 10th edition, April 2005.

[28] "A power-driven heavier-than-air aircraft, deriving its lift in flight chiefly from aerodynamic reactions on surfaces which remain fixed under given conditions of flight." See definitions in note 226.

[29] "A heavier-than-air aircraft supported in flight chiefly by the reactions of the air on one or more power driven rotors on substantially vertical axes." See definitions in note 226.

[30] See Shawcross and Beaumont (1977) Paras. 207–209. There are three other freedoms of the air that have been added since the Chicago Convention was signed: The Sixth Freedom provides that an airline has the right to carry traffic between two foreign States *via* its own State or registry. This freedom can also be considered a combination of third and fourth freedoms secured by the State of registry from two different States producing the same effect as the fifth freedom *vis a vis* both foreign States; The Seventh Freedom allows an airline operating air services entirely outside the territory of its State of registry, to fly into the territory of another State and there discharge, or take on, traffic coming from, or destined for, a third State or States; and, the Eighth Freedom is Cabotage, as referred to in Article 7 of the Chicago Convention. See Dempsey (1987), at 50, part IV, standard 1.1.2 of Annex 1 (wording identical to 9th edition).

amended to be applicable for helicopters with a certificated take-off mass over 750 kg only. In terms of licensing it has to be noted that Annex 1[31] to the Chicago convention, defines SARPs for personnel licensing, in that a person shall not act as an air crew member unless a valid license is held[32] by that person. Pilots are considered not only flight crew but as well flight navigators, flight engineers and radiotelephone operators.[33] Implicitly, this means that not only is the remote pilot of RPAS subject to licensing, but also personnel who are involved in the navigation and technical operation of RPAS should be licensed as well. Furthermore mechanics of RPAS be should also be licensed according the provisions in chapter 4.1 and 4.2 of Annex 1 to the Chicago Convention. Article 29 of the Chicago Convention requires the carriage of documents in aircraft such certificates of registration and airworthiness but also the appropriate licenses for each member of the crew. Although certificates of airworthiness can be carries in an aircraft in the manner required, the carriage of other documents may pose difficulties as some RPAS are designed to operate over extended periods of time, up to several months, and the specific operators who would operate for such long periods may not be known at the initial stage of the flight. One potential solution could be to electronically store the data and electronic licenses (be it in the form of scanned documents or other forms) of the current crew on board of the vehicle, but this would need in depth assessment in regards to the legal validity of such a form.

Annex 2 to the Chicago Convention, detailing the rules of the air referred to in Article 12 of the Convention, states *inter alia* that the rules of the air shall apply to aircraft bearing the nationality and registration marks of a Contracting State.[34] These rules applicable to RPAS as well. Two main categories of rules of the air exist: visual flight rules and instrument flight rules.[35] The note to article 2.2. of Annex 2 states inter alia that a pilot may elect to fly in accordance with instrument flight rules in visual meteorological conditions. The rules of the air adhered to are thus distinct and separate from the metrological conditions prevailing in the area of operation, except for instrument metrological conditions, requiring instrument flight rules to be applied. Chapter 3.1 of Annex 2, contains an article on unmanned free balloons, stating that they shall be operated in such a manner as to minimize hazards to persons, property or other aircraft and in accordance with the conditions specified in Appendix 4. Appendix 4 states inter alia that heavy balloons[36] need to comply with similar provisions like normal aeroplanes, inter alia minimum height over "congested areas of cities, towns or settlements or an open-air assembly of

[31] Annex 1 to the Convention on International Civil Aviation, "Personnel Licensing", 10th edition, July 2006.

[32] *Supra* note 30, Standard 1.2.1 of Annex 1.

[33] *Supra* note 30, Chapter 3.

[34] *Resolutions and Recommendations of the Assembly 1–9th Sessions* (1947–1955) Part II, Doc 7670 at 78, Article 2.1.1.

[35] *Supra*, note 30, Article 2.2.

[36] ICAO *Doc 4510, A1-EC/72*, May 1947, 35, Appendix 4, article 1 (c).

persons not associated with the operation",[37] SSR equipment,[38] and lightening.[39] Article 3.3 of Appendix 4 to the Annex 2 to the Chicago convention contains a remarkable requirement to unmanned balloons. Such vehicles shall be equipped with at least two payload flight termination devices or systems. It may well be argued that such devises or systems are required for RPAS as well. An analogy to the operation of RPAS exists in Annex 2 which requires obliges pilots-in-command to take action as will best avert collision. The Annex also requires that vigilance for the purpose of detecting potential collisions be exercised on board an aircraft, regardless of the type of flight or the class of airspace in which the aircraft is operating. Therefore, it can be concluded that pilots flying according instrument flight rules are required to scan the environment visually in order to detect potentially conflicting traffic. This task may prove difficult in the case of RPAS in that although many RPAS are equipped with video cameras, it would be difficult for RPAS operators to detect vehicles nearby, to assess the potential for conflict and to initiate appropriate actions. This inability might result in infringement of article 3.2.1 of Annex 2, which provides that an aircraft shall not be operated in such proximity to other aircraft as to create a collision hazard. A potential solution to this problem could be that movement sensors, based on radar or ultrasound devices, similar to parking assistants for cars, are built into RPAS. The drawback of such a measure would be the cost involved and the additional weight that has to be carried by the RPAS.

2 Work of ICAO

In early 2011, as a result of sustained work carried out on RPAS[40] by ICAO, the Organization released a circular entitled Unmanned Aircraft Systems (RPAS)[41] the purposes of which was to: apprise States of the emerging ICAO perspective on the integration of RPAS into non-segregated airspace and at aerodromes; consider the fundamental differences from manned aviation that such integration will involve; and encourage States to help with the development of ICAO policy on RPAS by providing information on their own experiences associated with RPAS.[42] The fundamental premise that ICAO follows in this regard is that, since unmanned

[37]Id, article 3.2.

[38]Id, article 3.4.

[39]Id, article 3.6.

[40]In November 2007 the Air Navigation Commission of ICAO established the unmanned Aircraft Systems Study Group comprising Australia, Austria, Brazil, China, Czech Republic, France, Germany, Italy, Netherlands, New Zealand, Russian Federation, Singapore, South Africa, Sweden, U.K., U.S., CANSO, EASA, EUROCAE, EUROCONTROL, IAOPA, ICCAIA, IFALPA, IFATCA, UVS Intl.

[41]*Unmanned Aircraft Systems* (RPAS) Cir 328- AN/190.

[42]*Id.* Paragraph 1.6.

aircraft fall within the definition of "aircraft" all SARPs of the Annexes to the Chicago Convention applicable to aircraft would apply to RPAS as well.[43]

It must be underscored that the preliminary aim of the guidance material in the ICAO Circular is to ensure aviation safety[44] based on the fact that the risk of mid-air collisions between RPAS and aircraft manned by pilots on board is a critical safety concern for RPAS operations worldwide. Accident investigation therefore becomes crucial both in cases where accidents cause death or injury to persons and damage to property and in instances where no collision occurs between RPAS and manned aircraft. In order to determine what aspect of the operation failed, whether additional, previously unanticipated hazards were contributory, and what deficiencies need to be corrected to prevent such an event from progressing to a more serious outcome in the future.

To begin with, the Circular makes explicit mention of Article 8 of the Chicago Convention, which, as earlier discussed, requires special authorization by the State flown over by an aircraft capable of being flown without a pilot which is in fact flown without a pilot. In this context the Circular clarifies any obfuscation that may arise as to what a "pilotless aircraft" is by quoting another ICAO document which states that Article 8 refers to an aircraft which is flown without a pilot in command on board the aircraft but which is either remotely or fully controlled from another place (ground, another aircraft or space).[45]

One of the main issues addressed by the Circular is that aircraft operating without a pilot on board present a wide array of hazards to the civil aviation system. These hazards must be identified and the safety risks mitigated,[46] just as with introduction of an airspace redesign, new equipment or procedures. In this regard, States are required to establish a State Security Programme (SSP) to include safety

---

[43]*Id.* Paragraph 1.7. Model aircraft, which are outside the purview of the Chicago Convention are not included in within this principle. *Id.* 2.4. It must be noted that a number of Civil Aviation Administrations (CAAs) have adopted the policy that RPAS must meet the equivalent levels of safety as manned aircraft. RPAS operations must be as safe as manned aircraft insofar as they must not present a hazard to persons or property on the ground or in the air that is any greater than that attributable to the operation of manned aircraft of equivalent class or category. In general, RPAS should be operated in accordance with the rules governing the flight of manned aircraft and meet equipment requirements applicable to the class of airspace within which they intend to operate. RPAS must be able to comply with ATC instructions.

[44]Safety is defined as: "The state in which the possibility of harm to persons or of property is reduced to, and maintained at or below, an acceptable level through a continuing process of hazard identification and safety risk management". See Cir 328-AN/190, C-WP/9781 Appendix for the Secretary General's Report at p. 5.

[45]*The Global Air Traffic Management Concept*, Doc 9854 referred to in Cir 328-AN/190 in 2.2. It must be noted that ICAO recognizes many categories of aircraft, including but not limited to balloons, gliders, aeroplanes and rotorcraft whether they operate from land or water.

[46]The term "safety management" includes two key concepts. First is the concept of a State safety programme (SSP), which is an integrated set of regulations and activities aimed at improving safety. Second is the concept of a safety management system (SMS) which is a systematic approach to managing safety, including the necessary organizational structures, accountabilities, policies and procedures.

rulemaking, policy development and oversight. The operation of RPAS in desegregated airspace would not only affect operations carried by commercial air carriers but would also affect general aviation. The International Council of Aircraft Owner and Pilot Association (IAOPA) has commented on the issue of RPAS that operating rules for RPAS must take into account their potential impact on general aviation aircraft operating in un-segregated airspace. IAOPA added that while segregated airspace contains operations subject to air traffic control, un-segregated airspace depends almost entirely on certain Annex 2 (to the Chicago Convention) cruising altitude conventions and mutual self-separation methods. Because self-separation methods for RPAS are still in the conceptual stage and will likely require some time to perfect, there will be a temptation to impose un-segregated airspace restrictions on manned aircraft to accommodate RPA. Since un-segregated airspace is almost entirely the domain of general aviation, we do not want this to occur.

The second point raised by IAOPA is that State or military RPAS must abide by whatever RPAS operating rules are devised to ensure safe, hazard-free operations. Because non-civil RPAS operations may wish to use lower altitude un-segregated airspace, there could be a tendency for States and the military to pre-empt conventional flight rules in these areas, either on a temporary or permanent basis.[47]

The answer to these two points seemingly rests with the fact that RPAS will operate in accordance with ICAO Standards that exist for manned aircraft as well as any special and specific standards that address the operational, legal and safety differences between manned and unmanned aircraft operations.[48] This includes applicable environmental rules and guidelines as well.[49] In order for RPAS to integrate into non-segregated airspace and at non-segregated aerodromes, there will be a pilot responsible for the RPAS operation. Pilots may utilize equipment such as an autopilot to assist in the performance of their duties; however, under no circumstances will the pilot responsibility be replaced by technologies in the foreseeable future. For greater fluidity in identifying who the pilot is, the Circular has introduced the concept of "remotely-piloted aircraft" (RPA)—which is a subset

[47]See http://www.iaopa.org/news/RPAS.html.

[48]The principles so applicable in the context of the Chicago Convention are: Article 12 pertaining to rules of the air; Article 15 in the context of airport and air navigation services charges; Article 29 on documents carried on board aircraft; Article 31 which stipulates that every aircraft engaged in international navigation shall be provided with a certificate of airworthiness issued or rendered valid by the State in which it is registered; Article 32 which provides that the pilot of every aircraft and the other members of the operating crew of every aircraft engaged in international navigation shall be provided with certificates of competency and licenses issued or rendered valid by the State in which the aircraft is registered; and Article 33 which requires that Certificates of airworthiness and certificates of competency and licenses issued or rendered valid by the contracting State in which the aircraft is registered, shall be recognized as valid by the other contracting States, provided that the requirements under which such certificates or licences were issued or rendered valid are equal to or above the minimum standards which may be established from time to time pursuant to the Convention.

[49]*Supra*, note 5 paragraphs 6.48–6.51.

of unmanned aircraft—into the lexicon. An RPA[50] is an aircraft piloted by a licensed "remote pilot" situated at a "remote pilot station" located external to the aircraft (i.e. ground, ship, another aircraft, space) who monitors the aircraft at all times and can respond to instructions issued by ATC, communicates via voice or data link as appropriate to the airspace or operation, and has direct responsibility for the safe conduct of the aircraft throughout its flight. An RPA may possess various types of auto-pilot technology but at any time the remote pilot can intervene in the management of the flight. This equates to the ability of the pilot of a manned aircraft being flown by its auto flight system to take prompt control of the aircraft.

From a legal perspective, and in accordance with Article 3 *bis* of the Chicago Convention which stipulates *inter alia* that Contracting States recognize that every State, in the exercise of its sovereignty, is entitled to require the landing at some designated airport of a civil aircraft flying above its territory without authority, Contracting States are entitled, in certain circumstances, to require civil aircraft flying above their territory to land at designated aerodromes, Therefore the pilot of the RPA will have to be able to comply with instructions provided by the State, including through electronic or visual means, and have the ability to divert to the specified airport at the State's request. The requirement to respond to instructions based on such visual means may place significant requirements on certification of RPAS detection systems for international flight operations.

In terms of collision avoidance, the Circular makes the pilot in command of a RPAS as responsible as a pilot of a manned aircraft for detecting and avoiding potential collisions and other hazards. Furthermore it provides that technology to provide the remote pilot with sufficient knowledge of the aircraft's environment to fulfil the responsibility must be incorporated into the aircraft with counterpart components located at the remote pilot station. Also, remote pilots, despite not being on board the aircraft, will be subject to the same requirements as aircraft pilots who are required to observe, interpret and heed a diverse range of visual signals intended to attract their attention and/or convey information. Such signals can range from lights and pyrotechnic signals for aerodrome traffic to signals used by intercepting aircraft. This would necessitate development and approval of alternate means of compliance with this requirement.

With regard to air traffic management (ATM) the Circular prescribes that whether the aircraft is piloted from on board or remotely, the provision of air traffic services (ATS) should, to the greatest practicable extent, be one and the same.[51] It further states that the introduction of RPA must not increase the risk to other aircraft or third parties and should not prevent or restrict access to airspace. ATM

---

[50]RPA may have the same phases of flight—taxi, departure, en-route and arrival—as manned aircraft or they may be launched/recovered and/or conduct aerial work. The aircraft performance characteristics may be significantly different from traditional manned aircraft. Regardless, the remote pilot will operate the aircraft in accordance with the rules of the air for the State and airspace in which the RPA is operating. This will include complying with directions and instructions provided by the air traffic services (ATS) unit.

[51]*Supra*, note 41.

procedures for handling RPA should mirror those for manned aircraft whenever possible. There will be some instances where the remote pilot cannot respond in the same manner as could an on-board pilot and the Circular calls for ATM procedures to be able to take account of these differences. For this purpose, ATS/remote pilot communication requirements must be assessed in the context of an ATM function, taking into account human interactions, procedures and environmental characteristics. A safety management system (SMS) approach should be employed to determine the adequacy of any communications solutions.[52] The information exchange between ATC and the remote pilot will likely require the same levels of reliability, continuity and integrity, referred to as QOS, that are required to support operations with manned aircraft in the airspace in which a UA is intended to operate.[53]

The exchange of control information between the aircraft and its remote pilot station will require an extremely high level of availability, reliability, continuity and integrity. The determination of required communication performance and associated QOS levels will be based on functionality considering the level of ATS being provided.[54]

In terms of aerodrome operations and RPAS, the Circular recognizes that integration of RPA into aerodrome operations will prove to be among the greatest challenges. At issue are provisions for the remote pilot to identify, in real-time, the physical layout of the aerodrome and associated equipment such as aerodrome lighting and markings so as to manoeuvre the aircraft safely and correctly. The Circular provides that RPA must be able to work within existing aerodrome parameters. Aerodrome standards should not be significantly changed, and the equipment developed for RPA must be able to comply with existing provisions to the greatest extent practicable. Moreover, where RPA are operated alongside manned aircraft, there needs to be harmonization in the provision of ATS.[55]

Meteorology is another important element that needs to be properly coordinated in the operation of RPAS. The Circular provides that meteorological information plays a role in the safety, regularity and efficiency of international air navigation and is provided to users as required for the performance of their respective functions. Meteorological information supplied to operators and flight/remote crew members covers the flight in respect of time, altitude, and geographical area. Accordingly, the information relates to appropriate fixed times, or periods of time, and extends to the aerodrome of intended landing. It also covers meteorological conditions expected between the aerodrome of intended landing and alternate aerodromes designated by the operator.[56]

---

[52]*Id.* Paragraph 5.14.
[53]*Id.* Paragraph 6.33.
[54]*Id.* Paragraph 6.34.
[55]*Id.* Paragraph 5.23.
[56]*Id.* Paragraph 5.27.

Meteorological services are critical for the planning, execution and safe operation of international aviation. Since the remote pilot is not on board the aircraft and may not be able to determine meteorological conditions and their real-time effects on the aircraft, obtaining meteorological information from appropriate sources prior to and during flight will be especially critical for the safe operation of these aircraft.[57]

The Circular recognizes the Annex 3 to the Chicago Convention—*Meteorological Service for International Air Navigation* has a requirement for aircraft on its registry operating on international air routes to make automated routine observations, if so equipped. RPA may not be so equipped. Likewise, there is a requirement for all aircraft to make special observations whenever severe turbulence, severe icing, severe mountain wave, thunderstorms, hail, dust, stone and volcanic ash are encountered during a flight. However, RPA may not be able to comply with these provisions as the pilot is remote from the aircraft, and the aircraft may not have the sensors to detect these phenomena.[58]

It is also recognized that conversely, the RPA specifically equipped for such purposes may in fact be used to monitor meteorological conditions, relaying information back to ground sensors. These aircraft could potentially be used in conditions and locations where manned aircraft cannot safely operate such as in hurricanes, convective weather or in the vicinity of volcanic ash/gases.[59]

One of the critical elements in RPAS operations is the security of the system as security is a vital issue for RPA with aspects that are both similar and unique when compared with manned aircraft. As a remote pilot station is similar in purpose and design to a cockpit, it must likewise be secure from sabotage or unlawful malicious interference. Chapter 13 of Annex 6 to the Chicago Convention—Part I—*International Commercial Air Transport—Aeroplanes* contains SARPs to secure the flight crew compartment. However, due to the fixed and exposed nature of the remote pilot station (as opposed to the restricted nature of a commercial aeroplane where the intrusion and use of heavier weapons is less likely) further consideration should be given to the potential vulnerability of the premises against unlawful interference.[60] Similarly, the aircraft itself must be stored and prepared for flight in a manner that will prevent and detect tampering and ensure the integrity of vital components.

3 Operations Over the High Seas

The Circular prescribes that operators must have approval from the State of the operator before conducting operations in high seas airspace. They must likewise coordinate their operations with the ATS provider responsible for the airspace

[57] Id. Paragraph 5.28.

[58] Id. Paragraph 5.29.

[59] Id. Paragraph 5.30.

[60] Id. Paragraph 5.32.

concerned.[61] Article 12 of the Chicago Convention unambiguously states that over the high seas, the rules in force shall be those under the Convention and each Contracting State undertakes to insure the prosecution of all persons violating the regulations applicable. This peremptory principle,[62] of adherence by States and aircraft bearing their nationality to any Standards and Recommended Practices (SARPs) adopted in regard to the high seas, effectively precludes any possible reliance by States on Article 38 of the Convention which allows States to deviate from SARPs in general. In other words, Annex 2 on Rules of the Air, which contains provisions relating to the operation of aircraft over the high seas, is sacrosanct and inviolable. The first legal issue that would emerge from this clear principle is the question of applicability of Annexes (other than Annex 2) to the high seas and whether their provisions, if directly related to the principles of manoeuvre and navigation of aircraft over the high seas, would be binding with no flexibility offered by Article 38 of the Convention. Kaiser offers the opinion:

Over the high seas, the rules of the air have binding effect under Article 12, Sentence 3 of the Chicago Convention. It should be clarified that rules of the air have a broader meaning than Annex 2 and encompass the Standards and Recommended Practices of all other Annexes as far as their application makes sense over the high seas.[63]

Kaiser is of course referring mainly to Annexes 10 and 11 to the Chicago Convention relating to air traffic services and air traffic management, while at the same time drawing the example of Annex 16 (on environmental) protection being applicable in a future date if extended beyond noise and engine emissions to the high seas under Article 12 of the Chicago Convention.[64] This argument, which would ascribe to the ICAO Council wider control over larger spans of the world's air space, would be acceptable only if provisions of other Annexes (other than those of Annex 2) would directly have a bearing on the manoeuvre and navigation of aircraft over the high seas, as exclusively provided for by Article 12 of the Chicago Convention.

The provision of air navigation services are mainly regulated by three Annexes to the Chicago Convention, namely Annex 2 (Rules of the Air), Annex 3 (Meteorological Service for International Air Navigation) and Annex 11 (Air Traffic

---

[61]*Id.* at paragraph 3.19.

[62]Bin Cheng confirms that over the high seas there is absolutely no option for States to deviate from rules established under the Chicago Convention for the manoeuvre and operations of aircraft. See Cheng (1962), at 148.

[63]Kaiser (1995) at 455. Bin Cheng states that contracting States are expected to be able to exercise control over all that takes place within their territories, but outside their respective territories only over aircraft bearing their nationality. Cheng (1962), for an extended discussion on this issue see Milde (2012), at 110.

[64]*Ibid.*

Services).[65] Of these, compliance with Annex 2 is mandatory[66] and does not give the States the flexibility provided in Article 38 of the Chicago Convention to register differences from any provisions of the Annex.

With regard to maritime navigation, the *United Nations Convention on the Law of the Sea (UNCLOS)*, Article 39, lays down the duties of ships and aircraft involved in transit navigation to the effect that ships and aircraft, while exercising the right of transit passage, should : proceed without delay through or over the strait; refrain from any threat or use of force against the sovereignty, territorial integrity or political independence of States bordering the strait, or in any other manner in violation of the principles of international law embodied in the Charter of the United Nations; refrain from any activities other than those incident to their normal modes of continuous and expeditious transit unless rendered necessary by force majeure or by distress; and comply with the relevant provisions of the Convention. Article 39 (3) explicitly states that aircraft in transit passage shall observe the Rules of the Air established by ICAO as they apply to civil aircraft and that state aircraft will normally comply with such safety measures and will at all times operate with due regard for the safety of navigation. The provision further states that at all times aircraft shall monitor the radio frequency assigned by the competent internationally designated air traffic control authority or the appropriate international distress radio frequency.

Standard 2.1.1 of Annex 2 to the Chicago Convention provides that the rules of the air shall apply to aircraft bearing the nationality and registration marks of a Contracting State, wherever they may be, to the extent that they do not conflict with

---

[65]Article 54 l) of the Chicago Convention stipulates as a mandatory function of the Council the act of adopting, in accordance with Chapter VI of the Convention, international standards and recommended practices (SARPs) and for convenience designate them as Annexes to the Convention. Article 37 of the Convention reflects the areas in which SARPs should be developed and Annexes formed. Article 38 obliges contracting States to notify ICAO of any differences between their own regulations and practices and those established by international standards or procedures. The notification of differences however, does not absolve States from their continuing obligation under Article 37 to collaborate in securing the highest practicable degree of uniformity in international regulations, standards, and procedures.

[66]In October 1945, the Rules of the Air and Air Traffic Control (RAC) Division at its first session made recommendations for Standards, Practices and Procedures for the Rules of the Air. These were reviewed by the then Air Navigation Committee and approved by the Council on 25 February 1946. They were published as *Recommendations for Standards, Practices and Procedures—Rules of the Air* in the first part of Doc 2010, published in February 1946. The RAC Division, at its second session in December 1946–January 1947, reviewed Doc 2010 and proposed Standards and Recommended Practices for the Rules of the Air. These were adopted by the Council as Standards and Recommended Practices relating to Rules of the Air on 15 April 1948, pursuant to Article 37 of the Convention on International Civil Aviation (Chicago, 1944) and designated as Annex 2 to the Convention with the title *International Standards and Recommended Practices—Rules of the Air*. They became effective on 15 September 1948. On 27 November 1951, the Council adopted a complete new text of the Annex, which no longer contained Recommended Practices. The Standards of the amended Annex 2 (Amendment 1) became effective on 1 April 1952 and applicable on 1 September 1952.

the rules published by the State having jurisdiction over the territory over-flown.[67] The operation of an aircraft either in flight or on the movement area of an aerodrome shall be in compliance with the general rules and, in addition, when in flight, either with: visual flight rules (VFR); or the instrument flight rules (IFR).[68] Standard 2.3.1 further provides that the pilot-in-command of an aircraft shall, whether manipulating the controls or not, be responsible for the operation of the aircraft in accordance with the rules of the air, except that the pilot-in-command may depart from these rules in circumstances that render such departure absolutely necessary in the interests of safety.

4 Air Traffic Services

The provision of air traffic services[69] is addressed in Annex 11 to the Chicago Convention which provides *in limine* that Contracting States shall determine, in accordance with the provisions of the Annex and for the territories over which they have jurisdiction, those portions of the airspace and those aerodromes where air traffic services will be provided. They shall thereafter arrange for such services to be established and provided in accordance with the provisions of this Annex, except that, by mutual agreement, a State may delegate to another State the responsibility for establishing and providing air traffic services in flight information regions, control areas or control zones extending over the territories of the former.[70]

---

[67]The Council of the International Civil Aviation Organization resolved, in adopting Annex 2 in April 1948 and Amendment 1 to the said Annex in November 1951, that the Annex constitutes Rules relating to the flight and manoeuvre of aircraft within the meaning of Article 12 of the Convention. Over the high seas, therefore, these rules apply without exception.

[68]Information relevant to the services provided to aircraft operating in accordance with both visual flight rules and instrument flight rules in the seven ATS airspace classes is contained in 2.6.1 and 2.6.3 of Annex 11. A pilot may elect to fly in accordance with instrument flight rules in visual meteorological conditions or may be required to do so by the appropriate ATS authority.

[69]According to Paragraph 2.2 of the Annex, The objectives of the air traffic services shall be to: (a) prevent collisions between aircraft; (b) prevent collisions between aircraft on the manoeuvring area and obstructions on that area; (c) expedite and maintain an orderly flow of air traffic; (d) provide advice and information useful for the safe and efficient conduct of flights; (e) notify appropriate organizations regarding aircraft in need of search and rescue aid, and assist such organizations as required.

[70]24 Standard 2.1.1. It is also provided in the Annex that if one State delegates to another State the responsibility for the provision of air traffic services over its territory, it does so without derogation of its national sovereignty. Similarly, the providing State's responsibility is limited to technical and operational considerations and does not extend beyond those pertaining to the safety and expedition of aircraft using the concerned airspace. Furthermore, the providing State in providing air traffic services within the territory of the delegating State will do so in accordance with the requirements of the latter which is expected to establish such facilities and services for the use of the providing State as are jointly agreed to be necessary. It is further expected that the delegating State would not withdraw or modify such facilities and services without prior consultation with the providing State. Both the delegating and providing States may terminate the agreement between them at any time.

The Standards and Recommended Practices in Annex 11, together with the Standards in Annex 2, govern the application of the *Procedures for Air Navigation Services—Air Traffic Management*[71] and the *Regional Supplementary Procedures*—Rules of the Air and Air Traffic Services, contained in Doc 7030, Annex 11 pertains to the establishment of airspace, units and services necessary to promote a safe, orderly and expeditious flow of air traffic. A clear distinction is made between air traffic control service, flight information service and alerting service. Its purpose, together with Annex 2, is to ensure that flying on international air routes is carried out under uniform conditions designed to improve the safety and efficiency of air operation.

The Standards and Recommended Practices in Annex 11 apply in those parts of the airspace under the jurisdiction of a Contracting State wherein air traffic services are provided and also wherever a Contracting State accepts the responsibility of providing air traffic services over the high seas or in airspace of undetermined sovereignty. A Contracting State accepting such responsibility may apply the Standards and Recommended Practices in a manner consistent with that adopted for airspace under its jurisdiction.

Standard 2.1.2 of the Annex stipulates that those portions of the airspace over the high seas or in airspace of undetermined sovereignty where air traffic services will be provided shall be determined on the basis of regional air navigation agreements. A Contracting State having accepted the responsibility to provide air traffic services in such portions of airspace shall thereafter arrange for the services to be established and provided in accordance with the provisions of the Annex.[72] The Annex goes on to say that when it has been determined that air traffic services will be provided, the States concerned shall designate the authority[73] responsible for providing such services.[74] Situations which arise in respect of the establishment and provision of air traffic services to either part or whole of an international flight are as follows:

Situation 1: A route, or portion of a route, contained within airspace under the sovereignty of a State establishing and providing its own air traffic services.

Situation 2: A route, or portion of a route, contained within airspace under the sovereignty of a State which has, by mutual agreement, delegated to another State, responsibility for the establishment and provision of air traffic services.

---

[71]Doc 4444, PANS-ATM.

[72]21 The phrase "regional air navigation agreements" refers to the agreements approved by the Council of ICAO normally on the advice of Regional Air Navigation Meetings. The Council, when approving the Foreword to this Annex, indicated that a Contracting State accepting the responsibility for providing air traffic services over the high seas or in airspace of undetermined sovereignty may apply the Standards and Recommended Practices in a manner consistent with that adopted for airspace under its jurisdiction.

[73]The authority responsible for establishing and providing the services may be a State or a suitable Agency.

[74]Standard 2.1.3.

Situation 3: A portion of a route contained within airspace over the high seas or in airspace of undetermined sovereignty for which a State has accepted the responsibility for the establishment and provision of air traffic services.

For the purpose of the Annex, the State which designates the authority responsible for establishing and providing the air traffic services is:

In *Situation 1*: the State having sovereignty over the relevant portion of the airspace;

In *Situation 2*: the State to whom responsibility for the establishment and provision of air traffic services has been delegated;

In *Situation 3*: the State which has accepted the responsibility for the establishment and provision of air traffic services.

References

Cheng B (1962) The law of international air transport. Oceania Publications, London

Dempsey PS (1987) Law and foreign policy in international aviation. Transnational Publishers Inc, Dobbs Ferry/New York

Kaiser SA (1995) Infrastructure, airspace and automation - air navigation issues for the 21st century. Ann Air Space Law XX-I:447

Milde M (2012) International air law and ICAO. In: Benko M (ed) Essential air and space law. Eleven International Publishing, The Hague

Shawcross, Beaumont (1977) Air law, vol. 1, 4th edn. Butterworths, London

ter Kuille A. RPASD and the ATM community: the CANSO policy of engagement, UAV systems, the global perspective, 2006/2007. Blyenburgh & Co, France

ter Kuille A. UAV and the ATM community: the CANSO policy of engagement, RPAS systems, the global perspective, 2006/2007. Blyenburgh & Co, France

Article 9
Prohibited Areas

(a) **Each contracting State may, for reasons of military necessity or public safety, restrict or prohibit uniformly the aircraft of other States from flying over certain areas of its territory, provided that no distinction in this respect is made between the aircraft of the State whose territory is involved, engaged in international scheduled airline services, and the aircraft of the other contracting States likewise engaged. Such prohibited areas shall be of reasonable extent and location so as not to interfere unnecessarily wit air navigation. Descriptions of such prohibited areas in the territory of a contracting State, as well as any subsequent alterations therein, shall be communicated as soon as possible to the other contracting States and to the International Civil Aviation Organization.**

(b) **Each contracting State reserves also the right, in exceptional circumstances or during a period of emergency, or in the interest of public safety, and with immediate effect, temporarily to restrict or prohibit flying over the whole or any part of its territory, on condition that such restriction or prohibition shall be applicable without distinction of nationality to aircraft of all other States.**

(c) **Each contracting State, under such regulations as it may prescribe, may require any aircraft entering the areas contemplated in subparagraphs (a) or (b) above to effect a landing as soon as practicable, thereafter at some designated airport within its territory.**

Contents

1 No Fly Zones

On 12 December 1950 the ICAO Council, at the seventeenth meeting of its eleventh session, approved the following definitions of prohibited, restricted and danger areas:

Prohibited Area: A specified area within the land areas of a State or territory waters adjacent thereto over which the flight of aircraft is prohibited.[1]

[1]It will be noted that Annex 15 to the Chicago Convention defines a prohibited area as; an airspace of defined dimensions, above the land areas or territorial waters of a State, within which the flight of aircraft is prohibited.

DOI 10.1007/978-3-319-00068-8_10,

Restricted Area: A specified area within the land areas of a State or territorial waters adjacent thereto, designated for other than air traffic control purposes, over which the flight of aircraft is restricted in accordance with certain specified conditions[2];

Danger Area: A specified area within or over which there may exist activities constituting a potential danger to aircraft flying over it.[3]

On July 22 1951 the Council, at the ninth meeting of its thirteenth session decided to ask all ICAO member States to issue NOTAMs[4] on the establishment or change of boundaries of any prohibited, restricted or danger area well in advance of such establishment or change and to include in the NOTAMs an indication of the period when a hazard in any danger area would exist.

On 1 April 1957 the Council of ICAO, at the eighth meeting of its thirtieth session sought the views of the Air Navigation Commission

> that consideration might be given to the possibility of establishing a procedure by which airlines would give advance notification of flights in the vicinity of restricted airspace to States controlling such airspace, when these are not States that would be automatically informed through the filing of the flight plan.

The Air Navigation Commission, in its Report to the Council[5] saw more disadvantages than advantages in such a procedure and informed the Council that it was unable to offer any technical solution for the problem of avoiding attacks on civilian aircraft inadvertently infringing restricted airspace.

Fast track to 2011, during the "Arab Spring"[6] of 2011, the Libyan military, its equipment and personnel who were launching military attacks[7] on protesting civilians came under heavy attack from NATO as a consequence of United Nations Security Council Resolution 1973. This Resolution, which was adopted on 17 March 2011, demanded an immediate ceasefire in Libya, including an end to the current attacks against civilians, which it said might constitute "crimes against

---

[2]It will be noted that Annex 15 to the Chicago Convention defines a Restricted Area as: an airspace of defined dimensions, above the land areas or territorial waters of a State, within which the flight of aircraft is restricted in accordance with certain specified conditions.

[3]Doc 7188-C/828, Part II Proc. Of Council, 11th S. p. 29. Annex 15 to the Chicago Convention defines a Danger Area as: an airspace of defined dimensions within which activities dangerous to the flight of aircraft may exist at specified times

[4]A notice distributed by means of telecommunication containing information concerning the establishment, condition or change in any aeronautical facility, service, procedure or hazard, the timely knowledge of which is essential to personnel concerned with flight operations. See Annex 15: Aeronautical Information Services of the Chicago Convention, Definitions.

[5]C-WP/2552.

[6]The 2010–11 "Arab Spring" is a revolutionary wave of demonstrations and protests and protests that has been taking place in the Middle East and North Africa. Since 18 December 2010. Prior to this period Sudan was the only Arab country to have successfully toppled dictatorial regimes.

[7]The uprising in Libya instantly became violent when the Libyan government reacted harshly towards peaceful protests. On February 18, 3 days after the protests began; the country erupted into an armed conflict when protesters executed policemen and men loyal to Colonel Muammar Gaddafi for killing protesters.

humanity", the Security Council this evening imposed a ban on all flights in the country's airspace—a no-fly zone—and tightened sanctions on the Qadhafi regime and its supporters.

2 Military Perspectives

The Resolution invoked an earlier Resolution RES 1970 which provided *inter alia* that all Member States shall take the necessary measures to prevent the entry into or transit through their territories of individuals provided that nothing in the Resolution would oblige a State to refuse its own nationals entry into its territory.

Resolution 1970 also decided that the measures imposed by the above paragraph shall not apply:

- Where the Committee determines on a case-by-case basis that such travel is justified on the grounds of humanitarian need, including religious obligation;
- Where entry or transit is necessary for the fulfilment of a judicial process;
- Where the Committee determines on a case-by-case basis that an exemption would further the objectives of peace and national reconciliation in the Libyan Arab Jamahiriya and stability in the region; or
- Where a State determines on a case-by-case basis that such entry or transit is required to advance peace and stability in the Libyan Arab Jamahiriya and the States subsequently notifies the Committee within 48 h after making such a determination

Adopting Resolution 1973 by a vote of ten in favour to none against, with five abstentions (Brazil, China, Germany, India, Russian Federation), the Security Council authorized Member States, acting nationally or through regional organizations or arrangements, to take all necessary measures to protect civilians under threat of attack in the country, including Benghazi, while excluding a foreign occupation force of any form on any part of Libyan territory—requesting them to immediately inform the Secretary-General of such measures.

Recognizing the important role of the League of Arab States in the maintenance of international peace and security in the region, and bearing in mind the United Nations Charter's Chapter VIII, the Council asked the League's member States to cooperate with other Member States in implementing the no-fly zone.

The Council stressed the need to intensify efforts to find a solution to the crisis that responded to the legitimate demands of the Libyan people, noting actions being taken on the diplomatic front in that regard. It further demanded that Libyan authorities comply with their obligations under international law and take all measures to protect civilians and meet their basic needs and to ensure the rapid and unimpeded passage of humanitarian assistance.

In that connection, the Council specified that the flight ban would not apply to flights that had as their sole purpose humanitarian aid, the evacuation of foreign nationals, enforcing the ban or other purposes "deemed necessary for the benefit of the Libyan people".

It further decided that all States should deny permission to any Libyan commercial aircraft to land in or take off from their territory unless a particular flight had been approved in advance by the committee that was established to monitor sanctions imposed by resolution 1970 (2011).

In tightening the asset freeze and arms embargo established by that resolution, the Council this evening further detailed conditions for inspections of transport suspected to be violating the embargo, requesting States enforcing the embargo to coordinate closely with each other and the Secretary-General on the measures they were taking towards implementation.

It requested the Secretary-Secretary to create an eight-member panel of experts to assist the Security Council committee in monitoring the sanctions.

Introducing the resolution, the Foreign Minister of France, Alain Juppé, said "the situation on the ground is more alarming than ever, marked by the violent re-conquest of cities that have been released". The Security Council could not stand by and "let the warmongers flout international legality". The world was experiencing "a wave of great revolutions that would change the course of history", but the will of the Libyan people had been "trampled under the feet of the Qadhafi regime". Earlier Council measures had been ignored and violence against Libyan civilians had redoubled.

He said that the urgent need to protect the civilian population had led to the elaboration of the current resolution, which authorized the Arab League and those Member States wishing to do so to take all measures to protect areas that were being threatened by the Qadhafi regime. "We have very little time left—perhaps only a matter of hours," he said, adding that each hour and day that passed "increased the weight" on the international community's shoulders.

Speaking after the vote, representatives who had supported the text agreed that the strong action was made necessary because the Qadhafi regime had not heeded the first actions of the Council and was on the verge of even greater violence against civilians as it closed in on areas previously dominated by opposition in the east of the country. They stressed that the objective was solely to protect civilians from further harm.

The text of Resolution 1973 (2011) relating to the no-fly zone imposed on Libya reads as follows:

"*The Security Council*,

"*Recalling* its resolution 1970 (2011) of 26 February 2011,

"*Deploring* the failure of the Libyan authorities to comply with resolution 1970 (2011),

"*Expressing* grave concern at the deteriorating situation, the escalation of violence, and the heavy civilian casualties,

"*Reiterating* the responsibility of the Libyan authorities to protect the Libyan population and *reaffirming* that parties to armed conflicts bear the primary responsibility to take all feasible steps to ensure the protection of civilians,

"*Condemning* the gross and systematic violation of human rights, including arbitrary detentions, enforced disappearances, torture and summary executions,

"*Further condemning* acts of violence and intimidation committed by the Libyan authorities against journalists, media professionals and associated personnel and *urging* these authorities to comply with their obligations under international humanitarian law as outlined in resolution 1738 (2006),

"*Considering* that the widespread and systematic attacks currently taking place in the Libyan Arab Jamahiriya against the civilian population may amount to crimes against humanity,

"*Recalling* paragraph 26 of resolution 1970 (2011) in which the Council expressed its readiness to consider taking additional appropriate measures, as necessary, to facilitate and support the return of humanitarian agencies and make available humanitarian and related assistance in the Libyan Arab Jamahiriya,

"*Expressing its determination* to ensure the protection of civilians and civilian populated areas and the rapid and unimpeded passage of humanitarian assistance and the safety of humanitarian personnel,

"*Recalling* the condemnation by the League of Arab States, the African Union and the Secretary-General of the Organization of the Islamic Conference of the serious violations of human rights and international humanitarian law that have been and are being committed in the Libyan Arab Jamahiriya,

"*Taking note* of the final communiqué of the Organization of the Islamic Conference of 8 March 2011, and the communiqué of the Peace and Security Council of the African Union of 10 March 2011 which established an ad hoc High-Level Committee on Libya,

"*Taking note also* of the decision of the Council of the League of Arab States of 12 March 2011 to call for the imposition of a no-fly zone on Libyan military aviation, and to establish safe areas in places exposed to shelling as a precautionary measure that allows the protection of the Libyan people and foreign nationals residing in the Libyan Arab Jamahiriya,

"*Taking note further* of the Secretary-General's call on 16 March 2011 for an immediate ceasefire,

"*Recalling* its decision to refer the situation in the Libyan Arab Jamahiriya since 15 February 2011 to the Prosecutor of the International Criminal Court, and *stressing* that those responsible for or complicit in attacks targeting the civilian population, including aerial and naval attacks, must be held to account,

"*Reiterating its concern* at the plight of refugees and foreign workers forced to flee the violence in the Libyan Arab Jamahiriya, *welcoming* the response of neighbouring States, in particular Tunisia and Egypt, to address the needs of those refugees and foreign workers, and *calling on* the international community to support those efforts,

"*Deploring* the continuing use of mercenaries by the Libyan authorities,

"*Considering* that the establishment of a ban on all flights in the airspace of the Libyan Arab Jamahiriya constitutes an important element for the protection of civilians as well as the safety of the delivery of humanitarian assistance and a decisive step for the cessation of hostilities in Libya,

"*Expressing concern* also for the safety of foreign nationals and their rights in the Libyan Arab Jamahiriya,

"*Welcoming* the appointment by the Secretary General of his Special Envoy to Libya, Mr. Abdul Ilah Mohamed Al-Khatib and supporting his efforts to find a sustainable and peaceful solution to the crisis in the Libyan Arab Jamahiriya,

"*Reaffirming* its strong commitment to the sovereignty, independence, territorial integrity and national unity of the Libyan Arab Jamahiriya,

"*Determining* that the situation in the Libyan Arab Jamahiriya continues to constitute a threat to international peace and security,

"*Acting* under Chapter VII of the Charter of the United Nations,

"1. *Demands* the immediate establishment of a ceasefire and a complete end to violence and all attacks against, and abuses of, civilians;

"2. *Stresses* the need to intensify efforts to find a solution to the crisis which responds to the legitimate demands of the Libyan people and *notes* the decisions of the Secretary-General to send his Special Envoy to Libya and of the Peace and Security Council of the

African Union to send its ad hoc High-Level Committee to Libya with the aim of facilitating dialogue to lead to the political reforms necessary to find a peaceful and sustainable solution;

3. *Demands* that the Libyan authorities comply with their obligations under international law, including international humanitarian law, human rights and refugee law and take all measures to protect civilians and meet their basic needs, and to ensure the rapid and unimpeded passage of humanitarian assistance;

"Protection of civilians

"4. *Authorizes* Member States that have notified the Secretary-General, acting nationally or through regional organizations or arrangements, and acting in cooperation with the Secretary-General, to take all necessary measures, notwithstanding paragraph 9 of resolution 1970 (2011), to protect civilians and civilian populated areas under threat of attack in the Libyan Arab Jamahiriya, including Benghazi, while excluding a foreign occupation force of any form on any part of Libyan territory, and *requests* the Member States concerned to inform the Secretary-General immediately of the measures they take pursuant to the authorization conferred by this paragraph which shall be immediately reported to the Security Council;

"5. *Recognizes* the important role of the League of Arab States in matters relating to the maintenance of international peace and security in the region, and bearing in mind Chapter VIII of the Charter of the United Nations, requests the Member States of the League of Arab States to cooperate with other Member States in the implementation of paragraph 4;

"No-fly zone

"6. *Decides* to establish a ban on all flights in the airspace of the Libyan Arab Jamahiriya in order to help protect civilians;

"7. *Decides further* that the ban imposed by paragraph 6 shall not apply to flights whose sole purpose is humanitarian, such as delivering or facilitating the delivery of assistance, including medical supplies, food, humanitarian workers and related assistance, or evacuating foreign nationals from the Libyan Arab Jamahiriya, nor shall it apply to flights authorised by paragraphs 4 or 8, nor other flights which are deemed necessary by States acting under the authorization conferred in paragraph 8 to be for the benefit of the Libyan people, and that these flights shall be coordinated with any mechanism established under paragraph 8;

"8. *Authorizes* Member States that have notified the Secretary-General and the Secretary-General of the League of Arab States, acting nationally or through regional organizations or arrangements, to take all necessary measures to enforce compliance with the ban on flights imposed by paragraph 6 above, as necessary, and *requests* the States concerned in cooperation with the League of Arab States to coordinate closely with the Secretary General on the measures they are taking to implement this ban, including by establishing an appropriate mechanism for implementing the provisions of paragraphs 6 and 7 above,

"9. *Calls upon* all Member States, acting nationally or through regional organizations or arrangements, to provide assistance, including any necessary overflight approvals, for the purposes of implementing paragraphs 4, 6, 7 and 8 above;

"10. *Requests* the Member States concerned to coordinate closely with each other and the Secretary-General on the measures they are taking to implement paragraphs 4, 6, 7 and 8 above, including practical measures for the monitoring and approval of authorised humanitarian or evacuation flights;

"11. *Decides* that the Member States concerned shall inform the Secretary-General and the Secretary-General of the League of Arab States immediately of measures taken in exercise of the authority conferred by paragraph 8 above, including to supply a concept of operations;

"12. *Requests* the Secretary-General to inform the Council immediately of any actions taken by the Member States concerned in exercise of the authority conferred by paragraph

8 above and to report to the Council within 7 days and every month thereafter on the implementation of this resolution, including information on any violations of the flight ban imposed by paragraph 6 above;

"Ban on flights

"17. *Decides* that all States shall deny permission to any aircraft registered in the Libyan Arab Jamahiriya or owned or operated by Libyan nationals or companies to take off from, land in or overfly their territory unless the particular flight has been approved in advance by the Committee, or in the case of an emergency landing;

"18. *Decides that* all States shall deny permission to any aircraft to take off from, land in or overfly their territory, if they have information that provides reasonable grounds to believe that the aircraft contains items the supply, sale, transfer, or export of which is prohibited by paragraphs 9 and 10 of resolution 1970 (2011) as modified by this resolution, including the provision of armed mercenary personnel, except in the case of an emergency landing;

Article 10
Landing at Customs Airport

Except in a case where, under the terms of this Convention or a special authorization, aircraft are permitted to cross the territory of a contracting State without landing, every aircraft which enters the territory of a contracting State shall, if the regulations of that State so require, land at an airport designated by that State for the purpose of customs and other examination. On departure from the territory of a contracting State, such aircraft shall depart from a similarly designated customs airport. Particulars of all designated customs airports shall be published by the State and transmitted to the International Civil Aviation Organization established under Part II of this Convention for communication to all other contracting States.

Contents

1 Regulations on Landing

Annex 9 (Facilitation) to the Chicago Convention defines an international airport as

> Any airport designated by the Contracting State in whose territory it is situated as an airport of entry and departure for international air traffic, where the formalities incident to customs, immigration, public health, animal and plant quarantine and similar procedures are carried out.

Annex 15 to the Chicago Convention has a similar definition. The main functions of an airport are activities concerned with: administration and finance; operation of airport facilities; engineering, construction works and maintenance; marketing and public relations; ground handling; air traffic operations; security, immigration, health and customs. The functions and responsibilities of an airport will vary according to its size, type of traffic and areas of responsibility. For example, some airports are responsible for air traffic control as well as for meteorological services, while at most other airports such services are provided by separate government entities. Many airports are involved in security functions in varying degrees and in providing facilities for customs, immigration and health authorities. Ground-handling services for the airlines, including terminal handling or ramp handling, or both, are provided by some airports, while at others they are provided by the airlines or by specialized agents or companies. Certain airports also

R. Abeyratne, *Convention on International Civil Aviation*,
DOI 10.1007/978-3-319-00068-8_11,

perform functions that exceed the scope of conventional airport activities, such as consultancy services, public works, construction, and real estate development.

The ICAO Assembly, at its 37th Session (Montréal, 28 September–8 October 2010) adopted Resolution A37-15 (Consolidated statement of continuing ICAO policies and associated practices related specifically to air navigation) Appendix P of which states that major improvements to the physical characteristics of aerodromes are required at many locations and in certain cases these improvements will involve considerable outlay and it would be inadvisable to plan such work without taking into account future developments. The Resolution also recognized that States and aerodrome authorities will continue to need to know the general trends in aerodrome requirements which succeeding generations of aircraft will most likely produce.

A significant development in recent times that would adversely affect the operation of aircraft into designated customs airports would be civil unrest. This was seen in the Arab Sprig of early 2011. What started on 17 December 2010 with an act of self-immolation by Mohammed Bouazizi, a 26-year-old man trying to support his family by selling fruits and vegetables in the central town of Sidi Bouzid in Tunisia, led to massive protests in the country, resulting in the overthrow of Zine El Abidine Ben Ali, the country's president on 14 January 2011. On 25 January 2011, protests, at least partly inspired by the toppling of the authoritarian government in Tunisia, erupted in Egypt[1] and grew increasingly worse. As a result, Hosni Mubarak, President of Egypt, was deposed within weeks of a virulent peoples' uprising. Contemporaneous protests went on other States such as in Algeria, Yemen, and Bahrain, the last of which held a "day of rage" on February 14, instigated by youths, and inspired by events in Egypt and Tunisia. Furthermore, at the time of writing, there was acute unrest in Libya as a result of mass civil unrest.[2]

In the context of the Libyan crisis, many airlines adopted a cautious approach in planning their flights to Libya while others cancelled scheduled flights.[3] Stocks of European airlines rapidly declined and airlines such as British Airways and KLM cancelled their flight to Tripoli.[4] An inevitable corollary to intensifying violence in

---

[1]Tourism and transport combined forms the largest industry in the world. Air transport is a significant driver of tourism and visitors arriving by air directly support approximately 6.7 million jobs worldwide in the tourism industry with the foreign exchange they spend during their travels. Both the tourism industry and air transport industry depend on the policies of governments and the individual stability of States for their sustenance and development. The unrest wrought by mass protests in North Africa and the Middle East in 2011 seriously disrupted tourism and air transport. Tourism earned Egypt more than 11 billion dollars in the last fiscal year. In the third quarter of 2010, Egypt was receiving about 280 million US dollars a week from tourism. See http://www.suite101.com/content/tourism-crisis-as-foreign-visitors-desert-egypt-a342840.

[2]Wikipedia identifies civil unrest with synonyms such as civil disorder, or civil strife, which are broad terms typically used by law enforcement to describe one or more forms of disturbance caused by a group of people.. Examples of civil disorder include, but are not necessarily limited to: illegal parades; sit ins; and other forms of obstructions; and other forms of crime. http://en.wikipedia.org/wiki/Civil_disorder.

[3]Airlines wary on operating to Libya, *Air Letter*, No. 17,180, Thursday 24 February 2011at p. 3.

[4]*Id.* at p. 4.

Libya, which is a large oil supplier to Europe, would be that airlines will be forced to charge higher fares. IATA observed that if the unrest were to continue in the various countries in the Middle East and North Africa, airlines would be forced to stop operating flights into those States, which would definitely result in significant losses to the airlines.[5]

In Libya, the runway at Benghazi airport was destroyed as a result of the continuing clashes between anti-government protesters and security forces.[6] It is reported that protesters against the government of Libya had surrounded the airport and the government of the United Kingdom, among others, was "urgently seeking landing permission from the Libyan Government" for a charter aircraft to airlift stranded British citizens out of the country.[7] The first point of contact of a tourist is the airport and if the airport premises is under severe civil unrest and attack, there will be no tourists visiting that country.

The security of a State is entirely dependent on the level of peace prevailing in its territory and any breach of that security, starting at the entry points to its territory, will also impact on loss of income for the State as the case is with tourism. Most, if not all countries affected by the civil unrest in the Middle East and North Africa are tourist intensive and their income will suffer immensely. Many States issued travel advisories on Tunisia, Egypt and Libya. At the time of the unrest in early 2011, Tunisia was recovering from the devastating effects on its tourism industry brought about by the terrorist attacks of 2001 and 2002 when the country lost a substantial number of tourists from its traditional markets of France, Germany, Italy and the United Kingdom. The 500,000 German tourists lost in the process was a big blow to Tunisia's tourism.[8] With regard to Egypt, hotel capacity increased by approximately 7,000 rooms between 2009 and 2010 to a total of 220,000 hotel rooms. In December 2010 The United Nations World Tourism Organization (UNWTO) increased its collaboration with Egypt in enhancing the country's tourist intake worked closely with Egypt in enhancing its capacity to measure the economic impact of tourism and provide consistent, internationally benchmarked tourism statistics.[9] With such an upsurge in tourism promotion, It is therefore heartening that tourism in Tunisia and Egypt, States that carried out a successful revolution in overthrowing their existing regimes, did not suffer for too long and recovered quickly. UNWTO has expressed its appreciation of proactive efforts by national

---

[5]Airlines set for losses as mid-east unrest continues, *Air Letter,* No. 17,181, Friday, 25 February 2011at p. 3.

[6]Reals (2011). Seehttp://www.flightglobal.com/articles/2011/02/22/353498/runway-at-libyas-benghazi-airport-destroyed-capita.html.

[7]*Ibid.*

[8]This loss gradually balanced from the new European markets and especially from Poland, Czech Republic and Hungary. See http://www.traveldailynews.com/pages/show_page/23601-Tunisia-unveils-new-tourism-plan.

[9]http://www.ameinfo.com/252453.html.

authorities to restore confidence among tourists and by foreign governments to update travel advisories accordingly. Tourism is a significant contributor to both countries' economies and, as tourism returns to normalcy, overall economic recovery can be stimulated.[10]

As the situation in both Egypt and Tunisian returns to normal, tourism stakeholders from the private and public sectors have reacted accordingly. Major tourism sites are open to the public, airlines have resumed flights, tour operators in many of the main source markets have restarted selling holidays and governments have updated their travel advisories to reflect the unfolding situation.

From an aviation and tourism perspective the unrest in these regions has impelled the markets to respond with oil prices shooting skywards to $119 a barrel for Brent crude. These higher oil prices is highly worrying for airlines. Having retrenched and cut back, airlines were hoping for a return to profitability in 2011 as growth returns following the downturn. However, the latest rise in oil prices could, as IATA forecasts extinguish any airline gains this year, causing a global domino effect on aviation. Airlines were hoping for a return to profitability in 2011 as growth returns following the downturn.[11]

2 Keeping Airports Open

To begin with, The 37th Session of the ICAO Assembly of the ICAO held from 28 September to 10 October 2010 officially recognized that ICAO has three Strategic Objectives: safety, security and environmental protection and sustainability of air transport. The last strategic objective, although relevant to the consequences of civil unrest on air transport by no means impels ICAO to intervene in the internal affairs of States or to ensure that amidst the clash of arms air transport carries on regardless. However, what it does is to draw a nexus between ICAO and the Chicago Convention which provides *inter alia* that an aim of ICAO is to foster the planning and development of international air transport so as to meet the needs of the peoples of the world for safe, regular, efficient and economical air transport.[12]

The Chicago Convention requires States to keep their airports open to all airlines operating into and out of their territories and provide meteorological, radio and other information as well as facilities such as ground services. Of course, one might argue that Article 89 of the Chicago Convention enables Contracting States to have freedom of action irrespective of the provisions of the Convention in case of war, whether belligerents or neutrals. It also allows a State which has declared a state of

[10] http://www.traveldailynews.com/pages/show_page/41810-UNWTO-welcomes-signs-of-tourism-recuperation-in-Egypt-and-Tunisia.

[11] http://www.aerosocietychannel.com/aerospace-insight/2011/02/shifting-sands/.

[12] *Id.* Article 44 (d).

national emergency (and notifies the ICAO Council of such) to have the same freedom of action notwithstanding the provisions of the Convention. Therefore, unless a State is at war (which the Convention does not define)[13] or has declared a state of national emergency, it would be bound by the provisions of the Convention.

The first duty of a Contracting State not falling within the purview of Article 89 of the Chicago Convention is to keep its airport open to all incoming aircraft. Article 15 of the Convention requires *inter alia* that, uniform conditions shall apply to the use, by aircraft of every Contracting State, of all air navigation facilities, including radio and meteorological services, which may be provided for public use for the safety and expedition of air navigation. This condition is subject to Article 9 which stipulates that each Contracting State may, for reasons of military necessity or public safety, restrict or prohibit uniformly the aircraft of other States from flying over certain areas of its territory, provided that no distinction in this respect is made between the aircraft of the State whose territory is involved, engaged in international scheduled airline services, and the aircraft of the other Contracting States likewise engaged. The provision goes on to say that Each Contracting State reserves also the right, in exceptional circumstances or during a period of emergency, or in the interest of public safety, and with immediate effect, temporarily to restrict or prohibit flying over the whole or any part of its territory, on condition that such restriction or prohibition will be applicable without distinction of nationality to aircraft of all other States.

The question arises as to whether a State in which there is acute civil unrest is bound to follow the abovementioned principles of the Chicago Convention. States or international organizations which are parties to such treaties have to apply the treaties they have signed and therefore have to interpret them. Although the conclusion of a treaty is generally governed by international customary law to accord with accepted rules and practices of national constitutional law of the signatory States, the application of treaties are governed by principles of international law. If however, the application or performance of a requirement in an international treaty poses problems to a State, the constitutional law of that State would be applied by courts of that State to settle the problem. Although Article 27 of the *Vienna Convention*[14] requires States not to invoke provisions of their internal laws as justification for failure to comply with the provisions of a treaty, States are free to choose the means of implementation they see fit according to their traditions

[13]Article 31.1 of the *Vienna Convention on the Law of Treatie* provides that "a treaty shall be interpreted in good faith in accordance with the ordinary meaning to be given to the terms of the treaty in their context and in the light of its object and purpose". See *Vienna Convention on the Law of Treaties 1969*, done at Vienna on 23 May 1969. The Convention entered into force on 27 January 1980. United Nations, *Treaty Series*, vol. 1155, p. 331. The ordinary meaning of war can be considered as a behavior pattern of organized violent conflict typified by extreme aggression, societal disruption, and high mortality. This behavior pattern involves two or more organized groups. http://en.wikipedia.org/wiki/War.

[14]*Id.*

and political organization.[15] The overriding rule is that treaties are juristic acts and have to be performed.

3 Airport and Aviation Security

The biggest threat to security in the vicinity of the airport, where aircraft landing and takeoff are at their lowest altitude, is Man Portable Air Defence Systems (MANPADS). Since the events of 11 September 2001, there have been several attempts against the security of aircraft in flight through the misuse of Man Portable Air Defense Systems (MANPADS).[16] The threat of MANPADS to aviation security is by far the most ominous and the international aviation community has made some efforts through ICAO. MANPADS have posed a serious threat to aviation security. On 5 January 1974, 220 soldiers and 200 police sealed off five square miles around Heathrow International airport in London after receiving reports that terrorists had smuggled SA-7s into Britain in the diplomatic pouches of Middle-Eastern embassies and were planning to shoot down an El Al airliner.[17]

Another significant incident occurred on 13 January 1975 when an attempt by terrorists to shoot down an El Al plane with a missile was believed to have brought civil aviation to the brink of disaster. Two terrorists drove their car onto the apron at Orly airport, where they set up a rocket launcher and fired at an El Al airliner which was about to take off for New York with 136 passengers. The first round missed the target thanks to the pilot's evasive action and hit the fuselage of a Yugoslav DC-9 aeroplane waiting nearby to embark passengers for Zagreb. The rocket failed to explode and no serious casualties were reported. After firing again and hitting an administration building, which caused some damage, the terrorists escaped by car? A phone call from an individual claiming responsibility for the attack was received at Reuters. The caller clearly implied that there would be another such operation, saying 'Next time we will hit the target'.

In fact, six days later another dramatic though unsuccessful attempt did occur at Orly airport. The French authorities traced the attack to the PFLP Venezuelan terrorist, and leader of the PFLP group in Europe, Carlos.[18] It is also known that once again an El Al airliner had been deliberately chosen as a target by Gadafi in an attempt to avenge the loss of the Libyan airliner shot down by Israel over the Sinai Desert.[19]

MANPADS are extremely effective weapons which are prolific in their availability worldwide. The significance of the abuse of MANPADS as a threat to civil aviation in the airport context is that MANPADS could be used in the vicinity of the

---

[15]Reuter (1989), at 16.

[16]The use of SAMs and anti-tank rockets by terrorists goes back to 1973. On 5 September 1973 Italian police arrested five Middle-Eastern terrorists armed with SA-7s. The terrorists had rented an apartment under the flight path to Rome Fumicino Airport and were planning to shoot down an El Al airliner coming in to land at the airport. See Dobson and Payne (1987), p. 366.

[17]Mickolus (1980), p. 428.

[18]Dobson and Payne (1987), *supra*, note 16, p. 53.

[19]*Ibid.*

perimeter of the airport or in the airport premises itself in view of the short range needed to hit an aircraft approaching an airport or departing from one. Introduced in the 1950s and originally meant to deter terror attacks from air to ground to be used by State authorities and other protection agencies, these weapons have got into the wrong hands and are being used against civil and military aviation. The surface to air MANPAD is a light weapon which offers very little warning before impact, and is often destructive and lethal.[20] They are cheap, easily carried, handled and concealed. It is claimed that there are at least 100,000 and possibly in excess of 500,000 systems in inventories around the world and several thousands of these are vulnerable to theft from State authorities.[21] It is also claimed that there is a 70 % chance that a civil aircraft will be destroyed if hit by a MANPAD.[22] A study conducted and published in early 2005 by the Rand Corporation concludes that, based on the effects of the attacks of September 11 2001, it is plausible for air travel in the United States to fall by 15–20 % after a successful MANPADS attack on a commercial airliner in the United States.[23] The international aviation community is aware that civil aircraft are particularly vulnerable to hand held ground to air missiles and that susceptibility avoidance techniques (calculated to avoid being hit) and vulnerability avoidance (survival after being hit) systems must be in place. This is particularly so since tracking the proliferation of MANPADS is difficult since any intelligence gathered on this particular threat is usually *ex post facto*, through the recovery of launchers or fragments from expended missiles. Contrary to popular belief, the MANPAD is considerably durable and can be used several years after inactivity, with recharged batteries.

The World's attention to the deadly threat posed by MANPADS was further drawn in November 2002 when there was an unsuccessful attempt to bring down a civilian aircraft leaving Mombasa, Kenya. Over the past 35 years, significant developments have taken place in dangerous weapons systems creating more opportunities for terrorists. The ready acceptance of new modern technologies by the international community and our growing dependence on them have created many targets, such as nuclear and civil aircraft in flight. Similarly, developments in electronics and microelectronics, and the trend towards miniaturization and simplification have resulted in a greater availability of tactical weapons with longer ranges and more accuracy that are also simpler to operate. One of the most effective developments in individual weaponry is portable, precision-guided munitions (PGMs), which are lightweight and easy to operate. They can usually be carried and operated by a single person. The United States-made Stinger, the British-made

---

[20]The lethality of the weapon can be reflected by the 340 MANPADS used by Afghan Mujahedeen rebels to successfully hit 269 soviet aircraft. See http://www.janes.com/security/international_security/news/.

[21]MANPADS, *Ploughshares Monitor* Autumn 2004, at 83.

[22]*Ibid.* The deadly accuracy and ease of handling of MANPADS were demonstrated when Somali gunmen shot down two US MH-60 Black Hawk helicopters in October 1993.

[23]Infrastructure Safety and the Environment, *Protecting Commercial Aviation against the Shoulder-Fired Missile Threat*, Rand Corporation, 2005, at 9.

Blowpipe and the Russian-made SA-7 missiles are examples of these smaller weapons. These are shoulder-fired, anti-aircraft missiles that have infra-red, heat-seeking sensors in the projectile that guide it to the heat emitted from an aircraft engine. It is known that more than 60 States possess SA-7 missiles and there is no doubt that most of them maintain strict security measures to prevent the outflow of the weapons. However, it has been alleged that some States, including Libya, have supplied PGMs to terrorist organizations. It is incontrovertible that in the hands of terrorists these missiles are not likely to be used against conventional targets such as tanks and military fighter aircraft. Of particular concern is the prospect of civilian airliners being shot at by SAMs and anti-tank rockets as they land at or take off from airports[24] Dr. Richard Clutterbuck subsumes the great threat of missile attacks:

> Recent years have seen increasing use of expensive and sophisticated surface-to-surface and surface-to-air missiles (SSM and SAM) by terrorists, generally of Russian or East European origin and redirected by Arab Governments, notably Colonel Gadafi's. Continuing development of these weapons for use by regular armies will ensure that new and more efficient versions will become available for terrorists.[25]

With increased airport security, the possibility of placing explosive devices on civil aircraft is becoming more difficult, but now the same destructive result can be achieved far more easily by using modern missiles or rockets.

Perimeter security at the airport is a vital element in ensuring security of the airport itself as well as the security of incoming and outgoing aircraft. For a successful missile attack against aircraft, the firing position has to be located within range of the flight path. A missile's guidance system is such that the weapon has to be fired within a few degrees of the flight path if the infra-red guidance is to locate the target. Accordingly, a possible preventive measure would be to prevent terrorists from getting into a firing position with their missiles. However, it would be very difficult to cut off areas of up to 6 km wide that lie in the paths of aircraft as they land and take off. This measure is therefore impracticable if not impossible.[26] This difficulty can be overcome to an extent by patrolling the outer areas of airports in times of stringent security conditions might prevent such attacks. Even in times when no specific threat has been received, it is within the capacity of most States to monitor those strips of land from which a SAM could be launched and thus minimize the risk. At the same time, these security operations would deter terrorists from spending vital resources on buying SAMs given the limited possibilities for their use.

Although the success rate so far of Western States in preventing terrorist missile attacks against civil aviation is satisfactory, and security forces, with the help of good intelligence, have been successful in tracking down and capturing missiles before they could be used, it is not unlikely that there will be attempts to use surface-to-air missiles to attack civil aviation in the near future. As some targets are

---

[24]Hanle (1989), p. 185; Ofri (1984), p. 49; Pierre (1975–6), p. 1256; Dorey (1983), p. 142.

[25]Clutterbuck (1991), at 175.

[26]Dorey (1983), p. 142.

becoming more difficult for terrorists to attack it can be anticipated that they will make efforts to overcome the enhanced security systems as well as redirecting their efforts towards less secure targets. The displacement of the increasingly ineffective system of hijacking by missile attacks against civil aviation is a real threat.

Another aspect in securing aviation in times of civil unrest is diplomacy and the meaning and purpose of aviation as interpreted by the founding fathers of the Chicago Convention. Given its strategic objective on sustainability of air transport and its compelling diplomatic role which ICAO has played over the past 66 years with aplomb and competence, member States of ICAO could well consider the role of aviation in bringing about peace. The importance of aviation toward maintaining peace has been accepted since World War 2 and is aptly reflected in the Statement of the British at that time, that civil aviation holds the key to power and importance of a nation and therefore it must be regulated or controlled by international authority. Lord Beaverbrook for the British Government of that time stated in Parliament:

> Our first concern will be to gain general acceptance of certain broad principles whereby civil aviation can be made into a benign influence for welding the nations of the world together into a closer cooperation...it will be our aim to make civil aviation a guarantee of international solidarity, a mainstay of world peace.[27]

The intensely political overtones that moulded the incipient civil aviation system of the world immediately after the War, thereby incontrovertibly establishing the relevance of diplomacy, international politics and international relations in civil aviation, is borne out by the statement of the first President of the ICAO Council, Edward Warner, when he said:

> It is well that we should be reminded...if the extent of the part which diplomatic and military considerations have played in international air transport, even in periods of undisturbed peace. We shall have a false idea of air transport's history, and a very false view of the problems of planning its future, if we think of it purely as a commercial enterprise, or neglect the extent to which political considerations have been controlling in shaping its course.[28]

In retrospect, it must be noted that this statement is a true reflection of what civil aviation stood for at that time, and, more importantly, that the statement has weathered the passage of time and is true even in the present context. A more recent commentator correctly observes that over the past decades, civil aviation has had to serve the political and economic interests of States and that, in this regard, ICAO has alternated between two positions, in its unobtrusive diplomatic role and its more pronounced regulatory role.[29]

---

[27] *Flight*, Vol. XLV No. 1331, January 27, 1944, at pp. 97–98.

[28] Warner (1942), p. V.

[29] Sochor (1991), t xvi.

An inherent characteristic of aviation is its ability to forge inroads into human affairs and promote international discourse. It also promotes international goodwill and develops "a feeling of brotherhood among the peoples of the world".[30] Therefore, it has been claimed that problems of international civil aviation constitute an integral part of the universal political problems of world organization and therefore aviation problems cannot be solved without involving the world political and diplomatic machinery.[31] It is at these crossroads that one encounters the profound involvement of the United Nations mechanism in general and ICAO in particular.

References

Clutterbuck R (1991) Living with terrorism. Butterworths, London

Dobson C, Payne R (1987) Appendix B: the chronology of terror: 1968–1987. In: War without end: the terrorists: an intelligence dossier. Sphere Books, London

Dorey FC (1983) Aviation security. Granada, London

Hanle DJ (1989) Terrorism: the newest face of warfare. Pergamon-Brassey's, New York

Mickolus EF (1980) Transnational terrorism: a chronology of events, 1969–1979. Aldwych, London

Ofri A (1984) Intelligence and counterterrorism. ORBIS 28:49

Pierre AJ (1975–6) The politics of international terrorism. ORBIS 19:1256

Reals K (2011) Runway at Benghazi airport destroyed: Capita Symonds. Air Transport Intelligence News (22 February 2011)

Reuter P (1989) Introduction to the law of treaties. Pinter Publishers, London

Schenkman J (1955) International civil aviation organization. Librairie E. Droz, Geneve

Sochor E (1991) The politics of international aviation. Macmillan, London

Warner E (1942) Foreword to international air transport and national policy by Lissitzyn OJ. Council on Foreign Relations, New York

---

[30] Schenkman (1955) at p. 6

[31] *Id.* Vi.

Article 11
Applicability of Air Regulations

Subject to the provisions of this Convention, the laws and regulations of a contracting State relating to the admission to or departure from its territory of aircraft engaged in international air navigation, or to the operation and navigation of such aircraft while within its territory, shall be applied to the aircraft of all contracting States without distinction as to nationality, and shall be complied with by such aircraft upon entering or departing from or while within the territory of that State.

Contents

1 Sovereignty Extended

At its 2nd Assembly (Geneva, 1–21 June 1948) the Council was directed through Resolution A2-45 which revised an earlier Resolution A1-30 to establish procedures for the reporting of individual cases of alleged breaches of, or non-compliance with national aeronautical regulations and the corrective action taken by States in such cases, including adoption of measures to ensure that airmen of their respective States would be familiar with the regulations of other Contracting States and with rules in force over the high seas. The Council, in its Annual Report to the 7th Session of the Assembly (Brighton, 16 June–6 July 1953) in formed the Assembly that:

> After further study of the difficulties involved in finding a practicable and effective procedure, the subject has been dropped, at least for the time being[1]".

[1]Doc 7367, A7-P/1.

DOI 10.1007/978-3-319-00068-8_12,

Resolution A2-45 was eventually declared to be no longer in force by Resolution A16-1 at the 16th Session of the Assembly (Buenos Aires, 3–26 September 1968). It must be noted that Annex 17 (security) to the Chicago Convention has an analogous provision (presumably stemming from Article 11) where Recommended Practice 6.1.2 provides that each ICAO member State should require the operators of aircraft of its registry to conform to the international civil aviation security requirements of those States into which they operate. Article 11 therefore has a strong link to security.

The carriage of dangerous goods into a country is a case in point, examples being the air cargo security and the carriage by air of human remains and dangerous substances.

2 Air Cargo Security

The air cargo industry is a 60 billion dollar business and passenger-accompanied air cargo is a major profit maker accounting for approximately 15 % of the industry's overall revenue. At a high level aviation security conference held by the International Civil Aviation Organization in September 2012, ICAO advised that civil aviation was estimated to grow by 6 % in 2013; 6.4 % in 2014; and 4.9 % on an annual basis until 2020, such growth being encouraging to the air transport industry from an economic perspective. At the same time, ICAO cautioned that this growth may not necessarily halt the persisting threat of unlawful interference with civil aviation and that the threat will evolve into a complex web of activity as new innovative methods of attack are conceived. This brought to bear the compelling need to address the growing vulnerabilities of the air transport system and the sustainability of aviation security.

In this context the conference noted that air cargo security, its sustainability and the need for innovative approaches to risk management were considered paramount and that it was essential that a global legal and regulatory system should govern the carriage of air cargo, where carriers would observe and fulfil requirements for security of each State they operated air services into and out of.

The High Level Conference on Aviation Security, convened by the International Civil Aviation Organization from 12 to 14 September 2012 in Montreal was attended by over 700 participants representing 132 of ICAO's 191 member States, and 23 Intergovernmental and industry international and regional organizations. The Conference focused on three overarching themes on cargo security, sustainability of security measures and innovation in addressing threats to aviation security.

The Conference recalled the events of 29 October 2010 when terrorists exploited vulnerabilities in the air cargo security system to introduce improvised explosive devices intended to destroy aircraft in flight, and endorsed certain principles on air cargo and mail security. One principle was that a strong, sustainable and resilient air cargo security system is essential and that the threat is to the air cargo system as a whole and risk-based consideration must be given to strengthening security measures across all aspects of the system, including enhancing the ability to recover from a major disruption. Another was that appropriate security controls should be implemented at the point of origin. Cargo and mail should come from a secure supply chain or be screened and, in either case, protected throughout the entire journey, including at transfer and transit points. At points of transfer, States should satisfy themselves that security controls previously applied to cargo and mail meet ICAO Standards. In doing so, they should avoid unnecessary duplication of security controls.

An important milestone in this regard was the recognition that air cargo advanced information for security risk assessment is a developing area that enhances air cargo security, particularly in the context of express delivery carriers such as FEDEX, UPS, DHL Express and TNT Express who carry around thirty million shipments daily, which typically contain high-value added, time-sensitive cargo. These carriers guarantee the timely delivery of these vast volumes of shipments, ranging from same-day delivery to 72 h after pick-up, virtually anywhere in the world. They operate in 220 countries and territories.

The conference noted that a real risk in the area of cargo and mail security would arise when an express delivery carrier experiences a technical problem in an aircraft and is forced to transfer cargo to a passenger carrier, in which instance strict supply chain standards should be adhered so that the risk in the transfer of cargo could be obviated.

Participants agreed that it was essential that solid standards and mutual recognition programmes be in place in order to make sure that States all along an air cargo supply chain satisfy themselves that air cargo is secure, and so let it flow unimpeded. Such standards and recommended practices should allow for the speedy transit and transhipment of legitimate air cargo worldwide, through any combination of air routes and transit or transhipment points.

Another risk factor considered by the Conference was the insider threat and the need to implement 100 % screening of persons other than passengers. This includes personnel at the airport, visitors and others who do not carry a boarding card. The Conference was reminded that States needed to acknowledge that the roles of people other than passengers that are working in civil aviation can present particular vulnerabilities that should be addressed.

It was also noted that there is already an ICAO Standard which requires each State to ensure that persons other than passengers, together with items carried, being granted access to security restricted areas are screened; however, if the principle of 100 % screening cannot be accomplished, other security controls, including but not limited to proportional screening, randomness and unpredictability, should be applied in accordance with a risk assessment carried out by the relevant national authorities.

One of the recommendations put forward at the Conference was to establish more airport authorities with increased aviation security expertise. In this regard it was noted that comprehensive background checks of all personnel selected for hire/employment at an airport need to be carried out by the relevant State's security agencies based on risk assessment. In addition, re-vetting of airport workers such as cleaners, duty free shop personnel, catering staff and concessionaires must be carried out frequently in order to mitigate collusion to commit acts of unlawful interference.

The author is of the view that if aviation security concerning cargo and mail were to be addressed in a results-based manner, there are two key areas that need to be enforced: sustainability and innovation. "Sustainable aviation security" can be defined as the detection and prevention of, and response to and recovery from, acts or attempted acts of unlawful interference with civil aviation, utilizing means that can be sustained by the entity or entities responsible for the period of time required. It is worth noting a number of important inter-related policy principles and practices that can contribute to the achievement of sustainable aviation security. These and other means can, more broadly, support the development of a sound and economically-viable civil aviation system. The starting point for consideration of any security measure must be a risk assessment. Such risk assessments, carried out objectively by appropriate security authorities on a continuous basis and informed by available and relevant information, including security intelligence, help assure that new or revised security measures are justified, aligned with actual needs and are proportionate to the level of risk.

The Conference recognized that the sustainability of aviation security measures and arrangements is an important strategic issue for all entities with aviation security-related responsibilities. It noted that risk-based security measures, outcomes-focused security measures, rationalization of security measures, optimization of technology, mutual recognition of equivalence and one-stop security, harmonization, and preparedness for crisis events are policy principles and practices whose implementation can contribute significantly to the sustainability of aviation security measures and arrangements.

From the time aviation was used as a weapon of mass destruction on 11 September 2001, there have been 75 terrorist attacks on aircraft and airports worldwide which have resulted in 157 deaths up to the end of 2011. When one compares this statistic to other modes of transport, such as trains and buses, one notes that there have been approximately 2,000 attacks and about 4,000 deaths resulting. On this basis, aviation has been fortunate. However, one cannot be complacent. The terrorist anchors himself on the Displacement Theory, moving from one mode of attack to another when the going gets bad. The 9/11 attacks on buildings turned to attacks on airports and then onwards to cargo. Examples go back to the 1980s where, in the early 1980s, aircraft were attacked through the cargo hold (recalling just three instances when aircraft of Air India and PANAM were blown up in midair as well as the attack on the Airlanka aircraft) to using aircraft as weapons of mass destruction in 2001, to attacking airports in the 1990s. Recalling the events of 29 October 2011 discussed above, one could argue that unlawful interference with civil aviation has turned full 360 degrees and has seemingly returned to attacks on cargo.

The Conference viewed with approval the ICAO Risk Content Statement and endorsed the Declaration adopted at the 37th ICAO Assembly Session convened by ICAO from 28 September to 8 October 2010, which, inter alia called for the strengthening of security screening procedures, enhancement of human factors and utilization of modern technologies to detect prohibited articles and support of research and development of technology for the detection of explosives, weapons and prohibited articles in order to prevent acts of unlawful interference. The Declaration also calls on all Member States to share best practices and information on a range of key aviation security matters, including threat-based risk assessments.

2.1 The Risk Content Statement

The Risk Content Statement submitted to the Conference by ICAO was entirely based on a global risk based approach which advocated a robust methodology for national risk assessment. It aims at providing a description of the global risk picture; assisting States in their efforts to protect air transportation and prevent its use for unlawful acts; presenting high-level statements for an improved approach in creating and maintaining State national civil aviation security programmes; and assisting ICAO in improving Standards and Recommended Practices (SARPs) and guidance material. The Statement focused the attention of ICAO member States to Standard 3.1.3 of Annex 17 to the Convention on International Civil Aviation (Chicago Convention) which requires each State to keep under constant review the level of threat to civil aviation within its territory, and establish and implement policies and procedures to adjust relevant elements of its national civil aviation security programme, based on a security risk assessment carried out by relevant national authorities. The Conference, based on this fundamental premise, recognized that a reasonably designed risk-based approach is one by which States identify the criteria to measure potential criminal activities, principally from terrorism. The identification of risks permits States to determine and implement proportionate measures and controls to mitigate against each risk type.

The Statement exhorted member States of ICAO to share information based on the premise that in conducting a risk assessment, it is necessary to assemble information about the threat. Such information may come from a variety of sources, such as those relating to: actual incidents, including successful or unsuccessful attacks on aviation, which provide information on proven terrorist methodologies; closed sources, from primarily counter-terrorist intelligence, which may be gathered by intelligence, law enforcement and other agencies of States; and open sources, which may include publicly available information on unusual or suspicious occurrences and the availability of items that could be used for terrorist purposes, and any other information that may contribute to the threat picture.

The Statement also highlights the insider threat as being of compelling significance. It emphasizes the danger of vulnerability associated with insiders which may be considered greater if they have access to the last layer of security in a way that a passenger does not, and points out that the likelihood associated with insiders might be less if they have already been subject to vetting and selection procedures and/or screening. It cautions that the consequence of a threat associated with insiders might be greater if an insider has access deeper within the system. For instance, an insider could perpetrate a more credible and thus more disruptive hoax. In summary, the Statement explains that the methodology involves considering each role within the system and whether it offers a particular tactical advantage in relation to each threat type or whether it poses the same issues as passengers, and that, in applying this methodology, it is possible to consider insider vulnerabilities as part of an integrated risk assessment.

Finally, the Risk Content Statement identifies risk assessment as a process which evaluates risk by threat identification, i.e. identifying the threat scenario, consisting of a defined target (e.g. airport terminal or aircraft), as well as the means and method of possible attack (e.g. attack by passenger using an improvised explosive device, or attack by an insider using weapon, etc.); likelihood i.e. considering the probability of the threat occurring; consequence—assessing the nature and scale of likely impacts associated with a successful attack, including consideration of human, economic, political and reputational factors (based on a reasonable worst-case scenario); vulnerabilities, i.e. evaluating the effectiveness and vulnerabilities of current security measures (i.e. security strengths and weaknesses of SARPs) in mitigating the potential threat scenario identified; and residual risk i.e. assessing the remaining risk of that type of attack being successfully carried out against that target, to enable a judgement to be made as to whether that is acceptable in risk management terms.

2.2 Capacity Building

The High Level Security Conference noted that capacity building on an international scale was critical to a risk based approach for air cargo security. In this regard it was recognized that an international capacity building strategy for air cargo and mail security would draw on the ICAO Assistance and Capacity Building Strategy for Aviation Security , and allow for targeted assistance for States in need. This Strategy would be guided by the ICAO security audit results, where air cargo and mail security has been identified as a priority need. The Strategy would include a proposal to coordinate bilateral and multilateral capacity building initiatives regionally, as well as amongst international organizations such as the World Customs Organization (WCO) and the Universal Postal Union (UPU) in order to align such initiatives, maximize limited resources, and avoid duplication of efforts. The development of such a strategy would also be in line with the ICAO Comprehensive Aviation Security Strategy (ICASS) 2011–2016 that was endorsed at the 37th ICAO Assembly in 2010.

In this regard the Conference viewed favourably the idea that any new arrangements must recognize that many donor states engage in aviation security capacity building for specific national interest reasons, generally related to the nature of flights into donor states. This is understandable and, in fact, is a concept which drives many bilateral aviation security efforts across the globe. In encouraging this capacity building to continue, the proposed framework seeks to better coordinate and inform its development by building on existing and future Government-to-Government arrangements with targeted industry-to-industry capacity building efforts, and using ICAO-sponsored capacity building where regional "gaps" in bilateral, multilateral and industry capacity building efforts are identified.

The reality that capacity building requires a long-term commitment and should be focused on "regular/repeated engagement", rather than the provision of one-off courses on an irregular basis, was recognized, together with the fact that effective capacity building takes years and will only succeed when issues of trust, mutual respect and culture are addressed and fostered on an on-going basis. The Conference endorsed the development of an International Capacity Building Strategy specific to air cargo and mail security, to aid those ICAO member States that require assistance to adequately implement enhanced ICAO air cargo and mail security standards. This Strategy was to be aligned with the ICAO Assistance and Capacity Building Strategy for Aviation Security, and avoid duplication of efforts. It also encouraged all ICAO Contracting States to further support the Secretariat in its efforts to provide capacity building assistance based on USAP audit results, subject to the consent of the State(s) receiving assistance, focusing on air cargo and mail where it has been identified as a priority; and urged other entities within the air cargo environment to continue taking action to effectively secure those sections of the supply chain in which they operate.

2.3 Insider Threats

A third element addressed within the parameters of the risk based approach was the insider threat and the need for screening of persons other than passengers. The Conference considered as the basis for discussion Standard 4.2.6 to Annex 17 to the Chicago Convention which, through Amendment 12 to the Annex (which became applicable on 1 July 2011), states that each Contracting State is required to ensure that persons other than passengers, together with items carried, being granted access to security restricted areas are screened; however, if the principle of 100 % screening cannot be accomplished, other security controls, including but not limited to proportional screening, randomness and unpredictability, shall be applied in accordance with a risk assessment carried out by the relevant national authorities.

The above provision notwithstanding, the Conference recognized that it was indeed very difficult to preclude, detect and face an act of unlawful interference carried out with the internal support of persons who have access to security-restricted areas, even though such persons may have had their records verified. The danger and risk were compounded by the fact that such persons usually have access to sterile lounges and other security-restricted areas where they have the opportunity to mingle with passengers and therefore could well interfere with their carry-on baggage and/or the checked baggage already inspected. They also have access to aircraft during ground and pre-flight services. One participating State suggested that in all access control points, conditions should be created securing that 100 % of persons who are not passengers, as well as the articles transported, are subject to security inspections with whatever of the different means available for this purpose, including manual inspection.

Some national practices submitted to the Conference were: (a) supervise or accompany daily/seasonal workers in the restricted security area; (b) closely examine all officials, employees or staff entering the restricted security area; closely examine all Janitors before permitting them to enter the aircraft; oversee the restricted security areas and facilities related to flight operations by patrolling periodically or continuous surveillance using CCTV; inspect all cabin carry-on, baggage and cargo as well as food (catering items) and equipment required and sold in flight (stores) and watch them before and during the loading onto the aircraft; oversee the process of boarding passengers and loading of goods; aircraft security check before departure (pre-flight security check); supervise, control and update the permit issuance and use of appropriate entry of airport, including applying background checks and stop list procedure; implement security awareness training for all airport pass applicants; carry out internal and external quality control regularly based on risk assessment; and be aware of religious, social and cultural approach among stakeholders.

2.4 Sustainability and Innovation

The Conference recognized that the sustainability of aviation security measures and arrangements is an important strategic issue for all entities with aviation security-related responsibilities and that risk-based security measures, outcomes-focused security measures, rationalization of security measures, optimization of technology, mutual recognition of equivalence and one-stop security, harmonization, and preparedness for crisis events are policy principles and practices whose implementation can contribute significantly to the sustainability of aviation security measures and arrangements.

The need for each State to carry out continuous risk assessments as a preliminary measure was considered paramount for the sustainability of security. One view was that aviation security has to be sustained in a balanced manner so that, on the one hand, applying security measures to mitigate identified threats, and on the other hand, the essential task of facilitating operations, passengers' experience and trade could be ensured. Security should not accumulate layer upon layer of controls and associated costs, but should rather ensure the sustainability of the system from the perspectives of cost, efficiency, and acceptability by passengers and air transport operators, which should be a central consideration when designing security processes. Another means of achieving sustainability is at transfer points where security controls are known to have been performed effectively at the point of origin. The Conference took note of the fact that, in such instances, the concept of "One Stop Security" should be advanced, where ICAO Member States, by virtue of recognising the equivalence of each other's aviation security regimes, can allow incoming passengers, baggage and cargo to transfer onto a connecting flight without being subjected, once again, to the same security controls as at the point of origin. The conclusion of such "One Stop Security" arrangements remains an issue to be addressed Member State to Member State.

One proactive suggestion towards achieving and retaining sustainability was to follow the practice of reciprocal acceptance of equivalent security measures across the board, with due regard to the principle of host State responsibility, as envisaged by the Chicago Convention. In that respect, it was suggested that the need for any one State to require extra security measures of another State can be avoided by working together to align international requirements to the global threat environment. A further recommendation was that this approach should be reflected in Annex 17 to the Chicago Convention which deals with the subject of security.

The Conference was called upon to endorse a coordinated response to security incidents and threats whereby States could collectively accept, without derogating a State's freedom to take its own measures, the measures adopted by one State as a global norm if that norm were to be consistent with the Standards and Recommended Practices of Annex 17 and provided such recognition was accepted as such by ICAO.

Incontrovertibly, innovation in air cargo security lies in two areas: advancement of technology; and intelligence. In the field of air cargo security the ICAO Conference showed a marked deficiency of discussion. It is submitted that, critical to a discussion of technology and innovation is the subject of supply chain security. Preeminent among technological progress is the need to establish basic security packaging mandates for shippers. Cargo is either being flown or stored at any given point in time and therefore both phases must be covered in the tracking and identification of cargo. Hoffer recommends:

Courier boxes and envelopes supplied by carriers should be required to have an original number and (if possible) a tamper-evident seal and markings (tied to the bill of lading), so that it is harder to replace a package with a similar box. Recipients would have the ultimate responsibility to compare manifest numbers with packages before accepting them.

Air cargo can be loaded individually or in bulk form from 1 kg to a weight of several tons and can be loaded on various platforms such as unit load devices (ULDs) crates and assembled pallets. Several technologies are used at present in ensuring cargo security. These may vary between explosive detection devices, explosive trace detection computer aided tomography and X-ray, in addition to certified canine teams. The Transportation Security Administration (TSA) has identified such advanced technologies as XR/PFNA X-ray systems with pulsed fast neuron analysis; pressure activated sampling systems, quadruple resonance and miniature explosive and toxic chemical detector utilizer sensors.

In the context of military intelligence, the author submits that as a mirror reflection of the "known shipper" and "known consignor" practice, military intelligence be employed to track and identify unknown consignors as well as insiders. Taking into consideration the aircraft bombings that have taken place (some of which have been discussed in this article) it is fair to conclude that most of these attacks were perpetrated by groups of incendiary persons. Military intelligence, which essentially is information relating to the armed forces of a foreign country that is significant to the planning and conduct of another country's military doctrine, policy, and operations, largely penetrates such groups and could be effectively used to take pre-emptive and preventive measures against threats to air cargo security.

It is eminently clear that the glue that binds the elements discussed above, including those that relate to the global supply chain, is law and practice. These are already in place in principle. For instance, Standard 4.6.1 of Annex 17 requires each Contracting State to ensure that appropriate security controls, including screening where applicable, are applied to cargo and mail, prior to their being loaded onto an aircraft engaged in passenger commercial air transport operations. The operative words here are "security controls" which brings to bear the reality that different States could have different security controls and that they should be harmonized in ensuring supply chain security and global security standards. Screening and examination of cargo and mail are paramount to this consideration.

Standard 4.6.2 requires that each Contracting State establish a supply chain security process, which includes the approval of regulated agents and/or known consignors, if such entities are involved in implementing screening or other security controls of cargo and mail. A regulated agent is defined in the Annex as an agent, freight forwarder or any other entity who conducts business with an operator and provides security controls that are accepted or required by the appropriate authority in respect of cargo or mail. There are five other provisions under Chapter 4.6 of Annex 17 pertaining to: the protection of cargo and mail from unauthorized interference from the point of screening or other security controls being applied until the departure of the aircraft; the non-acceptance of cargo or mail by operators unless it is confirmed that screening or other procedures have been applied and conformed by a regular agent; the appropriate screening of catering, stores and supplies intended for carriage by air; the appropriate screening of merchandise and supplies introduced into security restricted areas; and the fact that the security controls mentioned above have been implemented on the basis of a security risk assessment carried out by relevant national authorities.

If laws and practices are the glue that keeps air cargo security together, political will is the fuel which will ignites its progress and development. The thrust of political will essentially lies in a security culture that must be visible in every State. A security culture would make States aware of their rights and duties, and, more importantly, enable States to assert them. Those who belong to a security culture also know which conduct would compromise security and they are quick to educate and caution those who, out of ignorance, forgetfulness, or personal weakness, partake in insecure conduct. This security consciousness becomes a "culture" when all the 191 member States of ICAO as a whole make security violations socially and morally unacceptable within the group. In building a security culture within ICAO member States it is imperative that consideration should also be given to the development of a process for ensuring that all Member States are notified when deficiencies identified during the course of an ICAO security audit conducted under the Universal Security Audit Programme (USAP) remain unaddressed for a sustained period of time. A notification process could involve the use of information which does not divulge specific vulnerabilities but enables States to initiate consultations with the State of interest to ensure the continued protection of aviation assets on a bilateral basis. States have to adopt a security culture that admits of an overall approach to the threat as a potential harm to humanity. This should inevitably include strict adherence by States to the provisions of Annex 17.

3 Carriage by Air of Human Remains

If a person dies in a country other than his own, there are no global rules or guidance that dictates the manner in which his remains could be transported back to his country, with dignity and care. This matter was highlighted in 2003 before the European Parliament with a real example of a British national who died while on holiday in Greece. The Greek authorities had carried out an autopsy which concluded that the deceased tourist had died of a heart attack. When the body was transported back home the deceased's family had requested a second autopsy, only to find that most of the deceased's organs had been removed in Greece after the autopsy and destroyed, according to Greek law. This had caused severe mental distress to the deceased's kin.

There are three dimensions to this subject: the health and sanitation aspects of carrying human remains by air; and the rights of close relatives of the deceased to bring his remains back home with speedy dispatch; and the risk of such carriage to aviation security. The former may have security implications as well as safety implications in that security clearance for this type of cargo has to be carefully ascertained and ensured. As this article will discuss, international standards for packaging bring to bear certain vulnerabilities in the acceptance of human remains for carriage by air.

There have been some concerns about human remains being used to transport explosives. The transportation Security Administration of the United States cautions consignors and consignees:

> Out of respect to the deceased and their family and friends, under no circumstances will an officer open the container even if the passenger requests this be done. Documentation from the funeral home is not sufficient to carry a crematory container through security and onto a plane without screening. You may transport the urn as checked baggage provided that it is successfully screened. We will screen the urn for explosive materials/devices using a variety of techniques; if cleared, it will be permitted as checked baggage only. Crematory containers are made from many different types of materials, all with varying thickness. At present, we cannot state for certain whether your particular crematory container can successfully pass through an X-ray machine. However, we suggest that you purchase a temporary or permanent crematory container made of a lighter weight material such as wood or plastic that can be successfully X-rayed.[2]

There is a serious lacuna in regulatory consistency in the carriage by air of human remains which lays the thrust of Article 11 wide open. There have been some attempts by the international community to address the subject. The bottom line is that that, although several attempts have been made at international level in the past—some clear and some unclear—they lack unification and stand fragmented and ambivalent. To their credit, airlines, under the guidance of the International

[2] See Transportation Security Administration, *Transporting the Deceased*, at http://www.tsa.gov/travelers/airtravel/specialneeds/editorial_1296.shtm.

Air Transport Association (IATA), have adopted their own principles in carrying human remains with compassion and dedication. The conclusion suggests a way forward in binding the threads of this issue in a harmonious manner.

Human dignity is an international concept which is extended both to the living and the dead. The 1948 *Universal Declaration of Human Rights* of the United Nations—the cornerstone of human dignity—declares that the inherent dignity and the equal and inalienable rights of all members of the human family are the foundations of freedom, justice and peace in the world and that all human beings are born free and equal in dignity and rights. This statement establishes human dignity as the conceptual basis for human rights. 75 % of the constitutions of ICAO's 191 member States use the concepts of "human dignity" or "personal dignity" explicitly.[3] It follows therefore that if the remains of a human being are not given equal respect and dignity, the moral imperative of the doctrine of human dignity[4] would be rendered destitute of meaning and purpose.

From an aviation perspective, most airlines in the world offer services for the transportation of human remains and cremated remains. These services are varied according to the policies of each airline, but all share a common thread of dedication and compassion in offering the service in the transportation of funeral shipments. Usually, airlines employ specially trained staff to address all the travel-related issues that may arise when shipping such very sensitive cargo. The tasks assigned to these staff include providing advice to those seeking the airlines' services on applicable regulations, taking into account the delicateness of the responsibility that devolves upon the carrier.

In terms of property rights pertaining to a cadaver or other remains, such rights do not exist at common law. However, for the purpose of transportation—whether it be for embalming, cremation or internment—the corpse or cremated remains of a human being is considered to be property or quasi-property, the rights to which are held by the surviving spouse or next of kin. This right cannot be transferred and does not exist while the deceased is living. A corpse or urn carrying cremated remains may not be retained by either an undertaker or a carrier as security for unpaid funeral expenses, particularly if such were kept without authorization and payment was demanded as a condition precedent to its release. Upon burial the body accrues to the ground and any appurtenant property such as jewelry which was on the corpse on burial accrue to their rightful owner as determined by applicable principles of property laws and wills and testaments as they might exist.

The purpose of this article is to discuss *de lege lata* the fragmented regime applicable to the carriage by air of human remains. Two antiquated multilateral agreements, one Resolution and one Regulation all in Europe; some maundering by the ICAO Council decades ago; three Annexes to the Chicago Convention which may

---

[3] http://www.constitution.org; http://www.oefre.unibe.ch/law/icl; http://www.psr.keele.ac.uk.

[4] Human dignity has not been comprehensively defined and has remained a somewhat squishy subject, often explained theologically. However, the dictionary definition of dignity is that it is *inter alia* "the quality or state of being worthy of esteem or respect". See http://www.thefreedictionary.com/dignity.

have applicability to this subject; some proactive guidelines by the International Air Transport Association and the World Health Organization and procedures and policy of individual air carriers comprise the history of this subject. Against this backdrop, this article will inquire into the need for a global regulatory process that would properly address this esoteric but important area of carriage by air.

3.1 The Berlin Agreement of 1937

The *International Arrangement Concerning the Conveyance of Corpses*[5] (Berlin Agreement), signed at Berlin on 10 February 1937 was the first recorded attempt at the unification of rules relating to the carriage of human remains. The agreement, which applied to the international transport of corpses immediately after decease or exhumation, was designed to avoid the difficulties resulting from differences in the regulations concerning the conveyance of corpses, and recognized the necessity and the convenience of laying down uniform regulations in this area of transportation. Accordingly, the signatory States[6] undertook to accept the entry into their territory, or the passage in transit through their territory, of the corpses of persons deceased in the territory of any one of the other Contracting countries upon certain conditions, which were incorporated in the Agreement.

The initial condition, as laid out in Article 1 of the Agreement was that, for the conveyance of any corpse by any means and under any conditions, a special laissez-passer be issued for a corpse which would state the surname, first name and age of the deceased person, and the place, date and cause of decease. The competent authority for the place of decease or the place of burial in the case of corpses exhumed had to issue the laissez-passer and it was recommended that the laissez-passer should be made out, not only in the language of the country issuing it, but also in at least one of the languages most frequently used in international relations.

The Berlin Agreement further stated that neither the country of destination nor the countries of transit shall require, over and above such papers as are required under international conventions for the purpose of transports in general, any document other than the laissez-passer referred to in Article 1. The following had to be presented to the competent authority for the issuance of laissez-passer: a certified true copy of the death certificate; and official certificates to the effect that conveyance of the corpse is not open to objection from the point of view of health or from the medico-legal point of view, and evidence that the corpse has been placed in a coffin in accordance with the regulations laid down in the Agreement.[7]

[5]League of Nations, Treaty Series 1938, No. 439r at 315–325.

[6]Germany, Belgium, Chile Denmark, Egypt, France, Italy, the Netherlands, Switzerland, Czechoslovakia and Turkey.

[7]Berlin Agreement, Airlines set for losses as mid-east unrest continues, *Air Letter*, No. 17,181, Friday, 25 February 2011at p. 3, Article 2.

As for packaging the human remains, the Agreement, in Article 3 provided that corpses must be placed in a metal coffin, the bottom of which has been covered with a layer approximately 5 cm. of absorbent matter such as peat, sawdust, powdered charcoal or the like with the addition of an antiseptic substance. Where the cause of decease was a contagious disease, the corpse itself was required to be wrapped in a shroud soaked in an antiseptic solution. A further requirement was that the metal coffin must thereupon be hermetically closed (soldered) and fitted into a wooden coffin in such a manner as to preclude movement. The wooden coffin was required to be of a thickness of not less than 3 cm. and its joints must be completely watertight. It was also required that the coffin be closed by means of screws not more than 20 cm. distant from one another, and strengthened by metal hoops. In the case of transport by air, The Agreement, in Article 7, required that coffins must be conveyed either in an aircraft specially and solely used for the purpose or in a special compartment solely reserved for the purpose in an ordinary aircraft.

The Agreement precluded bodies of persons who had died as a cause of plague, cholera, small-pox or typhus from being conveyed between the territories of the Contracting parties until the lapse of at least 1 year after the demise. No articles were permitted to be transported along with the coffin in the same aircraft or in the same compartment, other than wreaths, bunches of flowers and the like.[8]

3.2 Agreement on the Transfer of Corpses (Strasbourg—1973)

The second international agreement was in 1973 called the *Agreement on the Transfer of Corpses*, and it was drawn up within the Council of Europe by the European Public Health Committee. The Strasbourg Agreement was opened for signature by the member States of the Council of Europe on 26 October 1973. This agreement was designed to adapt the provisions of the Berlin Agreement concerning the conveyance of corpses, to the new situation arising from developments in the field of communications systems, international relations and commercial and tourist activities. A proposal to examine anew the problem of the transfer of corpses with a view to drawing up a new instrument was approved by the Committee of Ministers of the Council of Europe in 1967 and this task was entrusted to the European Public Health Committee which, in the course of its work, gave due consideration to the observations, among others, of the European Federation of Funeral Directors (Brussels) and the European Funeral Directors Association (Vienna). The text of the draft Agreement was submitted to the European Committee on Legal Co-operation (CCJ) before its final adoption by the Committee of Ministers of the Council of Europe in April 1973. It was opened for signature by member States of the Council of Europe on 26 October 1973.

[8] *Id.* Article 4.

The Strasbourg Agreement defines the transfer of corpses as the international transport of human remains from the State of departure to the State of destination. Accordingly, the State of departure is that in which the transfer began; in the case of exhumed remains, it is that in which burial had taken place; the State of destination is that in which the corpse is to be buried or cremated after the transport. The Agreement does not apply to the international transport of ashes. Article 3 of the Agreement states that during the transfer, any corpse is required to be accompanied by a special document (laissez-passer for a corpse) issued by the competent authority of the State of departure. The laissez-passer has to include at least the information set out in the model annexed to the Agreement; and be made out in the official language or one of the official languages of the State in which it was issued and in one of the official languages of the Council of Europe.

Article 4 provides that, with the exception of the documents required under international conventions and agreements relating to transport in general, or future conventions or arrangements on the transfer of corpses, neither the State of destination nor the transit State shall require any documents other than the laissez-passer for a corpse. The laissez-passer is issued by the competent authority referred to in Article 8 of the Agreement,[9] after it has been ascertained that: all the medical, health, administrative and legal requirements of the regulations in force in the State of departure relating to the transfer of corpses and, where appropriate, burial and exhumation have been complied with; the remains have been placed in a coffin which complies with the requirements laid down in Articles 6 and 7 of the Agreement; and that the coffin only contains the remains of the person named in the laissez-passer and such personal effects as are to be buried or cremated with the corpse.

Article 6 requires that the coffin must be impervious and that the inside must contain absorbent material. If the competent authority of the State of departure consider it necessary the coffin must be provided with a purifying device to balance the internal and external pressures. It may consist of: either an outer coffin in wood with sides at least 20 mm thick and an inner coffin of zinc carefully soldered or of any other material which is self-destroying; or a single coffin in wood with sides at least 30 mm thick lined with a sheet of zinc or of any other material which is self-destroying. If the cause of death is a contagious disease, the body itself is required to be wrapped in a shroud impregnated with an antiseptic solution.

Article 6 further provides that the coffin, if it is to be transferred by air, has to be provided with a purifying device or, failing this, present such guarantees of resistance as are recognised to be adequate by the competent authority of the State of departure. If the coffin is to be transported like an ordinary consignment, it has to be packaged so that it no longer resembles a coffin, and it shall be indicated that it be handled with care.[10]

[9] Article 8 states that each Contracting Party shall communicate to the Secretary General of the Council of Europe the designation of the competent authority referred to in Article 3, paragraph 1, Article 5 and Article 6, paragraphs 1 and 3 of the Agreement.

[10] Article 7 of the Strasbourg Agreement.

3.3 Resolution 2003/2032 (INI)

The European Community was dissatisfied with both the Berlin Agreement and the Strasbourg Agreement (which only some member States had signed), claiming that these Agreements advocated indirect discrimination by providing for non-European Community residents. Also it was claimed that these two agreements imposed strict rules on the cross-border transfer of mortal remains, applied essentially to "non-nationals" and hence ran counter to the Community scheme of things. Accordingly, and with a view to addressing the case where a Community citizen expired in a Community country other than his own and his remains had to be repatriated to his country, a Committee was appointed by the European Parliament to consider an instrument that addressed the conveyance of mortal remains suggested in 2003 Resolution 2003/2032 (INI). This Resolution noted that, on account of the above agreements, the death of a Community citizen in a Member State other than his country of origin results in more complex procedures, a longer period of time before burial or cremation takes place and higher costs than if the death had occurred in the deceased person's country of origin,

Another compelling reason for this Resolution was the recognition that, in view of the growth in intra-Community tourism, the increasing numbers of retired people who choose to live in a country other than their own and, more generally, greater intra-Community mobility which is actually encouraged, the number of Community citizens who die in a country other than their country of origin was bound to increase. This was considered against the backdrop that Community citizens should, *mutatis mutandis*, be able to move between and reside in Member States in similar conditions to nationals of a Member State moving around or changing their place of residence in their own country, and that exercising the right to freedom of movement and freedom of residence should be facilitated to the utmost by reducing administrative formalities to an absolute minimum.

The European Community was of the view that, at the time the Resolution was proposed, it was still far from true that a Community citizen who dies in a Member State other than his own is treated in the same way as a national who dies in his home country. For example, the fact that a zinc coffin is required for the repatriation of a corpse from Salzburg to Freilassing (a distance of 10 km) but not for the transfer of a body from Ivalo to Helsinki (a distance of 1,120 km) (2).

Therefore it was pointed out that the repatriation of mortal remains without excessive cost or bureaucracy in the event of the death of a European Community citizen in a country other than the one in which either burial or cremation was to take place may be regarded as a corollary of the right of each EU citizen to move and reside freely within the territory of the Member States.

The Resolution called upon the Commission to see that the standards and the procedures applied in the cross-border transportation of corpses were harmonized throughout the Community and to endeavor to ensure that, as far as possible, Community citizens were treated in the same way as nationals in their home country.

A Regulation, covering intra-community transport of bodies according to the European Standard CEN/BT/TF 139 on Funeral Services and approved on 27 July 2005 goes on to say in Article 1 that the identification of the deceased must be performed before the body is placed in the coffin by the funeral enterprise or operator of the country of departure. The elements of identification relate to the civil status of the deceased and are indicated on the laissez-passer for the body. For identification, the body must be provided with: an identification bracelet attached to the body part (wrist, ankle...,); and a non-removable and tamper-proof identification tag attached to the coffin and its wrapping, if any. The information required on the bracelet were: surname and first name(s); sex; date and place of birth; date and place of death; and nationality. The information required on the identification tag were to be: surname and family name(s); date of birth; and date of death.

Article 2 of the Regulation required that the coffin or casket that carried the remains must be made of solid material—the main material used in Europe being wood (excluding the use of carton or chipboard). The material used for the coffin must be biodegradable. It also required that the coffin must be impervious; the products used to make it impervious must be biodegradable and in conformity with the standards applicable to crematorium emissions. In particular, the coffin must be impervious to decomposition liquids and fitted with absorbent material. The out cover of the coffin/casket was required to meet necessary sanitary requirements.

The Regulation had chemical requirements that were not contained in the 1937 Berlin Agreement and the 1973 Strasbourg Agreement. For instance, Articles 2.3 and 2.5, specified conditions for international carriage of corpses by providing that if the cause of death was a contagious disease (as per the WHO official list), the outer container (usually wooden) used for the transport of the body may be lined with a hermetically sealed container. The hermetically sealed container must be provided with a purifying filter. If the consecutive treatments (thanatopraxy) have been performed within 36 h after the death the body must be encoffined within 6 days. The transport must be done not more than 48 h after encoffining and sealing. The conditions required for long distance international transport outside Europe under the Agreement were: hermetically sealed container; and/or embalming/thanatopractical treatment; and/or refrigeration. In the case of refrigeration at no time shall the temperature inside the container exceed 80 °C during transport.

The Regulation requires two types of documents for carriage of corpses: medical certificate upon death; and a laissez passer. The medical certificate is required to be drawn up, on the one hand, in the language of the country of departure in which the death had occurred and, on the other hand, in one of the following languages: English, German or French. It must contain information relating to the deceased such as: surname and maiden name in the case of a married woman; first name(s); date and place of birth; date and place of death; sex; and cause of death.

3.4 ICAO Initiatives

3.4.1 The Council

The Council of ICAO, at its Thirty Second Session in 1957 addressed the carriage under the heading "Carriage of Sick Persons, Pregnant Women, Live Animals and Coffins—Sanitation on Board Aircraft" at which IATA recommended that in addition to the prevailing requirement—that human remains be placed in hermetically-sealed coffins which are enclosed in outside cases—human remains should be embalmed prior to being placed in the coffin. IATA further suggested that acceptance of such coffins is dependent upon the type of aircraft, requirements of entry and clearance and prior approval of the countries of origin, transit and destination.[11] The Council noted that comments on the carriage of coffins had been received from twenty seven States (from a total of 72 member States at that time) and two overseas territories. Three of these States reported that they were bound by the provisions of the 1937 Berlin Agreement and Eight States advised ICAO that the carriage of corpses existed in their national legislations. Thirteen States commented that they had not, in their experience encountered serious difficulties in this area. The United States made the comment:

Because of known effects of rare atmosphere at high altitude on sealed caskets, such caskets should not be carried by aircraft.[12]

The ICAO Secretariat responded in assent:

> Differences in atmospheric pressure are known to have caused bursting of coffins, particularly when sealed hermetically (by welding) according to provisions of Articles 5 and 7 of the Berlin Arrangement, or similar provisions in national legislation. Prompted by rapid decomposition in flight, such transports occasionally arrive in appalling conditions; in some States (Australia, Philippines, Venezuela, Netherlands Antilles), therefore, it is required that corpses be embalmed prior to air transport, thus eliminating at least certain difficulties. If some pressure-relief system were applied to sealed caskets, the difficulties caused by pressure differences might disappear, but international transport would not permitted by existing laws.

It is noteworthy that during these discussions, cremated human remains were not mentioned, except by Belgium which said that "incinerated corpses are accepted without any restrictions and are carried on all types of aircraft".[13] The ICAO Council concluded that the difficulties reported by States were caused by variations of atmospheric pressure; a characteristic of transport by air, while for international transport coffins must be hermetically sealed.

ICAO has approached this subject from another dimension i.e. the carriage of human remains of an aircraft accident victim. In 2001 the Council released its *Guidance on Assistance to Aircraft Accident Victims and their Families*[14] where

[11] C-WP/2448, 5/6/57, Addendum and Corrigendum, 21/11/576 at 3.

[12] *Id.* Paragraph 20.1 at p. 10.

[13] C-WP/2448, 5/6/57, 42 tit.24, pr 22 S 4, Appendix "A" at 25.

[14] *Guidance on Assistance to Aircraft Accident Victims and their Families, ICAO Circular 285 –AN/166.*

ICAO recognizes that in an accident context the identification, custody and return of human remains are very important forms of family assistance but remains are often difficult to recover and identification can be an arduous and time consuming process. The ICAO guidance goes on to say that legislation often requires a post mortem examination of those killed in an accident and in some instances there may be remains that cannot be identified.[15] ICAO also calls for personal effects of the deceased to be correctly handled and returned to their lawful owners.[16] The Guidance also calls for the State of occurrence to provide for the return of human remains[17] while also devolving that burden—of the carriage of such remains—upon the aircraft operator involved in the accident.[18]

3.4.2 Annexes 9, 17 and 18 to the Chicago Convention

There are three Annexes to the Chicago Convention which bear some relevance to the carriage of human remains by air—Annex 9 (Facilitation), Annex 17 (Security) and Annex 18 (The Safe Transport of Dangerous Goods by Air). The Annex 9 definition of cargo implies that human remains could be categorized as cargo by giving the definition of cargo as "*any property carried on an aircraft other than mail, stores and accompanied or mishandled baggage*. This definition is slightly different from the one contained in another ICAO document—*Technical Instructions for the Safe Transport of Dangerous Goods by Air*[19] which defines "cargo" as "*any property carried on an aircraft other than mail and accompanied or mishandled baggage*. Annex 18 does not define the word "cargo" but defines "dangerous goods" as articles or substances which are capable of posing a risk to health, safety, property or the environment and which are shown in the list of dangerous goods in the Technical Instructions or which are classified according to those instructions. The Technical Instructions do not list human remains as being dangerous cargo. However, it behooves the international aviation community to inquire, along the lines of ICAO discussions in the Council, whether human remains could be ruled out as not posing a risk to health or the environment under any circumstances of carriage by air or whether human remains, depending on the way it is packed for transport, could be considered as dangerous goods.[20]

Standard 4.6.1 of Annex 17 requires each Contracting State to ensure that appropriate security controls, including screening where applicable, are applied to cargo and mail, prior to their being loaded onto an aircraft engaged in passenger commercial air transport operations. The operative words here are "security

[15] *Id.* Paragraph 3.10.

[16] *Id.* Paragraph 3.11.

[17] *Id.* Paragraph 5.1.

[18] *Id.* Paragraph 5,7.

[19] ICAO Doc 9284, AN/905 (2011–2012 Edition).

[20] American Airlines requires that human remains packed in dry ice are subject to dangerous goods regulations. https://www.aacargo.com/shipping/humanremains.jhtml.

controls" which brings to bear the reality that different States could have different security controls[21] and that they should be harmonized in ensuring supply chain security and global security standards. Screening and examination of cargo and mail are paramount to this consideration.

Standard 4.6.2 requires that each Contracting State establish a supply chain security process, which includes the approval of regulated agents and/or known consignors, if such entities are involved in implementing screening or other security controls of cargo and mail. A regulated agent is defined in the Annex as an agent, freight forwarder or any other entity who conducts business with an operator and provides security controls that are accepted or required by the appropriate authority in respect of cargo or mail. There are five other provisions under Chapter 4.6 of Annex 17 pertaining to: the protection of cargo and mail from unauthorized interference from the point of screening or other security controls being applied until the departure of the aircraft[22]; the non-acceptance of cargo or mail by operators unless it is confirmed that screening or other procedures have been applied and conformed by a regular agent[23]; the appropriate screening of catering, stores and supplies intended for carriage by air[24]; the appropriate screening of merchandise and supplies introduced into security restricted areas[25]; and the fact that the security controls mentioned above have been implemented on the basis of a security risk assessment carried out by relevant national authorities.[26]

Getting back to Annex 9, there is a whole chapter in the Annex—Chapter 4—dedicated to the entry and departure of cargo and other articles. Surprisingly, there is no provision in the Annex for priority of clearance or transport of human remains over other cargo, despite the prominence given to the subject in *ICAO Circular 285—AN/166.*[27] Another surprise is that, although there is a *Recommended Practice* in the Annex which suggests that electronic information systems for the release and clearance of "goods" (my emphasis) should cover their transfer between air and other modes of transport,[28] there is no definition of "goods" in the Annex. Do corpses or cremated human remains come under the purview of "goods"? This question is valid in the context of Appendix 3 to the Annex which has a template for a cargo manifest where there exists a column for "Nature of Goods". There is no mention of the word "cargo" in this template.

---

[21] Annex 17 defines a security control as "a means by which the introduction of weapons, explosives or other dangerous devices, articles or substances which may be used to commit acts of unlawful interference can be prevented.

[22] Standard 4.6.3.

[23] Standard 4.6.4.

[24] Standard 4.6.5.

[25] Standard 4.6.6.

[26] Standard 4.6.7.

[27] Baldwin (1910), 97.

[28] Annex 9 to the Convention on International Civil Aviation, Thirteenth Edition: July 2011, *Recommended Practice* 4.18.

In view of the above discussion it might be worthwhile for a detailed discussion on the status of human remains in the global aviation context and a re-visit of the 1957 discussions in the ICAO Council. The added dimension of related ICAO documentation such as Circular 285—AN/166 makes it all the more compelling.

3.4.3 IATA, WHO and United States Guidelines

The International Air Transport Association has clear, cogent guidance on the carriage by air of human remains. In its *Airport Handling Manual* (AHM) IATA prescribes that for special cargo, such as valuable cargo, perishables, vulnerable cargo, human remains and shipments of special importance or urgency, particular points to be considered are: that all personnel concerned are made fully aware of the nature and handling requirements of all such shipments; suitable arrangements are made for the security of valuable and vulnerable cargo; perishables are handled in accordance with the requirements of the particular commodity and in particular the most recent edition of the Perishable Cargo Regulations Manual; that a check is made to ensure that the final load assembled for dispatch to the aircraft *does* include shipments of special importance or urgency; and that shipments considered as special cargo have "special consignment" labels visibly attached to each package.[29]

The *IATA Ground Operations Manual* (IGOM) provides that human remains should be carried in an aircraft only if accepted by the operating airline for transport. The IGOM requires the carrier to make sure that a Human Remains Acceptance Checklist has been used (if required by the operating airline). Carriers are required, according to the IGOM, not to accept any human remains that are consolidated with any cargo other than other human remains. With regard to cremated human remains the Manual requires that only urns or other suitable containers as cargo with no special restrictions are accepted for carriage and that the carrier should make sure that the urn or other container is packed in a neutral outer pack that will protect the urn from breakage and/spillage.[30] It also prescribes that human remains in coffins should not be stored next to food or live animals, adding that there appears to be no scientific or technical reason why live animals and human remains should be segregated in aircraft cargo compartments, except that it may be ethical for cultural reasons to segregate them.

IATA in AHM 333 states that, should a body fluid leakage occur while transporting dead bodies, the usual accepted guidelines endorsed by WHO for dealing with spilled body fluids should be followed and the handler is advised to: wear disposable gloves and, if available, a plastic apron. If the spillage has occurred on an aircraft, the AHM provision advises the handler to only use cleaning materials suitable for aircraft use. He should not try to clean the body fluids by hosing with water or air and should use material that will adsorb the body fluids and scrape the material into a biohazard bag. Afterwards, he should wash the area with water/ disinfectant after removal of the adsorbent material, dispose of gloves and apron in a biohazard bag and wash hands thoroughly with soap and water afterwards.

[29] IATA Airport Handling Manual, AHM 310 at 149.

[30] IATA IGOM, Chapter 3.

WHO has also some guidance pertaining to the handling of human remains, and recommends as a fundamental measure that the handling of human remains should be kept to a minimum. Additionally, WHO recommends, particularly in the case of deaths caused by infectious diseases that remains should not be sprayed, washed or embalmed and that only trained personnel should handle remains during the outbreak. Personnel handling remains should wear personal protective equipment (gloves, gowns, apron, surgical masks and eye protection) and closed shoes.[31]

In the United States, there are no requirements for importation into the country if human remains consist entirely of: clean, dry bones or bone fragments or human hair; teeth; fingernails or toenails; and human remains that are cremated before entry into the United States. Human remains intended for interment or subsequent cremation after entry into the United States must be accompanied by a death certificate stating the cause of death. If the death certificate is in a language other than English, then it should be accompanied by an English language translation.

If the cause of death was a quarantinable communicable disease (i.e., cholera, diphtheria, infectious tuberculosis, plague, smallpox, yellow fever, viral hemorrhagic fevers, SARS, or pandemic influenza), the remains must meet the applicable standards and may be cleared, released, and authorized for entry into the United States only if: the remains are cremated; or the remains are properly embalmed and placed in a hermetically sealed casket; or the remains are accompanied by a permit issued by the Director of the Centre for Disease Control and Prevention (CDC). The CDC permit (if applicable) must accompany the human remains at all times during shipment. If the cause of death was anything other than a quarantinable communicable disease, then the remains may be cleared, released, and authorized for entry into the United States if: the remains meet the standards for applicable or properly embalmed and placed in a hermetically sealed casket, or are accompanied by a permit issued by the CDC Director); or the remains are shipped in a leak-proof container.

Federal quarantine regulations (42 CFR Part 71) state that the remains of a person who is known or suspected to have died from a quarantinable communicable disease may not be brought into the United States unless the remains are; properly embalmed and placed in a hermetically sealed casket, cremated, or accompanied by a permit issued by the CDC Director. Quarantinable communicable diseases include cholera; diphtheria, infectious tuberculosis; plague; smallpox, yellow fever; viral hemorrhagic fevers (Lassa, Marburg, Ebola, Congo-Crimean, or others not yet isolated or named); severe acute respiratory syndrome (SARS); and influenza caused by novel or re-emergent influenza viruses that are causing or have the potential to cause a pandemic. A CDC permit may be required when the remains are not embalmed or cremated, especially if the person is suspected or known to have died from a communicable disease.

Persons wishing to import human remains, including cremated remains, into the United States must obtain clearance from CDC's Division of Global Migration and Quarantine (DGMQ). Clearance can be obtained by presenting copies of the foreign

[31] Interim Infection Control Recommendations for Care of Patients with Suspected or Confirmed Filovirus (Ebola, Marburg) Haemorrhagic Fever, BDP/EPR/WHO, Geneva March 2008.

death certificate and if needed, a CDC/DGMQ permit to the CDC Quarantine Station with jurisdiction for the U.S. port of entry. A CDC/DGMQ permit may be needed to import human remains if the deceased is known or suspected to have died from a quarantinable communicable disease. A copy of the foreign death certificate and the CDC/DGMQ permit must accompany the human remains at all times during shipment. The foreign death certificate should state the cause of death and must be translated into English.

The basic principle that should apply to the handling of human remains must be consistent with the policy which currently applies in case of aircraft accident investigations, in that the country in which the death occurred must act contemporaneously and in close consultation with the country of nationality. This would obviate the case of the British tourist who died in Greece. The second principle should be that the principles of *ICAO Circular 285—AN/166* should be incorporated into Annex 9 along with a Standard in Chapter 4 that human remains should be accorded priority and dignity and that specially reduced rates should be promulgated by States on their airlines for this purpose. This Standard should be adopted in accordance with the basic philosophy of Article 44 (d) of the Chicago Convention which states that ICAO should strive to meet the needs of the people of the world for safe, regular, efficient and economical air transport.

Annex 9 should contain a separate Appendix for the carriage of human remains by air, which would lay down global principles for the handling, care and commitment that States could ensure. This Appendix should have a cross reference to Annex 18 and the *Technical Instructions* contained in Doc 9284[32] with appropriate linkages that ensure the harmonious application of both Annexes to this sensitive subject.

As for Annex 18, a study should be undertaken to determine as to when a cadaver or cremated remains would, if at all, become a dangerous good. The focus area would be both on the condition the human remains are at the point of acceptance for carriage, and the manner in which they are packaged. In the ultimate analysis, there has to be core global rules in place for this important area of air transportation. It cannot be left for individual States or airlines to decide.

Enhancing global civil aviation security and facilitation is one of ICAO's Strategic Objectives as adopted by the Council in May 2012. This is the first time facilitation has been mentioned in ICAO's strategic language and it should be a harbinger of new studies and new cooperation with the international community between ICAO and its member States o the carriage by air of human remains.

References

Baldwin SE (1910) The law of the airship. Am J Int Law 4(95):97

Dobson C, Payne R (1987) Appendix B: the chronology of terror: 1968–1987. In: War without end: the terrorists: an intelligence dossier. Sphere Books, London

[32] Dobson and Payne (1987), p. 53.

Article 12
Rules of the Air

Each contracting State undertakes to adopt measures to insure that every aircraft flying over or manoeuvring within its territory and that every aircraft carrying its nationality mark, wherever such aircraft may be, shall comply with the rules and regulations relating to the flight and manoeuvre of aircraft there in force. Each contracting State undertakes to keep its own regulations in these respects uniform, to the greatest possible extent, with those established from time to time under this Convention. Over the high seas, the rules in force shall be those established under this Convention.

Each contracting State undertakes to insure the prosecution of all persons violating the regulations applicable.

Contents

1 Mandatory Rules of Navigation

There are three facets to Article 12. The first reflects that aircraft flying over or manoeuvring within the territory of a State are required to comply with the rules and regulations relating to the flight and manoeuvre of aircraft there in force. The second is that each ICAO member State undertakes to keep its own regulations in these respects uniform, to the greatest possible extent, with those established from time to time under this Convention. The third is that over the high seas, the rules in force shall be those established under this Convention. Each Contracting State undertakes to insure the prosecution of all persons violating the regulations applicable. All these three aspects inextricably link themselves to the action of the Council in pursuance of the Chicago Convention with regard to rules of the air and this means that this provision is heavily reliant upon the Annex that would address the subject. This is Annex 2 (Rules of the Air), a discussion on which follows.

The ICAO Council on 20 June 1947 suggested that, pending the formal adoption of Standards and Recommended Practices under the terms of the Chicago Convention ICAO member States should continue to apply in their national civil aviation practices the recommendations for Standards and Recommended Practices which had been endorsed by the Interim Council of the Provisional International Civil

R. Abeyratne, *Convention on International Civil Aviation*,
DOI 10.1007/978-3-319-00068-8_13, © Springer International Publishing Switzerland 2014

Aviation Organization (PICAO).[1] The Interim Council of PICAO had, on 25 February 1946 adopted "Recommendations for Standards, Practices and Procedures—rules of the Air and Air Traffic Control "of which Part 1 contained Rules of the Air. These Recommendations were developed during the first session of the RAC Division[2] and submitted with a number of amendments to the Interim Council by the Air Navigation Committee in it twelfth Interim Report.[3] The Council requested that States also introduce into their national practices, in so far as they individually consider it advisable and appropriate, the Recommendations presented by Divisions in the Reports not yet formally acted upon by the Council.[4] The Council, on 15 April 1948 at the twenty second meeting of its third session considered the draft Standards and Recommended Practices for Rules of the Air[5] and decided that "Rules of the Air" and "Air Traffic Control" were sufficiently distinctive to be issued as separate Annexes, and adopted Standards and Recommended Practices for Rules of the Air and designated them as "Annex 2—Rules of the Air" to the Chicago Convention.

Another resolution of the Council was to the effect that States should, in complying with ICAO Standards which re of a regulatory character, to introduce the text of such standards into their national regulations, as nearly as possible, in the wording and arrangement employed by ICAO and that States be notified that

> the Rules of the Air Annex constitutes rules relating to the flight and manoeuvre of aircraft within the meaning of Article 12 of the Convention and deviations from these rules may not be made in so far as they relate to flight over the high seas.[6]

It must be noted that while States have the flexibility, by virtue of Article 38[7] of the Chicago Convention, not to adhere to any Standards in an Annex to the Convention, the only exception, as ruled by Council lies in Annex 2.

---

[1]Doc 6808-C/791, Proc. Of Council, 1st S., p. 34.

[2]Doc806-RAC/97 and 824-RAC/100.

[3]Doc 1360-AN/177.

[4]The First Assembly of ICAO, held earlier in May 1947, had expressed general satisfaction with the Recommendations for Standards, Practices and Procedures proposed by the Divisions, but decided to rename them henceforth *International Standards and Recommended Practices.* It then resolved that those "Standards" on which substantial agreement had been reached should be adopted by the Council as soon as possible.

[5]Doc 5300-AN/604.

[6]Doc 7310-C/846, Proc. Of Council, 3rd S., pp. 26 and 28.

[7]Article 38 provides: "Any State which finds it impracticable to comply in all respects with any such international standard or procedure, or to bring its own regulations or practices into full accord with any international standard or procedure after amendment of the latter, or which deems it necessary to adopt regulations or practices differing in any particular respect from those established by an international standard, shall give immediate notification to the International Civil Aviation Organization of the differences between its own practice and that established by the international standard. In the case of amendments to international standards, any State which does not make the appropriate amendments to its own regulations or practices shall give notice to the Council within sixty days of the adoption of the amendment to the international standard, or indicate the action which it proposes to take. In any such case, the Council shall make immediate

The Council, at the second meeting of its seventeenth session on 16 September 1952, in response to a request of the United States for interpretation of paragraph 3.1.2 of Annex 2 declared that:

> In applying paragraph 3.1.2 of the International Standards and Recommended Practices for Rules of the Air (Annex 2 to the Convention on International Civil Aviation), the "appropriate authority" in the case of flight over the high seas should be interpreted as the appropriate authority of the State of Registry and, in all other cases, as the appropriate authority of the State having sovereignty over the area over which the flight takes place.[8]

This was confirmed by the Council on 15 November 1972 at the sixth meeting of its seventy seventh session, when it adopted Amendment 14 to Annex 2.[9]

The ICAO Council, in adopting Annex 2 in April 1948 and subsequently in November 1951 when Amendment 1 to the Annex was adopted, resolved that the Annex constitutes *rules* relating to the flight and manoeuvre of aircraft within the meaning of Article 12 of the Convention. Therefore, the Council explicitly recognized that the rules in the Annex applied to the manoeuvre and operation of aircraft without exception. Annex 2, in its Foreword, states that the Standards in the Annex, together with the Standards and Recommended Practices of Annex 11, govern the application of the Procedures for Air Navigation Services Rules of the Air and Air Traffic Services, and the Regional Supplementary Procedures. The Regional Supplementary Procedures are subsidiary procedures of regional applicability. It is clear that by this introduction, there is established a distinct disparity between Annex 2 and Annex 11 where the provisions of the former remain unquestionably mandatory, and the provisions of the latter remain subject to Article 38 of the Chicago Convention and capable of being deviated from. However, it is clear that the purpose of Annex 11 is to ensure that flying on international routes is carried out under uniform conditions designed to improve the safety and efficiency of air operation and, therefore, provisions relating to air traffic control services, flight information services, and alerting services of Annex 11 when linked to the provisions of Annex 2, have a coercive effect that may in certain circumstances, transcend the parameters set in Article 38 of the Convention.

2 High Seas

The second issue in determining the legal status of rules of the air over the high seas in relation to sovereignty is the element of control exercised by a State over aircraft operations over the high seas. Article 2.1.2 of Annex 2 provides that a Contracting

notification to all other states of the difference which exists between one or more features of an international standard and the corresponding national practice of that State".

[8]Doc 7353-C/856, Action of the Council, 17th S., p. 26.

[9]Doc 9078-C/1012, Action of the Council, 77th S., p. 25.

State may deem to have accepted (unless ICAO is otherwise advised) that it provides air traffic services through an appropriate ATS Authority as designated to be responsible for providing air traffic services over parts of the high seas. An appropriate ATS Authority is defined in the foreword of the Annex as the relevant authority designated by the State responsible for providing air traffic services in the air space concerned. A Contracting party accepts an Appropriate ATS Authority pursuant to a regional air navigation agreement, which is an agreement approved by the Council of ICAO usually based on the outcome of the findings of Regional Air Navigation Meetings.

It is somewhat disconcerting that neither the legal status of the regional air navigation plan or agreement, nor its definition is clear. In November 1996, at the 38th meeting of the European Air Navigation Planning Group, it was recorded that an Air Navigation Plan consisted of an authoritative internationally agreed reference document, which corresponded to a contract between States covered by the Plan regarding air navigation facilities to be provided, to be approved by the ICAO Council in accordance with the provisions of the Chicago Convention.[10] It was deemed that the Council, in any given instance, would be acting on behalf of all Contracting States, including those not covered by the Plan. There is a marked dichotomy in the terminology used, which refers to the Plan on the one hand as a contract between parties and on the other hand as a reference document. Buergenthal offers a more coherent view, by saying that ICAO Annexes, Plans, SUPPS[11] and Regional Air Navigation Plans constitute an integral body of aviation legislation comparable both in structure and content to comprehensive domestic air navigation codes.[12] Yet another view is that the Regional Air Navigation Plan, not involving the process of ratification, signature or adoption, is a technical and operational document.[13] Confusion is further confounded by the fact that there is no direct consequence for any State which does not perform its obligations under a Regional Air Navigation Plan.

The provision of air navigation services are mainly regulated by three Annexes to the Chicago Convention, namely Annex 2 (Rules of the Air), Annex 3 (Meteorological Service for International Air Navigation) and Annex 11 (Air Traffic

[10]ICAO Doc. EANPG COG/2-WP/6, 12/03/1996 at 3.

[11]The ICAO Regional Supplementary Procedures (SUPPS) form the procedural part of the Air Navigation Plan developed by Regional Air Navigation (RAN) Meetings to meet those needs of specific areas which are not covered in the worldwide provisions. They complement the statement of requirements for facilities and services contained in the Air Navigation Plan publications. Procedures of worldwide applicability are included either in the Annexes to the Convention on International Civil Aviation as Standards or Recommended Practices, or in the Procedures for Air Navigation Services (PANS). See *ICAO Doc 7030.*

[12]Buergenthal (1969), at 121.

[13]Milde (2002) at 192.

Services).[14] Of these, compliance with Annex 2 is mandatory[15] and does not give the States the flexibility provided in Article 38 of the Chicago Convention to register differences from any provisions of the Annex.

With regard to navigation over the high seas, the United Nations Convention on the Law of the Sea *UNCLOS*, Article 39, lays down the duties of ships and aircraft involved in transit navigation to the effect that ships and aircraft, while exercising the right of transit passage, should : proceed without delay through or over the strait; refrain from any threat or use of force against the sovereignty, territorial integrity or political independence of States bordering the strait, or in any other manner in violation of the principles of international law embodied in the Charter of the United Nations; refrain from any activities other than those incident to their normal modes of continuous and expeditious transit unless rendered necessary by force majeure or by distress; and comply with the relevant provisions of the Convention. Article 39 (3) explicitly states that aircraft in transit passage shall observe the Rules of the Air established by ICAO as they apply to civil aircraft and that state aircraft will normally comply with such safety measures and will at all times operate with due regard for the safety of navigation. The provision further states that at all times aircraft shall monitor the radio frequency assigned by the competent internationally designated air traffic control authority or the appropriate international distress radio frequency.

Standard 2.1.1 of Annex 2 to the Chicago Convention provides that the rules of the air shall apply to aircraft bearing the nationality and registration marks of a Contracting State, wherever they may be, to the extent that they do not conflict with

---

[14]Article 54 (l) of the Chicago Convention stipulates as a mandatory function of the Council the act of adopting, in accordance with Chapter VI of the Convention, international standards and recommended practices (SARPs) and for convenience designate them as Annexes to the Convention. Article 37 of the Convention reflects the areas in which SARPs should be developed and Annexes formed. Article 38 obliges contracting States to notify ICAO of any differences between their own regulations and practices and those established by international standards or procedures. The notification of differences however, does not absolve States from their continuing obligation under Article 37 to collaborate in securing the highest practicable degree of uniformity in international regulations, standards, and procedures.

[15]In October 1945, the Rules of the Air and Air Traffic Control (RAC) Division at its first session made recommendations for Standards, Practices and Procedures for the Rules of the Air. These were reviewed by the then Air Navigation Committee and approved by the Council on 25 February 1946. They were published as *Recommendations for Standards, Practices and Procedures—Rules of the Air* in the first part of Doc 2010, published in February 1946. The RAC Division, at its second session in December 1946–January 1947, reviewed Doc 2010 and proposed Standards and Recommended Practices for the Rules of the Air. These were adopted by the Council as Standards and Recommended Practices relating to Rules of the Air on 15 April 1948, pursuant to Article 37 of the Convention on International Civil Aviation (Chicago, 1944) and designated as Annex 2 to the Convention with the title *International Standards and Recommended Practices—Rules of the Air*. They became effective on 15 September 1948. On 27 November 1951, the Council adopted a complete new text of the Annex, which no longer contained Recommended Practices. The Standards of the amended Annex 2 (Amendment 1) became effective on 1 April 1952 and applicable on 1 September 1952.

the rules published by the State having jurisdiction over the territory over-flown.[16] The operation of an aircraft either in flight or on the movement area of an aerodrome shall be in compliance with the general rules and, in addition, when in flight, either with: visual flight rules (VFR); or the instrument flight rules (IFR).[17] Standard 2.3.1 further provides that the pilot-in-command of an aircraft shall, whether manipulating the controls or not, be responsible for the operation of the aircraft in accordance with the rules of the air, except that the pilot-in-command may depart from these rules in circumstances that render such departure absolutely necessary in the interests of safety.

References

Buergenthal T (1969) Law making in the international civil aviation organization. Syracuse University Press, Syracuse

Milde M (2002) Legal aspects of airports constructed in the sea. In: Milde M, Khadjavi H (eds) Public international air law, vol 2. McGill University Faculty of Law, Montreal

[16]The Council of the International Civil Aviation Organization resolved, in adopting Annex 2 in April 1948 and Amendment 1 to the said Annex in November 1951, that the Annex constitutes Rules relating to the flight and manoeuvre of aircraft within the meaning of Article 12 of the Convention. Over the high seas, therefore, these rules apply without exception.

[17]Information relevant to the services provided to aircraft operating in accordance with both visual flight rules and instrument flight rules in the seven ATS airspace classes is contained in 2.6.1 and 2.6.3 of Annex 11. A pilot may elect to fly in accordance with instrument flight rules in visual meteorological conditions or may be required to do so by the appropriate ATS authority.

Article 13
Entry and Clearance Regulations

The laws and regulations of a contracting State as to the admission to or departure from its territory of passengers, crew or cargo of aircraft, such as regulations relating to entry, clearance, immigration, passports, customs, and quarantine shall be complied with by or on behalf of such passengers, crew or cargo upon entrance into or departure from, or while within the territory of that State.

Contents

1 Control of Air Transport

This article forms the basis of ICAO's facilitation programme. The Assembly, at its 37th Session (Montreal, September/October 2010) adopted Resolution A37-20 (Consolidated statement of continuing ICAO policies in the air transport field) Appendix D of which was on development and implantation of facilitation provisions. The Resolution stated:

> *Whereas* Annex 9 — Facilitation, was developed as a means of articulating the obligations of Contracting States under Articles 22, 23 and 24 of the Convention and standardizing procedures for meeting the legal requirements referred to in Articles 10, 13, 14, 29 and 35;
>
> *Whereas* implementation of the Standards and Recommended Practices in Annex 9 is essential to facilitate the clearance of aircraft, passengers and their baggage, cargo and mail and manage challenges in border controls and airport processes so as to maintain the efficiency of air transport operations;
>
> *Whereas* it is essential that Contracting States continue to pursue the objective of maximizing efficiency and security in such clearance operations;
>
> *Whereas* the Convention on the Rights of Persons with Disabilities and its Optional Protocol, that had been adopted in December 2006 by the United Nations General Assembly, entered into force on 3 May 2008;

R. Abeyratne, *Convention on International Civil Aviation*,
DOI 10.1007/978-3-319-00068-8_14, © Springer International Publishing Switzerland 2014

Whereas the development of specifications for machine readable travel documents by the Organization has proved effective in the development of systems that expedite the movement of international passengers and crew members through clearance control at airports while enhancing immigration compliance programmes; and

Whereas the development of a set of standard signs to facilitate the efficient use of airport terminals by travellers and other users has proved effective and beneficial;

The Assembly:

1. *Urges* Contracting States to give special attention to increasing their efforts to implement Annex 9 Standards and Recommended Practices;

2. Requests the Council to ensure that Annex 9 — *Facilitation,* is current and addresses the contemporary requirements of Contracting States with respect to administration of border controls, cargo and passengers, the protection of passenger and crew health and the accessibility to air transport by persons with disabilities;

3. Requests the Council to ensure that the provisions of Annex 9 — *Facilitation,* and Annex 17 — *Security,* are compatible with and complementary to each other;

4. *Requests* the Council to ensure that its specifications and guidance material in Doc 9303, *Machine Readable Travel Documents,* remain up to date in the light of technological advances and to continue to explore technological solutions aimed at improving clearance procedures; and

5. *Requests* the Council to ensure that Doc 9636, *International Signs to Provide Guidance to Persons at Airports and Marine Terminals,* is current and responsive to the requirements of Contracting States[1]

On machine readable travel documents (particularly passports and visas) the Resolution's Appendix D goes on to say the following:

Whereas the passport is the basic official document that denotes a person's identity and citizenship and is intended to inform the State of transit or destination that the bearer can return to the State which issued the passport;

Whereas international confidence in the integrity of the passport is essential to the functioning of the international travel system;

Whereas the veracity and validity of machine readable travel documents (MRTDs) depends on the documentation used to establish identity, confirm citizenship or nationality and assess entitlement of the passport applicant (i.e. "breeder" documentation);

Whereas Member States of the United Nations have resolved, under the Global Counter-Terrorism Strategy adopted on 8 September 2006, to step up efforts and cooperation at every level, as appropriate, to improve the security of manufacturing and issuing identity and travel documents and to prevent and detect their alteration or fraudulent use;

Whereas Resolution 1373 adopted by the United Nations Security Council on 28 September 2001 decided that all States shall prevent the movement of terrorists or terrorist groups by effective border controls and controls on issuance of identity papers and travel documents, and through measures for preventing counterfeiting, forgery or fraudulent use of identity papers and travel documents;

Whereas high-level cooperation among States is required in order to strengthen resistance to passport fraud, including the forgery or counterfeiting of passports, the use of forged or counterfeit passports, the use of valid passports by impostors, the use of expired or revoked passports, and the use of fraudulently obtained passports;

Whereas the use of stolen blank passports, by those attempting to enter a country under a false identity, is increasing worldwide; and

Whereas ICAO provides assistance to States in all matters related to MRTDs including project planning, implementation, education, training and system evaluation services, and

[1]ICAO Doc 9958, Assembly Resolutions in Force (as of October 2010) at III-9 to III-10.

has set up the Public Key Directory (PKD) to strengthen the security of biometrically-enhanced MRPs (ePassports);

The Assembly:

1. *Urges* Contracting States to intensify their efforts to safeguard the security and integrity of the breeder documentation;

2. *Urges* Contracting States to intensify their efforts to safeguard the security and integrity of their passports, to protect their passports against passport fraud, and to assist one another in these matters;

3. *Urges* those Contracting States that have not already done so, to issue machine readable passports in accordance with the specifications of Doc 9303, Part 1;

4. *Urges* Contracting States to ensure that the expiration date of non-machine readable passports falls before 24 November 2015;

5. *Urges* those Contracting States requiring assistance in implementing MRTD standards and specifications to contact ICAO without delay;

6. *Requests* the Council to take appropriate measures to establish guidance on breeder documentation;

7. *Requests* the Council to continue the work on enhancing the effectiveness of controls on passport fraud by implementing the related SARPs of Annex 9 and developing guidance material to assist Contracting States in maintaining the integrity and security of their passports and other travel documents;

8. *Urges* those States issuing ePassports to join the ICAO PKD; and all receiving States to verify the digital signatures associated with the passports; and

9. *Urges* those Contracting States that are not already doing so to provide routine and timely submissions of lost and stolen passport data to Interpol's Automated Search Facility/Stolen and Lost Travel Document Database.

On national and international cooperation on facilitation matters the Resolution states:

Whereas there is a need for continuing action by Contracting States to improve the effectiveness and efficiency of clearance control formalities;

Whereas the establishment and active operation of national facilitation committees is a proven means of effecting needed improvements;

Whereas cooperation on facilitation matters amongst Contracting States and with the various national and international parties interested in facilitation matters has brought benefits to all concerned; and

Whereas such cooperation has become vital in the light of the proliferation of non-uniform passenger data exchange systems that adversely affect the viability of the air transport industry;

The Assembly:

1. *Urges* Contracting States to establish and utilize national facilitation committees and adopt policies of cooperation on a regional basis among neighbouring States;

2. *Urges* Contracting States to participate in regional and subregional facilitation programmes of other intergovernmental aviation organizations;

3. *Urges* Contracting States to take all necessary steps, through national facilitation committees or other appropriate means, for:

a) regularly calling the attention of all interested departments of their governments to the need for:

1) making the national regulations and practices conform to the provisions and intent of Annex 9; and

2) working out satisfactory solutions for day-to-day problems in the facilitation field; and

b) taking the initiative in any follow-up action required;

4. *Urges* Contracting States to encourage the study of facilitation problems by their national and other facilitation committees and to coordinate the findings of their committees on facilitation problems with those of other Contracting States with which they have air links;

5. *Urges* neighbouring and bordering States to consult one another about common problems that they may have in the facilitation field whenever it appears that these consultations may lead to a uniform solution of such problems;

6. *Urges* Contracting States to encourage their aircraft operators to continue to cooperate intensively with their governments as regards:

a) identification and solution of facilitation problems; and

b) developing cooperative arrangements for the prevention of illicit narcotics trafficking, illegal immigration and other threats to national interests;

7. *Urges* Contracting States to call upon international operators and their associations to participate to the extent possible in electronic data interchange systems in order to achieve maximum efficiency levels in the processing of passenger and cargo traffic at international terminals;

8. *Urges* Contracting States, in their use of electronic data interchange systems, to ensure that their passenger data requirements conform to international standards adopted by relevant United Nations agencies for this purpose; and

9. *Urges* States and operators, in cooperation with interested international organizations, to make all possible efforts to speed up the handling and clearance of air cargo, while ensuring the security of the international supply chain.

Illegality or irregularity in migratory movements, although generally viewed from the perspective of the destination country, may occur in either one of the countries of origin, transit or destination or all of them. The country of origin becomes involved when a person leaves that country without a valid passport or an equivalent document of identity as required by national legislation. As regards the country through which an illegal migrant travels—or the transit country, its laws may be infringed upon if the person being smuggled did not have a transit visa to travel through that country. From the perspective of the country of destination, the main characteristics of illegal or irregular migration are more readily discernible. They are:

- The arrival by a person in a country and attempt thereafter to enter without compliance with required formalities or without authorization required by law for admission or stay in that country; or
- The cessation by a person of meeting conditions to which that person's stay or activity is subject.

Human migration is a compelling social indicator of the future and is driven by economic and social factors. Economically motivated migration characteristically takes place between countries which show a marked disparity in labour market situations (typically when one country shows shortage of labour and the other shows a surplus of labour) and sharply different levels of income.

The magnitude of the problem naturally spawns opportunities for profit making which in turn have given rise to the multi billion dollar human smuggling industry. In 1995, the Economist recorded:

> Around the world, smuggling organizations ranging in size and degree of sophistication are smuggling tens of thousands of people from poorer to richer countries. In the process, they are earning at least $7 billion per year, with the potential for even greater future profits.

The risk element involved in the smuggling of humans is far less than the smuggling of narcotics and therefore some gangs of notoriety have reportedly abandoned the drug trade in favour of smuggling humans. Human smuggling, to be defined as such, has to be composed of two different kinds of activity: the exchange of money or other form of payment between the would-be illegal migrant and smuggler; and the arrangement by a "facilitator" of passage across an international border. The movement has to be voluntary and the passage from one border to another should be illegal.

Smuggling is a popular recourse and an easy "way out" to the illegal immigrant who faces huge distances to travel; difficulties imposed by States' anti-immigrant restrictions; and difficulties adjusting to life in the host country.

Ironically, the last reason—difficulty in adjusting to life in the host country—works to the advantage of the smuggler. The process of payment, which is the essential part of the transaction of smuggling, is not always concluded with the act of smuggling. It may extend well beyond that point. The form and method of payment often reach insidious proportions of exploitation where the smuggler may discourage payment at the outset, making the illegal migrant an easy victim to exploitation and abuse. In many cases, the person seeking to be smuggled into a country illegally welcomes such deferral of payment making the smugglers' business flourish through easier recruitment of candidates for illegal migration. This situation also strengthens the position of the smuggler to tighten his grip on the migrants and use their virtually indentured services for unlawful and criminal activities.

A 1995 study on smuggled women carried out in Belgium suggests that although most women were not required to pay the smuggler in advance, a high proportion of them had found themselves on arrival to be required to perform various services to the smuggling network. The Vienna-based International Centre for Migration Policy Development (ICMPD) has revealed that 15–30 % of those who illegally went into Western Europe for purposes of employment or residence, estimated in 1993 to be a figure of 250,000–300,000 persons, used the services of smugglers.

2 United Nations Initiatives

Concerned that the activities of criminal organizations that profit illicitly by smuggling human beings were becoming a threat to the world community and recognizing that international criminal groups often convince individuals to migrate illegally by various means for enormous profit, the United Nations, at its 48th General Assembly in December 1993, adopted Resolution 48/102 on the prevention of smuggling aliens. This Resolution makes mention of smuggling of illegal migrants as an activity that endangers the lives of those smuggled and imposes severe costs on the international community. The United Nations noted that smuggling of aliens can involve criminal elements in many States and condemned the practice of smuggling aliens in violation of national and international law. Resolution 48/102, while urging States to adopt measures to frustrate the objectives and activities of smugglers of illegal migrants, identifies the International Civil Aviation Organization (ICAO) as one of the specialized agencies of the United Nations that could consider ways and means to enhance international co-operation to combat the smuggling of aliens. At the same meeting, the General Assembly of the United Nations adopted Resolution 48/103 on crime prevention and international justice which reaffirmed the importance of the United Nations crime prevention and criminal justice programme and the crucial role the Organization has to play in promoting international co-operation in crime prevention and criminal justice. Resolution 48/103, *inter alia*, invites governments to lend their full support to the United Nations crime prevention and criminal justice programme.

In a subsequent 1997 resolution the United Nations General Assembly, at its fifty-first session, *inter alia*, recognized that international criminal groups often convinced individuals to migrate illegally by various means for enormous profits and that socio-economic factors influenced the problem of the smuggling of aliens and also contributed to the complexity of current international migration. The Assembly requested States to cooperate bilaterally and on a multilateral basis to prevent the use of fraudulent documents and reaffirmed the need to fully observe international and national law in dealing with the smuggling of aliens, including the provision of humane treatment and strict observance of all human rights of migrants.

Following up on its early initiatives, the United Nations General Assembly, at its 53rd Session, held in December 1998, adopted Resolution 53/111 establishing an open-ended inter-governmental *ad hoc* committee for the purpose of elaborating a comprehensive international convention against transnational organized crime. This Resolution gave rise to United Nations Convention against Transnational Organized Crime (which was still in draft form at the time of writing). The purpose of the Convention is to promote co-operation to prevent and combat transnational organized crime more effectively.

The Convention, by Article 3, requires Contracting States to establish as criminal offences, when committed internationally, the act of agreeing with one or more persons to commit a serious crime for any purpose relating directly or indirectly to the obtaining of financial or other material benefit and conducted by a person with knowledge which is tantamount to participating in criminal activities of an organized criminal group.

The Convention also considers organizing, aiding, abetting, facilitating or counselling the Commission of serious crime involving an organized criminal group to be a criminal offence. The Convention attributes to any person who aids and abets a crime, knowledge, intent, aim and purpose to commit such crime as provided for in Article 3, which makes the conduct by a person who knowingly aids and abets a criminal or criminal organizations a crime. The provision does not impute liability to any person who aids and abets a crime if that person ought to have known that his conduct would aid and abet a crime. In other words, any person who commits a crime under the Convention should essentially have the knowledge that his conduct would facilitate, aid or abet a crime.

The Protocol which supplements the Convention (which was also in draft form at the time of writing) specifically addresses the smuggling of migrants by land, air and sea and records *in limine*, the concern of the States Parties to the Protocol that the smuggling of migrants may lead to the misuse of established procedures for immigration, including those for seeking asylum.

The Protocol defines the smuggling of migrants as follows:

> "Smuggling of migrants" shall mean the procurement of the illegal entry into or illegal residence of a person in [a] [any] State Party of which the person is not a national or a permanent resident in order to obtain directly or indirectly, a financial or other benefit.

The Protocol defines illegal entry as crossing of borders without complying with the necessary requirements for legal entry into the receiving State.

The purposes of the Protocol are to prevent, investigate and prosecute the smuggling of migrants, when involving an organized criminal group, as defined in the Convention and to promote international cooperation to meet these objectives. The Protocol excludes the prosecution of persons who are smuggled, thus exclusively applying only to those responsible, directly or indirectly, in carrying out the act of smuggling.

Article 4 of the Protocol requires States Parties to enact domestic legislation that would criminalize, *inter alia*, the act of producing a fraudulent travel or identity document. This would also mean that a person who aids and abets such an act would be deemed to be criminally liable under the Convention. Article 12 of the Protocol provides that State Parties shall adopt such measures as may be necessary, in accordance with available means, to ensure that travel or identity documents issued by them are of such quality that they cannot easily be misused and cannot readily be unlawfully altered, replicated, falsified or issued. The provision also calls for States Parties to ensure the integrity and security of travel or identity documents issued by or on behalf of the States Parties and to prevent their unlawful creation, issuance and use.

3 Work of ICAO

The International Civil Aviation Organization has been making sustained efforts at adopting technical specifications for machine readable travel documents which are aimed at making illegal migration more difficult and facilitating air transport. Specifications for the machine readable passport and visa are already published and Sri Lanka is one of the countries which produce machine readable passports and actively participates in ICAO meetings.

4 The Machine Readable Passport (MRP)

The machine readable passport (MRP) is a passport that has both a machine readable zone and a visual zone in the page that has descriptive details of the owner. The machine readable zone enables rapid machine clearance, quick verification and instantaneous recording of personal data. Besides these advantages, the

MRP also has decided security benefits, such as the possibility of matching very quickly the identity of the MRP owner against the identities of undesirable persons, whilst at the same time, offering strong safeguards against alteration, counterfeit or forgery. Another advantage of the MRP is the fact that the document obviates the need for the passenger to lodge embarkation or disembarkation cards, on the assumption that countries installing automatic reader equipment would accept the data on the passport as sufficient for their clearance purposes. Of course, the MRP had to offer safeguards equal to or better than those of conventional passports and satisfy those control requirements already set by conventional passports and other travel documents in use throughout the world. Also, since it was only natural that a certain number of States would not have wished to issue the MRP or adopt new procedures related thereto, it was expected that a machine readable system and conventional passport procedures would operate side by side for some time.

Although the MRP may be produced and used as a single and separate card, it has to take booklet-form since most States still insist on entrance visas, which have to be accommodated in the passport. The MRP's dimensions are smaller than those of most traditional passports, its overall dimensions being 88.75 mm × 125.75 mm. The page has two areas—with the visual-inspection zone on top of the page as the first area, and the machine-readable zone at the bottom, as the second area. The visual inspection zone contains the photograph and personal data of the owner. At the bottom—in the machine readable zone, are prescribed data elements printed in machine readable form in a prescribed sequence and position. When being used, the MRP is opened at the page containing the visual-inspection and machine-readable zones and placed face down on a glass surface in the reading machine, thus activating an electro optical-scanning mechanism. The mechanism illustrates the two machine readable lines and surrounding background using a light source. The whole process operates on a principle of "light absorption" where the mechanism in the reader uses an optical sensor to measure the presence or absence of light reflecting off the page. The cumulative efforts of this "imagery" process and a computer installed in the reader then produces on the screen of the reader, all the information that the inspecting officer requires, such as the passport number, date of expiry of the passport, name of the issuing State, passport-owner's name, nationality, sex, date of birth, and optionally, national ID number. The computer also interrogates simultaneously, a data base containing a list of persons considered undesirable by the State of entry and the results thereof are displayed on the screen momentarily, enabling the inspecting officer to decide whether the bearer of the passport can be admitted to the country. This process takes a mere 10 s, and adds a tremendous impetus to the facilitation efforts of ICAO.

ICAO recommends the use of the MRP by all States, even if meagre traffic flows may not justify the use of reading equipment. The first MRP was issued in the United States of America in 1981 and since then, well over thirty five million MRPs have been issued world-wide, while millions more are being issued every year in countries such as Canada, Australia, Germany. Technical specifications for MRP's were first published by ICAO in the First Edition of Doc 9303 in 1980, based on Standard 7501 of the International Organization for Standardization (ISO).

Developments in technology and modern exigencies of prolific air travel however, dictated that the specifications contained in Doc 9303 be improved. As the ICAO Panel on Passport Cards had been extinct with the publication of its Fifth Report in 1978, the International Organization for Standardization (ISO) took it upon itself to establish a working group to update the provisions of Doc 9303 so that the outcome of their deliberations would be published as a new ISO Standard. The care taken by the ISO working group led to considerable time being taken by the working group, compelling ICAO to establish a new group to succeed its Panel on Passport Cards, namely, the Technical Advisory Group on Machine Readable Passports (TAG/MRP) which met for the first time in 1986. ICAO thus regained its lead in developing MRP specifications and co-ordinating with other organizations the task of developing a single set of specifications for machine readable travel documents. On the strength of ICAO's new leading role, the Air Transport Committee of ICAO widened the scope of the terms of reference of ICAO to include the development of specifications for machine readable visas and to provide for the Group's membership to include participation by the authorities responsible for visas. Accordingly, the ISO Technical Committee, in the light of ICAO's new role in updating specifications in Doc 9303, adopted a proposal to withdraw ISO Standard 7501, so that there would not be confusion by the introduction of double standards or "overlapping" of specifications of ICAO and ISO.

ICAO's terms of reference in the development of specifications for MRPs stem from the Chicago Convention itself which provides for ICAO's adoption of international Standards and Recommended Practices dealing inter alia with customs and immigration procedures. It is interesting that passports apply to other modes of international travel as well, and the fact that ICAO has been singly designated to adopt specifications speaks for the uniqueness of its facilitation programme.

5 The Machine Readable Visa

With the terms of reference of ICAO having been expanded to cover the development of machine readable visas (MRV), the Technical Advisory Group changed its name to read as the Technical Advisory Group on Machine Readable Travel Documents (TAG/MRTD), and, in 1992 released specifications relating to a machine readable visa. Since the visa is inextricably linked to the passport, both the MRP and the MRV go side by side, and are not considered in isolation. As long as visas were required and needed to be attached to the passport, it was difficult to envisage the development of a Passport Card, and a booklet type passport was required for visual inspection. The MRV therefore is the main reason for the retention of the MRP concept in its present form. Since MRVs can be accommodated in any type of passport (whether machine-readable or conventional) the placement of an MRV in a conventional, non machine readable passport would make that passport machine readable, worldwide.

6 Official Travel Documents

The Technical Advisory Group on Machine Readable Travel Documents has also developed the Size 1 and Size 2 Official Travel Documents (TD-1) which are cards conforming with Specification 7810 of the International Standards Organization (ISO) and is designed to be read by machine similar to the machine readable passport and visa. The specifications stipulate standards for official documents of identity which can be used for travel purposes in terms of acceptance of a person by a receiving State. In the same year during which the official travel document specifications were released by ICAO—1996—the Organization released specifications for machine readability of the crew member certificate.

At its eleventh meeting held in September 1999, the Technical Advisory Group on Machine Readable Travel Documents reiterated its support for continued work to develop a set of indicative, probably short term, test methods that could emulate failure modes commonly found in travel documents. The Group also approved in principle that the future direction of the Group's work should include inter alia, the development of specifications for an electronic visa; an integrated automated border clearance system; a survey of user requirements and current applicability of machine readable travel documents; and specifications for a logical record format for use with optional capacity exPANsion technologies.

The United Nations is forging ahead with preventive measures against the smuggling of illegal migrants through an ongoing effort. The rapid development of information technology has placed in the hands of States such sophisticated security tools as the machine readable travel document. A noteworthy corollary to this trend is that airlines will now be required to exercise more vigilance in the future, particularly with the introduction of the excellent initiative of the International Air Transport Association (IATA) for Simplified Passenger Travel (SPT) which has now gained momentum. The SPT concept, which is calculated to be essentially a tool for facilitating air travel, uses a smart card which confirms a traveller's identity through trip related information and biometric data which is encoded. The check-in takes less than a minute with the SPT card. Reportedly, a number of airlines are already well into the process of developing smart card technology. This could only mean that such a process, when developed by some, would have a coercive effect on other airlines which are able to follow suit. Failure to follow such industry practices may have negative implications on an air carrier's security record and may result in uncalled for legal liability.

6.1 The ePassport

Over 104 States are currently producing and using ePassports and there are approximately four hundred million in circulation. This accounts for 33 % of all passports used globally. The additional feature that the ePassport carries in the conventional machine readable passport is a chip containing biometric and biographic information which have to be validated accurately, efficiently and quickly while retaining the security and integrity of the information. Ideally, an ePassport should be issued in accordance with the technical specifications approved by ICAO. However, this does not happen in all cases of issuance of ePassports. This lapse could seriously compromise global security. The nuances of this threat are described and discussed in this article against their legal background.

At a Symposium on machine readable travel documents, biometrics and security standards held at ICAO on 10–12 October 2012, experts addressed ICAO machine readable travel documents (MRTD) standards and specifications, identity management best practices and related border security issues. Foremost among these discussions was the ePassport, which is defined by ICAO as a passport which has a contactless integrated circuit (IC) chip within which is stored data from the machine readable passport page, a biometric measure of the passport and a security object to protect the public key infrastructure (PKI) cryptographic technology, and which conforms to the specifications of Doc 9303 part 1.[2] The ICAO Facilitation Manual defines the ePassport as a machine readable passport that has a contactless integrated circuit embedded in it and the capability of being used for biometric identification of the machine readable passport holder in accordance with the Standards specified in the relevant part of ICAO document 9303 (Machine Readable Travel Documents).[3] ePassports are easily recognised by the international ePassport symbol on the front cover.[4]

6.2 Biometric Identification

It is important to note that the operative terms in the definition of the ePassport are "biometric identification" and "public key infrastructure (PKI) cryptographic technology". Biometric technology involves a measurable, physical characteristic or

[2]Machine Readable Travel Documents Part 1 Volume 2 ICAO Doc 9303 Sixth Edition: 2006, at Page II-3 at Paragraph 6.1, Definitions.

[3]See The Facilitation Manual, Doc 9957, ICAO: Montreal, First Edition 2011, Definitions at X. ICAO has been working on the development of passports since 1968. The Seventh Session of the ICAO Facilitation Division in 1968 recommended that a small panel of qualified experts including representatives of the passports and/or other border control authorities, be established: to determine the establishment of an appropriate document such as a passport card, a normal passport or an identity document with electronically or mechanically readable inscriptions that meet the requirements of document control; the best type of procedures, systems (electronic or mechanical) and equipment for use with the above documents that are within the resources and ability of Member States; the feasibility of standardizing the requisite control information and methods of providing this information through automated processes, provided that these processes would meet the requirements of security, speed of handling and economy of operation.

[4]http://www.dhs.gov/xtrvlsec/programs/content_multi_image_0021.shtm.

personal behavioral trait used to recognize the identity, or verify[5] the claimed identity of a person. Biometric identification has been defined as

> a generic term used to describe automated means of recognizing a living person through the measurement of distinguishing physiological or behavioural traits.[6]

Biometrics target the distinguishing physiological or behavioral traits of the individual by measuring them and placing them in an automated repository such as machine encoded representations created by computer software algorithms that could make comparisons with the actual features. Physiological biometrics that have been found to successfully accommodate this scientific process are facial recognition, fingerprinting and iris-recognition which have been selected by ICAO as being the most appropriate. The biometric identification process is four-fold: firstly involving the capture or acquisition of the biometric sample; secondly extracting or converting the raw biometric sample obtained into an intermediate form; and thirdly creating templates of the intermediate data is converted into a template for storage; and finally the comparison stage where the information offered by the travel document with that which is stored in the reference template.

Biometric identification gets into gear each time an MRTD holder (traveler) enters or exists the territory[7] of a State and when the State verifies his identity against the images or templates created at the time his travel document was issued. This measure not only ensures that the holder of the document is the legitimate claimant to that document and to whom it was issued, but also enhances the efficacy of any advance passenger information (API)[8] system used by the State to pre-determine the arrivals to its territory. Furthermore, matching biometric data presented in the form of the traveler with the data contained in the template accurately ascertains as to whether the travel document has been tampered with or not. A three way check, which matches the traveler's biometrics with those stored in the template carried in the document and a central database, is an even more efficacious

[5]To "verify" means to perform a one-to-one match between proffered biometric data obtained from the holder of the travel document at the time of inquiry with the details of a biometric template created when the holder enrolled in the system.

[6]Machine Readable Travel Documents Part 1 Volume 2, Preamble (*supra* note 3) at Page II-3 at Paragraph 4.1.

[7]The Chicago Convention, Preamble (*supra* note 1), defines, in Article 2, "territory of a State" as the land areas and territorial waters adjacent to the State under the sovereignty, suzerainty, protection and mandate of such State.

[8]API involves exchange of data information between airlines and customs authorities, where an incoming passenger's essential details are notified electronically by the airline carrying that passenger prior to his arrival. The data for API would be stored in the passenger's machine readable passport, in its machine readable zone. This process enables customs authorities to process passengers quickly, thus ensuring a smoother and faster clearance at the customs barriers at airports. One of the drawbacks of this system, which generally works well and has proven to be effective, is that it is quite demanding in terms of the high level of accuracy required. One of the major advantages, on the other hand, is the potential carried by the API process in enhancing aviation security at airports and during flight. See Abeyratne (2002a).

way of determining the genuineness of a travel document. The final and most efficient biometric check is when a four way determine is effected, were the digitized photograph is visually matched (non electronically) with the three way check described above.[9] In this context, it is always recommended that the traveler's facial image (conventional photograph) should be incorporated in the travel document along with the biometric templates in order to ensure that his identity could be verified at locations where there is no direct access to a central database or where the biometric identification process has not entered into the legal process of that location.

6.3 Public Key Infrastructure (PKI) Cryptographic Technology

PKI Cryptographic technology uses a brand new technique known as quantum cryptography, designed to eliminate the terrifying vulnerabilities that arise in the way digitally stored data are exposed to fraudulent use. This new technique uses polarized photons instead of electronic signals to transmit information along cables. Photons are tiny particles of light that are so sensitive that when intercepted, they immediately become corrupted. This renders the message unintelligible and alerts both the sender and recipient to the fraudulent or spying attempt. The public key directory—designed and proposed to be used by customs and immigration authorities who check biometric details in an electronic passport, is based on cryptography—and is already a viable tool being actively considered by the aviation community as a fail-safe method for ensuring the accuracy and integrity of passport information.

In order to assure inspecting authorities (receiving States) that they would know when the authenticity and integrity of the biometric data stored in the MRTD, which they inspect, are compromised and tampered with, the Public Key Infrastructure (PKI) scheme was developed by the TAG/MRTD, which has been pioneering work on the MRTD for over a decade.[10] The scheme is not calculated to prescribe global implementation of public key encryption, but rather acts as a facilitator enabling States to make choices in areas such as active or passive authentication,

[9]Issuing States must ensure the accuracy of the biometric matching technology used and functions of the systems employed if the integrity of the conducted checks are to be maintained. They must also have realistic and efficient criteria regarding the number of travel documents checked per minute in a border control situation and follow a regular biometric identification approach such as facial recognition, fingerprint examination or iris identification system.

[10]ICAO's terms of reference in the development of specifications for machine readable passports stem from the Chicago Convention which provides for ICAO's adoption of international Standards and Recommended Practices dealing, *inter alia,* with customs and immigration procedures. Chicago Convention, Preamble (*supra* note 2), Article 37(j). It is interesting that, although passports apply to other modes of international travel as well, ICAO has been singly recognized as the appropriate body to adopt specifications for MRTDs. This alone speaks for the uniqueness of ICAO's facilitation programme. See Machine Readable Travel Documents, *ICAO Doc 9303/6* Sixth Edition 2006, 1-1 to 1-3.

anti-skimming and access control and automated border crossing, among other facilitative methods. The establishment of a public key directory, through means of public key cryptology and in a PKI environment, is consistent with ICAO's ultimate aim and vision for the application of biometric technology on the fundamental postulate that there must be a primary interoperable form of biometric technology for use at border control with facilities for verification, as well as by carriers and the issuers of documents. This initial premise is inevitably followed by the assumption that biometric technologies used by document issuers must have certain specifications, particularly for purposes of identification, verification and the creation of watch lists. It is also ICAO's vision that States, to the extent possible, are protected against changing infrastructure and changing suppliers, and that a technology, once put in place, must be operable or at least retrievable for a period of 10 years.

6.4 Features and Purpose of the ePassport

The story of the passport—the precursor of the ePassport—starts with the birth of an individual and his birth certificate, which records the event of birth and time and place thereof. The Civil Registry is able, with this document to primarily establish the identity of the person at birth and inform his country of his details for purposes of maintaining census and vital statistics. The passport, which uses this information, gives a person a name and nationality that is required for him to travel internationally. The passport is a basic document in the transport by air of persons. Its use therefore is of fundamental importance as a travel document, not only because it reflects the importance of the sovereignty of a State and the nationality of its citizens but also because it stands for the inviolability of relations between States that are linked through air transport.

The key consideration of an ePassport is *Global Interoperability*—the crucial need to specify a system for biometrics deployment that is universally interoperable. a Logical Data Structure (LDS) for ePassports required for global interoperability. It defines the specifications for the standardized organization of data recorded to a contactless integrated circuit capacity exPANsion technology of an MRP when selected by an issuing State or organization so that the data is accessible by receiving States. This requires the identification of all mandatory and optional Data Elements and a prescriptive ordering and/or grouping of Data Elements that must be followed to achieve global interoperability for reading of details (Data Elements) recorded in the capacity exPANsion technology optionally included on an MRP (ePassport). The other considerations are *Uniformity*—the need to minimize via specific standard setting, to the extent practical, the different solution variations that may potentially be deployed by member States; *Technical reliability*—the need to provide guidelines and parameters to ensure member States deploy technologies that have been proven to provide a high level of confidence from an identity confirmation viewpoint; and that States reading data encoded by other States can be sure that the data supplied to them is of sufficient quality and integrity to enable accurate verification in their own systems; *Practicality*—the need to ensure that specifications can be operationalized and implemented by States without their having to introduce a plethora of disparate systems and equipment to ensure they meet all possible variations and interpretations of the standards; and *Durability*—the requirement that the systems introduced will last the maximum 10-year life of a travel document, and that future updates will be backward compatible.

The major components of a biometric system are: *Capture*—acquisition of a raw biometric sample; *Extract*—conversion of the raw biometric sample data to an intermediate form; *Create template*—conversion of the intermediate data into a template for storage; and *Compare*—comparison with the information in a stored reference template.

In terms of security and privacy of the stored data, both the issuing and any receiving States need to be satisfied that the data stored on the IC has not been altered since it was recorded at the time of issue of the document. In addition, the privacy laws or practice of the issuing State may require that the data cannot be accessed except by an authorized person or organization. Accordingly ICAO has developed specifications in Section IV regarding the application and usage of modern encryption techniques, particularly interoperable public key infrastructure (PKI) schemes, to be used by States with their machine readable travel documents as made in accordance with the specifications set out in Doc 9303. The intent is primarily to augment security through automated means of authentication of MRPs and their legitimate holders internationally. In addition, ways and means are recommended to implement international ePassport authentication and to provide a path to the use of ePassports to facilitate biometric or e-commerce applications.

Annex 9[11] to the Convention on International Civil Aviation (Facilitation of Air Transport), in Standard 3.7 requires ICAO member States to regularly update security features in new versions of their travel documents, to guard against their misuse and to facilitate detection of cases where such documents have been unlawfully altered, replicated or issued. Recommended Practice 3.9 suggests that member States incorporate biometric data in their machine readable passports, visas and other official travel documents, using one or more optional data storage technologies to supplement the machine readable zone, as specified in Doc 9303, Machine Readable Travel Documents. The required data stored on the integrated circuit chip is the same as that printed on the data page, that is, the data contained in the machine-readable zone plus the digitized photographic image. Fingerprint image(s) and/or iris image(s) are optional biometrics for member States wishing to supplement the facial image with another biometric in the passport. Member States incorporating biometric data in their Machine Readable Passports are to store the data in a contactless integrated circuit chip complying with ISO/IEC 14443 and programmed according to the Logical Data Structure as specified by ICAO.

[11] Annex 9 to the Convention on International Civil Aviation, 12th Edition, 2006.

6.5 Legal Issues

The basic legal issues encompassing the issuance of ePassports are privacy of the individual[12]; and the internal security of a State. Ensuring both these are intrinsically and exclusively the responsibility of the State. As for privacy, The Chicago Convention, which established the regulatory framework for international civil aviation, underscores the fundamental aim of States in the context of civil aviation to exchange privileges which friendly nations have a right to expect from each other. In his message to the Conference in Chicago, President Roosevelt said:

> the Conference is a great attempt to build enduring institutions of peace, which cannot be endangered by petty considerations or weakened by groundless fears.[13]

6.6 Privacy

The Chicago Convention, in Article 13 of the Convention provides that the laws and regulations of a Contracting State as to the admission to and departure from its territory of passengers, crew or cargo of aircraft, such as regulations relating to entry, clearance, immigration, passports, customs and quarantine shall be complied with by or on behalf of such passengers, crew or cargo upon entrance into or departure from, or while within the territory of that State. This provision ensures that a Contracting State has the right to prescribe its own internal laws with regard to passenger clearance and leaves room for a State to enact laws, rules and regulations to ensure the security of that State and its people at the airport. However, this absolute right is qualified so as to preclude unfettered and arbitrary power of a State, by Article 22 which makes each Contracting State agree to adopt all practicable measures , through the issuance of special regulations or otherwise, to facilitate and expedite navigation of aircraft between the countries.

The above notwithstanding, there are three rights of privacy relating to the display and storage and use of personal data:

- The right of an individual to determine what information about oneself to share with others, and to control the disclosure of personal data;
- The right of an individual to know what data is disclosed, and what data is collected and where such is stored when the data in question pertains to that individual; the right to dispute incomplete or inaccurate data; and

[12]See Abeyratne (2002a). Also by the same author, The Exchange of Airline Passenger Information - Issues of Privacy, *Communication Law*, Vol. 6, No. 5; 2001: pp. 153–162, and also by Abeyratne (2003).

[13]Proceedings of the International Civil Aviation Conference, Chicago, Illinois, November 1–December 7 1944 The Department of State, Vol. 1 at p. 43.

- The right of people who have a legitimate right to know in order to maintain the health and safety of society and to monitor and evaluate the activities of government.[14]

It is incontrovertible that the data subject has a right to decide what information about oneself to share with others and more importantly, to know what data is collected about him. This right is balanced by the right of a society to collect data about individuals that belong to it so that the orderly running of government is ensured.

The data subject, like any other person, has an inherent right to his privacy.[15] The subject of privacy has been identified as an intriguing and emotive one.[16] The right to privacy is inherent in the right to liberty, and is the most comprehensive of rights and the right most valued by civilized man.[17] This right is susceptible to being eroded, as modern technology is capable of easily recording and storing dossiers on every man, woman and child in the world.[18] The data subject's right to privacy, when applied to the context of the full body scanner is brought into focus by Alan Westin who says:

> Privacy is the claim of individuals, groups or institutions to determine for themselves when, how, and to what extent information is communicated to others.[19]

The role played by technology in modern day commercial transactions has affected a large number of activities pertaining to human interaction. The emergence of the information superhighway and the concomitant evolution of automation have inevitably transformed the social and personal life styles and value systems of individuals, created unexpected business opportunities, reduced operating costs, accelerated transaction times, facilitated accessibility to communications, shortened distances, and removed bureaucratic formalities.[20] Progress notwithstanding, technology has bestowed on humanity its corollaries in the nature of automated mechanisms, devices, features, and procedures which intrude into personal lives of individuals. For instance, when a credit card is used, it is possible to track purchases, discovering numerous aspects about that particular individual, including, food inclination, leisure activities, and consumer credit behaviour.[21] In similar vein, computer records of an air carrier's reservation system may give out details of the passenger's travel preferences, *inter alia*, seat selection, destination

---

[14]Hoffman (1980), 142.

[15]Abeyratne (2001, 2002b).

[16]Young (1978) at 1.

[17]Warren and Brandies (1890–1891), at 193.

[18]As far back as in 1973 it was claimed that ten reels, each containing 1,500 m of tape 2.5 cm wide, could store a twenty page dossier on every man, woman, and child in the world. See Jones (1973).

[19]Westin (1970), at 124.

[20]Orwell (1984).

[21]For a detailed analysis of the implications of credit cards with respect to the right of privacy see Nock (1993) at 43.

fondness, ticket purchasing dossier, lodging keenness, temporary address and telephone contacts, attendance at theatres and sport activities, and whether the passenger travels alone or with someone else.[22] In similar vein, does it follow that a full body scanning exercise would reveal imperfections of the human body which person would desire to keep private? This scheme of things may well give the outward perception of surveillance attributable to computer devices monitoring individuals' most intimate activities, preferences and physical attributes, leading to the formation of a genuine "traceable society".[23]

The main feature of this complex web of technological activity is that an enormous amount of personal information handled by such varied players from the public and private sector, may bring about concerns of possible "data leaks" in the system, a risk that could have drastic legal consequences affecting an individual's rights to privacy.

At the international level, privacy was first recognized as a fundamental freedom in the *Universal Declaration of Human Rights*.[24] Thereafter, several other human rights conventions followed the same trend, granting to individuals the fundamental right of privacy.[25] The pre-eminent concern of these international instruments was to establish a necessary legal framework to protect the individual and his rights inherent to the enjoyment of a private life.

---

[22]The paramount importance of airline computer reservation system records is reflected in the world-renowned cases *Libyan Arab Jamahiriya* v. *United Kingdom* and *Libyan Arab Jamahiriya* v. *United States of America* regarding the PANAM 103 accident at Lockerbie, Scotland in 1988, where the International Court of Justice requested air carriers to submit to the Court the defendants' flight information and reservation details. See International Court of Justice. News Release 99/36, "Questions of Interpretation and Application of the 1971 Montreal Convention arising from the Aerial Incident at Lockerbie" (1 July 1999), online: http://www.icj-cij.org/icjwww/idocket/iluk/iluk2frame.html (date accessed: 14 July 2000). In a similar vein, Arthur R. Miller describes the significance of airline computer reservation system records when dealing with federal, state, local, and other types of investigations where these dossiers could provide valuable information. See also Miller (1971) at 42.

[23]See Scott (1995) at 307; Burnham (1983) at 20. *A contrario* to the argument supported in this thesis that the advancement of technology directly affects the intimacy of individuals. U.S. Circuit Judge Richard Posner favours the idea that other factors, such as urbanisation, income, and mobility development have particularly weakened the information control that, for instance, the government has over individuals: this denotes that individuals' privacy has increased. See Posner (1978) at 409.

[24]The text reads: "No one shall be subjected to arbitrary interference with his privacy, family, home or correspondence, nor to attacks upon his honor and reputation. Everyone has the right to the protection of the law against such interference or attacks". See *Universal Declaration of Human Rights*. GA Res. 217(III), 10 December 1948, Art. 12.

[25]See *International Covenant on Civil and Political Rights,* GA Res. 2200 (XXI), 16 December 1966, Art. 17; *American Declaration on the Rights and Duties of the Man* (1948), Art. 5; *American Convention on Human Rights,* 22 November 1969, San Jose, Costa Rica, Art. 11; *Convention for the Protection of Human Nations Convention on Migrant Workers*, A/RES/45/158, 25 February 1991, Art. 14; *United Nations Convention on Protection of the Child,* GA Res. 44/25, 12 December 1989, Art. 16.

Privacy represents different things for different people.[26] The concept *per se* has evolved throughout the history of mankind, from the original non-intrusion approach, which defended an individual's property and physical body against unwanted invasions and intrusions, then manifesting in whom to associate with, later enlarging its scope to include privacy as the individual's decision-making right,[27] and culminating in the control over one's personal information.[28] Thus, the conceptual evolution of privacy is directly related to the technological advancement of each particular period in history.

The right of privacy, as enunciated by the United States Judge Thomas M. Cooley, was the right "to be let alone" as a part of a more general right to one's personality. This idea was given further impetus by two prominent young lawyers, Samuel D. Warren and Louis D. Brandeis,[29] in 1890.[30] Before this idea was introduced, the concept of privacy reflected primarily a somewhat physical property or life. The foundations of "information privacy", whereby the individuals would determine when, how, and to what extent information about themselves would be communicated to others, inextricably drawing the right of control of information about oneself,[31] is a cornerstone of privacy. With the development of computer capabilities to handle large amounts of data, privacy has been enlarged to include the collection, storage, use, and disclosure of personal information.[32] The notion of informational privacy protection, a typically American usage, has been particularly popular both in the United States and Europe, where the term "data protection" is used.[33]

---

[26]See Regan (1995) at 33; Freund (1971) at 182.

[27]In this case, the US Supreme Court acknowledged the right of women to have abortions based on the grounds that the federal government could not interfere within her "decisional privacy" sphere. See *Roe* v. *Wade*, 410 U.S. 113 (1973). See also Cate (1997) at 49. See also Zelermyer (1959) at 16.

[28]In a remarkable case concerning the legality of a national census scheduled by the authorities, the German Constitutional court connected the individual's liberty and the personal data processing of the intended census, to rule that if the individuals do not know for what purposes and who is collecting the data, that situation will eventually create an abdication of the individual's rights to the processor's command, "which cannot be tolerated in a democratic society". See Simitis (1995) at 447–448. See also Hoffer (2000) at 8.1; Gavison (1980).

[29]See Cooley (1888), as cited in Warren and Brandeis (1980) at 195.

[30]The definition of privacy as the "Right to be Alone" is often erroneously attributed to Warren and Brandeis. See Warren & Brandeis. See Cooley (1888) as cited in Warren and Brandeis (1980) at 195. Additionally the concept of privacy as "the right to be let alone", and "the right most valued by civilized man: was embraced by US courts in the landmark dissenting opinion of Justice Louis D. Brandeis in *Olmsted* v. *United States.* See *Olmsted* v. *United States*, 277 U.S. 438, 478 (1928) [hereinafter *Olmstead.*]

[31]See Westin (1967) at 368. For a similar conceptualisation of privacy, see Fried (1978) at 425.

[32]See Reidenberg (1995) at 498.

[33]The former Privacy Commissioner of British Columbia, Canada, has asserted that privacy was originally a "non-legal concept". See Flaherty (1991) at 833–834. The term "data protection" has been translated from the German word *Datenschutz*, referring to a set of policies seeking to regulate the collection, storage, use, and transfer of personal information. See Bennet (1992) at 13.

Self-determination in the right to protect one's privacy was first judicially embraced by the German Bundesverfassungsgericht in 1983.[34] The US Supreme court followed this trend by adopting the principle of privacy self-determination in *DOJ* v. *Reporters Comm. for Freedom of the Press*.[35]

It must be borne in mind that privacy is not an absolute, unlimited right that operates and applies in isolation.[36] It is not an absolute right, applied unreservedly, to the exclusion of other rights. Hence there is frequently the necessity to balance privacy rights with other conflictive rights, such as the freedom of speech and the right to access information when examining individuals' rights *vis-à-vis* the interest of society.[37] This multiplicity of interests will prompt courts to adopt a balanced approach when adjudicating on a person's rights, particularly whose interests of a State are involved.

Since the data contained in equipment such as body scanners may be subject to trans-border storage, there is a compelling need to consider the introduction of uniform privacy laws in order that the interests of the data subject and the data seeker are protected. Although complete uniformity in privacy legislation may be a difficult objective to attain[38] (as has been the attempt to make other aspects of legislation uniform), it will be well worth the while of the international community to at least formulate international Standards and Recommend Practices (in the lines of the various ICAO Annexes) to serve as guidelines of State conduct. After all, as Collin Mellors pointed out:

Under international agreements, privacy is now well established as a universal, natural, moral and human right. Article 12 of the Universal Declaration of Human Rights, Article 17 of the United Nations Covenant on Civil and Political Rights and Article 8 of the European Convention for the Protection of Human Rights and Fundamental Freedoms, all specify this basic right to privacy. Man everywhere has occasion to seek temporary "seclusion or withdrawal from society" and such arrangements cannot define the precise area of the right to privacy.[39]

It is such a definition that is now needed so that the two requirements of ensuring respect for information about individuals and their privacy on the one hand, and the encouragement of free and open dissemination of trans-border data flows on the other, are reconciled.

In the provision of biometric data, the provider of the information and the receiver thereof are both under obligation to ensure that the data is not used for

---

[34]WHO Global Influenza Preparation Plan.

[35]See *DOJ* v. *Reporters Comm. for Freedom of the Press*, 489 U.S. 749 AT 763 (1988).

[36]See Simmel (1971) at 71.

[37]See Halpin (1997) at 111. See also Foschio (1990) at 35. For a comprehensive study on the conflictive interest on privacy and the mass media and the Freedom of Speech, see Pember (1972) at 227; Prowda (1995) at 769. See also J. Montgomery Curtis Memorial Seminar, *The Public, Privacy and the Press: Have the Media Gone Too Far? (American Press Institute, 1992) at 2.*

[38]*Computers and Privacy in the Next Decade*, Lance J. Hoffman ed. op. cit. at 146.

[39]Collin Mellors, *Governments and the Individual- Their Secrecy and His Privacy*, cited in, *A Look at Privacy*, Young (1978), at 94.

any purpose other than clearance of the owner of the information through customs barriers. This information may not later be used for commercial or other gain for instance for advertising purposes (such as using the physical profile of a prominent actor or actress whose biometric information originally given for customs clearance).[40]

The protection of human rights is the most significant and important task for a modern State, particularly since multi ethnic States are the norm in today's world. Globalization and increased migration across borders is gradually putting an end to the concept of the nation State, although resistance to reality can be still seen in instances where majority or dominant cultures impose their identity and interests on groups with whom they share a territory. In such instances, minorities frequently intensify their efforts to preserve and protect their identity, in order to avoid marginalization. Polarization between the opposite forces of assimilation on the one hand and protection of minority identity on the other inevitably causes increased intolerance and eventual armed ethnic conflict. In such a scenario, the first duty of governance is to ensure that the rights of a minority society are protected.

The foregoing discussion addressed the right of privacy of the individual which is paramount over most legal considerations. The only factor that would override this would be the security of State. Inherent to the concept of security of State is State responsibility[41] to its citizens and others who are in its territory. The fundamental issue in the context of State responsibility for the purposes of this article is to consider whether a State should be considered responsible for its own failure or non-feasance to prevent a private act of terrorism against civil aviation or whether the conduct of the State itself can be impugned by identifying a nexus between the perpetrator's conduct and the State. One view is that an agency paradigm, which may in some circumstances impute to a state reprehensibility on the ground that a principal-agent relationship between the State and the perpetrator existed, can obfuscate the issue and preclude one from conducting a meaningful legal study of the State's conduct.[42]

7 Security

It is incontrovertible that in issuing an ePassport, the State concerned ensures aviation security not only in its own territory but also in the territory of the State to which the ePassport holder travels. New and emerging threats to civil aviation are a constant cause for concern to the aviation community. Grave threats such as those posed by the carriage of explosives and dangerous pathogens on board, are real and have to be addressed with vigour and regularity. The leakage of dangerous

[40]See *Gould Estate* v. *Stoddart Publishing Company* (1996) O.J. No. 3288 (Gen. Div)

[41]For an in-depth discussion of State Responsibility see Abeyratne (2009).

[42]Caron (1998) 109, at 153–54 cited in Becker (2006), at 155.

pathogens[43] from laboratories also presents an ominous analogy to the aviation sector in that the same could well occur in the carriage of such dangerous goods by air.[44] Although past instances of the escape of dangerous pathogens are small in number, nonetheless their occurrence and the threat posed to the wellbeing of humanity cannot be underestimated. In 2002 when Anthrax spores escaped from two military laboratories in the United States, the authorities agreed that the leakage was due to a security lapse.[45] In 2003 a string of such leakages occurred in Asia, this time of the SARS virus.[46]

ICAO has been addressing these threats for some time and continues to do so on a global basis, particularly with regard to the impact of unpredictable security measures on passenger confidence in aviation security. There has been much support for this approach because of its value as a deterrent. It has been suggested that States adopt an approach providing for a baseline regime, but with the addition of unpredictable measures, thus achieving a balance between certainty and unpredictability.

The security ensured by the introduction of the ePassport undoubtedly has its genesis in the maintenance of international peace and security is an important objective of the United Nations,[47] which recognizes one of its purposes as being *inter alia*:

To maintain international peace and security, and to that end: take effective collective measures for the prevention and removal of threats to the peace, and for the suppression of acts of aggression or other breaches of peace, and to bring about by peaceful means, and in conformity with the principles of justice and international law, adjustment or settlement of international disputes or situations which might lead to a breach of the peace.[48]

It is clear that the United Nations has recognized the application of the principles of international law as an integral part of maintaining international peace and security and avoiding situations which may lead to a breach of the peace.

Liability of the manufacturer of the electronic chip, often a private entity, and the State is a significant factor in this equation. Ultimately, even though the chip

---

[43]Pathogens are microorganisms (including bacteria, viruses, rickettsia, parasites, fungi) or recombinant microorganisms (hybrid or mutant) that are known or are reasonably expected to cause infectious disease in humans or animals.

[44]See Abeyratne (2007).

[45]An year earlier, a covert event occurred in October 2001 when anthrax spores were sent through the mail exposing persons in the eastern USA to contaminated mail resulting in deaths, illnesses and identified exposures to Anthrax. Overt, announced events, in which persons are warned that an exposure has occurred, have taken place in the United States, although most of these were determined to have been hoaxes, that is, there were no true exposures to infectious agents.

[46]The leakages occurred in China, Taiwan and Singapore. See Air-Tight Security, *Intersec*, June 2007 33–35 at 34. See also International Responsibility in Preventing the Spread of Communicable Diseases through Air Carriage - The SARS Crisis. Abeyratne (2002c).

[47]Charter of the United Nations and Statute of the International Court of Justice, Department of Public Information, United Nations, New York, DPI/511 – 40108 (3-90), 100M at 1.

[48]*Id.* at 3.

incorporated in the ePassport is the creation of a public or private entity, is so incorporated in a State document—i.e. the passport, and therefore the State is ultimately liable for defects in the passport. State liability under administrative law can in turn be divided into two limbs: liability for acts of instrumentalities of State; and liability for privatized service providers for whose acts, relating to the provision of technical services, the State would still bear responsibility. The traditional model of administrative responsibility and accountability of the administrative State is based on the premise that Parliament controlled the executive but was in turn controlled by the people. Added to this, the fundamental postulate lay in the overarching principle that the judiciary played its role in keeping instrumentalities and agencies of the State intact. Accountability of the State for its agencies' actions was twofold: one stemming from a statutory power given to that agency by the State; and the other arising from delegation of authority by the State to the agency concerned. In the latter instance, however, the legislature could intervene and share some control of the agency. This gave rise to the inexorable principle that administrative law and judgments of courts on such agencies could be involved only in the former instance, when the State had provided a statutory base for a private agency or entity. In the 1983 British case of *O'Reilly & Mackman*,[49] the House of Lords limited the circumstances in which a public law remedy, such as a declaratory judgment or injunction, could be brought outside of Section 31 of the *Supreme Court Act* of 1918, which prescribed instances of legal actions to be brought against the State for an act of its statutory agent. This is notwithstanding the fact that Section 21 of the Crown Proceedings Act 1949 allows the Court in civil proceedings to issue a declaratory order against the State, although there could be no injunction specific performance orders against a State. Furthermore, a later case established that although the claim for judicial review might be brought against the Crown, the Crown's involvement is merely nominal and the ultimate dispute would be between the claimant and the defendant.[50] It is with the 1990 decision in the *Factorframe* Case[51] where Lord Bridge stated that injunctive relief against the Crown or its officers was not possible.

In the instance of a privatized service provider, the situation calls for a discussion of the reasons for privatization leading to the legal nature of a privatized entity.[52] The reasons for privatization could well range from improvement of efficiency to reducing government involvement in industrial decision making. The corollaries to privatization are often the widening of share ownership; encouraging share ownership by employees; providing more flexibility to pay policies; and enhancement of economic freedom. There could be two types of privatized service providers: the first being large companies which were once instrumentalities

---

[49][1983] 2. A.C. 237.

[50]*R.* (on the application of *Ben-Abdelaziz and Kugwa*) v. *London Borough of Hackney and the Secretary of State for the Home Department* [2001] 1 W.L.R. 1485, para 29.

[51]*R.* v. *Secretary of State ex parte Factorframe Ltd.* [1990] 2 A.C. 85.

[52]For a detailed discussion of the legal liability of States and of a privatized service provider see Abeyratne (2004).

of state, which, even after privatization, do not possess potential for undue competition in the market. These would easily transit to a position in which large companies had been private in the first place, and would not be subject to principles of public law. The second category of the privatized service provider is one which has market power and consequent potential for untrammelled competition. In such cases, the State may regulate the provider by bringing it under the administrative purview of a State agency. These privatized bodies may be vulnerable under public law through the agencies having administrative control over them.

One of the analogies in the United Kingdom of a privatization of a utility can be observed in the legislative initiative of 1984 with the adoption of the *Telecommunications Act* which brought about the privatization of a major public utility.[53] The 1984 legislation privatized the public corporation *British Telecom* (BT) and abolished BT's monopoly in providing telecom services, thus opening the doors to competition. The Director General of Telecommunications, established by the Act, can grant licenses to operators of telecom systems. The Director General is also empowered to refer a matter to the *Monopolies and Mergers Commission,* particularly on issues related to public interest such as pricing. If this particular feature were to be applicable to a privatized air navigation service provider appointed under Statute, there would be the interesting consideration under public law whether that provider complied with Article 15 of the Chicago Convention[54] on charges for services.

The operation of the administrative process in a State becomes somewhat complex when viewed in the context of competition policy where the State takes measures to curb the ill-effects on society of monopolies and cartels. An initial difficulty that arose was the nineteenth Century control of trade, which was aimed at promoting competition proved counterproductive, resulting in controlling competition. This difficulty was compounded by the early twentieth Century State policy of reluctance to interfere with citizens striking bargains for their benefit.[55] However, after World War 1, some British Governmental measures introduced comprehensive control of market power.[56]

British legislators can be proud of three legislative stages of unfair competition control. The first came in the form of the 1948 *Monopolies and Restrictive Practices (Inquiry and Control), Act* which devolved regulatory responsibility on an

---

[53]From 1912 until 1981 telecommunications are the responsibility of the Post Office. The 1981 legislation represented telecommunications from KP. Services and established British Telecom as a public corporation.

[54]Article 15 provides that every airport in an ICAO contracting State which is open to public use by its national aircraft shall likewise be open under uniform conditions to aircraft of all other Contracting States. The like uniform conditions shall apply to the use, by aircraft of every Contracting State, of all air navigation facilities, including radio and meteorological services, which may be provided for public use for the safety and expedition of air navigation services. Article 15 also provides that charges applicable to a foreign carrier for the provision of the air navigation services shall not be higher than those imposed on a carrier bearing the service provider State's nationality.

[55]*Mogul SS. Co. Ltd.* v. *McGregor Gow* [1892] A.C. 25. See also *Sorrell* v. *Smith* [1925] A.C. 700.

[56]Committee on Trusts Cmd. 9236 (1918).

agency—the Monopolies and Restriction Practices Commission (MRPC)—a body outside the normal departmental framework. The second stage commenced with the 1956 *Restrictive Trade Practices Act* which addressed the competitive threat of cartels and the *Restrictive Practices Court* was established to adjudicate an anti-competitive and privy issues. The third stage took on with the exPANsion of the *Monopolies Commission* which investigates monopolies issues. Merger irregularities were added to the jurisdiction of the Commission with the *Monopolies and Mergers Act* of 1968. The 1980 *Competition Act* which followed gave the Commission power to investigate particular anti-competitive practices. The final stage of the evaluation demarcates choice of institutions to investigate and adjudicate on anti-competitive practices. From an administrative perspective, the citizen has been known to challenge these State instrumentalities,[57] the most notable of which has been the challenge offered to the various governmental institutions created under Statute to define their extent of duty to give reasons for competition legislation.[58]

A Government's approach to regulation of a public utility, whether public or privatized, is usually based on the public interest rationale where individual consumer choice will determine the demand and supply for goods and their pricing and quantity.[59] In the United Kingdom, these factors are intrinsically related to transparency, accountability, proportionality, consistency and targeting.[60]

The foremost necessity is to establish a strong security culture in every State. For this, there must be a clear definition of State responsibility and accountability brought to bear by a close and unbreakable link between government and industry stakeholders. A security culture would make States aware of their rights and duties, and, more importantly, enable States to assert them. Those who belong to a security culture also know which conduct would compromise security and they are quick to educate and caution those who, out of ignorance, forgetfulness, or personal weakness, partake in insecure conduct. An ePassport must necessarily be the result of efficient and fail-safe organizational arrangements. It should be tested at border control by trained professionals.

eGovernment and eID are the bare essentials for State security. The digital economy has also brought much facilitation that helps the world move to paperless processes which result in greater economy and streamlined processes. However, there must essentially be global harmonization in this process. In this regard ICAO has made remarkable progress in advancing its MRTD programme to the level it is at now. If harmonization means ensuring consistency between global practices, standardization means compliance with international Standards. There is no room for doubt that both harmonization and globalization are needed in this context.

---

[57]See *R.* v. *Monopolies and Mergers Commission Exp. Elders* 1XL Ltd. [1987] 1. W.L.R. 1121. Also *R.V.M. & M. C Exp. Mathew Brown plc* [1987] 1 W.L.R. 1235.

[58]*R.* v. *Secretary of State for Trade Industry* Ex parte Lonrho plc [1989] 1 W.L.R. 325.

[59]Ogus (1994) Charter.

[60]See Better Regulation Guide, UK Cabinet Office (1998).

References

Abeyratne RIR (2001) The exchange of airline passenger information - issues of privacy. Commun Law 6(5):153–162
Abeyratne RIR (2002a) Intellectual property rights and privacy issues: the aviation experience in API and biometric identification. J World Intellect Property 5(4):631–650
Abeyratne RIR (2002) Attacks on America - privacy implications of heightened security measures in the United States, Europe, and Canada. J Air Law Commerce 67(1)
Abeyratne RIR (2002c) Transportation Law J 30(1):53–80
Abeyratne RIR (2003) Profiling of passengers at airports - imperatives and discretions. European Transport Law XXXVIII(3):297–311
Abeyratne RIR (2004) Privatization of Hong Kong international airport: some legal and economic issues. Asia Pacific Law Rev 12(1):31–51
Abeyratne RIR (2007) The safe carriage of dangerous pathogens by air: legal and regulatory issues. Eur Transp Law XLII(6):689–704
Abeyratne R (2009) Principles of responsibility for private acts of terrorism. Bar Assoc Law J XV:55–64
Becker T (2006) Terrorism and the state, Hart monographs in transnational and international law. Hart Publishing, Oxford
Bennet CJ (1992) Regulating privacy. Cornell University Press, Ithaca
Burnham D (1983) The rise of the computer state. Random House, New York
Caron DD (1998) The basis of responsibility: attribution and other trans-substantive rules. In: Lillich RB, Magraw DB (eds) The Iran-United States claims tribunal: its conclusions to state responsibility. Transnational Publishers, Hudson
Cate FH (1997) Privacy in the information age. Brookings Institution Press, Washington, DC
Cooley TM (1888) A treatise on the law of torts, 2nd edn. Callaghan, Chicago
Flaherty DH (1991) On the utility of constitutional rights to privacy and data protection. Case W Res 41:831
Foschio LG (1990) Motor vehicle records: balancing individual privacy and the public's legitimate need to know. In: Kuferman TR (ed) Privacy and publicity. Meckler, London
Freund PA (1971) Privacy: one concept or many. In: Pennnock JR, Chapman JW (eds) Privacy. Atherton, New York
Fried C (1978) Privacy: economics and ethics a comment on Posner. GA Law Rev 12:423
Gavison R (1980) Privacy and the limit s of the law. Yale Law J 89:421
Halpin A (1997) Rights & law analysis & theory. Hart Publishing, Oxford
Hoffer S (2000) World cyberspace law. Juris Publishing, Huntington
Hoffman LJ (ed) (1980) Computers and privacy in the next decade. Academic, New York
Jones RV (1973) Some threats of technology to privacy, privacy and human rights. In: Robertson AH (ed) Presented at the third colloquy about the European convention on human rights, Manchester University Press, Brussels, 30 September–3 October 1970
Miller AR (1971) The assault on privacy. The University of Michigan Press, Ann Arbor
Nock SL (1993) The costs of privacy. Aldine De Gryter, New York
Ogus A (1994) Regulation, legal form and economic theory. Oxford University Press, Oxford
Orwell G (1984) Nineteen eighty-four. Clarendon, Oxford
Pember DR (1972) Privacy and the press. University of Washington Press, Seattle
Posner R (1978) The right of privacy. GA Law Rev 12(3):393
Prowda JB (1995) A layer's ramble down the information superhighway: privacy and security of data. Fordham Law Rev 64:738
Regan PM (1995) Legislating privacy. The University of North Caroline Press, Chapel Hill
Reidenberg JR (1995) Data protection law and the European Union's directive: the challenge for the United States: setting standards for fair information practice in the U.S. private sector. Iowa Law Rev 80:497

Scott GG (1995) Mind your own business – the battle for personal privacy. Insight Books, New York
Simitis S (1995) From the market to the polis: the EC directive on the protection for personal data. Iowa Law Rev 80:445
Simmel A (1971) Privacy is not an isolated freedom. In: Pennnock JR, Chapman JW (eds) Privacy. Atherton, New York
Warren SD, Brandeis LD (1980) The right of privacy. Harv Law Rev 4(5):193
Warren SD, Brandies LD (1890–1891) The right to privacy. Harv Law Rev 4:193
Westin A (1967) Privacy and freedom. Atheneum, New York
Westin AF (1970) Privacy and freedom. Bodley Head, London
Young JB (ed) (1978) A look at privacy. Privacy, Willey, New York
Zelermyer W (1959) Invasion of privacy. Syracuse University Press, Syracuse

Article 14
Prevention of Spread of Disease

Each contracting State agrees to take effective measures to prevent the spread by means of air navigation of cholera, typhus (epidemic), smallpox, yellow fever, plague, and such other communicable diseases as the contracting States shall from time to time decide to designate, and to that end contracting States will keep in close consultation with the agencies concerned with international regulations relating to sanitary measures applicable to aircraft. Such consultation shall be without prejudice to the application of any existing international convention on this subject to which the contracting States may be parties.

Contents

1 Foresight Regarding Airborne Diseases

This provision imposes a responsibility on States to take measures in preventing the spread of disease through air transport. It is a very resilient provision that was designed to address communicable diseases prevalent in 1944 but leaves the door open to be inclusive to modern day diseases, two of which that spread were SARS and the Avian Flu pandemic. This provision explicitly devolves primary responsibility on States to take effective measures to prevent airborne diseases in aircraft and implicitly requires States to issue guidelines for airlines, by liaising with the international agencies concerned. *Non obstante,* airlines have to face certain legal issues themselves in terms of their conduct. Primarily, airlines are expected to conform to applicable international health regulations and the laws of the countries in which their aircraft land.[1] Furthermore, the airline owes its passengers a duty of

[1]World Health Organization, International Health Regulations, Third Annotated Edition, WHO: Geneva 1983.

R. Abeyratne, *Convention on International Civil Aviation*,
DOI 10.1007/978-3-319-00068-8_15, © Springer International Publishing Switzerland 2014

care[2] to exercise all caution in protecting their rights, so that a blatant instance of a person who looks sickly and coughs incessantly at the check-in counter cannot be ignored. Common law principles of tort law vigorously distinguish between negligence, recklessness and wilful blindness. Of these elements of liability, wilful blindness is particularly relevant since it brings to bear the need for an airline to be vigilant in observing passenger profiles in potentially dangerous or threatening situations

2 SARS

Civil aviation has traditionally been used not only as the speediest means of communication and commercial transport between and beyond national boundaries, but also as a means of solace, particularly in providing relief to communities in distress whether from natural disaster, famine and ill health or war. Unfortunately, aviation has also been used as a weapon of mass destruction, particularly in the context of the catastrophic events of 11 September 2001. The latest concern of the international community may well be that, although aviation cannot be matched by other means of transportation in view of the speed inherent in air transport, it nonetheless portends certain threats to human health which may emerge as a result of its very nature, requiring the cloistering of a large number of humans in a limited space where ventilation and air pressure have to be provided in a contrived manner.

In this regard, the compelling concern in 2002 was the spread of Severe Acute Respiratory Syndrome (SARS) which has an alarmingly high and increasing morbidity rate, which approached 6 % at its peak. Although the spread of the disease abated to a manageable degree in a few months of its outbreak, the threat of a pandemic lingered on and it would not be surprising if it were to resurface after a lapse of time. A vaccine against this dreaded disease may be several years away and the prospects of a cure are not in sight. Some experts on communicable diseases have gone to the extent of predicting a global pandemic, along the lines of the Influenza which afflicted the world in 1919–20 killing 20 million worldwide despite its low morbidity rate of 1 %. The threat posed by SARS is compounded by the fact that already large countries such as China are severely affected, along with countries that have a high rate of trans-border communication such as Hong Kong and Singapore. It could be envisioned that, unless contained, the disease could spread to other large countries such as Australia, Canada (which has already shown susceptibility) and the United States, along with the States of Europe. Stringent measures have already been taken by the countries afflicted such as enforcement of quarantines on thousands of hospital employees and patients, together with isolation of those not ill but have had some contact with afflicted individuals.

[2] Air carrier liability in negligence is extensively discussed in Abeyratne (2002b), *R.* v. *Secretary of State for Trade Industry* Ex parte Lonrho plc [1989] 1 W.L.R. 325 at 65–78.

2.1 Health Implications of SARS

From an aviation perspective, it was important to be aware of the grave risk that may be posed by the SARS virus in an in-flight situation. To have full realization, the nature of the disease and the manner in which it spreads has to be fully understood. In general, SARS begins with a fever greater than 100.4 °F (>38.0 °C). Other symptoms may include headache, an overall feeling of discomfort, and body aches. Some people also experience mild respiratory symptoms. After 2–7 days, SARS patients may develop a dry cough and have trouble breathing.

The primary way through which SARS appeared to spread is by close person-to-person contact. Most cases of SARS have involved people who cared for or lived with someone with SARS, or had direct contact with infectious material (for example, respiratory secretions) from a person who has SARS. Potential ways in which SARS can be spread include touching the skin of other people or objects that are contaminated with infectious droplets and then touching your eye(s), nose, or mouth. This can happen when someone who is sick with SARS coughs or sneezes droplets onto themselves, other people, or nearby surfaces. It also is possible that SARS can be spread more broadly through the air or by other ways that are currently not known. Thus, the aircraft cabin environment is highly conducive to the spread of the SARS virus.

Cases of SARS continued to be reported mainly among people who have had direct close contact with an infected person, such as those sharing a household with a SARS patient and healthcare workers who did not use infection control procedures while taking care of a SARS patient. Any airborne disease such as SARS is impacted by the environment, particularly if such were to be an enclosed one as in an aircraft cabin. The ventilation system plays a critical part in this regard and therefore, it is crucial to an air carrier's conduct to determine the manner in which an air carrier decides on ventilation systems in its aircraft. For instance, early jet aircraft until the last decade offered 100 % fresh air in the cabin. However, in the 1990s, ironically with more evolved technology, ventilation systems in aircraft were built in such a way as to recycle stale air, thus increasing the chances of survival of bacteria in the aircraft cabin. Even if such a practice were ineluctable, in that recycling is a universal practice which is calculated to conserve fuel, a prudent airline would take other measures, such as change of air filters through which ventilation is provided.

Air in the cabin is usually dry and lacking in humidity since the outside air at cruising altitudes has an extremely low water content. The humidity level in the air of an aircraft cabin at cruising level has been recognized as being of 10–20 % humidity which is approximately the same as desert air. The lack of humidity *per se* does not facilitate the transmission of airborne vectors, but makes breathing difficult, particularly for persons suffering from respiratory diseases, such as Asthma. When dry air becomes stale through recycling, the chances of removing droplets of air which is usually accomplished by fresh air becomes remote. A suggested solution for a prudent airline to take in this regard is to reintroduce 100 % fresh air which is humidified.

One of the major preoccupations of the World Health Organization (WHO) is to ensure the international prevention of disease. Quarantine regulations, which was the first step toward this aim, has a long history, having been introduced during the tenth Century. WHO adopted International Health Regulations in 1951,[3] the philosophy of which was recognized subsequently as:

The purpose of International Health Regulations is to prevent international spread of disease, and in the context of international travel, to do so with the minimum of inconvenience to the passenger. This requires international collaboration in the detection, reduction or elimination of the sources from which infection spreads rather than attempts to prevent the introduction of disease by legalistic barriers that over the years have proved to be ineffective.[4]

Of course, the purpose of this philosophy will be defeated if individual States have no willingness or the political will to notify the outbreak of communicable diseases to WHO, particularly in the absence of a monitoring body, incentives for States to notify or sanctions. Therefore the preeminent obligation of States is to ensure that the outbreak of any communicable disease is notified in a manner that would benefit the world and help prevent the spread of the disease across national boundaries. Regrettably, there have been instances recorded where WHO reports that no new instances of a communicable disease has been recorded while the news media give contrary information simultaneously.[5] One of the reasons adduced for the lack of interest on the part of States to report the incidence of communicable diseases to a world body such as WHO has been identified as the lack of importance attributed to International Health Regulations (IHR) by States who consider the regulations as an obsolete relic.[6]

The international health dimension of SARS involves human rights issues as well. International human rights law has laid down two critical aspects relating to public health: that protection of public health constitutes legitimate grounds for limiting human rights in certain circumstances (such as detention of persons or house arrest tantamount to quarantine exercises would be justified in order to contain a disease); and individuals have an inherent right to health. In this context it is not only the State or nation that has an obligation to notify WHO of communicable disease but the human concerned as well, who has an abiding moral and legal obligation. In 1975, WHO issued a policy statement which subsumed its philosophy on health and human rights which stating:

---

[3]See Gear and Deutschman (1981), pp. 273–343.

[4]World Health Organization, Vaccination Certificate Requirements an Health Advice for International Travel, Geneva, 1988, p. 7.

[5]See World Health Organization—Functioning of the International Health Regulations for the period January 1 to December 31 1984, 60 *Weekly Epidemiological Record*, December 13, 1985 at p. 386. It is also interesting to not that WHO reports that only 40 % of diagnosed cases of AIDS are reported to the Organization. See. Schachter and Joyner (1995), p. 865.

[6]See World Health Organization - Functioning of the International Health Regulations for the period January 1 to December 31 1985, Part 1, 61 *Weekly Epidemiological Record*, December 12, 1986 p. 303.

The individual is obliged to notify the health authorities when he is suffering from a communicable disease (including venereal diseases) or has been exposed to infection, and must undergo examination, treatment, surveillance, isolation or hospitalization. In particular, obligatory isolation or hospitalization in such cases constitutes a limitation on freedom of movement and the right to liberty and security of person.[7]

It is critical for an evaluation of the health and aeronautical implications of SARS that the term "health" be defined in context. While the WHO Constitution identifies as an objective of the Organization "attainment of the highest possible level of health", the state of health is defined as "a state of complete physical, mental and social well-being and not merely the absence of disease or infirmity".[8] In an aeronautical perspective, as will be seen later in this chapter, this is a tough act to follow, as international responsibility in the carriage of persons extends only as far as the obligation to prevent injury, wounding or death, and not to the physical or mental well-being of a person.

2.2 Aeronautical Implications of SARS

During the period November 1 2002–22 April 2003, the WHO had recorded 78 SARS related deaths and 2,223 suspected cases of SARS in 18 countries.[9] Following these statistics WHO declared that passengers with symptoms of SARS or those who may have been exposed to the virus should not be allowed to fly.[10] Some countries took immediate action, one of the first being the United States which advised its citizens to defer non essential travel to affected regions. Canada declared a health emergency and Taiwan advised against travel to the mainland.[11] It was reported that Airbus Industries had revealed in early May 2003 that several airlines hit by the SARS crisis requested formal postponement of aircraft deliveries.[12] According to this report the SARS crisis was persistent and affected traffic figures adversely, compounding problems already caused by the war in Iraq. The enormity of the problem is brought to bear by the response of the International Civil Aviation Organization (ICAO) which issued guidelines on 2 May 2003 urging member States to:

[7]The Individual's Duty to the Community and the Limitations on Human Rights and Freedoms Under Article 20 of the Universal Declaration of Human Rights, 100 UN Sales No. E.82.XIV.1 (1983)

[8]World Health Organization, Basic Documents, 1–2 (37th Ed. 1988).

[9]Fiorino (2003), p. 59.

[10]WHO Urges Screening of Air Pasengers for SARS on Some Flights, *Washington Aviation Summary*, April 2003 Edition (April 1 2003), Baker and Hostetler, LLP at p. 1.

[11]. *Ibid.*

[12]Some SARS Hit Airlines Want Deliveries Postponed, *Air Letter*, Friday, 2 May 2003, No. 15,232, p. 3.

- Implement pre-boarding medical screening of passengers at check-in;
- Provide all incoming passengers with a detailed information leaflet on SARS;
- Implement medical screening of passengers arriving directly from or via affected areas;
- Advise pilots to radio ahead if someone on board exhibits SARS symptoms;
- Instruct crew on dealing with suspected SARS-patients in flight; and
- Disinfect aircraft on which a suspected SARS-patient has travelled.[13]

The International Air Transport Association (IATA) as part of its response to the crisis, set up a SARS operations Centre in Singapore, one of the worst hit States, in order to help coordinate efforts in the region in containing the disease.[14] IATA's aim was to assist in the establishment of effective and efficient screening processes that could be the result of combined public health expertise offered by Governments along with operational expertise of airports and airlines.[15] Furthermore, IATA and WHO met in Bangkok in April 2003 to coordinate and refine plans to curb the possibility of the disease affecting air transport, where IATA identified the disease as a

> global problem, requiring a global solution, needing the coordinated support and understanding of governments. . .which meant that the imposition of reactionary and inefficient countermeasures must be avoided [16]

IATA's official view pertaining to the effects of SARS on the air transport industry was that the virus posed the biggest threat the airlines have ever faced and that SARS related airline losses would overtake the $10 billion loss suffered as a result of the Iraq war.[17] According to IATA, passenger loads on all airlines plunged as a result directly or indirectly of SARS and Hong Kong carriers such as Cathay Pacific and Dragonair had suffered losses as much as 70 %.[18]

On the insurance front, the London underwriters were reported to have withdrawn aviation insurance coverage for travel to countries affected by SARS.[19] The Air Transport Association of the United States announced that

> the world situation continues to play havoc with the airline market place. . .and that for the week ending 6 April, systemwide traffic for the biggest US carriers had dropped by 17.4% and domestic travel had fallen almost 15% compared to the same period in 2002.[20]

[13]ICAO Issues Guidelines Regarding SARS, *PIO 07/03*, Montreal, 02 May 2003.

[14]IATA Sets up Regional SARS Centre, *Air Letter*, Wednesday 30 April, 2003, No. 15,230, p. 3.

[15]*Ibid.*

[16]Airlines Refine Battle Plans to Fight SARS, *Washington Aviation Summary* (1 May 2003) May 2003 Edition, Baker & Hostetler LL., p. 1.

[17]IATA Predicts Tough Six Months head for Aviation Industry, *Aviation Daily*, Friday, April 25 2003, p. 5.

[18]*Ibid.*

[19]Travel Insurers Take Fright Over SARS, *Air Letter*, Monday 28 April 2003, No. 15,328, p. 1.

[20]Iraq, SARS send Travel to New Low, *Air Letter*, Friday, 11 April 2003, No. 15,219, p. 1.

Elsewhere, there were at least two airlines which reduced scheduled flights or operations as a result of the crisis: KLM announced its reduction of flights to Asia and its intent to fly smaller aircraft with lesser capacity to Asian destinations, thus reducing its total capacity by 3 %[21]; and QANTAS delayed its aircraft orders and downsized its staff by 400.[22] Cathay Pacific announced the most comprehensive and aggressive cabin health programme ever launched by a commercial carrier in order to ensure the health of passengers and reassure air crews of cabin safety despite the SARS threat.[23]

3 The Avian Flu Crisis

The Avian flu crisis occurred in 2008. Although the World Health Organization (WHO) categorized the outbreak of the H5N1 avian flu virus as being in Phase 3[24] (which is by no means in the pandemic stage) the outbreak caused widespread fears of a human flu pandemic. Air carriers, particularly in Asia where the virus broke out in birds, have taken several measures, looking at specific countries where they consider health risks are high, and placing an embargo on the transport of live birds.[25] However, the above notwithstanding, Asian countries are not panicking.[26] Some have even cautioned that, in reality, the threat of an avian flu pandemic is no more real today than it was when the virus first broke out 8 years ago and that there is no cause to believe that a pandemic is upon the world.[27] At the time of writing, neither the WHO nor the United States Disease Control Centres had issued travel advisories urging the public to avoid travelling to infected areas, despite a request by President Bush of the US Congress for a $7.1 billion fund injection for an avian flu preparedness plan.[28]

Unfortunately for the air transport industry, although precipitate political action in a Phase 3 situation will be addressed by the international Organizations and governmental authorities with cautious restraint, the public at large will panic if the situation worsens, causing the sale of air transport to plummet worldwide. The

---

[21]KLM Cuts Flights to Asia due to SARS, *Air Letter*, Tuesday 29 April 2003, No 15,229, p. 3.

[22]Qantas May Delay Orders Due to SARS, *Air Letter*, Friday 25 April 2003, No. 15,227 p. 2.

[23]Risk of Deadly Respiratory Infection Fuels Fear of Air Travel, *Air Safety Week*, April 14, 2003: Washington DC, Vol. 17, No. 15 p. 1.

[24]According to WHO, Phase 3 represents human infections of a new sub type, but no human to human spread or at more rare instances of spread to a close contact. In Phase 3, the virus subsides in initially affected countries but not in other countries to which it has spread. See *WHO Global Influenza Preparedness Plan*, WHO/CD/CSR/GIP/2005.5, World Health Organization, Department of Communicable Disease Surveillance and Response, Global Influenza Programme WHO: Geneva, at 7.

[25]Mathews (2005) at 38.

[26]*Ibid.*

[27]May (2005) at 7.

[28]*Ibid.*

example of Thailand when the outbreak of avian flu was in its incipient stage is a good one. When Thailand, which is the fourth largest exporter of poultry in the world, announced on 17 January 2004 the presence of highly pathogenic H5N1 avian influenza in both humans and poultry, the population of the country became bewildered and dismayed, which brought immediate and drastic consequences on the economy and tourism. The announcement percolated overseas, having dramatic results on the international community.[29]

A rapidly spreading pandemic would be as disastrous to the industry as the aftermath of the events of 11 September 2001, where States rallied together under the auspices of ICAO as well as by themselves to keep the air transport industry afloat.[30] As during the SARS outbreak, aviation insurance underwriters may review the situation in terms of aircraft fleet coverage[31] calling for the protection of the air transport industry by the global aviation community. It is incontrovertible that responsibility for curbing the spread of contagious diseases primarily devolves upon States and international organizations concerned. They would at least have a moral obligation to watch the interests of the air transport industry in the face of a global pandemic of avian flu.

Avian influenza or "bird flu" is a contagious disease of animals caused by viruses that normally infect only birds and, less commonly pigs. These viruses are normally highly species specific, but on rare occasions have crossed the species barrier to infect humans.[32] When the avian flu virus infects domestic poultry, the birds could show low pathogenic mild symptoms (ruffled feathers or a drop in egg production) which may not be easily detected, or show highly pathogenic symptoms affecting the entire poultry run.

Bird flu viruses do not usually affect humans but several cases of human infection through different strains of viruses have been detected since 1997.[33] The main concern is that the virus could gain the ability to spread easily from one person to another. As these viruses do not commonly infect humans, there is little or no immune protection against them in the human population.

---

[29] *Outbreak Communication, Best Practices for Communicating With The Public During an Outbreak*, Report of the WHO Expert Consultation on Outbreak Communications, Held in Singapore, 21–23 September 2004, WHO: Geneva 2004 at 17.

[30] See Abeyratne (2002a) at 84.

[31] Abeyratne (2002b) at 62.

[32] *Weekly Epidemiological Record*, 4 November 2005, No. 44, 2005, 80:377–388, World Health Organization: Geneva, at 377. Laboratory confirmed human cases have been reported in 5 countries – Cambodia, Indonesia, Thailand, Vietnam and Turkey, *Id* 379.

[33] *Highly Pathogenic H5N1 Avian Influenza Outbreaks in Poultry and in Humans: Food Safety Implications,* World Health Organization: Geneva, 4 November 2005, INFOSAN Information Note No 7/2005 at 3. This virus first infected people in 1997 in Hong Kong. In late 2003 and early 2004, outbreaks of avian influenza were reported among poultry in eight countries in Asia (Cambodia, China, Indonesia, Japan, Laos, South Korea, Thailand and Vietnam), resulting in more than hundred million bird deaths in the affected countries either from disease or from culling to try to control the outbreak.

In the past century, there were three known major influenza pandemics in 1918, 1957 and 1968. Although no one can accurately predict when the next pandemic will occur, since the spread of avian flu is so extensive and appears to be growing, and in view of the proven durability of the virus' persistence over the past several years, concern about a pandemic is the highest since 1969. At the time of writing there were 150 confirmed case of humans affected, half of which were fatal. The exact way in which people get infected by influenza viruses is not clear, but most cases appear to involve direct contact between the person affected and the infected bird. Among birds, infected birds are thought to contaminate the environment by shedding the virus in their faeces. However, some avian viruses appear also to spread among birds by respiratory transmission.

The symptoms of avian flu have ranged from relatively mild and self resolving fever and respiratory illness to rapid respiratory failure and death. The currently active viruses are seemingly sensitive to two antiviral medications—Oseltamivir (Tamiflu) and Zanamavir, although these mediations may not take effect unless used very early when symptoms first appear. There is still very little direct experience with these drugs when used to treat people affected by avian flu. Efforts to develop a vaccine against the flu are under way However, at the time of writing, there was no commercially available vaccine to protect humans against the avian flu virus.

Governments are having trouble with stopping the spread of the avian flu virus, since the manner in which the virus spreads within and between domestic and wild bird populations is not entirely clear. This makes it very difficult to develop fully effective control strategies. What is known, however, is that the widespread persistence of H5N1 in poultry populations poses two risks for humans: the first being when the virus passes from poultry to humans; and the second being when the virus changes, if given opportunity enough, into a strain which will spread from person to person. The second type of risk is most threatening to air transport, since such a change would not only start a global outbreak (pandemic) but would also make air transport a conduit between nations for the global transmission of the disease across boundaries.

3.1 Consequences of an Influenza Pandemic

The greatest influenza pandemic occurred in 1918–1919 and caused an estimated forty to fifty million deaths worldwide.[34] Although healthcare has significantly improved in the past decades, epidemiological models developed in the US Centres

[34]*WHO Global Influenza Preparedness Plan*, *R.* v. *Secretary of State ex parte Factorframe Ltd.* [1990] 2 A.C. 85 at 13.

for Disease Control and Prevention in Atlanta project that an avian flu pandemic will likely cause 2–7.4 million deaths globally[35] WHO's estimates are consistent with these figures.[36]

As for the potential economic impact of an avian flu pandemic, any conjecture on the possible human and economic impact would be fundamentally flawed if the nature of the pandemic and possible economic fallout are not fully certain[37] It has been estimated that the gross attack rate (infection rate), which reflects the percentage of the population that will be infected and become clinically ill will be typical to influenza rates usually seen at 20–40 %[38] In the case of the Spanish Flu, the mortality rate was between 2.5 % and 5 %.[39]

From an economic perspective, a flu pandemic may have different consequences from the SARS outbreak which occurred earlier this century. Whereas the impact of SARS was on the demand side, in the form of consumption and the demand for services contracted,[40] a flu pandemic will also affect and impact the supply side, as members of the labour force fall sick and in some cases succumb to the disease. A flu pandemic will also destroy human and physical capital, reducing global growth potential and having a significant impact on the global economy. Furthermore, such a pandemic will make investment drop significantly and will not allow a revival for a long time. Deaths resulting from avian flu will reduce the work force drastically and a widespread pandemic could lower the world GDP by 3.6 points than in a case where there is no pandemic.

Another factor that would affect the global economy and in turn the air transport industry is the psychological factor. Regionally, a virulent global pandemic could have serious results on the confidence of Europe, North America and Asia which have built their economies on their growth potential. There will be a significant loss to business as importers, exporters and the service industry experience a serious drop in demand. A direct corollary to this trend would be the closure of many businesses, lowering future investment and employment.

3.2 What Can Be Done?

The major role in combating a possible avian flu pandemic should be played by both governments and international Organizations, by preventing and mitigating a flu pandemic. Such an effort would naturally require cooperation and coordination,

---

[35]Http://www.who.int/csr/dusease/influenza/pandemic/en/index.html. The forecast also includes the probability of 134 to 233 million outpatient visits to hospitals and a million hospital admissions.

[36]Bloom et al. (2005), at 2.

[37]*Ibid.*

[38]Taubenberger (2005) at 24.

[39]J. Barry, 1918 Revisited: Lessons and Suggestions for Further Inquiry, published in *The Threat of Pandemic Influenza: Are We Ready*? http://users.lmi.net/wfanca/pp_annan_on_sov.html, at 33.

[40]Fan (2003) at 5.

along with a concerted effort on the part of the international community to coordinate assistance with a view to ensuring support for all major areas while obviating duplication of efforts. A key support area would lie in financing, particularly poor countries and the provision of critical commodities to them. Needless to say, air transport would be playing a key role in this endeavour, which is all the more reason to have a contingency plan for the sustenance of global air transport in a crisis situation.

The Avian flu situation was different from earlier outbreaks of an influenza pandemic. Firstly, the world had been warned in advance. Secondly, this warning gave us ample opportunity to prepare for an outbreak. WHO observed that, since late 2003, the world had progressively moved closer to a pandemic since 1968 when the last pandemic of the twentieth century occurred. WHO also said that, during 2005, ominous changes have been observed in the epidemiology of the disease in animals.[41] WHO advised that, as a response to a pandemic threat, the world should take advantage of the gradual process of the adaptive mutation of the virus and implement early intervention with antiviral drugs, supported by other public health measures.[42]

In this regard, measures had already been proposed both by the Food and Agriculture Organization (FAO) and the World Organization for Animal Health (OIE)[43] along with a draft global strategy.[44] WHO had prepared a comprehensive avian influenza preparedness plan which recognizes that air travel might hasten the spread of a new virus and decrease the time available for preparing intervention.[45] At an international meeting of health ministers held in Ottawa in October 2005, it was stressed that there was a need for a multi sectoral approach calculated to: strengthen the capacity for surveillance; develop a global approach to vaccine and anti-viral policy for research and development; and, above all, achieve full transparency between countries and institutions involved in responding to the risk of a pandemic, while carrying out a global programme to conduct disease surveillance.[46]

3.3 The Air Transport Perspective

On 18 November 2005, temperature screening of people arriving at Hong Kong at Lowu and Lok Ma Chau were activated[47] using infra-red thermo imagery techniques. This measure amply demonstrates that, from an air transport perspective,

---

[41] *Responding to the Avian Influenza Pandemic Threat: Recommended Strategic Actions*, World Health Organization Communicable Disease Surveillance and Response to the Global Influenza Programme: Geneva, 2005, at 3.

[42] *Ibid.*

[43] http://www.fao.org/ag/againfo/subjects/en/health/diseases-cards/27septrecom.pdf.

[44] http://www.fao.org/ag/againfo/resources/documents/empres/AIglobalstrategy.pdf.

[45] *WHO Global Influenza Preparedness Plan*, WHO/CDS/CSR/GIP/2005.5, Preamble (*supra* note 5) at 3.

[46] *Ottawa 2005: Global Pandemic Influenza Readiness – An International Meeting of Health Ministers*, Health Canada News Release, October 25, 2005; at 7.

[47] *Temperature Screening for Incoming Travellers Activated in Phases*, Hong Kong Department of Health Bulletin 05117, 18 November 2005.

technology is available to combat an outbreak of flu around the world as States will find it increasingly easier to implement measures once used during the SARS crisis, particularly as both ICAO and IATA carried out an exhaustive programme of action when the SARS crisis erupted. Both Organizations worked closely with WHO during that crisis and are continuing their efforts at present in the context of the new threat to public health. IATA's Medical Advisory Group has worked with WHO to develop guidelines for check-in agents, cabin crew, cleaning staff and maintenance staff. ICAO has already put into action a systemic approach to a possible outbreak of communicable disease. At the 35th Session of the ICAO Assembly, held in September/October 2004, ICAO Contracting States adopted Resolution A 35–12,[48] which declares that the protection of the health of passengers and crews on international flights is an integral element of safe air travel and that conditions should be in place to ensure its preservation in a timely and cost effective manner. Through this Resolution, the Council has been requested to review existing Standards and Recommended Practices (SARPs) of relevant Annexes to the Chicago Convention and adopt new SARPs as necessary, while maintaining institutional arrangements to coordinate efforts by Contracting States and other members of the international civil aviation community.

It is quite evident that both ICAO and IATA are concentrating on protecting the health of passengers and crew on the basis that the spread of a communicable disease within the aircraft should be avoided. Much has already been done regarding this area of concern in a technological context so much so that it can now be reasonably assumed that there is little possibility of the spread of a communicable disease through the ventilation system of an aircraft. As one commentator has observed:

> ...there is nothing about an aircraft cabin that makes it easier to contract a communicable disease. In fact, quite the opposite appears to be true. The ventilation patterns on aircraft, combined with the circulation of air through High Efficiency Particulate Air (HEPA) filters reduces the spread of airborne pathogens, especially when compared with other public places[49]

While all this is well and good, the question is whether, as was experienced during the outbreak of SARS in Toronto, where two Toronto residents brought SARS from Hong Kong to Toronto after travelling by air, the international community should be more concerned with the transmission of the disease across boundaries, which is the real danger and not merely within the aircraft itself.

3.4 International Regime Relating to Public Health

The international health dimension of avian flu involves human rights issues as well. International human rights law has laid down two critical aspects relating to public health: that protection of public health constitutes legitimate grounds for

[48]*Protection of the health of the passengers and crews and prevention of the spread of communicable diseases through international travel*, Assembly Resolutions in Force (as of 8 October 2004), ICAO Doc 9848, at 1–50.

[49]May (2005) at 7.

limiting human rights in certain circumstances (such as detention of persons or house arrest tantamount to quarantine exercises would be justified in order to contain a disease); and individuals have an inherent right to health. In this context it is not only the State or nation that has an obligation to notify WHO of communicable disease but the human concerned as well, who has an abiding moral and legal obligation. In 1975, WHO issued a policy statement which subsumed its philosophy on health and human rights which stated:

The individual is obliged to notify the health authorities when he is suffering from a communicable disease (including venereal diseases) or has been exposed to infection, and must undergo examination, treatment, surveillance, isolation or hospitalization. In particular, obligatory isolation or hospitalization in such cases constitutes a limitation on freedom of movement and the right to liberty and security of person.[50]

It is critical for an evaluation of the health and aeronautical implications of avian flu that the term "health" be defined in context. While the WHO Constitution identifies as an objective of the Organization "attainment of the highest possible level of health", the state of health is defined as "a state of complete physical, mental and social well-being and not merely the absence of disease or infirmity".[51] In an aeronautical perspective, this is a tough act to follow, as international responsibility in the carriage of persons extends only as far as the obligation to prevent injury, wounding or death, and not to the physical or mental well-being of a person.[52]

References

Abeyratne RIR (1999) Mental injury caused in accidents during international air carriage - a point of view. Aviat Q 4:193–205

Abeyratne R (2002a) Attacks on America - privacy implications of heightened security measures in the United States, Europe, and Canada. J Air Law Commerce 67(1):83–115

Abeyratne RIR (2002b) International responsibility in preventing the spread of communicable diseases through air carriage - the SARS crisis. Transp Law J 30(1):53–80

Bloom E, de Wit V, San-Jose MJC (2005) Potential economic impact of an avian flu pandemic in Asia. ERDP policy brief no 42, Asian Development Bank

Fan E (2003) SARS: economic impact and implications. ERDP policy brief no 15. Economics and Research Department, Asian Development Bank, Manila

Fiorino F (2003) SARS: a new blow. Aviat Week Space Technol 59

Gear HS, Deutschman Z (1981) Disease control and international travel - a review of international sanitary regulations. 10 Chronicle of the World Health Organization, Spec. No. 9-10

---

[50]The Individual's Duty to the Community and the Limitations on Human Rights and Freedoms Under Article 20 of the Universal Declaration of Human Rights, 100 UN Sales No. E.82.XIV.1 (1983).

[51]World Health Organization, Basic Documents, 37th Ed. 1988, 1–2.

[52]See Abeyratne (1999).

Mathews N (2005) No panic, Asian carriers say risk of bird flu has to be kept in perspective. Aviat Week Space Technol 38

May J (2005) Avian flu: a call for precaution, not panic. Aviation Daily (Tuesday, 22 November 2005)

Schachter O, Joyner CC (eds) (1995) United Nations legal order, vol 2. Cambridge University Press, Cambridge

Taubenberger J (2005) Chasing the elusive 1918 virus: preparing for the future by examining the past. In: Knobler S, Mack A, Mahmiud A (eds) The threat of pandemic influenza: are we ready? National Academic Press, Washington, DC

Article 15
Airport and Similar Charges

Every airport in a contracting State which is open to public use by its national aircraft shall likewise, subject to the provisions of Article 68, be open under uniform conditions to the aircraft of all the other contracting States. The like uniform conditions shall apply to the use, by aircraft of every contracting State, of all air navigation facilities, including radio and meteorological services, which may be provided for public use for the safety and expedition of air navigation. Any charges that may be imposed or permitted to be imposed by a contracting State for the use of such airports and air navigation facilities by the aircraft of any other contracting State shall not be higher,

(a) As to aircraft not engaged in scheduled international air services, than those that would be paid by its national aircraft of the same class engaged in similar operations, and

(b) As to aircraft engaged in scheduled international air services, than those that would be paid by its national aircraft engaged in similar international air services.

All such charges shall be published and communicated to the International Civil Aviation Organization: provided that, upon representation by an interested contracting State, the charges imposed for the use of airports and other facilities shall be subject to review by the Council, which shall report and make recommendations thereon for the consideration of the State or States concerned. No fees, dues or other charges shall be imposed by any contracting State in respect solely of the right of transit over or entry into or exit from its territory of any aircraft of a contracting State or persons or property thereon.

Contents

1 Airport Economics

The global policy on airport and similar charges has been useful to the international aviation community as guidance in the steadily evolving practices on the imposition of charges and their use. As early as 1947 the First Assembly of ICAO (Montreal, 6–27 May 1947) adopted Resolution A1-66 (Studies to be conducted by the Council in the Joint Support Field) which requested the Council to continue conducting its

R. Abeyratne, *Convention on International Civil Aviation*,
DOI 10.1007/978-3-319-00068-8_16,

studies entitled …"International Air Transport Paying Economic Rates with respect to air navigation Facilities and Services".[1] The same Resolution requested ICAO member States to furnish ICAO with all necessary information relating to the cost of construction, operation and maintenance of air navigation facilities and services, to revenues derived, and to charges levied, such information to assist the Council's study.

The Assembly, at its 2nd session (Geneva, 1–21 June 1948) adopted Resolution A2-14 (Study on International Air Transport Paying Economic Rates with respect to air navigation Facilities and Services) in which the Council was reminded of Resolution A1-66 and additionally requested to ensure that its study of the issue duly take into account the economics of all phases of airport and air navigation services for international civil aviation, including telecommunications, radio and other aids to navigation, air traffic control, meteorological services and other ancillary services.

On the basis of its Study the Council was requested to formulate recommendations for the guidance of member States with regard to the principles on which providers of this services for international civil aviation may derive revenue therefrom with regard to the methods that may be employed in the collection of such revenue.

The Assembly, at its 12 Session (San Diego, 16 June–9 July 1959) adopted Resolution A12-19 (Study of Airports and Air Navigation Facilities) which recalled that the Council, after considering the recommendations of the Airport Charges Conference and Route Facility Charges Conference, adopted statements for the Guidance of member State with regard to charges for route navigation facilities and services.[2] Accordingly, the Resolution urged member States to make every effort to implement that part of Article 15 of the Chicago Convention providing that any charges that may be imposed or permitted to be imposed by a State for the use of air navigation facilities and airports by the aircraft of any other member State shall be published and communicated to ICAO.

The Assembly, at its 14th Session (Rome, 21 August–15 September 1962), adopted Resolution A14-31 (Study of the Problems charges and the economics of air ports and air navigation facilities) resolved that the Council should as soon as possible, carry out a thorough fact finding study (the scope of which was to be determined in consultation with member States and IATA) of the problems of charges in relation to the economic situation of airports and route facilities in international air transport. In this respect, governments, airport authorities and their international organizations and IATA were requested to provide statistical and other information to the Council.

More recently, at its 37th Session (September/October 2010) the ICAO Assembly adopted Resolution A37-20 (Consolidated statement of ICAO policy

[1]Doc 4026 A1-FA/3 Appendix 1 Part 3.

[2]Doc 7806-C/899 and Doc 7941-C/913.

in the air transport field) Appendix F (Airport and Air Navigation Services—Charging Policy) stated:

Whereas ICAO policies in Doc 9082, ICAO's Policies on Charges for Airports and Air Navigation Services make a conceptual distinction between a charge and a tax in that "a charge is a levy that is designed and applied specifically to recover the costs of providing facilities and services for civil aviation, and a tax is a levy that is designed to raise national or local government revenues which are generally not applied to civil aviation in their entirety or on a cost-specific basis";

Whereas the matter of aircraft engine emission-related levies and market-based measures is addressed separately in Assembly Resolution A37-18, Consolidated statement of continuing ICAO policies and practices related to environmental protection — General provisions, noise and local air quality (Appendix H, Aviation impact on local air quality), and in Assembly Resolution A37-19, Consolidated statement of continuing ICAO policies and practices related to environmental protection — Climate change;

Whereas Article 15 of the Convention establishes the basis for the application and disclosure of charges for airports and air navigation services;

Whereas the Council has been directed to formulate recommendations for the guidance of Contracting States with regard to the principles on which providers of airports and air navigation services for international civil aviation may charge to recover the costs of their provision and derive other revenue therefrom, and with regard to the methods that may be employed to that effect; and

Whereas the Council has adopted and revised, as necessary, and published in Doc 9082, ICAO's Policies on Charges for Airports and Air Navigation Services;

The Assembly:

1. *Urges* Contracting States to ensure that Article 15 of the Convention is fully respected;

2. *Urges* Contracting States to base the recovery of the costs of the airports and air navigation services they provide or share in providing for international civil aviation on the principles set forth in Article 15 of the Convention and additionally in Doc 9082, ICAO's Policies on Charges for Airports and Air Navigation Services, regardless of the organizational structure under which the airports and air navigation services are operated;

3. *Urges* Contracting States to ensure that airport and air navigation services charges are applied towards defraying the costs of providing facilities and services for civil aviation;

4. *Urges* Contracting States to make every effort pursuant to Article 15 of the Convention to publish and communicate to the Organization any charges that may be imposed or permitted to be imposed by a Contracting State for the use of air navigation facilities and airports by the aircraft of any other Contracting State;

5. *Encourages* Contracting States to adopt the principles of non-discrimination, cost-relatedness, transparency and consultation with users, as espoused in Doc 9082, in their national legislation, regulation or policies, as well as in air services agreements, to ensure compliance by airports and air navigation; and

6. *Requests* the Council to ensure that the guidance and advice contained in Doc 9082 are current and responsive to the requirements of Contracting States.

The vexed issue of the need to reach consensus on a just and equitable basis for the imposition of charges levied on airlines by airports for services rendered has been the subject of discussion at many ICAO conferences. Part of the problem has been that airports have, over the years, been privatized and commercialized, necessitating them to be operated in a businesslike manner. However, some core issues have remained unchanged, the first being that, after everything is said and done, there is only one product in the air transport business and that is air transport which is provided by the airlines. The second is that it is an immutable principle that the State is ultimately responsible for meeting the needs of the people of the world for safe, regular, economic and efficient air transport services, working through ICAO, as per the Convention on international Civil Aviation. The blurring of concepts that has arisen in meshing these fundamental principles brings to bear the need to critically appraise one area that exemplifies the confusion—airports charges for services provided to airlines. This article critically appraises the issue and identifies certain anomalies that exist.

At the very core of the rationale for charging airlines for services rendered to them by airports is the Universal Declaration of Human Rights of the United Nations.[3] Article 7 of the Declaration states that all are equal before the law and are entitled without any discrimination to equal protection of the law. The provision goes on to say that all are entitled to equal protection against any discrimination in violation of the Declaration and against any incitement to such discrimination. Article 17 provides that everyone has the right to own property alone as well as in association with others[4] and that no one shall be arbitrarily deprived of his property.[5] Since property includes money[6] and charges levied on airlines by airports comprise money, the Declaration could be legally construed as prohibiting arbitrary charging in excess of amounts that correspond to the services rendered.

It has to be noted that the Declaration is not a treaty and therefore not a binding source of formal law. Therefore it is not a self-executing document and persons relying on the enforcement of these principles would have to rely on the justiciability of treaties that implement the Declaration.[7] However, the Declaration remains a statement of moral principles that is calculated to have a coercive influence on the community of nations. With this in mind, the starting point, as the moral denominator for charging, would be to recognize that to impose charges which are not commensurate with the services rendered would be tantamount to the levy of a tax that unjustly enriches the airport and the State concerned. In this context, it is worthy of note that ICAO, for the purpose of its policy objectives, makes a distinction between a charge and a tax, in that charges are levies to defray the costs of providing facilities and services for civil aviation while taxes are levies to raise general national and local government revenues that are applied for non-aviation purposes.[8]

---

[3]Adopted and proclaimed by General Assembly Resolution 217 A (III) of 10 December 1948.

[4]Article 17(1).

[5]Article 17(2).

[6]See Qureshi (1994) at 295.

[7]For example, on 4 November 1950 the Council of Europe member States signed the Convention for the Protection of Human Rights and Fundamental Freedoms, also known as the Rome Convention, which implemented the principles of the Universal Declaration of Human Rights and, in Article 25 accorded an individual the right to complain if his rights enshrined in the Declaration were eroded.

[8]ICAO's Policies on Taxation in the Field of Air Transport, Doc 8632, Third Edition: 2000, at 3.

Specific regulatory provisions applicable to charges levied by airports have their genesis in ICAO and are contained in Doc 9082[9] which, in the Foreword has the curious opening: "ICAO's policies on charges for airports and air navigation services which follow contain the recommendations and conclusions of the Council". It must be mentioned that the earlier version of this document used the words "statements" instead of "recommendations and conclusions". The former is seemingly more appropriate since the Chicago Convention does not empower the Council to arrive at recommendations and conclusions either in Article 54—which contain the mandatory functions of the Council—and Article 55—which lays down the permissive functions of the Council. These two provisions do not require the Council to issue guidance to States on any matter pertaining to air transport. However, Article 54 (b) makes it a mandatory function of the Council to carry out the directions of the Assembly, although there is no clear definition of the word "directions". If this word were to be interpreted to include a request of the Assembly, one could apply resolving clause 5 of Assembly Resolution A36-15[10] which requested the Council to ensure that the guidance and advice (not conclusions and recommendations) contained in Doc 9082 are current and responsive to the requirements of Contracting States. In this context, one can only laud the ICAO Council for taking a leadership role and for taking the initiative to publish its recommendations and conclusions, however inconsistent the words may be between the Assembly Resolution and those used in Doc 9082.

The inconsistency of wording does not stop there. Paragraph 8 (i) of Doc 9082 states that the Council recommends that States permit the imposition of charges only for services and functions which are provided for, directly related to or ultimately beneficial for, civil aviation operations. The anomaly lies in the word "functions" which is not defined or elaborated anywhere in the document. Does this give airports the licence to levy charges on airlines for "functions" as defined or determined by them? Or, could one take the wording of Doc 8632[11] and apply the word "facilities" which is used therein as being meant by the word "functions" in this context?

Recent trends in regulatory control of the levy of airports charges dictate that a close look should be taken with view to evaluating whether there exists an environment for the imposition on airlines of airports charges on a just and equitable basis. This article will discuss this issue with emphasis on currently applicable regulatory provisions. Although the ensuing discussions will focus only on airports charges, it must be noted that the applicable ICAO regulatory policy also applies *ex aequo* to air navigation services.

[9]ICAO's Policies on Charges for Airports and Air Navigation Services Doc 9082/7 Seventh Edition-2004.

[10]Resolution A36-15, Consolidated Statement of Continuing ICAO Policies in the Air Transport Field, *Assembly Resolutions in Force* (as of 28 September 2007) Doc. 9902, III-1 at III-I3.

[11]*Supra*, note 8.

2 Current Regulatory Provisions

The current policies of ICAO on charges for airports and air navigation services stemmed from the recommendations of the Conference on the Economics of Airports and Air Navigation Services (ANSConf 2000) which were endorsed by the Council of ICAO.[12] ANSConf 2000, which was held in Montreal on 19–28 June 2000, came to the conclusion that the profile of basic cost recovery policy may need to be raised.[13] It was recommended by the Conference that this measure could be adopted within the parameters of existing policy calling for revenues from charges levied on international civil aviation and it would only be applied towards defraying the costs of facilities and services provided for international civil aviation. It was also recommended that revenues from other sources than charges on air traffic shall be taken into account before the cost basis for charges on air traffic is determined. ICAO advised the Conference that airports and air navigation services may produce sufficient revenues to exceed all operating costs and so provide for a reasonable return on assets to contribute towards necessary capital improvements. Of course, the governing principle would be that consultation with users shall take place before significant changes in charging systems or levels of charges are introduced.[14]

As already mentioned, the baseline of ICAO's policies on charges lies in Article 15 of the Chicago Convention, the basic philosophy of which is that every airport in a Contracting State which is open to public use by its national aircraft shall likewise be open under uniform conditions to the aircraft of all the other Contracting States. It also requires that uniform conditions shall apply to the use, by aircraft of every Contracting State, of all air navigation facilities, including radio and meteorological services,[15] which may be provided for public use for the safety and expedition of air navigation.[16] Article 15 subsumes three fundamental postulates:

---

[12]See Report of the conference on the economics of airports and air navigation services: air transport infrastructure for the 21st century. Montreal, 19–28 June 2000. Doc 9764, ANSConf 2000. ICAO: Montreal, 2000. For a discussion on ANSConf 2000 see Abeyratne (2001).

[13]*ANSConf-WP/4* at para. 5.1.

[14]*Id.* para. 5.3. ICAO's recommendations to ANSConf 2000 were both timely and practical, given the evolving fabric of economic forces which now govern airports and air navigation services. The recommendations also stimulate some reflection on the complexities of financing principles now applicable to the services provided by airports and air navigation services providers. In substance, the issue of costing and pricing of services would be dependent upon underlying practices and economic factors as the bunching of aviation and non-aviation revenues and their effect on the overall pricing policy relating to airports and air navigation services and a significant paradigm shift from Article 15 of the Chicago Convention.

[15]Article 28 of the Chicago Convention calls on each Contracting State, so far as it may find practicable, to provide airport and air navigation facilities, in accordance with the standards and practices recommended or established in pursuance of the Convention.

[16]Article 15 also provides that any charges that may be imposed or permitted to be imposed by a Contracting State for the use of such airports and air navigation facilities by the aircraft of any other Contracting State shall not be higher: as to aircraft not engaged in scheduled international air

- Uniform conditions should apply in the use of facilities provided by airports and air navigation services;
- Aircraft operators should be charged on a non-discriminatory basis; and
- No charges should be levied for the mere transit over, entry into or exit from the territory of a Contracting State.

Current ICAO policy also recognizes that the financial situation of airports and air navigation services are in a constant state of evolution and that the financial situation of the primary users, the scheduled airlines, generally fluctuates with the performance of national, regional and global economies.[17] Accordingly, the ICAO Council recommends that States permit the imposition of charges only for services and functions which are provided for, directly related to, or ultimately beneficial for, civil aviation operations. States are therefore encouraged to refrain from imposing charges which discriminate against international civil aviation in relation to other modes of transport.[18]

ICAO's policies are at best only authoritative in practice and, from a legal perspective, are rendered destitute of effect by the acknowledged lack of enforcement power afflicting them. In this context it is curious that, six decades after the establishment of ICAO some still refer to its powers and functions.[19] There are some others who allude to ICAO's mandate. The fact is that ICAO has only aims and objectives, recognized by the Chicago Convention[20] which established the Organization.[21] Broadly, those aims and objectives are to develop the principles and techniques of international air navigation and to foster the planning and development of international air transport. In effect, this bifurcation implicitly reflects the agreement of the international community of States which signed the Chicago Convention that ICAO could adopt Standards in the technical fields of air navigation and could only offer guidelines in the economic field.

In its basic documentation, the ICAO Council notes that with the rapidly growing autonomy in the provision of airports and air navigation services, many States may wish to establish an independent mechanism for the economic regulation of airports and air navigation services[22] To this end the Council recommends

services, than those that would be paid by its national aircraft of the same class engaged in similar operations; and as to aircraft engaged in scheduled international air services, than those that would be paid by its national aircraft engaged in similar international air services.

[17]*ICAO's Policies on Charges for Airports and Air Navigation Services,* Abeyratne (2002) at paragraph 7.

[18]*Id.* Paragraph 8. Paragraph 9 that follows states that the Council is concerned over the proliferation of charges on air traffic and notes that the imposition of charges in one jurisdiction can lead to the introduction of charges in another jurisdiction.

[19]MacKenzie (2008), Preface at 1.

[20]Preamble, *supra* note 1.

[21]*Id.* Article 43. This article provides that an organization to be named the International Civil Aviation Organization is formed by the Convention. It is made of an Assembly, a Council, and such other bodies as may be necessary.

[22]For an extended discussion on this issue see Milde (2012) at paragraph 15 (ii).

inter alia that States should ensure there is no overcharging or other anti-competitive practice or abuse of dominant position.[23] The Council further states that for the successful collection of charges for airports and air navigation services entities, it is essential that a collection policy be established by an airport or air navigation services entity, or where applicable by a State.[24] In this regard the cost basis for airport charges is an important issue and the Council considers that, as a general principle it is desirable, where an airport is provided for international use, that the users shall ultimately bear their full and final share of the cost of providing the airport. It is therefore considered important that airports maintain accounts which provide information adequate for the needs of both airports and users and that the facilities and services related to airport charges be identified as precisely as possible.[25] It is interesting, once again to note that the term "facilities" has been used and that there is no mention of the word "functions". The cost to be shared is the full cost of providing the airport and its essential ancillary services, including appropriate amounts for cost of capital and depreciation of assets, as well as the cost of maintenance, operation, management and administration. However, there is a caveat that the costs to be shared must allow for all aeronautical revenues plus contributions from non-aeronautical revenues accruing from the operation of the airport to its operators.[26] The Council also states that the proportion of costs allocable to various categories of users, including State aircraft, should be determined on an equitable basis, so that no users shall be burdened with costs not properly allocable to them according to sound economic principles.

The aforesaid provisions have the underlying requirement of economic oversight if the regulators were to ensure that charges are being levied in a just and equitable manner.

3 Current Trends in Airports Charges

A Conference of ICAO on the Economics of Airports and Air Navigation Services (CEANS), which was held in Montreal from 15 to 20 September 2008,[27] agreed to submit to the Council of ICAO crucial recommendations for international civil aviation which will take cooperation between the air transport, airport and air navigation services industries to a higher level and increase the efficiency and cost-effectiveness in the provision and operation of airports and air navigation services around the world. These recommendations are calculated to serve the aviation industry expeditiously in coping with the current challenges that air transport faces.

---

[23] *Id.* paragraph 15 (ii).

[24] *Id.* paragraph 18.

[25] Doc 9082, *op. cit.*, note 8 paragraph 21.

[26] *Id.* paragraph 22 (i).

[27] CEANS was attended by 520 delegates from 104 States and 19 international organizations.

- Cost basis for charges: On the basis that the cost basis of an airport for charging purposes has usually been established by taking into account the costs of operation and maintenance, cost of capital and depreciation of assets (based on historical value in most cases), and a "reasonable" return on assets, CEANS approached the issue by attempting to build a consensus on possible ways to assess what would constitute a "reasonable" rate of return, and explore the possibility of consolidating several airport cost bases into one cost base for charging purposes.
- Cost allocation and charging systems: It was recognized that ICAO's current policies and guidance material on airport charges have provisions dealing with how the costs of the various airport facilities and services should be allocated to different categories of users. However, the Conference took note of the fact that some new trends have emerged such as the allocation of costs on a per passenger basis, which includes all or most cost bases of the aeronautical activities. The Conference's discussions revolved around the issue as to whether such new approaches are consistent with ICAO's policies, and consideration was given to appropriate amendments to the policies and guidance material on cost allocation.
- Non-discrimination aspects: On the subject of non-discrimination, CEANS recognized that, in recent years, airport operators have developed certain differential charges to attract and retain new airline services, for example, discounts on passenger service charges and incentive schemes for particular airlines, including low-cost carriers. It was also noted that some of these differential charges might be non-transparent, discriminatory and anti-competitive, especially when they constitute a form of State aid. The Conference addressed the issue as to how to deal with the measures taken by airport operators that have the potential to create unfair treatment, as well as the issue of access to airport facilities.
- Financing and cost recovery of security measures: According to ICAO's policies on security charges, the costs of security functions performed by States such as general policing, intelligence gathering and national security should not be passed on to the airport users. However, it was noted that practices differ between regions and States, which have financial implications on users. The Conference reviewed the current policies and discussed how to achieve a more harmonized implementation of ICAO's policies regarding airport security charges.

4 Recommendations of CEANS

CEANS recommended that ICAO should amend Doc 9082 with a view to allowing more flexibility in setting airport charges. This calls for airports to maintain cost data in adequate detail which ensures transparency and oversight and the avoidance of discrimination in setting charges.

On rate of returns for airports and air navigation services it was recommended that within their economic oversight responsibilities, States should, where necessary and in the light of national circumstances, clearly define the methodology for determining what is a reasonable rate of return on assets for their service providers. ICAO was called upon to develop additional guidance material regarding possible methodologies to assess the risk element involved in cost recovery and the value of assets in the context of the determination of a reasonable rate of return.

On differential charges CEANS recommended that within their economic oversight responsibilities States should, where necessary, assess the positive and negative effects associated with specific forms of differential charges applied by airports on a case-by-case basis according to national circumstances. Furthermore, States were called upon to ensure that differential charges are offered on a non-discriminatory basis; that they are transparent in terms of their creation, purpose and the criteria on which they are offered; that, without prejudice to modulated charging schemes, costs associated with differential charges are not allocated, either directly or indirectly, to those other users not benefiting from them; and that, if the purpose is to attract and/or retain new air services, they are offered only on a temporary basis. ICAO was called upon to amend Doc 9082 to reflect the principles of transparency and time limitation for start-up aids in the application of differential charges.

It was the view of CEANS that the recommendations will make ICAO's policies on charges, which regulate the relationship between airports and air navigation services providers (ANSPs) on the one hand, and airlines and other airport and airspace users on the other, more authoritative in practice. The enhanced cooperation suggested by these recommendations would strengthen policies on States' economic oversight responsibility, requirements on implementation of performance management systems by all airports and ANSPs, and the establishment of a clearly defined, regular consultation process by all airports and ANSPs. At the same time, they recommend that States enshrine the main principles of non-discrimination, cost-relatedness, transparency and consultation with users in their national legislation, regulations or policies as well as all air services agreements between States.[28]

One of the fundamental premises addressed by CEANS is that the protection of users against the potential abuse of dominant position by airports and air navigation services providers is the primary responsibility of the State and could be discharged by the exercise of economic oversight. It was suggested during the discussions that such oversight could be effectively carried out by diligent monitoring by a State of the commercial and operational practices of these service providers.

There was discussion during CEANS where some delegations suggested that, in order to "give teeth" to ICAO policy, there be a recommendation in Doc 9082 to the effect that amendment to Doc 9082 should be incorporated by States in their national legislation. It is submitted that such a measure would be tantamount to treading uncharted and dangerous ground. While it is one thing to assert that the only way that ICAO policies could be implemented is for States to opt for

[28]Other essential features of the recommendations of the Conference are: more flexibility for commercialized airports and ANSPs in setting charges; support for separation of regulation from service provision; the application of good governance through best practices; and the efficient and cost-effective implementation of the global Air Traffic Management (ATM) concept.

incorporating such principles in their legislation, it is something quite different to recommend that States go ahead and do so.

As a necessary compromise and in order to reach a balance, the Conference broadly recognized the need for economic oversight in the increasingly commercialized and privatized environment for airports and air navigation services. It considered a number of suggestions that were made by the delegates for improving the proposed new text for Doc 9082. The following conclusions were reached by the Conference:

- States should bear in mind that economic oversight is the responsibility of States with the objectives, *inter alia*, to prevent the risk that a service provider could abuse its dominant position, to ensure non-discrimination and transparency in the application of charges, to encourage consultation with users, to ensure the development of appropriate performance management systems, and to ascertain that capacity meets current and future demand, in balance with the efforts of the autonomous/private entities to obtain the optimal effects of commercialization or privatization;
- States should select the appropriate form of economic oversight according to their specific circumstances, while keeping regulatory interventions at a minimum and as required. When deciding an appropriate form of economic oversight, the degree of competition, the costs and benefits related to alternative oversight forms, as well as the legal, institutional and governance frameworks should be taken into consideration;
- States should consider adoption of a regional approach to economic oversight where individual States lack the capacity to adequately perform economic oversight functions; and
- ICAO should amend Doc 9082 to clarify the purpose and scope of economic oversight for airports and air navigation services with reference to its different forms and the selection of the most appropriate form of oversight.[29]

5 The Legal Status of ICAO Policy

Although there was much discussion at CEANS on "giving teeth" to ICAO policy in order to ensure economic oversight by States of their airports and air navigation services providers, the Conference failed to arrive at a consensus on including text in the recommendations to the effect that States should incorporate the principles enunciated in ICAO document 9082 in their legislation or rules. The end result was a somewhat watered down recommendation that States should select the appropriate form of economic oversight according to their specific circumstances, while keeping regulatory interventions at a minimum and as required. Most delegations were, quite rightly, reluctant to agree to a recommendation that would impose upon

[29]Draft Report on Agenda Item 1.1., Economic Oversight, CEANS-WP/73, 16/9/08, Draft Report on Agenda Item 1.1.

States an obligation to incorporate policy guidelines into national legislation. Furthermore, the Conference correctly noted that States differed considerably in their economic circumstances and demand for services rendered by airports and air navigation services in their territories and therefore should be left to decide the best course of economic oversight to be taken in their territories.

ICAO's economic policies emanate from the States. However, these policies are no more than consensual principles that offer policy guidance and are at best left to the discretion of the States to follow. To require or recommend that States incorporate such policy in their national legislation or regulations is a reversal of the empowerment of ICAO by States upon which ICAO is founded, whereby ICAO is enabled by States to pursue its aims and objectives under Article 44 of the Chicago Convention.[30]

No international body or institution can legitimately expect a sovereign State to incorporate, as national legislation, policies that the former adopts. In this case, such a policy directive would come from the ICAO Council, the mandatory functions of which are stipulated in Article 54 of the Chicago Convention.[31] Nowhere in either the mandatory or permissive functions (contained in Article 55) is the Council given authority to act as legislator or regulator.[32] Even if such a function were to be elevated to the level of the ICAO Assembly, a resolution of the Assembly cannot require or even recommend that its principles be incorporated into national law or regulation. *Brownlie* has expressed the view that decisions by international conferences and organizations can in principle only bind those States accepting them.[33] Shaw, referring to the binding force of United Nations General; Assembly Resolutions states:

> ...one must be alive to the dangers in ascribing legal value to everything that emanates from the Assembly. Resolutions are often the results of political compromises and arrangements and, comprehended in that sense, never intended to constitute binding norms. Great care must be taken in moving from a plethora of practice to the identification of legal norms.[34]

---

[30]The overarching aims and objectives of ICAO, as contained in Article 44 of the Convention is to develop the principles and techniques of international air navigation and to foster the planning and development of international air transport so as to meet the needs of the peoples for safe, regular, efficient and economical air transport.

[31]The closest the Council comes in this respect is to adopt morally binding Standards and Recommended Practices, where, as per Article 54 (l) of the Chicago Convention the Council may adopt international standards and recommended practices and for convenience, designate them as Annexes to the Convention and notify all Contracting States of the action taken.

[32]For a discussion on the role of the Council in this context see Abeyratne (1992). Also by the same author, see The Settlement of Commercial Aviation Disputes Under the General Agreement on Trade in Services and the ICAO Council - A Comparative Analysis. *International Trade Law and the GATT/WTO Dispute Settlement System*; Kluwer Law International: London, 1997, 395–412.

[33]Brownlie (1990), 691.

[34]Shaw (2003), 110.

With regard to the practice of other international organizations, a little more caution might be required, as a resolution might create a custom. Non binding instruments form a special category that is sometimes referred to as "soft law" which is definitely not law in the sense of enforceability.[35]

The above discussion brings to bear some anomalies that exist in the field of airports charges. The first is that, according to ICAO policy, four elements are critical for prudent charges policy: non discrimination (as enshrined in Article 15 of the Chicago Convention); transparency; cost relatedness; and consultation. The first two elements are self explanatory. However, cost relatedness and consultation are open to interpretation. One could argue that cost related charges need not necessarily be restricted to actual costs but could be geared to earn profits for the airports as long as such costs are calculated in relation to the cost of services provided. With regard to consultation, there have been instances where the service provider has met with users and other stakeholders and merely informed them that certain charges were to be increased.

The second anomaly is that, the statement in Doc 9082—that autonomy and privatization of airports are preferred modes of operating airports[36]—has inadvertently resulted in the obfuscation of the fundamental principle that the ultimate responsibility for the setting and levying of charges rests with the State concerned. Although by and large States have been observed to follow ICAO policy in this area, there are many airports today which set charges and impose them, with a cursory and *pro forma* notice to the State concerned which invariably approves it. The Chicago Conference of 1944, which resulted in the adoption of the Chicago Convention, in its consideration of draft Article 15 of the Convention at that time, has explicitly recorded that

> "[E]ach Contracting State shall establish scales of charges for the use of such airports and air navigation facilities which shall be uniformly applicable to the aircraft of all other States. . ."[37]

Article 15 of the Chicago Convention follows this approach when it states: "Any charges that may be imposed or permitted to be imposed by a Contracting State. . ."clearly implying that it is the State which is responsible for the imposition of charges".

The third anomaly is that Doc 9082 which sets policy at a high level, is open to interpretation as some of its key provisions, as pointed out in the Introduction to this article, may open the possibility for interpretation and subjective treatment of critical principles concerned with ICAO's charging policy. This notwithstanding, there is no room for doubt that Doc 9082 is a generally clear policy statement which has served ICAO Contracting States well and provided guidance over the years.

---

[35] *Id.* 111. See also Tammes (1958) at 265.

[36] Doc 9082, Abeyratne (2002) at paragraph 10.

[37] *Proceedings of the International Civil Aviation Conference*, Chicago: Illinois November 1–December 7, 1944, Vol. 1, at 663.

Finally, this discussion brings one to the conclusion that, ultimately, the responsibility clearly lies with the States, which not only have to oversee airport charges but to provide the necessary economic oversight to ensure that charges are levied justly, equitably and in a prudent manner. One effective way of ensuring this is for States to include the four elements of transparency; non-discrimination; cost relatedness and consultation in their bilateral air services agreements.

References

Abeyratne RIR (1992) Law making and decision making powers of the ICAO council - a critical analysis. Zeitschrift Fur Luft-und Weltraumrecht 41(4):387–394

Abeyratne RIR (2001) Revenue and investment management of privatized airports and air navigation services: a regulatory perspective. J Air Transport Manag 7:217–230

Abeyratne RIR (2002) International responsibility in preventing the spread of communicable diseases through air carriage—the SARS crisis. Transp Law J 30(1):53–80

Brownlie I (1990) Principles of public international law, 4th edn. Clarendon Press, Oxford

MacKenzie D (2008) ICAO, a history of the international civil aviation organization. University of Toronto Press, Toronto

Milde M (2012) International air law and ICAO. In: Benko M (ed) Essential air and space law. Eleven International Publishing, The Hague

Qureshi AH (ed) (1994) The public international law of taxation, text, cases and materials. Graham & Trotman Ltd, London

Shaw MN (2003) International law, 5th edn. Cambridge University Press, Cambridge

Tammes AJP (1958) Decisions of international organs as a source of international law. HR 94:265

Article 16
Search of Aircraft

The appropriate authorities of each of the contracting States shall have the right, without unreasonable delay, to search aircraft of the other contracting States on landing or departure, and to inspect the certificates and other documents prescribed by this Convention.

Contents

1 Search and Rescue

It is not entirely clear as to whether this provision solely pertains to documents carried in an aircraft. One could argue that there are two activities of State permitted by this provision: Searching an aircraft and inspecting certificates and other documents. The first may mean both searching an aircraft for security and safety reasons and searching for documents. The second clearly pertains to inspection of documents.

With regard to searching an aircraft and the State's role Standard 2.1 of Annex 9 requires Contracting States to adopt appropriate measures for the clearance of aircraft arriving from or departing to another Contracting State and shall implement them in such a manner as to prevent unnecessary delays. This is followed by Standard 2.2 which provides that in developing procedures aimed at the efficient clearance of entering or departing aircraft, Contracting States shall take into account the application of aviation security and narcotics control measures, where appropriate.

Article 29 of the Chicago Convention prescribes the documents that should be carried in an aircraft. They are (a) the certificate of airworthiness; (b) the appropriate licenses for each member of the crew; (c) its journey log book; (d) if it is equipped with radio apparatus, the aircraft radio station license; (e) if it carries passengers, a list of their names and places of embarkation and destination; (f) if it carries cargo, a manifest and detailed declarations of the cargo.

Article 32 of the Chicago Convention provides that the pilot of every aircraft and the other members of the operating crew of every aircraft engaged in international navigation shall be provided with certificates of competency and licenses[1] issued or

[1]The expression "licence" used throughout this Annex has the same meaning as the expressions "certificate of competency and license", "license or certificate" and "license" used in the Convention. Similarly the expression "flight crew member" has the same meaning as the expressions

R. Abeyratne, *Convention on International Civil Aviation*,
DOI 10.1007/978-3-319-00068-8_17,

rendered valid by the State in which the aircraft operates. The provision also states that each ICAO member State reserves the right to refuse to recognize, for the purpose of flight above its own territory, certificates of competency and licences granted to any of its nationals by another member State. Member States of ICAO, at the 21st Session of the ICAO Assembly in Resolution A21-21, which has been alluded to earlier, in Appendix A resolves that certificates of airworthiness and certificates of competency and licenses of the crew of an aircraft issued or rendered valid by the ICAO member State in which the aircraft is registered shall be recognized as valid by the other States for the purpose of flight over their territories, including landings and take offs subject to the provisions of Articles 33 and 32 (b) of the Chicago Convention.

Article 33 provides that Certificates of airworthiness and certificates of competency and licenses issued or rendered valid by the Contracting State in which the aircraft is registered, shall be recognized as valid by the other Contracting States, provided that the requirements under which such certificates or licences were issued or rendered valid are equal to or above the minimum standards which may be established from time to time pursuant to the Convention. Article 32 (b) provides, as mentioned earlier that each ICAO member State reserves the right to refuse to recognize, for the purpose of flight above its own territory, certificates of competency and licences granted to any of its nationals by another member State.

Standards and Recommended Practices for Personnel Licensing were first adopted by the Council on 14 April 1948 pursuant to the provisions of Article 37 of the Convention on International Civil Aviation (Chicago 1944) and designated as Annex 1 to the Convention. They became effective on 15 September 1948. The most recent amendment to Annex 1, was Amendment 168 (Annex 1, Tenth Edition), adopted by the Council on 23 February 2007. The amendment concerns: the replacement of the approach and area radar control ratings by approach and area control surveillance ratings to reflect the fact that surveillance systems are not limited to radar; the harmonization of the Human Factors knowledge requirements for air traffic controllers with those recently adopted as part of Amendment 167 to Annex 1 for flight crew; the applicability of the existing Standards on approved training for flight crew (Annex 1, 1.2.8 and Appendix 2) to the approved training required for the air traffic controller licence and ratings; and new provisions for student air traffic controllers receiving instruction in an operational environment.

Annex 1 contains Standards and Recommended Practices adopted by the International Civil Aviation Organization as the minimum standards for personnel licensing. The Annex is applicable to all applicants for and, on renewal, to all holders of the licences and ratings specified herein. The ICAO Council has decided that, in principle, amendments affecting existing licensing specifications are applicable to all applicants for, and holders of, licences but, in considering their application to existing holders of licences, the assessment, if necessary, by re-examination

---

"member of the operating crew of an aircraft" and "operating personnel" used in the Convention while the expression "personnel other than flight crew members" includes the expression "mechanical personnel" used in the Convention.

of the knowledge, experience and proficiency of individual licence holders is left to the discretion of Contracting States.

As long as air travel cannot do without pilots and other air and ground personnel, their competence, skills and training will remain the essential guarantee for efficient and safe operations. Adequate personnel training and licensing also instil confidence between States, leading to international recognition and acceptance of personnel qualifications and licences and greater trust in aviation on the part of the traveller. Standards and Recommended Practices for the licensing of flight crew members (pilots, flight engineers and flight navigators), air traffic controllers, aeronautical station operators, maintenance technicians and flight dispatchers, are provided by Annex 1 to the Chicago Convention.

Related training manuals provide guidance to States for the scope and depth of training curricula which will ensure that the confidence in safe air navigation, as intended by the Convention and Annex 1, is maintained. These training manuals also provide guidance for the training of other aviation personnel such as aerodrome emergency crews, flight operations officers, radio operators and individuals involved in other related disciplines.

Today's aircraft operations are so diverse and complex that protection must be provided against the possibility, however remote, of total system breakdown due to either human error or failure of a system component. The human being is the vital link in the chain of aircraft operations but is also by nature the most flexible and variable. Proper training is necessary so as to minimize human error and provide able, skilful, proficient and competent personnel. Annex 1 and ICAO training manuals describe the skills necessary to build proficiency at various jobs, thereby contributing to occupational competency. The medical standards of the Annex, in requiring periodic health examinations, serve as an early warning for possible incapacitating medical conditions and contribute to the general health of flight crews and controllers.

The Human Factors programme addresses known human capabilities and limitations, providing States with basic information on this vital subject as well as the material necessary to design proper training programmes. ICAO's objective is to improve safety in aviation by making States more aware of, and responsive to, the importance of human factors in civil aviation operations. Licensing is the act of authorizing defined activities which should otherwise be prohibited due to the potentially serious results of such activities being performed improperly. An applicant for a licence must meet certain stated requirements proportional to the complexities of the task to be performed. The licensing examination serves as a regular test of physical fitness and performance ensuring independent control. As such, training and licensing together are critical for the achievement of overall competency.

One of ICAO's main tasks in the field of personnel licensing is to foster the resolution of differences in licensing requirements and to ensure that international licensing standards are kept in line with current practices and probable future developments. This is ever more crucial as the flight crew will be exposed to increasing traffic density and airspace congestion, highly complicated terminal area patterns and more sophisticated equipment. To accomplish this task, Annex I is regularly amended to reflect the rapidly changing environment.

Article 16, which seemingly gives blanket approval of a State's right to board an aircraft with a view to search it, has ramifications at public international, and as such should be viewed with caution as it brings to bear a State's responsibility *vis a vis* the property of another State and the conduct required of a State with regard to such responsibility. The principle of State Responsibility lies primarily in Article 24 of the United Nations Charter[2] which calls upon all members to refrain in their international relations from the threat or use of force against the territorial integrity or political independence of any state, or in any other manner inconsistent with the purposes of the United Nations.

2 Safety of Aircraft

One instance that a State could clearly search an aircraft is when, upon landing it has reason to believe that the crew and passengers are at risk of exposure to deleterious materials in the aircraft. The aerotoxic syndrome comes to mind, where, if there are complaints from either category, the State in which the aircraft lands could board with a view to searching an aircraft.

It is a platitude to say that air travel is the safest means of transport. Snug in this belief, a passenger takes comfort in the fact that he is just one of 2.2 billion passengers that are transported by air every year and that no danger would be lurking to threaten his life during the flight. The air traveller, and in particular the passenger, also believes that the modern aircraft is extremely sophisticated in design and structure and that it has passed rigorous and stringent tests before it has been certified to fly across oceans and continents. The glamorous captain, and the humble chap in overalls who performs the daily maintenance check on the aircraft also assure him that no ill can befall the aircraft in which he travels. It is hard to think otherwise, when billions of people have travelled by air comfortably over the years at 7 miles a minute, while the temperature outside is twice as cold as the freezer in one's fridge at home.

The only snag seems to be in what one breathes while travelling.[3] Under normal circumstances, the aircraft cabin environment is bad enough. An aircraft in flight is

[2]Charter of the United Nations and Statute of the International Court of Justice, United Nations, New York.

[3]The ventilation system plays a critical part in this regard and therefore, it is crucial to an airline's conduct to determine the manner in which that carrier decides on ventilation systems in its aircraft. For instance, early jet aircraft until the last decade offered 100 % fresh air in the cabin. However, in the 1990s, ironically with more evolved technology, ventilation systems in aircraft were built in such a way as to recycle stale air, thus increasing the chances of survival of bacteria and deleterious particles in the aircraft cabin. Even if such a practice were ineluctable, in that recycling is a universal practice which is calculated to conserve fuel, a prudent airline would take other measures, such as change of air filters through which ventilation is provided.

a pressurised, airborne, air-conditioned, densely populated tourist and business facility at a high altitude with a relative humidity similar to that of Antarctica.[4] Inside the aircraft, humans release "on occasion, hostile viruses and bacteria, shed dead skin particles, fungal spores and emit body odours".[5]

Additionally, materials used in the operation of aircraft may contain hazardous ingredients, some with significant toxicities. Aircraft material such as jet fuel, de-icing fluids, engine oil and hydraulic fluids contain a range of ingredients, some of which are toxic.[6] Engine oil, hydraulic fluids and other materials which have in common the chemical presence of toxic ingredients such as organo-phosphates are used as of necessity by the aviation industry. Although these chemicals are usually contained within the engines and equipment into which they have been added, they can sometimes find their way into cabin air where crew and passengers are located. Most common causes of leakages of this kind are engine oil leaks, fluid ingestion by the Auxiliary Power Unit (APU) and engines, and by failure of seals designed to preclude seepage and leakage through ventilation systems. It is a fact that the air inhaled by those in the aircraft in flight is a mixture of "bleed air" (fresh air taken in from the atmosphere through the jet engines, part of which is used for pressurization of the cabin and the other part of which is used for purposes of inhalation) and air that is re-circulated after use.[7] If the engine seals (that seal off the lubricating oil) are not working properly, there could be an oil leak contaminating the air.[8] The leaks do not usually occur spontaneously as the seals do not leak suddenly but take a while to give way, spilling oil into the engine over a period of time.[9] This can happen under certain phases of flight where there is a great load on the engine than at other phases or when the aircraft is descending where the load on the engine is less than when there is less load.

Researchers have carried out tests over several years and concluded that passengers and crew have, immediately after flights, suffered from a range of common and similar symptoms which leave them debilitated, weak and suffering from coughing and difficulty to breathe. This basket of symptoms, which is yet to be identified as a single disease, is called the aerotoxic syndrome.[10] The toxicity and irritation causing this syndrome is due to neurotoxic organophosphates which contaminate the air that circulates in the cabins of aircraft propelled by jet engines.[11] The cause for this contamination has been identified as the use of lubricating oils and hydraulic fluids in jet engines and flawed designing of the air intake mechanisms.[12]

---

[4]Crawford (1989) 12.

[5]Holcomb (1988) at 3.

[6]See Rayman and McNaughten (1983) at pp. 738–740, Smith et al. (1997) at pp. 625–632.

[7]Captain Susan Michaelis (2007) at iii.

[8]Lorraine (2007) at 19–20.

[9]Captain John Hoyte, Aerotoxic Syndrome—Aviation's Best Kept Secret, http://www.aerotoxic.org/download/docs/news_and_articles/NEXUS-Aerotoxic-Syndrome.pdf.

[10]The term "aerotoxic syndrome" was first suggested in 1999. See Winder et al. (2002), at 321–338.

[11]Hale and Al-Seffar (2008) at 107.

[12]S. Myhill, Aerotoxic Syndrome, www.aerotoxic.org/articles/20071118.

Pilots have also been affected by the ill-effects of this syndrome, causing widespread illness[13] and compelling some of them to retire prematurely.[14] Since the first known instance of a study on this subject in 1977[15] where a 34 year old pilot was examined for inhaling oil fumes and subsequently developing mental disorientation and neuromuscular discomfort, there have been several studies that have borne witness to illness in pilots due to the inhalation of contaminated air in the cockpit. For example, an inquiry into this issue in 2000 by the Australian Senate revealed that pilots had been disoriented and unable to concentrate on flying the aircraft, due to a feeling akin to being drunk.[16] In the same year there were four bulletins issued by the Civil Aviation Authority (CAA) of the United Kingdom, warning of the danger of pilot incapacitation by contaminated cabin air and suggesting procedures to counter the problem.[17] A Report published by the UK CAA in 2001 resulting from research into pilot incapacitation by contaminated air concluded that engine oil fumes could have probably caused the incapacitation.[18] In the same year, a similar conclusion was reached by the Swedish air safety authorities.[19] Three years later, in the United States, the Federal Aviation Administration issued a directive that required the operators of BAe 146 aircraft to preclude oil residue from accumulating in the air system ductwork of the aircraft.[20]

---

[13]The symptoms are said to vary from fatigue, sleep deprivation, blackouts, seizures, neuromuscular pain and weakness. See Winder and Balouet (2001), 471–483.

[14]J. Hoyte, Captain Hoyte's Account, www.aerotoxic.org/articles 20071114. Also, Toxic Free Airlines,(TFA) Poisoned Pilots Launch Campaign at Parliamentary Meeting: The Aerotoxic Association and Toxic Free Airlines to Expose Massive Public Health Scandal and Support Victims, www.toxicfreeairlines.com.

[15]Montgomery et al. (1977), 423–426.

[16]Technical Report on Air Safety and Cabin Air Quality in the BAe146 Aircraft, Parliament of the Commonwealth of Australia, Senate Rural and Regional Affairs and Transport Legislation Committee, Senate Printing Unit: Canberra, Australia, 2000 at p. 115–128.

[17]CAA (2008) 'Flight Operations Department Communications (FODCOM) 17/2008' UK Civil Aviation Authority, Safety Regulation Group, Aviation House, Gatwick, West Sussex, England; CAA (2002) 'Flight Operations Department Communications (FODCOM) 21/2002' UK Civil Aviation Authority, Safety Regulation Group, Aviation House, Gatwick, West Sussex, England; CAA (2001) 'Flight Operations Department Communication (FODCOM) 14/2001' UK Civil Aviation Authority, Safety Regulation Group, Aviation House, Gatwick, West Sussex, England. CAA (2000) 'Flight Operations Department Communication (FODCOM) 17/2000' UK Civil Aviation Authority, Safety Regulation Group, Aviation House, Gatwick, West Sussex, England.

[18]Cabin air quality' CAA Paper 2004/04, Research Management Department, Safety Regulation Group, UK Civil Aviation Authority, Aviation House, Gatwick Airport South, West Sussex, UK. 6. Cabin air quality' CAA Paper 2004/04, Research Management Department, Safety Regulation Group, UK Civil Aviation Authority, Aviation House, Gatwick Airport South, West Sussex, UK.

[19]'Report RL 2001:41e 'Accident investigation into incident onboard aircraft SE-DRE during flight between Stockholm and Malmo M County, Sweden,' Statens Haverikommission Board of Accident Investigation, Stockholm, Sweden.

[20]Airworthiness Directive 2004-12-05: BAE Systems (Operations) Limited Model BAe 146 Series Airplanes' Docket No. 2003-NR-94-AD, Federal Aviation Administration, Washington, DC.

Although it is generally accepted that this problem is not confined to any particular type of aircraft and that all jet aircraft remain vulnerable to the seepage of oil from the engines and possible contamination of bleed air, a view has been expressed that the Boeing 787, which will come into operation in 2010, will not have this problem since supply air in this aircraft is processed in electrically generated compressors independent of the jet engines.[21] This design, called the "no bleed architecture", relies on electrically driven compressors to provide cabin pressure where fresh air is brought on board dedicated cabin air inlets.[22]

The Chicago Convention in its Annex 8 (Airworthiness of Aircraft)[23] stipulates that a certificate of airworthiness shall be issued by an ICAO member State concerning an aircraft, conditional upon and based on satisfactory evidence being received that the aircraft complies with the design aspects of the appropriate requirements.[24] The Annex goes on to state that a Contracting State shall not issue or render valid a Certificate of Airworthiness for which it intends to claim recognition pursuant to Article 33[25] of the Convention on International Civil Aviation unless it has satisfactory evidence that the aircraft complies with the applicable Standards of the Annex through compliance with appropriate airworthiness requirements.[26] The Annex also requires States to establish a safety programme with a view to achieving an acceptable level of safety in civil aviation,[27] while going on to say that the design of the airplane shall take into consideration The design of the aeroplane shall take into consideration the flight crew operating environment including: (a) effect of aeromedical factors such as level of oxygen, temperature, humidity, noise and vibration; (b) effect of physical forces during normal flight; (c) effect of prolonged operation at high altitude; and (d) physical comfort.[28]

ICAO Assembly Resolution A35-12[29] declares that the protection of the health of passengers and crews on international flights is an integral element of safe air travel and that conditions should be in place to ensure its preservation in a timely

---

[21]Submission by Susan Michaelis (Capt): To accompany all sections of A-NPA comments made by EASA CRT, 8/1/10, RE: A-NPA No. 2009–10 'Cabin air quality onboard large aeroplanes' at p. 9.

[22]See Sinnett (2007) at p. 8.

[23]Annex 8 to the Convention on International Civil Aviation—*Airworthiness of Aircraft*—Tenth Edition, April 2005.

[24]*Id.* Standard 3.2.1.

[25]Article 33 provides that certificates of airworthiness and certificates of competency and licenses issued or rendered valid by the contracting State in which the aircraft is registered, shall be recognized as valid by the other contracting States, provided that the requirements under which such certificates or licences were issued or rendered valid are equal to or above the minimum standards which may be established from time to time pursuant to the Convention.

[26]*Id.* Standard 3.2.2.

[27]*Id.* Standard 5.1.

[28]Id. Part III B, Sub Part J, J-4 (Operating Environmental Factors).

[29]Resolution A35-12 *Protection of the Health of Passengers and Crews and Prevention of the Spread of Communicable Disease through International Travel*, Assembly Resolutions in Force (as of 28 September 2007), ICAO Doc 9902, at 1–77.

and cost effective manner. It also requests the Council of ICAO to support further research on the consequences of air transport on the health of passengers and crews. Resolution A 27-13[30] reaffirms the public-service character of the service provided by air transport operators, recognizing that the essential purpose of such a service is to satisfy the common good of peoples in whose development States, carriers and users are all equally interested.

The last mentioned regulatory requirement is especially relevant to some recent cases, the first being *Victoria Vaughn Holsted and Valerie Vaughn*, Petitioners, *vs. Southwest Airlines Co.*[31] The facts of this case were as follow: on January 27, 2009, petitioners Victoria and Valerie boarded Southwest Airlines Flight 1705, which departed at approximately 10:00 AM from Los Angeles International Airport with stops scheduled for Nashville International Airport in Nashville, TN, then Birmingham-Shuttlesworth International Airport in Birmingham, AL and finally Baltimore/Washington International Airport in Baltimore, MD. About 1-h into the flight they, along with the other passengers, began to experience hypoxia (oxygen deprivation), among other problems. Once alerted to the air quality problem in the cabin, the pilot then engaged the engines at full thrust and entered a steep ascent. As this occurred, super heated air began to surge out of the ventilation system and onto the passengers. Also present was the appearance of a mist.

In their petition to the Los Angeles Superior Court for the preservation of evidence against the defendant airline, the petitioners claimed that exposure to contaminated air had caused them to suffer serious and debilitating health problems, among them motor skill deficiencies, loss of balance, vision impairment and uncontrollable tremors. The plane in question was a Boeing 737–300 jet aircraft. They also claimed that, despite repeated requests, the defendant airline was ambivalent and vague in its responses to questions posed by the petitioners that were calculated to assist the physicians of the petitioners so that the doctors would know how to best treat them.

Accordingly, the petitioners prayed for damages on the ground that the defendant was negligent or wanton in that it failed to follow relevant safety, operation, maintenance, repair, service and inspection procedures with regard to the subject aircraft and that the defendant failed to provide its passengers with an aircraft that was in good mechanical condition and free of defects. The petitioners also claimed damages in breach of contract on the ground that the defendant agreed to transport them for compensation and the contract of carriage included the agreement express and/or implied, to transport them safely, in a non-negligent manner.[32]

[30]*Id.* 1–79.

[31]Case No. BS120400, http://www.finanznachrichten.de/nachrichten-2009-04/13758467-southwest-airlines-flight-1705-passengers-file-petition-against-the-airline-to-preserve-evidence-of-onboard-exposure-to-contaminated-air-causing-them-004.htm.

[32]Case No. CV-09-HGD-2193-s in the District Court of Northern District of Alabama, 28 October 2009. 46 N. W. 677.

The second case which is of relevance is *Turner* v. *Eastwest Airlines Limited*[33] which involved an action instituted by an employee of an airline who claimed that, on descent, there was smoke in the cabin of the aircraft in the nature of a thick cloud of smoke which she inhaled, causing her to cough and break out in sore eyes, a burning throat accompanied by a headache. The cough had persisted thereafter, causing periods of paroxysms of coughing. The tribunal hearing the case sought answers to such questions as "what was the plaintiff's condition and cause of her illness?" "Was the injury foreseeable" "did the airline have a reasonable response to the problem when it arose?" "Was there economic loss" and "is the plaintiff entitled to damages?" On various counts, the New South Wales Dust Diseases Tribunal, which heard the case, awarded the plaintiff $137,757, the counts being *inter alia*, non economic loss, loss of earnings, future loss of earnings, past out of pocket expenses, future out of pocket expenses.

The defendant appealed against the award. In September 2010, the High Court upheld the decision of the Dust and Diseases Tribunal.[34]

When there is incontrovertible evidence of a person Contracting a disease as a result of being contaminated in an aircraft whilst on board, liability issues pertaining to the airline arising from the incident may involve principles of private air carrier liability. The *Montreal Convention* of 1999[35] which emerged consequent to the Diplomatic Conference on Private Air Law of the International Civil Aviation Organization held from 10 to 28 May 1999, provides that the carrier is liable for damage sustained in the event of death or bodily injury of a passenger upon condition only that the accident which caused the damage so sustained took place on board the aircraft or in the course of any of the operations of embarking or disembarking. The *Warsaw Convention* of 1929[36] provides that the carrier is liable for damage sustained in the event of death or wounding of a passenger or any other bodily injury suffered by a passenger, if the accident which caused the damage so sustained took place on board the aircraft or in the course of any of the operations of embarking or disembarking. Both these Conventions have similar wording, admitting only of death or bodily injury or wounding. Of course, on the face of the provision, the words "wounding" and "bodily injury" do not necessarily lend themselves to be associated with infection. *A fortiori*, according to the Montreal

[33][2009] NSW DDT 10, 5 May 2009. Also, New South Wales Dust Diseases Tribunal 10 (5 May 2009) Matter Number 428 of 2001, discussed in *ZLW*, 58 Jg 4/2009, at pp. 705–717.

[34]See, Flight attendant wins toxic cabin air damages, *Air Letter*, No. 17,074 Thursday 16 September 2010 at p. 3. This article states that A University of New South Wales survey has found that about 25 % of pilots who flew the BAe 146 aircraft suffered long term health degradation that deprived them of their pilot licences and that an Australian Senate inquiry had found East–west and Ansett Airlines had been paid more than $2 million by BAe Systems (British Aerospace's successor) to drop complaints about the aircraft.

[35]Convention for the Unification of Certain Rules for International Carriage by Air, signed at Montreal on 28 May 1999, ICAO Doc 9740.

[36]Convention for the Unification of Certain Rules Relating to International Carriage by Air, signed at Warsaw on 12 October 1929.

Convention, the bodily injury must be caused as a result of an accident, and, according to the Warsaw Convention, the wounding or injury must be caused by accident which is not typically a synonym for "infection" in both cases. However, the recent decision in *El Al Isreal Airlines Limited* v. *Tseng*[37] introduced a new dimension to the word "accident" under the Warsaw Convention by giving it pervasive scope to include such acts as security body searches performed by the airlines. In this context, the word "accident" loses its fortuity and it becomes applicable to an expected or calculated act. Thus, if an airline knows or ought to have known that there could be oil leakage from its engines that would mix with bleed air and make the passengers sick, it may well mean that the act of the airline would be construed by the courts as an accident within the purview of the Warsaw Convention.

References

Captain Susan Michaelis (ed) (2007) Aviation contaminated air references manual. Published by Captain Susan Michaelis and printed by CPI Antony Rowe, Eastbourne, England

Crawford WA (1989) Environmental tobacco smoke and airlines - health issues. Aerospace 12

Hale M, Al-Seffar J (2008) Preliminary report on Aerotoxic Syndrome (AS) and the need for diagnostic neurophysiological tests. J Assoc Neurophysiol Sci 2:107–118

Holcomb LC (1988) Impact of environmental tobacco smoke on airline cabin air quality. Environ Technol Lett 3

Lorraine T (2007) Toxic airlines. DFT Enterprises, London

Montgomery MR, Wier GT, Zieve FJ et al (1977) Human intoxication following inhalation exposure to synthetic jet lubricating oil. Clin Toxicol II(4):423–426

Rayman RB, McNaughten CB (1983) Smoke/fumes in the cockpit. Aviat Space Environ Med 54:738–740

Sinnett M (2007) 787 no-bleed systems: saving fuel and enhancing operational efficiencies. Aero Q 8

Smith LB, Bhattacharya, Lemasters G, Succop P, Puhula E, Medvelovic M, Joyce J (1997) Effect of chronic low level exposure to jet fuel on postural balance of US air force personnel. J Occup Environ Med 39:625–632

Winder C, Balouet J-C (2001) Aircrew exposure to chemicals in aircraft: symptoms of irritation and toxicity. J Occup Health Saf Aust-NZ 17(5):471–483

Winder C, Fonteyn P, Balouet J-C (2002) Aerotoxic syndrome: a descriptive epidemiological survey of aircrew exposed to in-cabin airborne contaminants. J Occup Health Saf Aust-NZ 18(4):321–338

[37] 1999 Westlaw 7724 (January 12, 1999).

Article 17
Nationality of Aircraft

Aircraft have the nationality of the State in which they are registered.

DOI 10.1007/978-3-319-00068-8_18,

Article 18
Dual Registration

An aircraft cannot be validly registered in more than one State, but its registration may be changed from one State to another.

R. Abeyratne, *Convention on International Civil Aviation*,
DOI 10.1007/978-3-319-00068-8_19,

Article 19

National Laws Governing Registration

The registration or transfer of registration of aircraft in any contracting State shall be made in accordance with its laws and regulations.

R. Abeyratne, *Convention on International Civil Aviation*,
DOI 10.1007/978-3-319-00068-8_20,

Article 20
Display of Marks

Every aircraft engaged in international air navigation shall bear its appropriate nationality and registration marks.

R. Abeyratne, *Convention on International Civil Aviation*,
DOI 10.1007/978-3-319-00068-8_21,

Article 21
Report of Registrations

Each contracting State undertakes to supply to any other contracting State or to the International Civil Aviation Organization, on demand, information concerning the registration and ownership of any particular aircraft registered in that State. In addition, each contracting State shall furnish reports to the International Civil Aviation Organization, under such regulations as the latter may prescribe, giving such pertinent data as can be made available concerning the ownership and control of aircraft registered in that State and habitually engaged in international air navigation. The data thus obtained by the International Civil Aviation Organization shall be made available by it on request to the other contracting States.

Contents

1 Nationality and Registration

These Articles, which address the nationality of aircraft and the necessity of identification where such aircraft are engaged in international air navigation, will be discussed together since they involve such closely related subject that a separate treatment will be repetitive, tedious and may present a convoluted picture.

The ICAO Council, on 8 February 1949 at the sixth meeting of its sixth session adopted Standards for Aircraft Nationality and Registration Marks and Designated them as Annex 7 to the Chicago Convention. These Standards supplemented certain provisions of Articles 17–21 (inclusive) of the Convention.[1] Furthermore, The Council on 5 December 1950 at the fifteenth meeting of its eleventh session approved the insertion of a note in Annex 6—Operation of Aircraft which stated that the Convention specified in a number of respects the fundamental responsibility of a member State of ICAO for aircraft of its nationality. The responsibility of a

[1]Doc 6957-C/807 Proc. Of Council, 6th S. pp. 29–30.

DOI 10.1007/978-3-319-00068-8_22,

State of Registry was further expanded in the Annex by this note which recognized that methods of discharging such responsibility may vary with each State but no particular method could in any way relieve the State of Registry of its basic responsibility. Subject to this basic responsibility the Annex precluded:

- In the case of an aircraft being chartered and operated by an operator having the nationality of an ICAO member State other than the State of Registry, the latter State delegating to the former, in whole or in part, the exercise of the functions imposed by the Annex; and
- In the case of international operations offered jointly with aircraft, not al of which are registered in the same State, the States concerned entering into an agreement for the joint exercise of the functions placed upon the State of Registry by the provisions of the Annex.[2]

2 Registration of Aircraft

There is no specific requirement for registration of aircraft in terms of statutory provision. The Chicago Convention merely provides that aircraft have the nationality of the State in which they are registered.[3] However, it must be noted that the International Law Association (ILA), at its conference held in Helsinki in 1966 observed that:

First of all, it seems clear that every aircraft, in order to engage in international air navigation under the terms of the (Chicago) Convention must be registered, even though the Convention may not be entirely specific on the subject...[4]

Article 18 of the Chicago Convention stipulates that an aircraft cannot be validly registered[5] in more than one State, but its registration can be changed from one

[2]Doc 7057-C/817 (Minutes) p. 203, at paragraph 4.

[3]*Id.* Article 17.

[4]International Law Association Helsinki Conference (1966), *Report on Nationality and Registration of Aircraft with Special Reference to Article 77 of the 1944 Chicago Convention on International Civil Aviation*, at 29.

[5]The first use of aircraft registrations was based on the radio call signs allocated at the London International Radiotelegraphic Conference in 1913. This was modified by agreement and published on April 23 1913. Although initial allocations were not specifically or exclusively for aircraft but were for any radio user, the Convention Related to the Regulation of Aerial Navigation which was held in Paris in 1919 made allocations specifically for aircraft registrations based on the 1913 call sign list. The agreement stipulated that the nationality marks were to be followed by a hyphen, then a group of four letters that must include a vowel (and for the convention Y was considered to be a vowel). At the International Radiotelegraph Convention at Washington in 1927 the list of markings was revised and adopted from 1928, and these allocations are the basis of the currently used registrations. The marking have been amended and added to over the years and the allocations and standards are managed by the International Civil Aviation Organization.

State to another. Although *ex facie* this provision may be perceived as prohibiting joint registration[6] of aircraft, it is now clear that it is not so. The ILA Helsinki Conference went on to say:

> Dual or multiple registration of an aircraft ": in more than one State", is, however, forbidden by Article 18 of the Convention…[T]he position is different in the case of joint registration where two or more States maintain a joint register. Aircraft borne on such a register would also have dual or multiple nationality, but in this case the States maintaining the joint register would doubtless have taken measures in order to remove possible conflicts of jurisdiction.[7]

On 14 December 1967 the Council of ICAO adopted a resolution based on Article 77 of the Chicago Convention which provides *inter alia* that the Council shall determine in what manner the provisions of the Convention relating to nationality of aircraft shall apply to aircraft operated by international operating agencies. Subsequently, ICAO's *Air Transport Regulation Panel,* at its Ninth Meeting held in Montreal from 10 to 14 February 1997, recommended that:

> States wishing to accept broadened criteria for air carrier use of market access in their bilateral and multilateral air services agreements agree to authorize market access for a designated carrier which:
> a) has its principal place of business and permanent residence in the territory of the designating State, and
> b) has and maintains a strong link to the designating State…[8]

The Recommendation went on to say that in judging the existence of a strong link, States should take into account elements such as the designated air carrier establishing itself, and having a substantial amount of its operations and capital investment in physical facilities in the designating State, paying income tax and registering its aircraft there, and employing a significant number of nationals in managerial, technical and operational personnel. Where a State believes it requires conditions or exceptions concerning the use of the principal place of business and permanent residence criteria based on national security, strategic or commercial reasons this should be the subject of bilateral or multilateral negotiations or consultations, as appropriate.[9]

This Recommendation brings to bear the advantages of registering aircraft in a national register. There are several reasons to proceed with the registration within the public records of a specific state. One of these reasons is the intention of

[6]The expression "joint registration" indicates that system of registration of aircraft according to which the States constituting an international operating agency would establish a register other than the national register for the joint registration of aircraft to be operated by the agency. See Resolution Adopted by the ICAO Council on Nationality and Registration of Aircraft Operated by International Operating Agencies, 17th Meeting of the Council, Sixty-second Session, 14 December 1967, *Appendix 1* at p. 5.

[7]*Id.* 29–30.

[8]Air Transport Regulation Panel, Ninth Meeting, Montréal 10–14 February 1997, REPORT ATRP/9-4, 2–3.

[9]*Ibid.*

ascribing a nationality to an aircraft. Therefore, the procedure of registration is mainly a legal requirement which is calculated to obtain certain rights that the state will grant to aircraft registered in its registry. When an aircraft obtains a national character, naturally flowing corollaries to this right, which are implicit, are put in place, such as the right to claim the nationality of the country of registration as well as protection under international law.

One very important reason for aircraft owners to register their aircraft in a specific registry is that the registration in a nation's registry would effectively preclude third parties outside the nation of registry from claiming ownership of the aircraft. Put differently, aircraft registration serves as *prima facie* evidence of ownership of the aircraft.[10] Also, registration is important, particularly in respect of mortgages, which is a tradition deriving from maritime law, that required that ships must be registered in accordance with the law of the flag state of the vessel that will govern provisions regarding the ranking between mortgages, their effects with regard to third parties, and all procedure concerned with the enforcement of the mortgage instruments.[11]

The above discussion reflects that there is seemingly a common thread and recurrent theme that runs through the registration of aircraft. This theme is the fundamental postulate of the Preamble to the Chicago Convention which calls for international civil aviation to be developed in a safe and orderly manner. Responsibility in this regard devolves upon ICAO and in turn upon its Council. Registration by States in their registries of aircraft which do not meet minimum safety standards is a safety deficiency that is picked up by the ICAO safety audits and there have been instances where ICAO has recommended that such aircraft be taken off a country's register. Failure to comply with this recommendation may give rise to possible action on the part of the ICAO Council. This brings one to the issue of legal responsibility of the Council to invoke Article 54 (j) of the Chicago Convention in reporting to Contracting States of the failure of a Contracting State to carry out the recommendations of the Council to de-register aircraft in its register that are registered without proper safety checks and licensing procedures.

The 35th Session of the Assembly, when it addressed the issue of expanding the audits from a limited Annex basis to a comprehensive systems approach, instructed the Secretary General to make the final safety audit reports available to all Contracting States and also to provide access to all relevant information derived from the Audit Findings and Differences Database (AFDD) maintained by ICAO.[12] Furthermore, in Resolution A36-2 (Unified Strategy to Resolve Safety Related Deficiencies) the Assembly, in operative Clause 6 of the Resolution, directs the

---

[10]Hill (1998), at p. 24. It is noteworthy that Standard 3.1 of Annex 7 to the Chicago Convention requires that the nationality or common mark and registration mark shall be painted on the aircraft or shall be affixed by any other means ensuring a similar degree or permanence. This Standard also requires that marks be kept clean and visible by the operator at all times. See Annex 7 to the Convention on International Civil Aviation, Aircraft Nationality and Registration Marks. Fifth Edition: July 2003, at 2.

[11]*International Convention on Maritime Liens and Mortgages* 1993, Articles 1 (a) and 2.

[12]Resolution A 35–6, Operative Clause 7.

Council to apply and review, as necessary, the procedures to inform Contracting States, within the scope of Article 54 (j) of the Chicago Convention, in the case of a State having significant shortcomings with respect to ICAO safety related SARPs in order for other Contracting States to take action in an adequate and timely manner.

The discussion to follow will address some legal issues concerned with the registration of aircraft. In order to do so, some reliance will be placed on the analogy of maritime practice, the basic principles of which, particularly in terms of registration, also apply to aviation.

3 Principles of Registration

3.1 The Maritime Analogy

In maritime parlance, the principle of conferring nationality to ships is today considered a sovereign right granted to any state, regardless of whether it is a coastal or land-locked country.[13]

The decision in the 1905 *Muscat Dhows* case[14] initially established the international jurisprudence with regard to the sovereign right associated with the conferral of nationality to vessels. In this case, the Permanent Court of Arbitration pronounced that "it belongs to every sovereign to decide to whom it will accord the right to fly his flag and to prescribe the rules governing such grants."[15]

The *Muscat Dhows* principle was reiterated in 1953 by the United States Supreme Court in the decision in *Lauritzen* v. *Larsen*, when the court pronounced that: "each state under international law may determine for itself the conditions on which it will grant its nationality to a merchant vessel."[16]

Consequently, the principle entrenched in customary international jurisprudence, recognized by the two cases mentioned above, that sovereign States can decide to whom the right to fly a national flag can be granted and what rules should apply to such grant, was codified in such international conventions as the Geneva Convention on the High Seas,[17] United Nations Convention on the Law of the Sea (UNCLOS),[18] and lately the United Nations Convention on Conditions for

---

[13]For a maritime analogy see Sohn and Gustafson (1984), at p. 3.

[14]*France* v. *Great Britain, Muscat Dhows Case* (1916) Hague Court Reports 93, Permanent Court of Arbitration, 1916. See also Coles (2002), at 3 for a detailed discussion on the *Muscat Dhows* case.

[15]*Ibid.*

[16]*Lauritzen* v. *Larsen*, 345 U.S. 571 (1953).

[17]http://www.intfish.net/treaties/genevahs.htm.

[18]The Law of the Sea, Official Text of the United Nations Convention on the Law of the Sea with Annexes and Index, United Nations: New York, 1983. The United Nations Convention on the Law of the Sea (UNCLOS) comprises 320 articles and nine annexes, governing all aspects of ocean space, such as delimitation, environmental control, marine scientific research, economic and commercial activities, transfer of technology and the settlement of disputes relating to ocean matters. See also *infra*, note 28 in Article 1.

Registration of Ships,[19] all of whom were consistent with the principle enunciated by both the *Muscat Dhows Case* and *Lauritzen* v. *Larsen*, which are decisions rendered years before the emergence of that international codification process.[20]

Although at the present time the sovereign right to confer nationality to a ship that vested in a State is a widely recognized rule of international law, the rules determining the conditions for registration of ships were contained in perennial domestic legislations of States around the world years earlier than either the codification process or the jurisprudence alluded to above. In this regard, Article 5 of The Geneva Convention remains a critical key provision particularly since, by inference, the sovereign right of conferral of jurisdiction over a vessel could be attributed to the text of the Convention, which provides *inter alia* that each State shall fix the conditions for the grant of its nationality to ships, for the registration of ships in its territory, and for the right to fly its flag.[21]

The trend reflected in the Geneva Convention can be observed in UNCLOS, which by virtue of Article 91, confers on each state the flexibility to put in place requirements that vessels registered under its flag must comply with.[22] Furthermore, in the Preamble to the Convention on Registration of Ships, it is reaffirmed that each country is free to set the requirements that are considered necessary for the registration of vessels under its flag.[23] The principle of the exclusive jurisdiction of the flag state involves not only rights, but also responsibilities and obligations that the country of registration must comply with. These duties, initially considered customary international law, were however not totally uniform due to the differences among national legislations.

The above discussion brings to bear the fact that, as a consequence of the development and codification of the law of the sea, the duties of the flag state were incorporated and codified into treaty law in a general way through the Geneva Convention on the High Seas, and then, more specifically in UNCLOS. Seeking harmonizing with these Conventions, the United Nations Convention on Conditions for Registration of Ships introduced specific obligations of the flag State. They are however inapplicable as the Registration of Ships Convention has not yet entered into force.

It must be noted that the text of the Geneva Convention does not contain specific provision prescribing the duties of a flag state. However there is a general mandate by implication in Article 5 (1) requiring the flag State to carry out certain legal responsibilities. Article 5 provides as follows:

---

[19]Geneva, 7 February 1986.

[20]Matlin (1990). The author emphasises that the *Muscat Dhowns case* and *Lauritzen* v. *Larsen* are compelling precedents leading to the principle that each state shall determine whether it will grant its nationality to a ship.

[21]See the Geneva Convention, *supra* note 19, Article 5 (1).

[22]See UNCLOS, *supra* note 18, Article 91.

[23]*United Nations Convention on Conditions for Registration of Ships Id, Preamble.*

> *Article 5: 1.*
> Each State shall fix the conditions for the grant of its nationality to ships, for the registration of ships in its territory, and for the right to fly its flag. Ships have the nationality of the State whose flag they are entitled to fly. There must be a genuine link between the State and the ship; in particular, *the State must effectively exercise its jurisdiction and control in administrative, technical and social matters over ships flying its flag....*"[24]

Sovereignty of the flag state over a ship and its users confers upon that State an exclusive mandate to exercise its sovereignty on board the vessel, particularly in instances where it become necessary to apply principles of international law. A flag State can be held responsible under principles of State responsibility for non compliance with applicable principles of international law.[25]

This principle was formally adopted in 1975 by the International Law Commission (ILC) in its Articles of State Responsibility which affirms in Article 1 that there is a general rule of public international law, widely supported by practice that every intentionally wrongful act of a State entails responsibility. Article 2 provides that an intentionally wrongful act of a State carried out through conduct involving omission or commission is directly attributable to the State and will be considered as constituting a breach of an international obligation.[26] This principle, which forms a cornerstone of international conduct by States, provides the basis for strengthening international comity and regulating the conduct of States both internally—within their territories—and externally, towards other States. States are effectively precluded by this principle of pursuing their own interests untrammelled and with disregard to principles established by international law.

It must be noted that it is international law, and not municipal law that determines as to what constitutes an intentionally wrongful act. Article 12 of the ILCs *Articles of State Responsibility* makes the act of a State which does not conform to what it is required to do under an obligation a breach of an international obligation.

A cardinal principle with regard to the legal duties of the flag state is that the country of registration must exercise its jurisdiction properly, enforcing its domestic law in an effective way in pursuance of its main aim of exercising full control on all administrative, technical and social aspects over all vessels flying its flag. It must be noted that the duties of the flag state are not limited to the vessel as a movable property. Another important consideration is that a key technical responsibility

---

[24]See the Geneva Convention, *supra* note 19, Article 5 (1).

[25]In its Report to the General Assembly, the International Law Commission in 1949 recommended a draft provision which required:

> Every State has the duty to conduct its relations with other States in accordance with international law and with the principle that the sovereignty of each State is subject to the supremacy of international law.

Report of the International Law Commission to the General Assembly on the Work of the 1st Session, A/CN.4/13, June 9 1949, at 21.

[26]Yearbook of the International Law Commission, 1976, Vol. II at 75 ff and ILC Commentary 2001 at 68. This principle has been accepted and affirmed in the courts. See i*n Re. Chorzow Factory (Jurisdiction) Case (1927) PCIJ, Ser. A, no. 9* at 21. Also, *Rainbow Warrior Case*, 82 ILR at 499.

devolves upon the flag State under the Geneva Convention on the High Seas with regard to the seaworthiness of the ship and other measures taken with respect to the ship.[27] Conversely, when the Convention makes reference to social issues, it is understood that it is referring to the manning of ships and labor considerations in relation to the master, officers and crew.

Article 10 of the Geneva Convention is also important in terms of the duties of a flag State. This Article provides that every State shall take such measures for ships under its flag as are necessary to ensure safety at sea with regard *inter alia* to: the use of signals, the maintenance of communications and the prevention of collisions; manning of ships and labour conditions for crews taking into account the applicable international labour instruments; and construction, equipment and seaworthiness of ships. Article 10 also goes on to say that, in taking such measures, each State is required to conform to generally accepted international standards and to take any steps which may be necessary to ensure their observance. With this provision, the Geneva Convention has effectively subsumed the key responsibilities with which a flag state must comply. Article 10 also requires that measures ensuring safety at sea, which is a compelling objective and aim in maritime practice, should largely be regulated by international law.[28] In pursuance of the objective of ensuring safety at sea, the provision also addresses technical aspects such as the seaworthiness of ships, their construction and equipment, as well as the prevention of collisions. Furthermore, it takes into consideration social aspects such as manning and labor conditions for crews. The final paragraph of Article 10 requires each flag state to ensure that its domestic legislation conform with international standards and that international rules be observed.

The duties of flag states are intrinsically linked to the performance of duties of those who are employed on the ship, including the master. Under the Geneva Convention, certain responsibilities devolve upon the master.[29] The Geneva Convention also contains provisions related to the prevention and punishment of slavery and piracy and prescribes penal sanctions.[30] These implicitly become additional duties of the flag state.

[27]See *Tetley's Glossary of Maritime Law, Abbreviations, Definitions, Terms, Links and Odds'N Ends*, which could be accessed at Prof. William Tetley's homepage at http://tetley.law.mcgill.ca! maritime/glossarymaritime.htm#letter_s (last visit July 24, 2003). Regarding seaworthiness, it is important to note Prof. Tetley's view that seaworthiness and importance to the law of the sea is a consistent thread in the fabric of maritime law. As a consequence, the issue of seaworthiness has a bearing on all maritime issues.

[28]The issue of safety at sea is mainly covered by the International Convention for the Safety of Life at Sea, amended 1974, in force May 25, 1980.

[29]Geneva Convention, *supra* note 19, Article 12 which provides that every State shall require the master of a ship sailing under its flag, in so far as he can do so without serious danger to the ship, the crew or the passengers: to render assistance to any person found at sea in danger of being lost; to proceed with all possible speed to the rescue of persons in distress if informed of their need of assistance, in so far as such action may reasonably be expected of him; and after a collision, to render assistance to the other ship, her crew and her passengers and, where possible, to inform the other ship of the name of his own ship, her port of registry and the nearest port at which she will call.

[30]*Id.*, Articles. 13 and 14.

Finally, it must be said that the Geneva Convention contains specific provisions in relation to the prevention of pollution of the seas by the discharge of oil from ships and from the dumping of radioactive waste.[31] These provisions are particularly of analogical relevance to the field of aviation, which contains no equivalent environmental provisions in the Chicago Convention, although Annex 16 to the Convention has in-depth regulatory provisions on noise and engine emissions. Another noteworthy issue is that issues pertaining to the protection of the marine environment are of the utmost importance to the world of maritime practice world and therefore these issues are closely linked to the registration of ships. It is the flag State that must take measures to ensure that ships flying under its flag are in compliance with the rules for the protection of the marine environment.

One significant provision which brings out the duties of the flag State is Article 94 of the United Nations Convention on the Law of the Sea (UNCLOS)[32] which reflects a detailed list of the flag state's responsibilities. Taking off where Article 5 (1) of the Geneva Convention leaves off, UNCLOS makes the following statement in Article 94:

> "Article 94:
> Duties of the Flag State
> Every State shall effectively exercise its jurisdiction and control in administrative, technical and social matters over ships flying its flag. . .[33]

Following this general statement Article 94 provides a detailed explanation that covers different aspects of the registration of vessels. Firstly, it states that, in order to maintain administrative control over all vessels flying its flag, the flag state must keep a register of ships, including information such as the names and particulars of those vessels registered.[34]

Arguably, the main responsibility of the flag state is identified in Article 94 (2) (b) with respect to the assumption of jurisdiction, over each ship flying its flag, over its master, and over its officers and crew on administrative, technical and social

---

[31]*Id.*, Articles 24 and 25.

[32]The Law of the Sea, Official Text of the United Nations Convention on the Law of the Sea with Indexes and Annex, Final Act of the Third United Nations Convention on the Law of the Sea, United Nations: New York, 1983. Also called the Law of the Sea Convention and the Law of the Sea Treaty, UNCLOS is the international agreement that resulted from the third United Nations Convention (Conference) on the Law of the Sea, which took place from 1973 through 1982. The Law of the Sea Convention defines the rights and responsibilities of nations in their use of the world's oceans, establishing guidelines for businesses, the environment, and the management of marine natural resources. The Convention concluded in 1982 replaced four 1958 treaties. UNCLOS came into force in 1994, a year after Guyana became the 60th state to sign the treaty. To date 155 countries and the European Community have joined in the Convention. The United States has signed the treaty, but the Senate has not ratified it.

[33]See UNCLOS, *Id.*, Article 94 (1).

[34]Article 94 (2) provides that in particular every State shall: (a) maintain a register of ships containing the names and particulars of ships flying its flag, except those which are excluded from generally accepted international regulations on account of their small size; and (b) assume jurisdiction under its internal law over each ship flying its flag and its master, officers and crew in respect of administrative, technical and social matters concerning the ship.

matters concerning the ship.[35] This jurisdictional duty is presumed to be performed when the flag state adopts in its own domestic legislation all the regulations regarding registration of ships, concerning all its practical aspects, including, as recognized by UNCLOS, administrative, technical and social matters.

UNCLOS in its Article 94 (3) uses words similar to those used in the Geneva Convention on the High Seas and prescribes the duties of the flag state pertaining to safety at sea,[36] listing the main points that should be addressed regarding this important maritime issue. UNCLOS clarifies those duties by prescribing the measures that should be followed by the flag state in order to ensure the safety at sea. These measures are contained in article 94 (4) which include the practice of regular surveys of the ships, the requirement to check mainly the seaworthiness of the vessel, and the proper manning of the ship, taking into account the qualification of the master, officers and crew of the ship, in a vast number of maritime issues that are crucial to marine safety, such as prevention of collisions and marine pollution.[37] It is also noted that Article 94 (5) provides that in taking the measures called for in paragraphs 3 and 4 each State is required to conform to generally accepted international regulations, procedures and practices and to take any steps which may be necessary to secure their observance. Article 94 (6) follows, by prescribing that a State which has clear grounds to believe that proper jurisdiction and control with respect to a ship have not been exercised may report the facts to the flag State. Upon receiving such a report, the flag State is obligated to investigate the matter and, if appropriate, take any action necessary to remedy the situation.

Finally, Article 94 (7) of UNCLOS provides that Each State shall cause an inquiry to be held by or before a suitably qualified person or persons into every marine casualty or incident of navigation on the high seas involving a ship flying its flag and causing loss of life or serious injury to nationals of another State or serious damage to ships or installations of another State or to the marine environment. The flag State and the other State are required to cooperate in the conduct of any inquiry held by that other State into any such marine casualty or incident of navigation.

The Chicago Convention is replete with provisions that impact the issue of registration of aircraft. Besides the fundamental provisions in Articles 17 and 18 alluded to earlier, Article 20 provides that every aircraft engaged in international air navigation shall bear its appropriate nationality and registration marks. This is

---

[35] *Ibid.* Article 94 (2) (b).

[36] Article 94 (3) provides that every State shall take such measures for ships flying its flag as are necessary to ensure safety at sea with regard, inter alia, to: (a) the construction, equipment and seaworthiness of ships; (b) the manning of ships, labour conditions and the training of crews, taking into account the applicable international instruments; and (c) the use of signals, the maintenance of communications and the prevention of collisions.

[37] Article 94 (4) provides that such measures shall include those necessary to ensure: (a) that each ship, before registration and thereafter at appropriate intervals, is surveyed by a qualified surveyor of ships, and has on board such charts, nautical publications and navigational equipment and instruments as are appropriate for the safe navigation of the ship (b) that each ship is in the charge of a master and officers.

further elaborated in Annex 7 to the Chicago Convention where the Convention requires that the nationality or common mark and registration mark shall be painted on the aircraft or shall be affixed by any other means ensuring a similar degree or permanence. This Standard also requires that marks be kept clean and visible by the operator at all times.[38] Article 21 requires each Contracting State to undertake to provide on demand to any other Contracting State or ICAO information concerning the registration and ownership of any particular aircraft registered in that State. Article 24 on customs duty provides *inter alia* that aircraft on a flight to, and from or across the territory of another State shall be admitted temporarily free of duty, subject to the customs regulations of that State. It also states that spare parts and equipment imported into the territory of a Contracting State for incorporation in or use on an aircraft of another Contracting State engaged in international air navigation shall be admitted free of customs duty.

Article 29 of the Chicago Convention requires every aircraft of a Contracting State to carry *inter alia*, its certificate of registration. This requirement implicitly recognizes the fact that under international law, the legal status of an aircraft could be determined by its registration and its affiliation, for purposes of jurisdiction, and attributed to the State in which such aircraft is registered. Article 33 of the Chicago Convention requires States to recognize as valid, certificates of airworthiness and certificates of competency and licences issued or rendered valid by the Contracting State in which the aircraft in question is registered provided such certificates or licences are rendered valid and equal to the minimum standards prescribed by the Convention.

Another important provision in the Chicago Convention is Article 12 on rules of the air, which requires each Contracting State to undertake to adopt measures to insure that every aircraft flying over or manoeuvring within its territory and that every aircraft carrying its nationality mark, wherever such aircraft may be, comply with the rules and regulations relating to the flight and manoeuvre of aircraft there in force. It also provides that each Contracting State undertakes to keep its own regulations in these respects uniform, to the greatest possible extent, with those established from time to time under the Convention. Article 12 goes on to say that over the high seas, the rules in force shall be those established under the Convention. Also, each Contracting State undertakes to insure the prosecution of all persons violating the regulations applicable.

Article 30 of the Convention, which is another provision pertaining to registration of aircraft, provides that aircraft of each Contracting State may, in or over the territory of other Contracting States, carry radio transmitting apparatus only if a license to install and operate such apparatus has been issued by the appropriate authorities of the State in which the aircraft is registered. The use of radio transmitting apparatus in the territory of the Contracting State whose territory is flown over is required to be in accordance with the regulations prescribed by that State. The next provision is Article 31 which states that every aircraft engaged in international

[38] See Annex 7 to the Convention on International Civil Aviation, Aircraft Nationality and Registration Marks, Fifth Edition: July 2003, at 2.

navigation shall be provided with a certificate of airworthiness issued or rendered valid by the State in which it is registered. Article 32 (a), with regard to the crew of an aircraft provides that the pilot of every aircraft and the other members of the operating crew of every aircraft engaged in international navigation will be provided with certificates of competency and licenses issued or rendered valid by the State in which the aircraft is registered.

3.2 Flags of Convenience

All the above provisions reflect the signal role played by the State of registry in ensuring aviation safety. This brings one to the discussion of an issue that has concerned the aviation community in relation to the registration of aircraft, which is flags of convenience. The term "flags of convenience" has not been defined in real terms in any international instrument.[39] However, its origin clearly lies in the maritime industry and it was obviously used to refer to open registries[40] maintained by States which offered distinct advantages to ship-owners who were seeking to circumvent the traditional national registries. Besides, the term is a misnomer when used in an aviation context as aircraft do not fly flags but merely carry the nationality of the State in which they are registered.[41]

"Flags of convenience"[42] associated with foreign registered aircraft could indeed be worthy of examination under the purview of registration of aircraft and aviation safety. When an aircraft rarely, if ever, returns to the State of Registry, its airworthiness oversight becomes an issue in the absence of safety oversight arrangements between the State of Registry and the State of the Operator. There are broadly two groups of foreign registered aircraft that can be deemed to operate under a flag of convenience: those done for fiscal purposes and those done to take advantage of a system with no or minimal economic or technical oversight. The first group may not pose a serious problem if arrangements are made between concerned States to ensure proper oversight, for example through bilateral agreements under Article 83 *bis*,[43] which permits States to transfer all or a part of certain safety oversight responsibilities under the Convention. Even for this group, the reality remains far from satisfactory in that relatively few bilateral agreements implementing Article 83 *bis* have been notified to ICAO and numerous aircraft of all types all

[39]Kasoulides (1989) at 551.

[40]UNCTAD defined an open registry as "the conferment of national charter upon ships regardless of ownership, control and manning. *Id.* 546.

[41]To register is to record formally and exactly in a book of public facts. The task of keeping a registry or record of such entries pertaining to ships is attributed to customs authorities. See *Blacks Law Dictionary*, Sixth Edition, West Publishing Co: St. Paul Minn: 1990 at 1283.

[42]"Flags of convenience" is a term derived from the maritime industry which denotes a situation in which commercial vessels owned by nationals of a State, but registered in another State, are allowed to operate freely between and among other States.

[43]*Infra*, note 49 and text in the article pertaining thereto.

over the world are still subject to split oversight responsibility. It is the second group that creates a major safety problem which needs to be addressed.[44]

An issue which requires some discussion is whether flags of convenience would raise a safety issue. There are some who argue that aviation safety would not necessarily be compromised by the practice of flags of convenience.[45] Lelieur argues that in the event of liberalization of ownership and control of airlines (which may lead to flags of convenience in some instances) there need not be a fear for safety since there would be a harmonization of safety and security measures worldwide, mainly through the ICAO umbrella of the Universal Safety Oversight Audit Programme (USOAP).[46] It must be noted that there is also the highly effective IATA Operational Safety Audit (IOSA) Programme[47] which is an internationally recognised and accepted evaluation system designed to assess the operational management and control systems of an airline. IOSA uses internationally recognised quality audit principles, and is designed so that audits are conducted in a standardised and consistent manner.[48]

Although theoretically the conclusion—that the safety oversight system could act as a buffer to obviating the possibility of flags of convenience—is seemingly logical, it remains to be seen whether such confidence is justified in a practical sense. ICAO's oversight system is anchored on measures that could be taken to remedy deficiencies discovered through the audit process. One such deficiency is the haphazard and arbitrary manner in which aircraft are registered in some States, which may give rise to aircraft of one State which do not meet minimum safety standards being admitted to another State's register. This process could be carried out through Article 83 *bis*[49] of the Chicago Convention with regard to registration

[44]This problem is currently being addressed separately by the ICAO Air Navigation Commission and the Council in association with the Unified Strategy to resolve safety-related deficiencies within the scope of Article 54 (j) of the Chicago Convention, which requires the Council to report to States any infraction of the Convention, as well as any failure to carry out recommendations or determinations of the Council. See ICAO Air Navigation Commission working paper AN-WP/ 8015.

[45]Lelieur (2003) at 83. For a contrary view see *infra*, text attached to notes 47 in Article 1 and 48 in Article 1 *infra*.

[46]*Id.* 108. For information on the ICAO USOAP programme and a discussion thereof, see Abeyratne (2007).

[47]It must be noted that in March 2006, ICAO and IATA agreed to share information from their respective audit programmes.

[48]Inherent in the IOSA Programme is a degree of quality, integrity and security such that mutually interested airlines and regulators can all comfortably accept IOSA audit reports. As a result, the industry will be in a position to achieve the benefits of cost-efficiency through a significant reduction in audit redundancy.

[49]Article 83 *bis* provides that, notwithstanding the provisions of Articles 12, 30, 31 and 32 (a) (which have been discussed earlier in this article), when an aircraft registered in a contracting State is operated pursuant to an agreement for the lease, charter or interchange of the aircraft or any similar arrangement by an operator who has his principal place of business or, if he has no such place of business, his permanent residence in another contracting State, the State of registry may, by agreement with such other State, transfer to it all or part of its functions and duties as State of

of aircraft is, which provides *inter alia* that, notwithstanding the provisions of Articles 12, 30, 31 and 32 (a) (which have been discussed earlier in this article), when an aircraft registered in a Contracting State is operated pursuant to an agreement for the lease, charter or interchange of the aircraft or any similar arrangement by an operator who has his principal place of business or, if he has no such place of business, his permanent residence in another Contracting State, the State of registry may, by agreement with such other State, transfer to it all or part of its functions and duties as State of registry in respect of that aircraft. The State of registry shall be relieved of responsibility in respect of the functions. The Chicago Convention requires that, when an aircraft possessing a valid Certificate of Airworthiness issued by a Contracting State is entered on the register of another Contracting State, the new State of Registry, when issuing its Certificate of Airworthiness may consider the previous Certificate of Airworthiness as satisfactory evidence, in whole or part thereof, that the aircraft complies with ICAO Standards.[50]

It has been argued that the objective of increased safety by the transfer of supervisory functions and duties provided by Article 83 *bis* could be adversely affected through the emergence of flags of convenience.[51] The main basis of this argument is that, in order to obtain financial gain, States could gain financially by entering aircraft of another State on their register, but be unable to provide supervisory functions and duties required to ensure that those aircraft are maintained according to the minimum safety standards required.[52] Another area that might impinge on the safety of flight concerns operations involving foreign flight crew. Split oversight problems could also occur in respect of foreign-licensed flight crew. For example, dry leases (i.e. the lease of an aircraft without crew) raise the problem of validation of foreign crew licences by the State of Registry. The issue becomes complicated when the rules and requirements for crew licences in the State of Registry are at variance with the corresponding rules in the State that initially issued the licences. Differences between the laws and regulations of the State of Registry and those of the State of the Operator may also exist in the case of wet leases (i.e. a lease of aircraft with crew). While the lessor usually remains the official operator in such cases, the lessee may already operate aircraft of a similar type under its Air Operators Certificate. It may happen then that the wet-leased aircraft are operated under the lessee's Air Operator's Certificate and, consequently, the State of the lessee becomes the State of the Operator. In such circumstances, proper surveillance of the operating crew may become difficult. The situation could become more complicated if the operation involves a mixed crew (e.g. the cabin crew from the lessee carrier and the cockpit crew from a foreign lessor carrier).

---

registry in respect of that aircraft. The State of registry shall be relieved of responsibility in respect of the functions The Protocol which amended the Chicago Convention with the introduction of Article 83 *bis* entered into force on 20 June 1997. By April 2008, 153 parties had deposited their instruments of ratification of Article 83 *bis* with the depository, ICAO.

[50]Annex 8 to the Convention on International Civil Aviation, *Airworthiness of Aircraft*, Tenth Edition: April 2005, Standard 3.2.4.

[51]Verhaegen (1997) at 273.

[52]*Ibid.*

Another interesting provision with regard to the registration of aircraft is found in Article 77 of the Chicago Convention which provides that two or more Contracting States could operate international agencies and pool their air services on any routes or in any regions. Such arrangements are strictly subject to the provisions of the Convention and the ICAO Council can determine the nationality of aircraft operated under such international operating agencies. The Council's duty to determine the nationality of aircraft operated under international operating agencies was subject to much consideration once in 1960 by a committee appointed by ICAO and subsequently by the ICAO Legal Committee. Consequently, in 1967 the Council adopted a resolution which provided that, in the context of Article 77 the words "joint registration" indicated a system of registration of aircraft according to which the States constituting an international operating agency would establish a register other than the national register for the joint registration of aircraft to be operated by the agency.[53] The Council also resolved that the expression "international registration" denoted the cases where the aircraft to be operated by an international operating agency would be registered not on a national basis but with an international organization having international legal personality, whether or not such international organization is composed of the same States as have constituted the international operating agency.[54]

In the case of joint registration, the Council resolved *inter alia* that the States constituting the international operating agency shall be jointly and severally bound to assume the obligation which, under the Chicago Convention, attach to a State of registry.[55] Furthermore it was resolved that the States constituting the international operating agency will identify for each aircraft an appropriate State from among themselves which would then be entrusted with the duty of receiving and replying to representations which might be made by other Contracting States to the Chicago Convention concerning that aircraft.[56] The resolution applies only when all the States constituting the international operating agency are and remain parties to the Chicago Convention. It does not apply to the case of an aircraft which, although operated by an international operating agency, is registered on a national basis.

3.3 Leasing of Aircraft

Another important commercial aviation practice which is impacted by the regulatory aspects pf registration is leasing. Leasing became a strategic commercial manoeuvre of airlines only in the past 20 years. Of these, the first decade—the 1980s—saw a boom in commercial aviation and therefore a corresponding upsurge

[53]*Resolution adopted by the Council on Nationality and Registration of Aircraft Operated by International Operating Agencies*, Preamble *supra* note 6, Appendix 1.

[54]*Ibid.*

[55]*Ibid.*, Appendix 2.

[56]*Ibid.*

in aircraft manufacture, and the 1990s saw a downturn of this trend, contributed in part by vacillations of the world economy which brought to bear regional economic crises such as the Asian slump in the late 1990s.

The downturn of the 1990s and the ensuring money market crisis underscored the value of juggling the most expensive singular asset of the aviation industry—aircraft. Aircraft financiers are quick to offer flexible investment options to airlines to obviate the burden of outright purchase of aircraft. Apart from traditional loans available, two of the most effective financial tools now available to the airline industry for the procurement of aircraft are finance leasing and operational leasing.[57] These leasing options are particularly beneficial to small airlines which are at high risk because of their limited asset bases.

Leasing of aircraft has effectively extended the operational life of aircraft to encompass second and third operators. The magnitude of this financial option is well borne out by the fact that at least 25 % of aircraft being used in the airline are leased.[58] Essentially the three most basic benefits bestowed to the lessor and lessee by a lease are reduction and spreading of risk of the asset; attendant tax benefits; and flexibility of operation. In terms of the nature of operation of a lease, the leasing process may either take the form of a wet lease—a lease where the air crew of the lessor is an integral part of the lease agreement—and a dry lease, where the lessor transfers possession of the aircraft without crew.

Very simplistically put, leasing is the transfer of possession without ownership. Legally speaking, however, the definition becomes somewhat longer in that a lease at law is essentially a commercial arrangement whereby a lessor (or equipment owner) conveys to the lessee (or operator of the equipment) for valuable consideration in the form of rentals over a period of time specified in the lease agreement, the right to use the equipment.[59] The lessee is legally obligated to return to the lessor the equipment he leases at the expiration of the term of the lease,[60] in reasonably good order, leaving a margin for wear and tear of normal usage.[61]

Donald Bunker, in his informative treatise[62] on aerospace financing, cites the 1960s paradigm of IBM and XEROX leases which typified the principles of the modern lease. Both companies utilized the lease of their equipment as a tool of marketing strategy which was calculated to maximize their profits over a standard sale, by amortizing the capital costs of the equipment and earning a profit over the

[57]A finance lease involves the substantial transfer of risks and rewards appurtenant to ownership, from lessor to lessee; and an operational lease keeps such risks and rewards within the lessor's scope of legal status. A finance lease is calculated to amortize the lessor's capital outlay and provide a profit at the end of the lease term with the lease payments received from the lessee. An operational lease does not amortize capital outlay at the end of the term and profits are derived usually after more than one lease term.

[58]Maria Wagland, A new Lease of Life, *Aerospace International*, March 1999 at p 22.

[59]*Beecham Foods Limited* v. *North Supplies (Edmonton) Ltd.*, [1959] 1 WLR 643.

[60]*Ballet* v. *Mingay* [1943] 1.K.B. 281.

[61]*Lang* v. *Brown* (1898) 34 N.B.R. 492.

[62]Bunker (1988), at p. 22.

sustenance of the lease period. Over and above this fundamental benefit, a lease effectively demarcates the market pricing between new and used equipment, thus allowing the resale market to flourish on its own by removing obsolete equipment from the market place. The blend of new and used equipment pricing policies balances an enterprise's cash flow and asset base while ensuring a more orderly growth of reported profits.[63] To the consumer, or operator, a lease offers maximum flexibility for selective use of a product, which, in lay terms would be the equivalent of walking into a baker's shop and being able to buy a slice of pie to allay one's hunger, without having to buy the whole pie. In the context of aircraft leasing, this financial principle is of paramount importance, since leased aircraft can meet seasonal demand for additional capacity without the operator having to incur the capital outlay involved in the outright purchase of an aircraft. Additionally, leased aircraft can be selected to fit into routes and meet specific measurements and requirements of certain routes for which an operator obtains air traffic rights but does not own the equipment to enjoy the rights. This is particularly applicable in the case of smaller air carriers who have traffic rights to operate on certain routes but do not have the appropriate equipment for the purpose.

The registration of the airline is a paramount legal consideration which has to be addressed when an airline uses leased aircraft. The most fundamental characteristic of an aircraft at international law is its nationality. Both the Paris Convention of 1919[64] and the Chicago Convention provide that the nationality of an aircraft is governed by the State in which such aircraft is registered. The Tokyo Convention on Offences Committed on Board Aircraft (1963)[65] provides that the State of registration has jurisdiction over offences and acts committed on board.[66] Therefore, it is reasonable to conclude that the national status of an aircraft would depend on the fact of its registration and to this extent is not dissimilar with the maritime law concept of nationality of ships. The most explicit pronouncement on nationality of vessels was given by the International Court of Justice in the famous *Nottebohm* case[67] were the Court held:

The character thus recognized on the international level as pertaining to nationality is in no way inconsistent with the fact that international law leaves it to each State to lay down the rules governing the grant of its own nationality. The reason for this is that the diversity of demographic conditions has thus far made it impossible for any general agreement to be reached on the rules relating to nationality, although the latter by its very nature affects international relations. It has been considered that the best way of making such rules accord with the varying demographic conditions in different countries is to leave the fixing of such rules to the competence of each State. On the other hand, a State cannot claim that the rules it

[63] *Ibid.*

[64] Convention for the Regulation of Aerial Navigation, Paris 1919, Articles 5-10.

[65] *Convention on Offences and Certain Other Acts Committed on Board Aircraft*, signed at Tokyo on 14 September 1963. See ICAO Doc 8364.

[66] *Id.*, Article 3.

[67] ICJ Reports (1955) at 1.

has thus laid down are entitled to recognition by another State unless it has acted in conformity with this general aim of making the legal bond of nationality accord with the individual's genuine connection with the State which assumes the defence of its citizens by means of protection as against other states.

> ... According to the practice of states, to arbitral and judicial decisions and to the opinions of writers, nationality is a legal bond having as its basis a social fact of attachment, a genuine connection of existence, interests and sentiments, together with the existence of reciprocal rights and duties. It may be said to constitute the juridical expression of the fact that the individual upon whom it is conferred, either directly by the law or as the result of an act of the authorities, is in fact more closely connected with the population of the State conferring nationality than with that of any other State. Conferred by a State, it only entitles that state to exercise protection vis-à-vis another State, if it constitutes a translation into juridical terms of the individual's connection with the State which has made him its national.[68]

In the particular instance of aircraft, the concept of registration and nationality has evolved with changing conditions of civil aeronautical activities relating to the development of airline contracts concerning the use of aircraft which brought in fiscal advantages to airlines. Specific contacts, such as leases, charters and interchange of aircraft are now assisting air carriers to obviate the need to find money to buy new aircraft. More carriers are now entering into short term lease agreements to keep their operations afloat and such dry or wet lease agreements necessitate a closer look at the requirements of registration and nationality as dictated to by the Chicago Convention.

In order to accord with commercial exigencies relating to lease and charters in the air transport industry, and as already mentioned, ICAO has introduced Article 83 *bis* to the Chicago Convention, which means that a State may lease aircraft registered in another State, and, by mutual agreement, take over responsibilities of the State of registration in respect of that aircraft. Under these circumstances, it may be reasonable to assume that in the event an aircraft leased by a State performs functions of a military nature for the lessee State, such State could be considered the State of registration if an agreement to that effect had been put into effect between the lessor and lessee.

Article 83 *bis* of the Convention was timely, in that it was adopted at a time when trade barriers were being rapidly obviated and many industries were being globalized. Instances of as many as ten multinational partners in one industry are not uncommon in today's commercial world. In particular, commercial trends in the United States and United Kingdom show new emergent large airlines with the participation of more than one nationality.

Although the current bilateral regulatory structure calls for substantial ownership and effective control of airlines by nationals or companies of a designating State—which essentially means that for Country A to designate its airline to operate commercial flights the airline must be substantially owned and effectively controlled by nationals or companies of Country A—this requirement is increasingly becoming impracticable to fulfil in various instances. In recognition of one such

---

[68] *Ibid.* at 3.

circumstance, the ICAO Assembly, at its 24th Session, adopted Resolution A24-12 which recognized the political reality of regional groupings of States into composite economic entities, forming a community of interest. The Assembly recognized that such a community of interest, when applicable to groups of developing States, would require their airlines to be identified on a common basis with regard to their substantial ownership and effective control in the context of bilateral regulation of air traffic rights. Therefore, the ICAO Assembly urged Contracting States by its Resolution to accept the designation of, and allow an airline substantially owned and effectively controlled by one or more developing State or Sates (or its or their nationals) belonging to a regional economic grouping to exercise the route rights and other air transport rights of any developing State or States within the same grouping under mutually acceptable terms.

There are other instances such as when airlines have had multinational ownership (involving ownership of one airline by several states such as in the instances of Gulf air, Air Afrique, SAS and LIAT); have ownership registered in one country but are being accepted as airlines of another (such as Britannia and Monarch whose ownership rested in Canada and Switzerland respectively but operated air services as designated carriers of the United Kingdom); and are owned by legal persons whose businesses are not domiciled in the country in which the carrier has its place of business (such as Cathay Pacific Airlines).

The "Third Package" of the European Union, which allowed for airlines within the Union to be owned by nationals or companies of any member State, gave further credence to the compelling need to consider the element of designation of airlines outside the purview of the Conventional philosophy of "substantial ownership and effective control" as required by the current bilateral regulatory regime.

In view of the above developments, the dictates of aircraft financing require financiers to be aware of the multitude of possibilities of litigation for ownership and control of aircraft financed by them and also the legal implications of aircraft leasing in the modern context. Donald Bunker states:

The concept of registration has now developed such that financiers of commercial aircraft for use internationally must be well aware of the effect that the country of registration could have on their rights. The relatively liquid world market in used aircraft makes aircraft financing quite attractive to many investors. However, most prudent financiers like to be assured of being able to obtain possession of their equipment, free and clear of a defaulting debtor's rights and deregistered by the operator's country so that an efficient realization of their security could be achieved.[69]

From the point of view of the airline which leases aircraft and sustains damage to the aircraft and to its passengers, the legal relationship between lessor and lessee of property would apply in common law jurisdictions. The lessor of the aircraft would usually be covered by his own insurance or by an indemnification agreement between the lessor and lessee. In a typical financial lease agreement of aircraft, the position of the lessor could be that of a lender at common law, and to that extent he would be

[69]Bunker (1988), at p. 157.

protected from the mere presumption *ipso facto* that he is liable by virtue of his ownership of the aircraft. However, this is not strictly an inflexible rule and different jurisdictions may impose strict liability in certain situations.[70] There is also the possibility that rules of negligence may apply in certain jurisdictions where an injured party—the lessee—may seek redress from the lessor of the aircraft. Such claims are often prompted by the favourable financial circumstances that lessors are in usually.

The protection of the lessee in instances of damage is usually assured by the liability insurance obtained by the lessee. The lessee could also qualify the indemnity agreement he signs with the lessor that the lessee's liability would be valid and effectual only in instances when the lessor is not negligent or in default of his agreement. The lessee would therefore be protected against such acts as arbitrary seizure of property by the lessor. Other legal measures available to the lessee are his capacity and legal right to insert a clause in the lease agreement that the leased property is accepted by the lessee on condition of warranty as to the quality of the property; and his ability to obtain warranty direct from the manufacturer.

As discussed, registration of aircraft is a critical issue which requires that, at least the minimum requirements set by the Chicago Convention and its relevant Annexes are met. However, the universality and transparency required to ensure that States which enter aircraft in their registries do so while adhering to international standards has to be tempered with caution and discretion as States are weary of information pertaining to their internal standards of safety being shared indiscriminately. This sets, for the regulator, a diplomatic tight rope that has to be treaded with an abundance of care and caution. Over the past decades, civil aviation has had to serve the political and economic interests of States and in this regard, ICAO has alternated between two positions, in its unobtrusive diplomatic role and in its more pronounced regulatory role.[71]

An aircraft registry of a State exudes the profile and character of that State and it is important to remember that, from the distant past, it has been recognized that a nation's air power is the sum total of all its civil and military aviation resources.[72] Nationality of aircraft, which is tied intrinsically to its registration by the Chicago Convention, is a matter of national pride and registration therefore becomes a political symbol of a State in the international arena. Registration also has a bearing on the application of Rules of the Air as Standard 2.1.1 of Annex 2 to the Chicago Convention provides that the rules of the air shall apply to aircraft bearing the nationality and registration marks of a Contracting State, wherever they may be, to the extent that they do not conflict with the rules published by the State having jurisdiction over the territory over-flown.[73]

---

[70] *Id.* at p. 288.

[71] Sochor (1991), xvi.

[72] van Zandt (1944) at pp. 28, 93.

[73] The Council of ICAO resolved, in adopting Annex 2 in April 1948 and Amendment 1 to the said Annex in November 1951, that the Annex constitutes Rules relating to the flight and manoeuvre of aircraft within the meaning of Article 12 of the Convention. Over the high seas, therefore, these rules apply without exception.

In this equation, the role of ICAO becomes an important one. Over its 60 years of service to the international civil aviation community, ICAO has, through its Assembly and Council adopted numerous resolutions. Additionally, the ICAO Council has taken several decisions and issued statements of policy guidance. An organization such as ICAO is tasked primarily to provide a certain predictability about its members by promulgating norms for the conduct of its Member States. Of course not all those norms are binding and not all of them are adopted with the same degree of formality. However, certainly all of them provide guidance to States. This situation has to mesh with the basic inquiry as to whether ICAO, as an international organization, has been given direct authority over individuals or States. Another issue is whether ICAO is primarily an instrument for cooperation among States.

Firstly, when one considers the background of ICAO and the statements of its founding fathers, and as discussed earlier, there is no room for doubt that ICAO is a specialized agency that has procedures to modify, without eliminating, the positivist principle that States are only bound by international rules to which they have consented. This approach admits of a process whereby ICAO adopts or amends rules after having given a designated period of time for its member States to examine such rules and decide whether they would accept them or not. Individual member States may object or mark their differences in practices to the ones ICAO suggests for adoption.[74] States objecting to a particular Standard and Recommended Practice (SARP) may choose if they wish to opt out of whole processes recommended by ICAO, even though general consensus is achieved to adopt them. There is no record of a single international Standard adopted through this process being disapproved by a majority of ICAO member States, although not all of ICAO's 190 member States have found it practicable to comply with all Standards[75] in the 18 Annexes to the Chicago Convention.[76]

The question arises as to whether a member State is formally bound by Standards contained in an Annex to the Chicago Convention, particularly when such a State has no convincing argument that it is impracticable to implement such Standards or when it has not notified the ICAO Council of differences as required. This is a

[74]Article 37 of the Chicago Convention confirms that each Member State undertakes to collaborate in securing the highest practicable degree of uniformity in regulations, standards, procedures and organization in relation to aircraft, personnel, airways and auxiliary services in all matters in which such uniformity will facilitate and improve air navigation. Article 38 gives any State the opportunity, if it finds it impracticable to adhere to ICAO's policy to file differences by giving notice to ICAO of the difference between what is recommended or required by ICAO and the practice prevalent in that State.

[75]The ICAO Assembly, at its 35th Session held in Montreal from 28 September to 8 October 2004, defined a Standard "as any specification. . .the uniform application of which is recognized as necessary for the safety or regularity of international air navigation and to which member States will conform in accordance with the Chicago Convention.; in the event of impossibility of compliance, notification to the Council is compulsory under Article 38 of the Convention. The same resolution describes a Recommended Practice as any specification for physical characteristics. . . which is recognised as desirable and one that member States will endeavour to conform to" See *Assembly Resolutions in Force* (As of 8 October 2004) ICAO Doc 9848, II-2 Appendix A.

[76]Buergenthal (1969), pp. 98–107.

vexed debate, particularly in the face of two blatant facts. The first is that the *travaux preparatoires* to the Convention contains a statement that "the Annexes are given no compulsory force".[77] The second is that in Article 54 of the Convention, which lays down the mandatory functions of the Council, it is provided that one of the mandatory functions is to

> Adopt, in accordance with the provisions of this Convention, international standards and recommended practices; for convenience (emphasis added) designate them as Annexes to this Convention; and notify all member States of action taken.[78]

One could argue therefore that the Annexes are not an integral part of the Convention by virtue of the statement in Article 54 and therefore do not form binding law.

There have been numerous views of legal scholars who have cautioned against this approach and advocated that the words of the Convention should not be taken literarily. One commentator is of the view that:

> The debate is largely academic. Whether or not ICAO standards are formally binding in the treaty law sense, they are highly authoritative in practice. This reflects their recognized importance for the safety and efficiency of civil air travel and the thorough process by which they are promulgated.[79]

All this leaves one with the inevitable question as to whether ICAO has sufficient clout to enforce its mandatory duties which appear under Article 54 of the Chicago Convention and in particular Article 54 (j). From a legal perspective, the above discussion points to a resolute "yes" with an additional qualifier that it is indeed ICAO's duty to do so. It is therefore largely left to ICAO to decide on the path it takes to ensure the legitimacy of its SARPs and the credibility of its Assembly and Council in the most diplomatic manner possible, in this defining point in the history of the Organization.

References

Abeyratne RIR (2007) State responsibility for safety management systems. J Aviat Manag 7–13

Buergenthal T (1969) Law making in the international civil aviation organization. Syracuse University Press, Syracuse

Bunker DH (1988) The law of aerospace finance in Canada. Institute and Centre of Air and Space Law, McGill

---

[77]See Whiteman (1968) at p. 404.

[78]Chicago Convention, Preamble *supra* note 1, Article 54 (l).

[79]Kirgis (1995), p. 109 at 126. There is a similar process in operation under the World Meteorological Organization, whereby a certain amount of decision making authority is given to the WMO Congress. Article 9 (a) of the WMO Convention provides that all members shall do their utmost to implement the decisions of the Congress. Article 9 (b) allows any member to opt out by notifying the Secretary General, with reasons if it finds it impracticable to give effect to the technical requirement in question. WMO Convention, reprinted in *International Organization and Integration* (Kapteyn et al., eds) 2nd Revised Edition, 1981, pt. I.B.1.9 a. Also in *WMO Basic Documents*, No. 1. WMO Doc. No. 15 at 9 1987.

Coles R (2002) Ship registration: law and practice. LLP, London
Hill C (1998) Maritime law, 5th edn. LLP, London
Kasoulides G (1989) The 1986 United Nations Convention on the conditions for registration of vessels and the question of open registry. Ocean Dev Int Law 20:543
Kirgis FL Jr (1995) Specialized law making processes. In: Schachter O, Joiner C (ed) United Nations legal order, vol 1. The American Society of International Law (Chapter 2)
Lelieur I (2003) Law, policy and substantial ownership and effective control of airlines, prospects for change. Aldershot, Ashgate
Matlin D (1990) Re-evaluating the status of flags of convenience under international law. Vand J Transnat'l J 23:1017–1031
Sochor E (1991) The politics of international aviation. Macmillan, London
Sohn LB, Gustafson K (1984) The law of the sea in a nutshell. West Publishing Co., Minnesota
van Zandt P (1944) Civil aviation and peace. Brookings Institution, Washington, DC
Verhaegen BM (1997) The entry into force of Article 83 *bis*: legal perspectives in terms of safety oversight. Ann Air Space Law XXII(Part II):269–283
Whiteman M (1968) Digest of international law. Kluwer, The Hague

Article 22
Facilitation of Formalities

Each contracting State agrees to adopt all practicable measures, through the issuance of special regulations or otherwise, to facilitate and expedite navigation by aircraft between the territories of contracting States, and to prevent unnecessary delays to aircraft, crews, passengers and cargo, especially in the administration of the laws relating to immigration, quarantine, customs and clearance:

Contents

1 Ensuring Expedition Inherent in Air Transport

Facilitation of formalities pertaining to the carriage by air of persons, mail and cargo is addressed through Annex 9 to the Chicago Convention. The Annex is a response to Article 22, 23 and 37 of the Convention.[1]

The story of facilitation is a long one, dating back to the first session of the ICAO Assembly held in 1947. The Assembly adopted Resolution A1-40 (Facilitation of International Air Transport) which recalled that the Interim PICAO Assembly at its meeting in May 1946 had endorsed the objective of PICAO's programme on facilitation which primarily involved the development of Standards and Recommended Practices. The Assembly reaffirmed this endorsement.[2]

[1]A number of other articles of the Convention have special pertinence to the provisions of the FAL Annex and have been taken into account in its preparation. In particular, persons responsible for the implementation of the provisions of this Annex should be familiar with the following articles in addition to Articles 22 and 23:

Article 10, Landing at customs airport; Article 11, Applicability of air regulations; Article 13, Entry and clearance regulations; Article 14, Prevention of spread of disease; Article 14, Prevention of spread of disease; Article 24, Customs duty; Article 29, Documents carried in aircraft; Article 35, Cargo restrictions.

[2]Doc 2005, FAL/40.

R. Abeyratne, *Convention on International Civil Aviation*, DOI 10.1007/978-3-319-00068-8_23,

Standards and Recommended Practices on Facilitation were first adopted by the Council on 25 March 1949, pursuant to the provisions of Article 37 of the Convention on International Civil Aviation (Chicago, 1944), and designated as Annex 9 to the Convention with the title "Standards and Recommended Practices—Facilitation". They became effective on 1 September 1949. The Standards and Recommended Practices were based on recommendations of the First and Second Sessions of the Facilitation Division, held at Montreal in February 1946 and at Geneva in June 1948. They were expanded and amended comprehensively as a result of subsequent Sessions of the Division, i.e., the Third Session, held at Buenos Aires in December 1951, the Fourth Session, held at Manila in October 1955, the Fifth Session, held at Rome in December 1959, the Sixth Session, held at Mexico City in March–April 1963, the Seventh Session, held at Montreal in May 1968, the Eighth Session, held at Dubrovnik in March 1973, the Ninth Session held at Montreal in April–May 1979, the Tenth Session held at Montreal in September 1988 and the Eleventh Session held in Montreal in April 1995, and the Third Meeting of the Facilitation (FAL) Panel held in Montreal in February 2001.

As a result of the Division's and FAL Panel's Recommendations for amendment of Annex 9 and Council's action thereon, the Second Edition of Annex 9 became effective on 1 March 1953, the Third Edition on 1 November 1956, the Fourth Edition on 1 November 1960, the Fifth Edition on 1 April 1964, the Sixth Edition on 1 April 1969, the Seventh Edition on 15 April 1974, the Eighth Edition on 15 July 1980, the Ninth Edition on 15 November 1990, the Tenth Edition on 30 April 1997 and the Eleventh Edition on 15 July 2002.

The Standards and Recommended Practices (SARPs) on Facilitation (FAL) are derived from several provisions of the Chicago Convention. *Article 37* obliges ICAO to adopt and amend from time to time international standards and recommended practices and procedures dealing with, *inter alia*, customs and immigration procedures. *Article 22* obliges each Contracting State to adopt all practicable measures to facilitate and expedite navigation by aircraft between the territories of Contracting States, and to prevent unnecessary delays to aircraft, crews, passengers, and cargo, especially in the administration of the laws relating to immigration, quarantine, customs and clearance. *Article 23* of the Convention expresses the undertaking of each Contracting State to establish customs and immigration procedures affecting international air navigation in accordance with the practices established or recommended pursuant to the Convention.

A number of other articles have special pertinence to the provisions of the FAL Annex and have been taken into account in its preparation. These include: *Article 10*, which requires all aircraft entering the territory of a Contracting State to land at, and depart from, an airport designated by that State for customs and other examination; *Article 13*, which require compliance of a Contracting State's entry, clearance, immigration, passports, customs and quarantine laws and regulations, by or on behalf of passengers, crew or cargo; *Article 14*, which obliges each Contracting State to take effective measures to prevent the spread by means of air navigation of communicable diseases; and *Article 24* (customs duty), *Article 29* (documents carried in aircraft) and *Article 35* (cargo restrictions).

These provisions of the Convention find practical expression in the SARPs of Annex 9, the first edition of which was adopted in 1949. The SARPs pertain specifically to facilitation of landside formalities for clearance of aircraft and commercial traffic through the requirements of customs, immigration, public health and agriculture authorities. The Annex is a wide-ranging document which reflects the flexibility of ICAO in keeping pace with international civil aviation. ICAO is recognized as being the first international body to make a real start on facilitation by developing Standards which bind its Contracting States.

The Annex provides a frame of reference for planners and managers of international airport operations, describing maximum limits on obligations of industry and minimum facilities to be provided by governments. In addition, Annex 9 specifies methods and procedures for carrying out clearance operations in such a manner as to meet the twin objectives of effective compliance with the laws of States and productivity for the operators, airports and government inspection agencies involved.

Initially, the main thrust of the Annex consisted of efforts to reduce paperwork, standardize internationally the documents that were to accompany traffic between States, and simplify the procedures required to clear aircraft, passengers and cargo. It was—as it still is—recognized that delays due to cumbersome formalities must be reduced, not just because they are unpleasant but, in practical terms, because they are costly to all of the "customer groups" in the community and because they interfere with the success of everyone.

Over the years, traffic volumes grew. States' resources for inspection regimes could not keep pace. The facilitation of landside clearance formalities became a much more complex issue. The focus of Annex 9 therefore changed. In its 11th edition (2002), the Annex 9 retained its original strategies, carried forward in all editions since the first, of reducing paperwork, standardizing documentation and simplifying procedures. However, it shifted its focus to inspection techniques based on risk management, with the objectives to increase efficiency, reduce congestion in airports and enhance security; to control abuses such as narcotics trafficking and travel document fraud; and to support the growth of international trade and tourism. In addition, new SARPs and guidance material were introduced to address certain high-profile issues of public interest such as the treatment of persons with disabilities.

More recently, the face of facilitation has been further shaped by major developments in the civil aviation environment which have occurred during the last 10 years (the mid-1990s and beyond). These phenomena include: technological progress, with the universal proliferation of the use of computers and electronic data interchange systems; massive increases in illegal migration which have become worldwide immigration and national security problems, with civil aviation the transport mode of choice and passport fraud a frequent tactic; and ongoing political and social upheaval, which has given rise to increased use of terrorism, in which unlawful interference with civil aviation is still a powerful technique for pursuing an objective.

These topics formed the basis of the agenda of the 12th Session of the Facilitation Division that was held in Cairo in early 2004 with the theme, "Managing Security Challenges to Facilitate Air Transport Operations." Discussions on the essential role that facilitation measures play in the improvement of security led to the Division making recommendations on the security of travel documents and border control formalities, on modernized provisions for facilitation and security in air cargo service operations, on controlling travel document fraud and illegal migration and on international health regulations and hygiene and sanitation in aviation.

The consequent 12th edition of Annex 9 (expected publication: 2005) reflects ICAO's contemporary FAL strategy. This is to advocate and support action by Contracting States in three principal areas: the standardization of travel documents, the rationalization of border clearance systems and procedures, and international cooperation to tackle security problems related to passengers and cargo. While the primary motivation of Annex 9 will continue to carry out the mandate in Article 22 of the Chicago Convention, "...to prevent unnecessary delays to aircraft, passengers and cargo...", numerous provisions, developed with the intent to increase efficiency in control processes, support also the objective to raise the level of general security.

Enhancing the security of travel documents and tackling illegal migration are among the major changes introduced into Annex 9 through its 12th edition. Most of the existing Chapters and Appendices of the Annex remain more-or-less unchanged from the 11th edition. Two Chapters, in particular, have been substantially amended to reflect new international realities.

Chapter 3, which deals with the entry and departure of persons and baggage, now contains a Standard obliging Contracting States to regularly update security features in new versions of their travel documents, to guard against their misuse and to facilitate detection of cases where such documents have been unlawfully altered, replicated or issued.

Another Standard requires States to establish controls on the lawful creation and issuance of travel documents. States are also now obliged to issue separate passports to all persons, regardless of age, and to issue them in machine readable form, in accordance with ICAO's specifications. States and airlines are required to collaborate in combatting travel document fraud. As for crew members, States are obliged to place adequate controls on the issuance of crew member certificates and other official crew identity documents.

Finally, an entirely new Chapter 5 is devoted to the growing problem of inadmissible persons and deportees. The SARPs of this Chapter set out in clear terms the obligations of States and airlines *vis-à-vis* transport of potentially illegal migrants and similar "problem" cases that the international air transport industry comes across in ever greater numbers daily. Strict adherence by Contracting States of the obligations to remove from circulation fraudulent travel documents or genuine documents used fraudulently will greatly help to constrict the flow of illegal migrants the world over.

The Assembly, at its 10th Session (Caracas, 19 June–16 July 1956), adopted Resolution A10-35 (General Programmes in the Facilitation Field) which resolved that each State will pay special attention to its obligations under Articles 22 and 23 and review any deviations from Annex 9 with regard to their regulations and practices.

The current 12th Edition contains *inter alia*, provisions arising from the A-type recommendations of the Twelfth Session of the Facilitation Division (FAL/12) (Cairo, Egypt, 22 March to 1 April 2004) on issues including Machine Readable

Travel Documents (MRTDs), the deployment of biometric technologies in travel documents, aviation security, travel document fraud and illegal immigration, advance passenger information, international health issues and regulations, and assistance to aircraft accident victims and their families. This again resulted in a comprehensive amendment of Annex 9. This 12th Edition of Annex 9 became effective on 11 July 2005 and is to become applicable on 24 November 2005.

The Standards and Recommended Practices on Facilitation are the outcome of Article 37 of the Convention, which provides, *inter alia,* that the

> International Civil Aviation Organization shall adopt and amend from time to time, as may be necessary, international standards and recommended practices and procedures dealing with . . . customs and immigration procedures . . . and such other matters concerned with the safety, regularity and efficiency of air navigation as may from time to time appear appropriate.

The policy with respect to the implementation by States of the Standards and Recommended Practices on Facilitation is strengthened by Article 22 of the Convention, which expresses the obligation accepted by each Contracting State

> to adopt all practicable measures, through the issuance of special regulations or otherwise, to facilitate and expedite navigation by aircraft between the territories of Contracting States, and to prevent unnecessary delays to aircraft, crews, passengers, and cargo, especially in the administration of the laws relating to immigration, quarantine, customs and clearance,

and by Article 23 of the Convention, which expresses the undertaking of each Contracting State

> so far as it may find practicable, to establish customs and immigration procedures affecting international air navigation in accordance with the practices which may be established or recommended from time to time pursuant to this Convention.

In addition to the Standards and Recommended Practices of Annex 9, the Organization's FAL Programme is based on the FAL Resolutions of the Assembly and B-type recommendations of FAL Division Sessions which are those recommendations which do not suggest amendments to the Annex provisions.

At its 16th Session (Buenos Aires, 3–26 September 1968) the Assembly adopted Resolution A16-27 (Implementation of the Provisions of Annex 9 and of ICAO Recommendations in the Field of Facilitation) reiterated its Resolution adopted at the 10th Session and requested the Council to review at least once every 3 years the status of implementation of Annex 9.

2 Border Crossing and State Responsibility

One of the key aspects of facilitation involves speedy, efficient and secure border crossing. The primary tool for this process is a valid travel document. A discussion of passports and visas has already take place in this book under Article 13. However, there remains one aspect to be discussed and that is State responsibility in protecting the integrity of the passport and the prevention of passport fraud.

The passport is a basic document in the transport by air of persons. Its use therefore is of fundamental importance as a travel document, not only because it reflects the importance of the sovereignty of a State and the nationality of its citizens but also because it stands for the inviolability of relations between States that are linked through air transport. The assassination of a leader of Hamas on 19 January 2010 by a group of individuals in Dubai who used forged passports belonging to various nations, raised a diplomatic outcry and brought to bear an important facet of air transport that is vulnerable to abuse and contention among States.

The fundamental issue that emerges is one that is critical to air law in the context of the integrity and ownership of the passport and its abuse in the course of criminal activity. There is also the issue, from a legal and diplomatic perspective as to whether a State or instrumentality of State, can, with impunity, use forged passports for travel of its staff on missions of espionage or assassination. *A fortiori*, an additional issue is whether a State could be complicit or condone or be seen to condone (in the absence of any action taken by the State to punish the miscreants) such abuse of travel documents belonging to other nations. In order to determine these issues, this article addresses two basic discussions: the first on complicity and condonation of a State and the second on the nature and integrity of the passport. Finally, it discusses issues of State responsibility, diplomacy and criminality.

On 19 January 2010, Mahmoud al-Mabhouh, considered to be a senior commander of Hamas, a radical Palestinian group, was assassinated at a hotel in Dubai in a manner usually employed by professionally trained military and secret service agencies. The killing was attributed to Mossad[3] The European Union, which considers Hamas a terrorist organization, nonetheless condemned the assassination of the Hamas leader and showed particular concern over the fact that the killers had used passports from Ireland, France, Germany and the UK—to coordinate their travel into Dubai from various parts of the world, synchronizing their arrival time from various flights into Dubai International Airport and checking into the hotel of the victim contemporaneously. The EU strongly condemned the fact that those involved in this action used fraudulent EU member states' passports and credit cards acquired through the theft of EU citizens' identities.[4]

Australia was another complainant who warned Israel that its friendly relations with Israel would be jeopardised if it were found to have condoned the suspected theft of three Australian citizens' identities which Mossad used to carry out its political assassination. The diplomatic impasse occurred when three Australians from Victoria living in Israel at the time were confirmed among 26 people from four nations whose tampered passports were allegedly used by a team of suspected Israeli Mossad agents who assassinated al-Mabhouh. Australian Prime Minister

[3]Mossad is responsible for the collection of intelligence and other covert activities including military operations. It is one of the most integral parts of the Israeli intelligence community and reports directly to the Prime Minister of Israel. See http://en.wikipedia.org/wiki/Mossad.

[4]Toby Vogel, EU Condemns Use of False Passports inn Hamas Killing, http://www.europeanvoice.com/article/2010/02/eu-condemns-use-of-false-passports-in-hamas-killing/67225.aspx.

Kevin Rudd is reported to have stated that Australia would be vocal in its contempt of any State if it were found that it

> . . . has been complicit in the use or abuse of the Australian passport system, let alone for the conduct of an assassination, and has treated Australia with contempt and there will therefore be action by the Australian government in response.[5]

Dubai authorities are reported to have said that they were virtually certain Israeli agents carried out the killing and had released the identities of 11 people who travelled on forged British, Irish, French and German passports to kill al-Mabhouh in a hotel.[6]

There is seemingly a history behind alleged Mossad involvement in the use of fake foreign passports in its activities. Reportedly, in 2004 New Zealand's prime minister imposed diplomatic sanctions—restricting visas and cancelling high level visits—after two Mossad agents were caught trying to acquire passports fraudulently—one in the name of a tetraplegic man. Seven years earlier, Mossad assassins carrying Canadian passports with assumed names attempted to murder the Hamas leader Khaled Meshaal by spraying nerve agent into his ear as he entered his office in Amman.[7]

The fundamental issue that emerges is one that is critical to air law in the context of the integrity and ownership of the passport and its abuse in criminal activity. There is also the issue, from a legal and diplomatic perspective is whether a State or instrumentality of State such as Mossad, can, with impunity, use forged passports for travel of its staff on missions of espionage or assassination. *A fortiori*, an additional issue is whether a State could be complicit or condone or be seen to condone (in the absence of any action taken by the State to punish the miscreants) such abuse of travel documents belonging to other nations.

---

[5]http://www.theaustralian.com.au/news/world/australians-caught-in-hit-on-hamas/story-e6frg6so-1225834538825. It is reported that in 1997, Mossad bungled the assassination of top Hamas leader Khalid Mishal, who was injected while in Jordan with a poison by Israeli agents travelling on Canadian documents. He survived after his assailants were captured by his bodyguards and Israel provided the antidote. In 2004, two Mossad agents were jailed in New Zealand after trying to obtain fake passports, one in the name of a cerebral palsy sufferer. *Ibid.*

[6]http://www.euractiv.com/en/foreign-affairs/eu-unhappy-israel-over-fake-passports-james-bond-killings-news-278602.

[7]David Sapsted, and Loveday Morris, Israel in the Dock Over Fake Passports, http://www.thenational.ae/apps/pbcs.dll/article?AID=/20100218/NATIONAL/702179796/1133/sport Hamas, which won 2006 legislative elections in the Palestinian territories, is shunned by the West for rejecting its calls to recognise Israel and renounce violence. Hit squads dispatched by Mossad have used foreign passports in the past, notably in 1997 when agents entered Jordan on Canadian passports and bungled an attempt to kill Meshaal with poison. In 1987, Britain protested to Israel about what London called the misuse by Israeli authorities of forged British passports and said it received assurances steps had been taken to prevent future occurrences. In 2003, the offices of several EU member countries in the Council's Justus Lipsius building, including France, Germany and the UK, were found to be bugged. Although the Union has been discrete over the incident, many consider Mossad to have been responsible for the wiretapping. *Ibid.*

3 Complicity

The fundamental issue in the context of State responsibility for the purposes of this chapter is to consider whether a State should be considered responsible for its own failure or non-feasance to prevent a private act that is a violation of its international responsibility towards a third State or whether the conduct of the State itself can be impugned by identifying a nexus between the perpetrator's conduct and the State. One view is that an agency paradigm, which may in some circumstances impute to a state reprehensibility on the ground that a principal-agent relationship between the State and the perpetrator existed, can obfuscate the issue and preclude one from conducting a meaningful legal study of the State's conduct.[8]

At the core of the principal-agent dilemma is the theory of complicity, which attributes liability to a State that was complicit in a private act. Hugo Grotius (1583–1645), founder of the modern natural law theory, first formulated this theory based on State responsibility that was not absolute. Grotius' theory was that although a State did not have absolute responsibility for a private offence, it could be considered complicit through the notion of *patienta* or *receptus*.[9] While the concept of *patienta* refers to a State's inability to prevent a wrongdoing, *receptus* pertains to the refusal to punish the offender.

The eighteenth century philosopher Emerich de Vattel was of similar view as Grotius, holding that responsibility could only be attributed to the State if a sovereign refuses to repair the evil done by its subjects or punish an offender or deliver him to justice whether by subjecting him to local justice or by extraditing him.[10] This view was to be followed and extended by the British jurist Blackstone a few years later who went on to say that a sovereign who failed to punish an offender could be considered as abetting the offence or of being an accomplice.[11]

A different view was put forward in an instance of adjudication involving a seminal instance where the Theory of Complicity and the responsibility of states for private acts of violence was tested in 1925. The case[12] involved the Mexico–United States General Claims Commission which considered the claim of the United States on behalf of the family of a United States national who was killed in a Mexican mining company where the deceased was working. The United States argued that the Mexican authorities had failed to exercise due care and diligence in apprehending and prosecuting the offender. The decision handed down by the Commission distinguished between complicity and the responsibility to punish and the Commission was of the view that Mexico could not be considered an accomplice in this case.

[8]Caron (1998) 109, at 153–54 cited in Becker (2006a), at 155.

[9]H Grotius, JB Scott, (tr), 2 *De Jure Belli Ac Pacis* (1646), 523–26.

[10]De Vattel and Fenwick (1916), 72.

[11]Blackstone and Morrison (2001), at 68.

[12]*Laura M.B. Janes (USA)* v. *United Mexican States* (1925) 4 R Intl Arb Awards 82.

The Complicity Theory, particularly from a Vattellian and Blackstonian point of view is merely assumptive unless put to the test through a judicial process of extradition. In this Context it becomes relevant to address the issue through a discussion of the remedy.

4 Condonation

The emergence of the Condonation Theory was almost concurrent with the *Jane* case[13] decided in 1925 which emerged through the opinions of scholars who belonged to a school of thought that believed that States became responsible for private acts of violence not through complicity as such but more so because their refusal or failure to bring offenders to justice, which was tantamount to ratification of the acts in question or their condonation.[14] The theory was based on the fact that it is not illogical or arbitrary to suggest that a State must be held liable for its failure to take appropriate steps to punish persons who cause injury or harm to others for the reason that such States can be considered guilty of condoning the criminal acts and therefore become responsible for them.[15] Another reason attributed by scholars in support of the theory is that during that time, arbitral tribunals were ordering States to award pecuniary damages to claimants harmed by private offenders, on the basis that the States were being considered responsible for the offences.[16]

The responsibility of governments in acting against offences committed by private individuals may sometimes involve condonation or ineptitude in taking effective action against terrorist acts, in particular with regard to the financing of terrorist acts. The United Nations General Assembly, on 9 December 1999, adopted the International Convention for the Suppression of the Financing of Terrorism,[17] aimed at enhancing international co-operation among States in devising and adopting effective measures for the prevention of the financing of terrorism, as well as for its suppression through the prosecution and punishment of its perpetrators.

The Convention, in its Article 2 recognizes that any person who by any means directly or indirectly, unlawfully or wilfully, provides or collects funds with the intention that they should be used or in the knowledge that they are to be used, in full or in part, in order to carry out any act which constitutes an offence under certain named treaties, commits an offence. One of the treaties cited by the Convention is the International Convention for the Suppression of Terrorist

[13] *Ibid.*

[14] *Black's Law Dictionary* defines condonation as "pardon of offense, voluntary overlooking implied forgiveness by treating offender as if offense had not been committed."

[15] Jane's case, *Supra* note 12, at 92.

[16] Hyde (1928) at 140–142.

[17] International Convention for the Suppression of the Financing of Terrorism, adopted by the General Assembly of the United Nations in resolution 54/109 of 9 December 1999.

Bombings, adopted by the General Assembly of the United Nations on 15 December 1997.[18]

The Convention for the Suppression of the Financing of Terrorism also provides that, over and above the acts mentioned, providing or collecting funds toward any other act intended to cause death or serious bodily injury to a civilian, or to any other person not taking an active part in the hostilities in the situation of armed conflict, when the purpose of such act, by its nature or context, is to intimidate a population, or to compel a government or an international organization to do or to abstain from doing any act, would be deemed an offence under the Convention.

The United Nations has given effect to this principle in 1970 when it proclaimed that:

> Every State has the duty to refrain from organizing or encouraging the organization of irregular forces or armed bands, including mercenaries, for incursion into the territory of another State. Every State has the duty to refrain from organizing, instigating, assisting or participating in acts of civil strife or terrorist acts in another State or acquiescing in organized activities within its territory directed towards the commission of such acts, when the acts referred to in the present paragraph involve a threat or use of force.[19]

Here, the words *encouraging* and *acquiescing in organized activities within its territory directed towards the commission of such acts* have a direct bearing on the concept of condonation and would call for a discussion about how States could overtly or covertly encourage the commission of such acts. One commentator[20] identifies three categories of such support: *Category I* support entails protection, logistics, training, intelligence, or equipment provided terrorists as a part of national policy or strategy; *Category II* support is not backing terrorism as an element of national policy but is the toleration of it; *Category III* support provides some terrorists a hospitable environment, growing from the presence of legal protections on privacy and freedom of movement, limits on internal surveillance and security organizations, well-developed infrastructure, and émigré communities.

Another commentator[21] discusses what he calls the *separate delict theory*' in State responsibility, whereby the only direct responsibility of the State is when it is responsible for its own wrongful conduct in the context of private acts, and not for the private acts themselves. He also contends that indirect State responsibility is occasioned by the State's own wrongdoing in reference to the private terrorist conduct. The State is not held responsible for the act of terrorism itself, but rather for its failure to prevent and/or punish such acts, or for its active support for or

---

[18]A/52/653, 25 November 1997.

[19]Declaration on Principles of International Law Concerning Friendly Relations and Co-operation Among States in Accordance with the Charter of the United Nations, UN General Assembly Resolution 2625 (XXV) 24 October 1970.

[20]Steven Metz, State Support for Terrorism, Defeating Terrorism, Strategic Issue Analysis, at http://www.911investigations.net/IMG/pdf/doc-140.pdf.

[21]Becker (2006b).

acquiescence in terrorism.[22] Arguably the most provocative and plausible feature in this approach is the introduction by the commentator of the desirability of determining State liability on the theory of causation. He emphasizes that:

The principal benefit of the causality based approach is that it avoids the automatic rejection of direct State responsibility merely because of the absence of an agency relationship. As a result, it potentially exposes the wrongdoing State to a greater range and intensity of remedies, as well as a higher degree of international attention and opprobrium for its contribution to the private terrorist activity.[23]

The causality principle is tied in with the rules of State Responsibility enunciated by the International Law Commission and Article 51 of the United Nations Charter which states that

> nothing in the Charter will impair the inherent right of individual or collective self-defense if an armed attack occurs against a Member of the United Nations, until the Security Council has taken measures necessary to maintain international peace and security. The provision goes on to say that measures taken by Members in the exercise of this right of self-defense will be immediately reported to the Security Council and will not in any way affect the authority and responsibility of the Security Council under the present Charter to take at any time such action as it deems necessary in order to maintain or restore international peace and security.

The International Law Commission has established that a crime against the peace and security of mankind entails individual responsibility, and is a crime of aggression.[24] A further link drawing civil aviation to the realm of international peace and security lies in the Rome Statute of the International Criminal court, which defines a war crime, *inter alia*, as intentionally directing attacks against civilian objects; attacking or bombarding, by whatever means, towns, villages, dwellings or buildings which are undefended and which are not military objects; employing weapons, projectiles, and material and methods of warfare that cause injury.[25] The Statute also defines as a war crime, any act which is intentionally directed at buildings, material, medical units and transport, and personnel using the distinctive emblems of the Geneva Conventions in conformity with international law.[26]

5 Knowledge

Another method of determining State responsibility lies in the determination whether a State had actual or presumed knowledge of acts of its instrumentalities, agents or private parties which could have alerted the State to take preventive

[22] *Id.* Chapter 2, 67.

[23] Becker (2006b), Chapter 2 at 335.

[24] Draft Code of Crimes Against the Peace and Security of Mankind, International Law Commission Report, 1996, Chapter II Article 2.

[25] Rome Statute of the International Criminal Court, Article 8.2 (b) (ii), (V) and (XX).

[26] Id. Article 8.2 (b) (XXIV).

action. International responsibility of a State cannot be denied merely on the strength of the claim of that State to sovereignty. Although the Chicago Convention in Article 1 stipulates that the Contracting States recognize that every State has complete and exclusive sovereignty over the airspace above its territory, the effect of this provision cannot be extended to apply to State immunity from responsibility to other States. Professor Huber in the *Island of Palmas* case[27] was of the view:

> Sovereignty in the relations between States signifies independence. Independence in regard to a portion of the globe is the right to exercise therein, to the exclusion of any other State, the functions of a State. . .Territorial sovereignty. . .involves the exclusive right to display the activities of a State.[28]

Professor Huber's definition, which is a simple statement of a State's rights, has been qualified by Starke as the residuum of power which a State possesses within the confines of international law.[29] Responsibility would devolve upon a State in whose territory an act of unlawful interference against civil aviation might occur, to other States that are threatened by such acts. The International Court of Justice (ICJ) recognised in the *Corfu Channel* Case:

> every State's obligation not to allow knowingly its territory to be used for acts contrary to the rights of other States.[30]

In the famous *Corfu Channel* case, the International Court of Justice applied the subjective test and applied the fault theory. The Court was of the view that:

> It cannot be concluded from the mere fact of the control exercised by a State over its territory and waters that the State necessarily knew, or ought to have known, of any unlawful act perpetrated therein, nor yet that it necessarily knew, or should have known the authors. This fact, by itself and apart from other circumstances, neither involves prima facie responsibility nor shifts the burden of proof.[31]

The Court, however, pointed out that exclusive control of its territory by a State had a bearing upon the methods of proof available to establish the involvement or knowledge of that State as to the events in question.

Apart from the direct attribution of responsibility to a State, particularly in instances where a State might be guilty of a breach of treaty provisions, or violate the territorial sovereignty of another State, there are instances where an act could be imputed to a State.[32] Imputability or attribution depends upon the link that exists

---

[27]The *Island of Palmas* Case (1928) 11 U.N.R. I.A.A. at 829.

[28]*Ibid.*

[29]Starke (1989) at 3.

[30](1949) *I.C.J.R.*1, 22.

[31]The *Corfu Channel* Case, ICJ Reports, 1949, p. 4.

[32]There are some examples of imputability, for example the incident in 1955 when an Israeli civil aircraft belonging to the national carrier El Al was shot down by Bulgarian fighter planes, and the consequent acceptance of liability by the USSR for death and injury caused which resulted in the payment of compensation to the victims and their families. See 91 *ILR* 287. Another example concerns the finding of the International Court of Justice that responsibility could have been be

between the State and the legal person or persons actually responsible for the act in question. The legal possibility of imposing liability upon a State wherever an official could be linked to that State encourages a State to be more cautious of its responsibility in controlling those responsible for carrying out tasks for which the State could be ultimately held responsible. In the same context, the responsibility of placing mines was attributed to Albania in the *Corfu Channel* case where the court attributed to Albania the responsibility, since Albania was known to have knowledge of the placement of mines although it did not know who exactly carried out the act. It is arguable that, in view of the responsibility imposed upon a State by the Chicago Convention on the provision of air navigation services, the principles of immutability in State responsibility could be applied to an instance of an act or omission of a public or private official providing air navigation services.

The sense of international responsibility that the United Nations ascribed to itself had reached a heady stage at this point, where the role of international law in international human conduct was perceived to be primary and above the authority of States.

The United Nations General Assembly, in its Resolution 56/83,[33] adopted as its Annex the International Law Commission's *Responsibility of States for Internationally Wrongful Acts* which recognizes that every internationally wrongful act of a State entails the international responsibility of that State[34] and that there is an internationally wrongful act of a State when conduct consisting of an action or omission is attributable to the State under international law and constitutes a breach of an international obligation of the State.[35] Article 5 of the ILC document provides that the conduct of a person or entity which is not an organ of State but which is empowered by the law of that State to exercise elements of the governmental authority shall be considered an act of State under international law, provided the person or entity is acting in that capacity in the particular instance.

In the *Pan Am* case,[36] where an aircraft was destroyed over Lockerbie the British allegation against Libya's involvement in the act of terrorism was that the accused individuals (Libyan nationals) had acted as part of a conspiracy to further the purposes of the Libyan Intelligence Services using criminal means that amounted to terrorism. The United Kingdom appeared to stress the point in the UN Security Council that Libya had failed to respond to the request for extradition of the

---

imputed to the United States in the *Nicaragua* case, where mines were laid in Nicaraguan waters and attacks were perpetrated on Nicaraguan ports, oil installations and a naval base by persons identified as agents of the United States. See *Nicaragua* v. *the United States*, ICJ Reports 1986, 14. Also, 76 *ILR* 349. There was also the instance when the Secretary General of the United Nations mediated a settlement in which a sum of $7 million was awarded to New Zealand for the violation of its sovereignty when a New Zealand vessel was destroyed by French agents in New Zealand. See the *Rainbow Warrior* case, 81 *AJIL*, 1987 at 325. Also in 74 *ILR* at 241.

[33] A/RES/56/83 Fifty Sixth Session, 28 January 2002.

[34] *Id.* Article 1.

[35] Id. Article 2.

[36] *Infra*, note 61 in Article 1.

implicated Libyan nationals, and arguably as a consequence, the Security Council adopted Resolution 731 on 21 January 1992 which expressed concerns over certain investigations which imputed reprehensibility to officials of the Libyan Government.[37]

5.1 Security of the Passport

Production of passport books and travel documents, including the personalization processes, should be undertaken in a secure, controlled environment with appropriate security measures in place to protect the premises against unauthorized access. If the personalization process is decentralized, or if personalization is carried out in a location geographically separated from where the travel document blanks are made, appropriate precautions should be taken when transporting the blank documents and any associated security materials to safeguard their security in transit.

There should be full accountability over all the security materials used in the production of good and spoiled travel documents and a full reconciliation at each stage of the production process with records maintained to account for all material usage. The audit trail should be to a sufficient level of detail to account for every unit of material used in the production and should be independently audited by persons who are not directly involved in the production. Certified records should be kept of the destruction of all security waste material and spoiled documents.

Materials used in the production of travel documents should be of controlled varieties and obtained only from bona fide security materials suppliers. Materials whose use is restricted to high security applications should be used, and materials that are available to the public on the open market should be avoided.

Sole dependence upon the use of publicly available graphics design software packages for originating the security backgrounds should be avoided. These software packages may however be used in conjunction with specialist security design software.

Security features and/or techniques should be included in travel documents to protect against unauthorized reproduction, alteration and other forms of tampering, including the removal and substitution of pages in the passport book, especially the biographical data page. In addition to those features included to protect blank documents from counterfeiting and forgery, special attention must be given to protect the biographical data from removal or alteration. A travel document should include adequate security features and/or techniques to make evident any attempt to tamper with it.

The combination of security features, materials and techniques must be well chosen to ensure full compatibility and protection for the lifetime of the document. There is another class of security features comprised of covert (secret) features, designed to be authenticated either by forensic examination or by specialist

[37] For a discussion on this point see Jorgensen (2000) at 249–254.

verification equipment. It is evident that knowledge of the precise substance and structure of such features should be restricted to very few people on a "need to know" basis. The purpose of these features is not to prevent counterfeiting but to enable authentication of documents where unequivocal proof of authenticity is a requirement (e.g. in a court of law). All travel documents should contain at least one covert security feature as a basic feature.

5.2 Threats to the Security of Passports

There are many threats to the security of passports such as: counterfeiting a complete travel document; photo-substitution; deletion/alteration of text in the visual or machine readable zone of the MRP data page; construction of a fraudulent document, or parts thereof, using materials from legitimate documents; removal and substitution of entire page(s) or visas; deletion of entries on visa pages and the observations page; theft of genuine document blanks; and impostors (assumed identity; altered appearance).

To provide protection against these threats and others, a travel document requires a range of security features and techniques combined in an appropriate way within the document. Although some features can offer protection against more than one type of threat, no single feature can offer protection against them all. Likewise, no security feature is 100 per cent effective in eliminating any one category of threat. The best protection is obtained from a balanced set of features and techniques providing multiple layers of security in the document that combine to deter or defeat fraudulent attack.[38]

Annex 9[39] to the Convention on International Civil Aviation, in Standard 3.7 requires ICAO member States to regularly update security features in new versions of their travel documents, to guard against their misuse and to facilitate detection of cases where such documents have been unlawfully altered, replicated or issued. Recommended Practice 3.9 suggests that member States incorporate biometric data in their machine readable passports, visas and other official travel documents, using one or more optional data storage technologies to supplement the machine readable zone, as specified in Doc 9303, Machine Readable Travel Documents. The required data stored on the integrated circuit chip is the same as that printed on the data page, that is, the data contained in the machine-readable zone plus the digitized photographic image. Fingerprint image(s) and/or iris image(s) are optional biometrics for member States wishing to supplement the facial image with another biometric in the passport. Member States incorporating biometric data in their Machine

[38] Machine Readable Travel Documents, ICAO Doc 9303 Part 1, Machine Readable Passports, Sixth Edition, 2006, III-4.

[39] Annex 9 to the Convention on International Civil Aviation, 12th Edition, 2006.

Readable Passports are to store the data in a contactless integrated circuit chip complying with ISO/IEC 14443 and programmed according to the Logical Data Structure as specified by ICAO.

5.3 The Diplomatic Fallout

Any diplomatic action in the context of the issues raised in this article must primarily be based on State responsibility. In turn, and as already discussed, the issue of responsibility hinges on knowledge, complicity and condonation of a State. The responsibility of a State is determined by the quantum of proof available that could establish intent or negligence of the State, which in turn would establish complicity or condonation on the part of the State concerned. One way to determine complicity or condonation is to establish the extent to which the State adhered to the obligation imposed upon it by international law and whether it breached its duty to others. In order to exculpate itself, the State concerned will have to demonstrate that either it did not tolerate the offence or that it ensured the punishment of the offender. *Brownlie* is of the view that proof of such breach would lie in the causal connection between the offender and the State.[40] In this context, the act or omission on the part of a State is a critical determinant particularly if there is no specific intent.[41] Generally, it is not the intent of the offender that is the determinant but the failure of a State to perform its legal duty in either preventing the offence (if such was within the purview of the State) or in taking necessary action with regard to punitive action or redress.[42]

There are a few principles that have to be taken into account when determining State responsibility. Firstly, there has to be either intent on the part of the State towards complicit or negligence reflected by act or omission. Secondly, where condonation is concerned, there has to be evidence of inaction on the part of the State in prosecuting the offender. Thirdly, since the State as an abstract entity cannot perform an act in itself, the imputability or attribution of State responsibility for acts of its agents has to be established through a causal nexus that points the finger at the State as being responsible. For example, The International Law Commission, in Article 4 of its Articles of State Responsibility states that the conduct of any State organ which exercises judicial, legislative or executive functions could be considered an act of State and as such the acts of such organ or instrumentality can be construed as being imputable to the State. This principle was endorsed in 1999 by the ICJ which said that according to well established principles

[40]Brownlie (1983) at 39.

[41]Report of the International Law Commission to the United Nations General Assembly, UNGOAR 56th Session, Supp. No. 10, *UN DOC A/56/10*, 2001 at 73.

[42]de Arechaga (1968), 531 at 535.

of international law, the conduct of any organ of a state must be regarded as an act of State.[43]

The law of State responsibility has evolved through the years, from being a straightforward determination of liability of the State and its agents to a rapidly widening gap between the State and non State parties. In today's world private entities and persons could wield power similar to that of a State, bringing to bear the compelling significance and modern relevance of the agency nexus between the State and such parties. This must indeed make States more aware of their own susceptibility.

The United Nations General Assembly, in 2002 adopted Resolution A 56/83[44] on the subject of Responsibility of States for internationally wrongful acts. The Resolution, which was the result of work of the International Law Commission on the subject, provides that every internationally wrongful act of a State entails the international responsibility of that State[45] and that such an act is attributable to that State under international law and constitutes a breach of an international obligation of that State. Article 5 to the Annex to the Resolution states that the conduct of a person or entity which is not an organ of a State but which is empowered by the law of that State to exercise elements of the governmental authority shall be considered an act of the State under international law, provided the person or entity is acting in that capacity in the particular instance. If, as alleged, the assassination of Mahmoud al-Mabhouh was carried out by Mossad, which reports to the Isreali Prime Minister, Article 8 of the Annex to the Resolution is particularly relevant as it provides that the conduct of a person or group of persons would be considered an act of a State under international law if the person or group of persons is in fact acting on the instructions of, or under the direction or control of that State in carrying out the conduct. The Resolution also recognizes that there is a breach of an international obligation by a State when an act of that State is not in conformity with what is required of it by that obligation, regardless of its origin or character.[46]

Diplomatic relations between States are intrinsically linked to State responsibility and are based on relations between States dependent on comity.[47] The fundamental fact in this context is that international society is not an unchanging entity, but is

[43]*Differences Relating to Immunity from Legal Process of a Special Rapporteur*, ICJ Reports 1999, 62 at 87.

[44]A/RES/56/83 Fifty sixty Session 28 January 2002.

[45]*Id.* Annex, Article 1.

[46]*Id*, Article 12.

[47]In law, comity specifically refers to legal reciprocity—the principle that one jurisdiction will extend certain courtesies to other nations (or other jurisdictions within the same nation), particularly by recognizing the validity and effect of their executive, legislative, and judicial acts. The term refers to the idea that courts should not act in a way that demeans the jurisdiction, laws, or judicial decisions of another jurisdiction. Part of the presumption of comity is that other jurisdictions will reciprocate the courtesy shown to them.

subject to the ebb and flow of political life and activity.[48] Analogies to diplomatic relations between States arise often in the context of terrorism as in the 1985 *Rainbow Warrior* case. The sinking of the ship *Rainbow Warrior* in Auckland Harbour in New Zealand as a result of an officially organized undercover French military operation calculated to obstruct Greenpeace protests against French nuclear operations in the South Pacific is a good example. The destruction of the ship, which resulted in the death of a Dutch seaman, was directly attributable to the placing of explosives by an agency reporting to the French Ministry of Defence. This was construed by the Government of New Zealand and by the High Court of New Zealand as a violation of the principles of international law against the sovereignty of the country.[49] The Secretary General of the United Nations ruled that France should offer New Zealand an apology and that compensation should follow to the amount of $7,000,000. The *Rainbow Warrior* incident goes down in the annals of diplomatic relations as an act of international delinquency resulting in the criminal responsibility of a State.

Another analogous diplomatic incident occurred in 1981 when, on May 6, the US Department of State announced at a special press briefing that the United States Government had decided to require the Socialist People's Libyan Arab Jamahiriya to close its People's Bureau at Washington immediately and to withdraw all personnel within five working days. The Department's official statement was that from the first days of the Administration, both the President and the Secretary of State had made known their very real concern about a wide range of Libyan provocations and misconduct, including support for international terrorism. The United States made it officially clear that it had been concerned by a general pattern of unacceptable conduct by the People's Bureau in Washington, which was contrary to internationally accepted standards of diplomatic behaviour. The United States therefore asked the Libyans to close their People's Bureau in Washington and have given them five working days starting today to withdraw their personnel. This action reduced US relations with Libya to the lowest level consistent with maintenance of diplomatic relations and was followed with a travel advisory which stated that due to unsettled relations between the United States and the Government of Libya, the Department of State warned American citizens against any travel to or residence in Libya. Travellers were also informed that the US Embassy in Tripoli was closed and the US was not in a position to provide consular protection and assistance to Americans presently in Libya.[50]

In 1999, the Clinton administration warned Russia to voluntarily reduce the large number of intelligence officers operating in the United States or face cutbacks in diplomatic positions or expulsions. U.S. Ambassador James Collins delivered the message in Moscow during a meeting with Vladimir Putin, the former

---

[48]See generally, Jennings and Watts (1992); Lauterpacht (1947); Chen (1951), and Shaw (2003), at 367.

[49]*R*. v. *Mafart and Prieur*, New Zealand High Court, Auckland Registry, 22 November 1985, Per Davison CJ reported in 74 *ILR* 241.

[50]Dept. of State File No. P81 0101–1084.

KGB domestic spying chief and currently Russia's top Security Council adviser, according to administration officials familiar with the issue. The warning followed two expulsions of Russian intelligence officers from the United States and the ouster of a U.S. Army attaché from Moscow a month earlier.[51]

The international treaty regulating diplomatic relations is the *Vienna Convention on Diplomatic Relations* of 1961.[52] The Convention does not explicitly make provision for the right to break diplomatic relations. It follows by implication from Article 2 which provides that the establishment of diplomatic relations takes place by mutual consent that if either State withdraws that consent diplomatic relations are broken. Breach therefore takes place normally in consequence of a unilateral act—even though it frequently follows a sequence of reciprocal or retaliatory moves between two States to downgrade their relations or a collective political decision by a number of States directed against another State whose conduct is regarded as unacceptable. Relations are broken from the moment of the initial action.[53] The other State has no option in the matter. There are no legal limitations on the right of a State to break diplomatic relations with another, but the action is now invariably taken for political reasons. Practical considerations will almost always favour the continuation of relations, though not necessarily the retention of a permanent mission. This has become more obvious in the light of some recent cases where diplomatic relations subsisted even while armed conflict was taking place between sending and receiving States—as between India and Pakistan in 1965 and 1971.

A breach of diplomatic relations generally precludes direct contact between sending and receiving States other than what is needed to effect orderly departure and some form of interim regime. It does not, however, preclude the sending and receiving of special missions (which may later herald a resumption of normal diplomatic relations), meetings between diplomatic representatives of the two States in a third State (for example the regular meetings in Warsaw over many years of representatives of the United States and of the People's Republic of China) or contacts between representatives of the two States to an international organization. Detailed rules on permissible contacts are usually provided in the internal diplomatic service regulations of each State. It is often a feature of modern diplomacy that those on occasion a much-advertised breach of relations may turn out to be only partially real. This occurs when two States, having broken off diplomatic relations, usually on the initiative of one of them, continue an active, if quiet, direct relationship despite the appointment of third States to protect the interests of each in the territory of the other State.[54]

---

[51]Bill Gertz, *The Washington Times*, 26 July 1999.

[52]Done at Vienna on 18 April 1961 and entered into force on 24 April 1964. United Nations Treaty Series, Vol. 5000 at p. 95.

[53]For an account of the series of incidents and complaints between France and Iran which led France to break diplomatic relations in July 1987, see *1987 AFDI 1000*. See also do Nascimento e Silva, *Diplomacy in International Law* p. 173–4.

[54]James (1991).

Whatever unilateral diplomatic action an aggrieved State might take, be it on grounds of sovereignty or the violation of its national property (passports) and the rights of its citizens who held the passports, there are certain legal nuances in the Mahmoud al-Mabhouh case which are incontrovertible. Falsification of passports and identity theft are serious criminal offences under most national laws. It could well be that these are also offences under the laws of Israel. Falsification of a national passport, whatever its country or nationality might be, by a member of the Israeli intelligence services brings to bear issues that are much more serious than mere breaches of diplomatic courtesy or relations. Under the theory of condonation any government involved would be seriously implicated, were it to turn out that it was aware that falsified travel documents were being used by its security agency as has been suggested by some.

The international community should therefore condemn the extra judicial killing of Mahmoud al-Mabhouh as a breach of international law and those involved must unequivocally declare as to whether they were aware that falsified travel documents were being used by Mossad in relation to this operation and/or any other. If there is cogent evidence implicating the Israeli Government, the international community must also require the former to confirm whether its intelligence services were involved in the murder of Mahmoud al-Mabhouh and demand that the Israeli government confirm whether or not their intelligence services used falsified passports for this or any other operation or whether they have done since any assurance that they would not do so. Furthermore the international community should seek an assurance from the Israeli government that their intelligence operatives will never falsify passports for use in operations and require the Israeli government to condemn the killing of Mahmoud al-Mabhouh as a breach of international law. Finally an assurance must be sought from the Israeli government that they will extradite to Dubai any of those identified by the Dubai authorities as having been involved in the killing to face trial for murder and to Ireland, Britain, France and/or Germany to face trial for offences arising out of the abuse of passports issued by those countries.

At present, the issue of extradition could be settled through the United Nations and its Organs such as the Security Council[55] and the International Court of Justice (ICJ).[56] Of noteworthy practical relevance with regard to the complicity theory, particularly on the issue of extradition and whether one State can demand the

[55]The Security Council is the branch of the United Nations charged with the maintenance of international peace and security. Its powers, outlined in the Charter of the United Nations, include the establishment of peacekeeping operations, the establishment of international sanctions, and the authorization for military action. The Security Council's power are exercised through its Resolutions. The Permanent members of the Security Council are the United States of America, United Kingdom, France, the Russian Federation and the Republic of China.

[56]The International Court of Justice (ICJ) is the principal judicial organ of the United Nations (UN). It was established in June 1945 by the Charter of the United Nations and began work in April 1946.The Court's role is to settle, in accordance with international law, legal disputes submitted to it by States and to give advisory opinions on legal questions referred to it by authorized United Nations organs and specialized agencies. The Court is composed of 15 judges, who are elected for terms of office of nine years by the United Nations General Assembly and the Security Council. It is assisted by a Registry, its administrative organ. Its official languages are English and French.

extradition of offenders harbored in another State is the opinion given by the ICJ[57] on the explosion over Lockerbie, Scotland on 21 December 1988 of PAN AM Flight 103. The explosion is believed to have been caused by the detonation of a plastic explosive concealed in a portable cassette player/radio. The ICJ noted that it was a general principle of international law that no State could be compelled to extradite its nationals and that the State concerned held the prerogative of trying the accused of a crime in its own territory. The ICJ was encumbered with the discussion as to whether the Court had jurisdiction over a United Nations Security Council Resolution on the issue. The essence of the views of the learned judges of the ICJ was that the complimentary roles played by the United Nations Security Council and the ICJ would devolve responsibility on States to respect both these organs on the subject of extradition of private offenders.

It appears that the question in The ICJ's was whether the Security Council, by its Resolution 748 (1992) which required Libya to extradite its nationals either to the United States or to the United Kingdom, had the authority to override an established principle of international law. The answer to this question was, in the view of one judge, in the affirmative.

If a State found reprehensible is unable or unwilling to make reparations as requested, there is nothing to prevent a State from unilaterally terminating diplomatic relations with any that State if the former wishes to do so.

References

Becker T (2006a) Terrorism and the state, Hart monographs in transnational and international law. Hart Publishing, Oxford

Becker T (2006b) Terrorism and the state; rethinking the rules of state responsibility. Hart Publishing, Portland

Blackstone W, Morrison W (eds) (2001) Commentaries on the laws of England (1765–69), vol 4. Cavendish, London

Brownlie I (1983) System of the law of nations: state responsibility, Part 1. Clarendon, Oxford

Caron DD (1998) The basis of responsibility: attribution and other trans-substantive rules. In: Lillich RB, Magraw DB (eds) The Iran-United States claims tribunal: its conclusions to state responsibility. Transnational Publishers, Hudson

Chen TC (1951) The international law of recognition. Stevens and Sons Ltd., London

de Arechaga EJ (1968) International responsibility. In: Sorenson M (ed) Manual of public international law. St. Martin's Press, New York

De Vattel E, Fenwick CG (tr) 2 (1916) The law of nations or, the principles of natural law: applied to the conduct and to the affairs of nations and sovereigns. Legal Classics Library, New York

Hyde C (1928) Concerning damages arising from neglect to prosecute. Am J Int L 22:140

James D (1991) Diplomatic relations and contacts. BYIL 375

Jennings RY, Watts AD (eds) (1992) Oppenheim's international law, 9th edn. Longman, London

Jorgensen NHB (2000) The responsibility of states for international crimes. Oxford University Press, Oxford

Lauterpacht H (1947) Recognition in international law. Cambridge University Press, Cambridge

Shaw MN (2003) International law, 5th edn. Cambridge University Press, Cambridge

Starke JG (1989) Introduction to international law, 10th edn. Butterworths, London

[57] I.C.J. Reports 1980, 116.

Article 23
Customs and Immigration Procedures

Each contracting State undertakes, so far as it may find practicable, to establish customs and immigration procedures affecting international air navigation in accordance with the practices which may be established or recommended from time to time, pursuant to this Convention. Nothing in this Convention shall be construed as preventing the establishment of customs-free airports

Contents

1 Entry and Exit Controls

As far as customs and immigration go, this is a straightforward provision, as States establish their own customs and immigration laws and procedures. Annex 9 offers assistance in this regard. For example, Standard 3.40 requires that in order to expedite inspections, Contracting States, with the cooperation of airport operators, shall use applicable technology and adopt a multi-channel immigration inspection system, or other means of streaming passengers, at international airports where the volume of passenger traffic justifies such measures. Standard 3.41 provides that, except in special circumstances, Contracting States shall not require that travel documents or other identity documents be collected from passengers or crew before they arrive at the passport control points. This is followed by Standard 3.42 which requires public authorities concerned to expeditiously accept passengers and crew for examination as to their admissibility into the State.

A passenger or crew member is "accepted for examination" when he makes his first appearance at the arrivals control point after disembarkation, to seek entry into the country concerned, at which time the control officer makes a determination whether he should be admitted or not. This does not include the sighting of travel documents, which may be carried out immediately upon disembarkation. Recommended Practice 3.47 states that where appropriate, Contracting States should introduce a system of advance passenger information which involves the capture of certain passport or visa details prior to departure, the transmission of the details by electronic means to their public authorities, and the analysis of such data for risk management purposes prior to arrival in order to expedite clearance. To minimize handling time during check-in, document reading devices should be used to capture the information in machine readable travel documents.

R. Abeyratne, *Convention on International Civil Aviation*,
DOI 10.1007/978-3-319-00068-8_24,

One of the compelling issues that States are confronted with is their policy on undocumented persons and refugees. Recommended Practice 3.44 suggests that after *acceptance*, the public authorities concerned should be responsible for the custody and care of passengers and crew members until they are admitted or found inadmissible. The following provision states that The responsibility of an aircraft operator for custody and care of passengers and crew members shall terminate from the moment such persons have been admitted into that State.

As for customs free airports, Airports are a complex, big business. The first element in the airport business equation is the customer and it is therefore a good starting point to determine who the customers of the airport are. It is incontrovertible that airline passengers generate the bulk of the concession revenue and that the airlines who bring them would normally generate most of the rental or lease income. However other market groups are by no means inconsequential. Rigas Doganis, whose analysis of the two airport models represented by Frankfurt and Athens airports respectively states:

> Frankfurt airport has had both the space and the commercial acumen to try to meet the needs of all these target groups in and around its main terminal. An analysis of concession revenues earned by the airport authority has shown that 76 per cent comes from passengers, 13 per cent from airport employees and the remaining 11 per cent from visitors of various kinds including meeters and greeters and local residents. This excludes rents from office space, land, etc. As a result of its strategy, Frankfurt had developed the extensive and very wide-ranging shopping, leisure, business and service facilities which were summarized earlier. Athens in contrast targets only two groups at its two terminals, the airlines and the passengers, and provides only the very minimum of facilities for both.
>
> Airports world-wide fall somewhere in the range represented by the two extremes of Athens and Frankfurt. Whether they have followed the traditional model or the commercial model will depend partly on the strategic options adopted by the management and in particular on which of the above customer groups they have decided to target. But the model adopted will also be dependent on the volume and composition of the traffic handled by each airport, by the terminal space and land available and the degree to which the management has been given the freedom to adopt commercially oriented policies.[1]

In view of the current trends in the airport business model, airports have to make strategic decisions to be financially self sufficient and be run like businesses. There should therefore be a commercial airport model which maximizes revenue, whether such is generated through aeronautical revenues (such as airport charges from airlines) or non- aeronautical revenues (such as from concessionaires). They should also have a reasoned view on who their target clients are and who their most attractive customers are.

[1] Doganis (2005) at 32.

2 Commercial Perspectives

Airports are subject to both external and internal factors with regard to revenue generation and their cost-benefit equation. The most significant external factors are current and projected traffic levels, global and local economic fluctuations and currency exchange rates, taxes and charges imposed on airports by governments and authorities as well as charges that can be exercised by airports on their users. Internally, strategic planning in terms of air and terminal space, slot allocation and the nature and effect of taxes. A large number of factors will influence an airports' ability to maximize its commercial revenues. There are first of all certain external factors, which are crucially important in affecting revenue generation or strategic options but which are largely outside the control of individual airport managements. These include the airport's traffic levels and its proportion of international traffic, exchange rates and the level of taxes or duties imposed on alcohol or tobacco. Then there are a variety of factors which can be influenced directly by management. These are the area and the location of terminal space allocated to commercial activities, the nature of the contracts negotiated with the concessionaires and the quality of the concessionaires themselves.

Airports have a range of goods, services and facilities to offer. Their income comes from rents and concession fees. Rental income is earned primarily from the renting or leasing of space and the customers are direct users of what airports have to offer, such as users airlines, freight forwarders and handling agents. The indirect users are hotels, catering firms, manufacturing companies and other similar business enterprises. Concession fees are earned from, payments made to the airport by the providers of various services for the right to offer their services in the airport premises. These fees are generally based on the volume of business generated by the concessionaires and not usually on the dimensions of space occupied, although it is not uncommon for some concession agreements to include a straightforward ground-rent criterion.

Duty-and tax-free shops are arguably the most attractive to airports in terms of non-aeronautical revenue. These shops are ready to pay competitive rents since consumers find them more attractive than city centre shops as the prices in the former are substantially low. An added attraction for concessionaires to pay high rents is that since the profit margins earned by the concessionaires are considerably high, airports can negotiate a contract that is advantageous in obtaining for the airport a large share of the profits earned by such shops.

Airports also recognize the need to provide duty and tax free shopping in the landside areas of the passenger terminal since duty-free shopping is only accessible to passengers who have entered the departure or transit lounges. A wide range of duty- and tax-paid shopping needs to be provided in the public landside areas of terminals. Such shops will be of three kinds. These shops could sell goods such as travel goods, tobacco, books sweets, drugs and toiletries, food, flower, shops, fashion goods souvenirs, glassware, clothing, videos and records. As I any place where humanity gathers, the airport also needs food and drink outlets.

The provision of these services would depend on market demand as determined by the type of passenger who frequents the area in question. Another revenue generating source is through bill board advertising and other types of promotional sources business enterprises might wish to have in the premises of an airport.

Reference

Doganis R (2005) The airport business. Routeledge, London

Article 24
Customs Duty

(a) **Aircraft on a flight to, from, or across the territory of another contracting State shall be admitted temporarily free of duty, subject to the customs regulations of the State. Fuel, lubricating oils, spare parts, regular equipment and aircraft stores on board an aircraft of a contracting State, on arrival in the territory of another contracting State and retained on board on leaving the territory of that State shall be exempt from customs duty, inspection fees or similar national or local duties and charges. This exemption shall not apply to any quantities or articles unloaded, except in accordance with the customs regulations of the State, which may require that they shall be kept under customs supervision.**

(b) **Spare parts and equipment imported into the territory of a contracting State for incorporation in or use on an aircraft of another contracting State engaged in international air navigation shall be admitted free of customs duty, subject to compliance with the regulations of the State concerned, which may provide that the articles shall be kept under customs supervision and control.**

Contents

1 Exemptions to Assist Air Transport

The Assembly, at its first session in 1947 adopted Resolution A1-42 (Onerous Economic Burdens on International Air Transport—Double Taxation and Similar Burdens) which identified that double taxation and taxation on fuel and equipment not consumed within the jurisdiction of the country imposing such tax, requested the Council to conduct a study. The Resolution also requested States to provide the Council with information pertaining to these two subjects. Pursuant to this Resolution, and in the course of its Study, the Council adopted several Resolutions on taxation starting in 1948. Of particular note was a Resolution dated 14 November 1966 which led to the adoption of Council Statements on taxation in the field of air transport.

R. Abeyratne, *Convention on International Civil Aviation*,
DOI 10.1007/978-3-319-00068-8_25, © Springer International Publishing Switzerland 2014

A **Tax** is a "pecuniary contribution made by persons liable, for the support of government".[1] Courts have adjudicated that a tax is "a pecuniary burden laid upon individuals or property to support the government and is a payment exacted by legislative authority". It has also been identified as "annual compensation paid to government for annual protection and for current support of government." An early American decision identified a tax as:

> A ratable portion of the produce of the property and labour of the individual citizens, taken by the nation, in the exercise of its sovereign rights, for the support of government, for the administration of the laws, and as the means for continuing in operation the various legitimate functions of the State.[2]

According to these definitions, a tax is a very general imposition, often described as a "once and for all" payment. Therefore, a tax could not be named, as a specific tax, such as "aviation fuel tax" or "aircraft equipment tax". The fact that a tax was levied "for the support of the government" makes its general nature more explicit.

In the 1956 case of *Heirs* v. *Mitche*,[3] the court held that a tax was:

> An enforced contribution of money or other property, assessed in accordance with some reasonable rule or apportionment by authority of some sovereign State on persons or property within its jurisdiction for the purpose of defraying the public expenses. Therefore, a tax came to be known as a "contribution" and was regarded in a general sense to be any contribution imposed by government upon individuals, for the use and service of the State, whether under the name of toll, tribute, tallage, gable, impost, duty, custom, excise, subsidy, supply, aid or *any other name*.

The legal definition of a tax is that it is an enforced contribution by the public or section thereof, introduced by legislative decree, for the purposes of defraying public expenses. Judicially, a tax has been identified as a "contribution", among other synonyms, including, quite disturbingly, with "any other word".[4] Experts in taxation maintain that the "efficiency" test in taxation calls for devising tax levies which cause minimal reduction in or disruption of, overall productivity of a society. It is in this perspective that the overall context of taxation in the field of international air transport should be viewed. In many instances "taxes" imposed on international air transport have been labeled as iniquitous.[5] It is strongly claimed that a tax which is "a compulsory contribution levied upon persons, property or

---

[1]Black's Law Dictionary, 1951.

[2]*New London* v. *Miller*, 1941 *Connecticut Reporter* at 112.

[3]1956 *Southern Reporter* at 81.

[4]*In re. Mytinger*, D:C: Tex., 31 *F. Supp.* 977 at 979.

[5]See *Aviation Daily*, 16 November 1990 at 328 where IATA Director General calls the $1 "Facilitation Fee" proposed by the US and which was to be imposed on aliens visiting the US a "tax". See also, *Aviation Daily*, 11 January 1990 at 71 where he makes a strong plea to US Secretary of Transport to urgently review the taxation of international transport. This tax was later withdrawn by the United States authorities, largely as a response to the effective lobbying against the imposition of this tax by such organizations as IATA. See Abeyratne (1993), at 450–460.

business for the support of government: any assessment"[6] is an onerous demand upon any one's person or resources and when imposed upon international air transport justifies its definition as a verb—"to subject to a severe strain".[7] While it is accepted that taxation must be for a public purpose[8] the amount of the tax charged must be compatible with principles of commerce and should be proportionate to the cost of the specific facility or services used rather than the cost of overall governmental services in general. The formula must admit of the tax being directly proportionate to the cost of the service or facility used.

2 Definition of Charge

A "charge" has reference to impositions for improvements which are specially beneficial to particular individuals or property, and which are imposed in proportion to the benefits supposed to be conferred. Charges are special and local impositions upon property in the immediate vicinity of municipal improvements and are laid with reference to the special benefit which the property is expected to have derived therefrom.

A charge levied upon the products of a particular industry is expected to be utilized in the improvement of that industry, while a tax is generally imposed in the national interest and is directed accordingly towards the national treasury. In concept, the former is not objectionable, since it is calculated to benefit a particular industry for which the charge is collected, while the latter, it is claimed, should be borne by States as part of their national responsibility. However, this clear demarcation has often been shrouded in anomalous terminology resulting in a passenger service charge being identified as a tax imposed for the national benefit.

In a broader sense, taxes, as have been judicially defined, could be considered as including assessments and charges. But practically, a "tax" is a public burden imposed generally on the inhabitants of a State or upon a division thereof for governmental purposes, without reference to particular or peculiar benefits to particular individuals or property. The main criticism of a tax is that it is a compulsory contribution levied upon persons ... for the support of government[9] and therefore is a heavy demand upon one's person or resources.

3 ICAO Council's Work on Taxation

ICAO Policies on Taxation in the Field of Air Transport (Doc 8632, 3 ed, 2000) recognizes that the imposition of national or local taxes on the acquisition of fuel, lubricants and consumable technical supplies for use by aircraft in connection with

[6]See generally, Hinshaw (1939) at 75–94.

[7]*Ibid.*

[8]Tell (1931) at 347. See also, *Lowell* v. *Boston* III Mass. 454 (1873).

[9]Hinshaw (1939), *supra*, note 6, *loq. cit.*

international air transport may have an adverse economic and competitive impact on international air transport operations. Under the ICAO Policies, the Council has resolved, *inter alia* on principles based on reciprocity that:

> "when an aircraft registered in one Contracting State, or leased or chartered by an operator of that State, is engaged in international air transport to, from or through a customs territory of another Contracting State its fuel, lubricants and other consumable technical supplies shall be exempt from customs or other duties on a reciprocal basis, or alternatively, in the cases of fuel, lubricants and other consumable technical supplies taken on board in sub-paragraphs ii) or iii) such duties shall be refunded, when:
>
> i) the fuel etc. is contained in the tanks or other receptacles on the aircraft on its arrival in the territory of the other State, provided that no quantity may be unloaded except temporarily and under customs control;
> ii) the fuel etc. is taken on board for consumption during the flight when the aircraft departs from an international airport of that other State either for another customs territory of that State or for the territory of any other State, provided that the aircraft has complied, before its departure from the customs territory concerned, with all customs and other clearance regulations in force in that territory; or
> iii) the fuel etc. is taken on board the aircraft at an international airport in one customs territory of another State and the aircraft makes successive stops at two or more international airports in that customs territory on its way to another customs territory of that State or to the territory of any other State;
>
> The provisions of sub-paragraphs i), ii) and iii) above apply whether the aircraft is engaged in an individual flight or in the operation of an air service and whether or not it is operating for remuneration.

No Contracting State complying with this Resolution is obliged to grant to aircraft registered in another Contracting State or aircraft leased or chartered by an operator of that State any treatment more favourable than its own aircraft are entitled to receive in the territory of that other State.

Notwithstanding the underlying principle of reciprocity, Contracting States are encouraged to apply the exemption, to the maximum extent possible, to all aircraft on their arrival from and departure for other States.

The expression "customs and other duties" include import, export, excise, sales, consumption and internal duties and taxes of all kinds levied upon the fuel, lubricants and other consumable technical supplies; and the duties and taxes described above shall include those levied by any taxing authority within a Contracting State, whether national or local. These duties and taxes shall not be or continue to be imposed on the acquisition of fuel, lubricants or consumable technical supplies used by aircraft in connection with international air services except to the extent that they are based on the actual costs of providing airports or air navigation facilities and services and used to finance the costs of providing them:

ICAO Assembly Resolution A 37–20 in Appendix E States that ICAO Doc 8632 makes a conceptual distinction between a "charge[10]" and a "tax[11]" in that "a charge is a levy that is designed and applied specifically to recover the costs of providing facilities and services for international civil aviation, and "a tax is a levy that is designed to raise national or local government revenues which are generally not applied to civil aviation in their entirety or on a cost-specific basis".

Resolution A37-20 APPENDIX E goes on to say that it is a matter of great concern that taxes are increasingly being imposed by some Contracting States in respect of certain aspects of international air transport and that charges on air traffic, several of which can be categorized as taxes on the sale or use of international air transport are proliferating.

References

Abeyratne RIR (1993) Air transport tax and its consequences on tourism. Ann Tourism Res 20(3):450–460

Hinshaw H (1939) The protection of aviation from inequitable taxation. J Air Law 9:75–94

Tell WK (1931) Taxation of aircraft motor fuel. J Air Law 2:342

[10]A "charge" has reference to impositions for improvements which are specially beneficial to particular individuals or property, and which are imposed in proportion to the benefits supposed to be conferred. Charges are special and local impositions upon property in the immediate vicinity of municipal improvements and are laid with reference to the special benefit which the property is expected to have derived therefrom.

[11]The legal definition of a tax is that it is an enforced contribution by the public or section thereof, introduced by legislative decree, for the purposes of defraying public expenses. Judicially, a tax has been identified as a 'contribution', among other synonyms, including, quite disturbingly, with 'any other word'. Experts in taxation maintain that the 'efficiency' test in taxation calls for devising tax levies which cause minimal reduction in or disruption of, overall productivity of a society. It is in this perspective that the overall context of taxation in the field of international air transport should be viewed.

Article 25
Aircraft in Distress

Each contracting State undertakes to provide such measures of assistance to aircraft in distress in its territory as it may find practicable, and to permit, subject to control by its own authorities, the owners of the aircraft or authorities of the State in which the aircraft is registered to provide such measures of assistance as may be necessitated by the circumstances. Each contracting State, when undertaking search for missing aircraft, will collaborate in coordinated measures which may be recommended from time to time pursuant to this Convention.

Contents

1 Regulation of Assistance to Aircraft

The ICAO Assembly, at its 37th Session (Montreal, September/October 2010) adopted Resolution A37-15 (Consolidated statement of continuing ICAO Policies and associated practices related specifically to air navigation) appendix N of which provides:

> *Whereas* in accordance with Article 25 of the Convention each Contracting State undertakes to provide such measuresof assistance to aircraft in distress in its territory as it may find practicable and to collaborate in coordinated measures which may be recommended from time to time pursuant to the Convention;
>
> *Whereas* Annex 12 to the Convention contains specifications relating to the establishment and provision of search and rescue services within the territories of Contracting States as well as within areas over the high seas;
>
> *Whereas* Annex 12 to the Convention specifies that those portions of the high seas where search and rescue serviceswill be provided shall be determined on the basis of regional air navigation agreements, which are agreements approved by the Council usually on the advice of regional air navigation meetings;
>
> *Whereas* Annex 12 to the Convention recommends that search and rescue regions should, insofar as practicable, be coincident with corresponding flight information regions and, with respect to those areas over the high seas, maritime search and rescue regions;
>
> *Whereas* Article 69 of the Convention specifies that, if the Council is of the opinion that the air navigation services of a Contracting State are not reasonably adequate for the safe operation of international air services, present or contemplated, the Council shall consult with the State directly concerned, and other States affected, with a view to finding means by

DOI 10.1007/978-3-319-00068-8_26, © Springer International Publishing Switzerland 2014

which the situation may be remedied, and may make recommendations for that purpose; and

Whereas the air navigation services referred to in Article 69 of the Convention include, inter alia, search and rescue services;

The Assembly resolves that:

1. search and rescue regions, whether over States' territories or, in accordance with regional air navigationagreement, over an area greater than a State's sovereign airspace or over the high seas, shall be delimited on the basis of technical and operational considerations, including the desirability of coincident flight information regions, search and rescue regions, and, with respect to areas over the high seas, maritime search and rescue regions, with the aim of ensuring safety, and optimizing efficiency with the least overall cost;

2. States shall ensure the closest practicable cooperation between maritime and aeronautical search and rescue services where they serve the same area and, where practical, establish joint rescue coordination centres to coordinate aeronautical and maritime search and rescue operations;

3. if any search and rescue regions need to extend over the territories of two or more States, or parts thereof, agreement thereon should be negotiated between the States concerned;

4. the providing State in implementing search and rescue services over the territory of the delegating State shall do so in accordance with the requirements of the delegating State, which shall establish and maintain in operation such facilities and services for the use of the providing State as are mutually agreed to be necessary;

5. any delegation of responsibility by one State to another or any assignment of responsibility over the high seas shall be limited to technical and operational functions pertaining to the provision of search and rescue services in the area concerned;

6. remedies to any inadequacies in the provision of efficient search and rescue services, including over the high seas, should be sought through negotiations with States which may be able to give operational or financial assistance in search and rescue operations, with a view to concluding agreements to that effect; and, *furthermore, declares* that:

7. any Contracting State which delegates to another State the responsibility for providing search and rescue services within its territory does so without derogation of its sovereignty; and

8. the approval by Council of regional air navigation agreements relating to the provision by a State of search and rescue services within areas over the high seas does not imply recognition of sovereignty of that State over the area concerned.

The Resolution also stated that Contracting States should, in cooperation with other States and the Organization, seek the most efficient delineation of search and rescue regions and consider, as necessary, pooling available resources or establishing jointly a single search and rescue organization to be responsible for the provision of search and rescue services within areas extending over the territories of two or more States or over the high seas. It also requested the Council to encourage States whose air coverage of the search and rescue regions for which they are responsible cannot be ensured because of a lack of adequate facilities, to request assistance from other States to remedy the situation and to negotiate agreements with appropriate States regarding the assistance to be provided during search and rescue operations.

After the events of 11 September 2001, it is only natural to assume that there is heightened awareness of the possibility of aircraft being used as weapons of destruction in the future. From a social and political perspective, the world has to prepare for eventualities leading up to search and rescue of aircraft that may need to be located without loss of time and the passengers and crew rescued. There are already two international treaties on the subject, although one—the Brussels Convention of 1938,[1] has unfortunately not been ratified by the requisite number of States and has therefore not come into effect. The Brussels Convention contemplated only assistance and salvage operations at sea. The other Convention is the Chicago Convention which requires member States of ICAO to fulfil their obligations under Article 25 of the which provides:

Each Contracting State undertakes to provide such measures of assistance to aircraft in distress in its territory as it may find practicable, and to permit, subject to control by its own authorities, the owners of aircraft or authorities of the State in which the aircraft is registered to provide such measures of assistance as may be necessitated by the circumstances. Each Contracting State, when undertaking search for missing aircraft, will collaborate in coordinated measures which may be recommended from time to time pursuant to this Convention.

Annex 12 to the Chicago Convention elaborates on this fundamental requirement by qualifying that Contracting States shall arrange for the establishment and provision of SAR services within their territories on a 24 h basis.[2] Contracting States are further requested to delineate the SAR process under the Annex on the basis of regional air navigation agreements[3] and provide such services on a regional basis without overlap.[4] A search and rescue region has been defined in the Annex as an area of defined dimensions within which SAR service is provided.

The dilemma facing many States extending both to airports and airlines, relates to the lack of rapid response, adequate equipment and well-trained crews, all of which are critical to passenger survival in the event of an aircraft disaster. Although most States are particularly mindful of these compelling needs, they are by no means confined to the a particular region. An example of this crisis can be cited with the 1980 incident of a Saudi Arabian Airlines L-1011 catching fire shortly after leaving Riyadh Airport. Although the pilot turned back for an emergency landing and made a perfect touchdown, nearly 30 min passed before firemen managed to go in, by which time all passengers and crew had perished. This could have been a survivable accident.[5] To the contrary, a hijacking incident involving a Boeing 767 aircraft on the shores of Comoros, in November 1996, when the aircraft crashed due to lack of fuel, showed how spontaneous reaction from even non-trained professionals at rescue

[1]Convention for the Unification of Certain Rules Relating to Assistance and Salvage of Aircraft at Sea, Brussels, September 29, 1938.

[2]Annex 12 to the Convention on International Civil Aviation, Search and Rescue, Sixth Edition March 1975, Standard 2.1.1.

[3]*Id.* Standard 2.1.1.1.

[4]*Id.* Standard 2.2.1 Boundaries of search and rescue regions should, insofar as practicable, be coincident with the boundaries of corresponding flight information regions. See Recommendation 2.2.1.1 of Annex 12.

[5]David Morrow, Preparing for Disaster, Airport Support, April 1995 at p. 29.

efforts could help. In this instance, the quick response of tourists at the scene ensured that 51 of the 175 passengers on board were saved.[6]

The discussion to follow will outline principles of responsibility of States and political, economic and humanitarian consequences pertaining to search and rescue of aircraft within their territorial boundaries.[7]

2 Political Issues

Annex 12 to the Chicago Convention requires Contracting States to coordinate their SAR organizations with those of neighbouring Contracting States[8] with a recommendation that such States should, whenever necessary, coordinate their SAR operations with those of neighbouring States[9] and develop common SAR procedures to facilitate coordination of SAR operations with those of neighbouring States.[10] These provisions collectively call upon all Contracting States to bond together in coordinating both their SAR organizations and operations.

At the 32nd Session of the Assembly, held in 1998, ICAO adopted Resolution A32-14, Appendix O which addresses the provision of SAR services. This Resolution refers to Article 25 of the Convention in which each Contracting State undertakes to provide such measures of assistance to aircraft in distress in its territory as it may find practicable and to collaborate in coordinated measures which may be recommended from time to time pursuant to the Convention.

The Resolution mentions Annex 12 to the Convention which contains specifications relating to the establishment and provision of SAR services within the territories of Contracting States as well as within areas over the high seas. The resolution recognizes that Annex 12 specifies that those portions of the high seas where SAR services will be provided shall be determined on the basis of regional air navigation agreements, which are agreements approved by the Council normally on the advice of regional air navigation meetings. Annex 12 also recommends that boundaries of SAR regions should, insofar as practicable, be coincident with the boundaries of corresponding flight information regions.

Article 69 of the Convention, which is also outlined in the Resolution, specifies that, if the Council is of the opinion that the air navigation services of a Contracting State are not reasonably adequate for the safe operation of international air services, present or contemplated, the Council shall consult with the State directly concerned, and other States affected, with a view to finding means by which the

---

[6]Report in FAZ No. 275/1996 (25 November 1996) at p.9.

[7]It is not the intent of this discussion to address issues pertaining to rights in recovery of costs incurred in search and rescue of aircraft and passengers. For this aspect of SAR, see Andreas Kadletz, Rescue and Salvage of Aircraft *ZLW* 46. Jg 2/1997, pp. 209–216.

[8]Standard 3.1.1.

[9]Recommendation 3.1.2.

[10]Recommendation 3.1.2.1.

situation may be remedied, and may make recommendations for that purpose; and the air navigation services referred to in Article 69 of the Convention include, *inter alia*, SAR services.

In taking into consideration the above facts, the Assembly resolves in A32-14 that the boundaries of SAR regions, whether over States? territories or over the high seas, shall be determined on the basis of technical and operational considerations, including the desirability of coincident flight information regions and SAR regions, with the aim of ensuring optimum efficiency with the least overall cost. If any SAR regions need to extend over the territories of two or more States, or parts thereof, agreement thereon should be negotiated between the States concerned.

The Resolution also calls upon the providing State, in implementing SAR services over the territory of the delegating State, to do so in accordance with the requirements of the delegating State, which shall establish and maintain in operation such facilities and services for the use of the providing State as are mutually agreed to be necessary. Any delegation of responsibility by one State to another or any assignment of responsibility over the high seas shall be limited to technical and operational functions pertaining to the provision of SAR services in the area concerned. Remedies to any inadequacies in the provision of efficient SAR services, particularly over the high seas, should be sought through negotiations with States which may be able to give operational or financial assistance in SAR operations, with a view to concluding agreements to that effect.

Furthermore, the Resolution declares that any Contracting State which delegates to another State the responsibility for providing SAR services within its territory does so without derogation of its sovereignty; and the approval by Council of regional air navigation agreements relating to the provision by a State of SAR services within areas over the high seas does not imply recognition of sovereignty of that State over the area concerned.

It is also stated in the Resolution that Contracting States should, in cooperation with other States and the Organization, seek the most efficient delineation of SAR regions and consider, as necessary, pooling available resources or establishing jointly a single SAR organization to be responsible for the provision of SAR services within areas extending over the territories of two or more States or over the high seas.

Finally, the Resolution calls on the Council to encourage States, whose air coverage of the SAR regions for which they are responsible cannot be ensured because of a lack of adequate facilities, to request assistance from other States to remedy the situation and to negotiate agreements with appropriate States regarding the assistance to be provided during SAR operations.

The legal validity of Resolution A32-14, as substantive law recognized under public international law, and therefore binding on States, is a relevant issue if the obligations of States in search and rescue are to be determined. All resolutions adopted within the United Nations framework embody declarations of principles and rules of international law. They are particularly compelling when adopted without dissent. Article 38 of the Statute of the International Court of Justice cites, as a source of public international law, general principles of law recognized by civilized nations into which category resolutions adopted by the United Nations could well fall.

Legal experts have consistently argued that resolutions could be authoritative evidence of binding international law on the grounds that such resolutions or declarations could be considered authentic interpretations of the United Nations Charter agreed by all parties. They have also adduced reasons for recognizing resolutions adopted within the United Nations system as affirmations of recognized customary law and as expressions of general principles of law recognized by States. Some confirmation of these arguments has been given by the International Court of Justice when the Court, over a period of years, recognized the force of several declarations adopted within the United Nations.[11]

In practical application however, non-observance by States purportedly bound by such resolutions would render such States destitute of the desired legal effect. This would essentially be the case if there are negative votes or reservations attached to an Assembly resolution. In the case of A32-14, however, there is no question of reservation as the Resolution was adopted by consensus.

The real utility of an Assembly resolution lies in the fact that primarily it supplements the absence of law in a given area by filling a legal lacuna that has not been filled by a formal legislative process. Treaty law making is often long winded and involves a cumbersome process. A resolution offers a quick fix while embodying principles in a declaration that introduces legitimacy and validity to a given principle or group of principles. In this context, it would be correct to assume that the ICAO Standards and Recommended Practices (SARPs) referred to earlier in this paper on the subject of the implementation of Annex 12 are of equal persuasion. Together, the resolution and SARPs have a clear and substantial impact, reflecting the meticulous and thoughtful work that have gone with the development of these instruments and recognized importance of safety and efficiency of civil aviation.[12]

In the case of the Africa–Indian Ocean Region, the ICAO Regional Air Navigation Plan,[13] in Part V addresses issues of Search and Rescue by pointing to the provisions of the ICAO Search and Rescue Manual (Doc 7333), referring in particular to the need for aircraft to carry specified equipment,[14] carry out paper and communications exercises[15] and, more importantly, for the need for States to pool their resources and provide mutual assistance in the case of SAR operations. The Plan calls for precise agreements between States to implement these measures.[16] The ICAO Regional Plan

[11]Advisory Opinions on Western Sahara, 1975 ICJ Rep 12 (October 16); Legal Consequences for States of the Continued Presence of South Africa in Namibia (South West Africa) notwithstanding Security Council Resolution 276 (1970), 1971 ICJ Rep 16 (June 21).

[12]See Joyner (1997) at p. 84.

[13]Air Navigation Plan Africa-Indian Ocean Region, Doc 7474.

[14]*Id.* 3.1.

[15]*Id.* 3.3 (a).

[16]*Id.* 4.1.

also calls upon States, in order to ensure compatibility between aeronautical and maritime search and rescue regions (SRRs), and aeronautical search and rescue authorities, to maintain close liaison with their maritime counterparts and the International Maritime Organization (IMO).

In 1985, ICAO signed a memorandum of understanding (MOU) with the IMO concerning cooperation in respect of safety of aircraft operations to and from ships and other marine vehicles and of aeronautical and maritime SAR activities. Both ICAO and IMO signed this understanding with a view to ensuring the best possible coordination of activities between the Organizations in matters concerned with the safety of aircraft operations to and from ships and other marine vehicles and with aeronautical and maritime search and rescue operations, agreeing to make arrangements for consultations between the Secretariats of the two Organizations in regard to these matters, with a view to ensuring consistency or compatibility between services and procedures in all cases where joint efforts or close cooperation may be required and in order to avoid any unnecessary duplication of efforts by them.

In determining the allocation of responsibilities of the two Organizations to ensure safety of aircraft operations to and from ships and other marine vehicles, the following principles are applied:

- All matters which are directly connected with the design, construction, equipment and operation of aircraft in general, and of helicopters in particular, should be regarded as falling primarily within the field of responsibility of ICAO.
- All matters which are directly connected with the design, construction and equipment of ships and other marine vehicles and their operation should be regarded as falling primarily within the field of responsibility of IMO.
- Matters which do not fall clearly within sub-paragraphs (a) or (b) above should be regarded as the responsibility of both Organizations and dealt with by appropriate collaboration between them.

In determining the allocation of responsibilities of the two Organizations in respect of search and rescue in maritime areas, the following principles are applied:

- All matters which are directly connected with search and rescue by aircraft in general, and with air search and rescue facilities and operating procedures in particular, should be regarded as falling primarily within the field of responsibility of ICAO.
- All matters which are directly connected with search and rescue by marine craft in general, and with marine search and rescue facilities and operating procedures in particular, should be regarded as falling primarily within the field of responsibility of IMO.
- Matters which do not fall clearly within sub-paragraphs (a) and (b) above should be regarded as the responsibility of both Organizations and dealt with by appropriate collaboration between them.

The MOU also provides that any draft amendment to Annex 12 Search and Rescue to the Convention on International Civil Aviation being considered by ICAO or any amendment to the Technical Annex to the International Convention on Maritime Search and Rescue, 1979, being considered by IMO and related to matters covered by this MOU will be communicated by the Organization proposing the amendment to the other Organization. Similarly, draft amendments to the ICAO SAR Manual or to the IMO SAR Manual which are related to matters covered by this MOU will be communicated in due time to the other Organization with a view to keeping both Manuals aligned as closely as possible.

The consultations referred to above should also take place in respect of matters falling primarily within the responsibility of one or the other Organization, so that each Organization may, when it deems it necessary, safeguard its responsibilities and interests in these matters and thereby ensure effective cooperative action whether carried out by one or the other or both Organizations.

In practice, the two Secretariats are required to take all available steps to ensure that the consultations referred to in paragraph 1 are undertaken before either Organization proceeds to take definitive action on matters subject to this MOU. The two Secretariats are also expected to make available to each other relevant information and documentation prepared for meetings at which matters covered by this MOU are to be considered.

Both Organizations have also agreed to take appropriate steps to ensure that relevant advice from other Organizations and bodies are made available in matters covered by this MOU, in accordance with the regulations and procedures of the respective signatory Organization.

All the above mentioned documents cited bring to bear the compelling need for the critical link between the legislative nature of the documentation and implementation State responsibility. All the law making and guidance material, declarations and resolutions would be destitute of effect if there was no element of State responsibility to give legitimacy to the instrument by complying with and adhering to the instruments.

When discussing principles of State responsibility in the field of search and rescue, it is an incontrollable fact that the provisions of the Chicago Convention, which is an international treaty are binding on Contracting States to the Convention and therefore are principles of public international law. The International Court of Justice (ICJ), in the *North Sea Continental Shelf Case*,[17] held that legal principles that are incorporated in Treaties, such as the "common interest" principle, become customary international law by virtue of Article 38 of the 1969 Vienna Convention on the Law of Treaties.[18] Article 38 recognizes that a rule set forth in a treaty would become binding upon a third State as a customary rule of international law if it is generally recognized by the States concerned as such. Obligations arising from *jus cogens* are considered applicable *erga omnes* which would mean that States using space technology owe a duty of care to the world at large in the provision of such technology. The ICJ in the *Barcelona Traction Case* held:

[17] *I.C.J. Reports* 1970, at 32.

[18] *Vienna Convention on the Law of Treaties*, United Nations General Assembly Document *A/CONF.39/27*, 23 May 1969.

> [A]n essential distinction should be drawn between the obligations of a State towards the international community as a whole, and those arising *vis a vis* another State in the field of diplomatic protection. By their very nature, the former are the concerns of all States. In view of the importance of the rights involved, all States can be held to have a legal interest in their protection; they are obligations *erga omnes*.[19]

The International Law Commission has observed of the ICJ decision:

> [I]n the Courts view, there are in fact a number, albeit limited, of international obligations which, by reason of their importance to the international community as a whole, are- unlike others - obligations in respect of which all States have legal interest.[20]

The views of the ICJ and the International Law Commission, which has supported the approach taken by the ICJ, give rise to two possible conclusions relating to *jus cogens* and its resultant obligations *erga omnes*:

- Obligations *erga omnes* affect all States and thus cannot be made inapplicable to a State or group of States by an exclusive clause in a treaty or other document reflecting legal obligations without the consent of the international community as a whole;
- Obligations *erga omnes* preempt other obligations which may be incompatible with them.

Some examples of obligations *erga omnes* cited by the ICJ are prohibition of acts of aggression, genocide, slavery and discrimination.[21] It is indeed worthy of note that all these obligations are derivatives of norms which are *jus cogens* at international law.

International responsibility relates both to breaches of treaty provisions and other breaches of legal duty. In the *Spanish Zone of Morocco Claims* case, Justice Huber observed:

> [R]esponsibility is the necessary corollary of a right. All rights of an international character involve international responsibility. If the obligation in question is not met, responsibility entails the duty to make reparation.[22]

It is also now recognized as a principle of international law that the breach of a duty involves an obligation to make reparation appropriately and adequately. This reparation is regarded as the indispensable complement of a failure to apply a convention and is applied as an inarticulate premise that need not be stated in the breached convention itself.[23] The ICJ affirmed this principle in 1949 in the *Corfu Channel* Case[24] by holding that Albania was responsible under international law to pay compensation to the United Kingdom for not warning that Albania had laid

---

[19] *Barcelona Traction, Light and Power Company Limited, I.C.J. Reports, 1974*, 253 at 269–270.

[20] *Yearbook of International Law Commission* 1976, Vol II, Part One at 29.

[21] *I.C.J. Reports*, 1970 at 32.

[22] *1925 RIAA ii* 615 at 641.

[23] *In Re. Chorzow Factory (Jurisdiction) Case, (1927) PCIJ, Ser. A, no. 9* at 21.

[24] *ICJ Reports (1949)*, 4 at 23.

mines in Albanian waters which caused explosions, damaging ships belonging to the United Kingdom. Since the treaty law provisions of liability and the general principles of international law as discussed complement each other in endorsing the liability of States to compensate for damage caused by space objects, there is no contention as to whether in the use of nuclear power sources in outer space, damage caused by the uses of space objects or use thereof would not go uncompensated. The rationale for the award of compensation is explicitly included in Article XII of the *Liability Convention* which requires that the person aggrieved or injured should be restored (by the award of compensation to him) to the condition in which he would have been if the damage had not occurred. Furthermore, under the principles of international law, moral damages based on pain, suffering and humiliation, as well as on other considerations, are considered recoverable.[25]

This principle, which forms a cornerstone of international conduct by States, provides the basis for strengthening international comity and regulating the conduct of States both internally—within their territories—and externally, towards other States. States are effectively precluded by this principle of pursuing their own interests untrammelled and with disregard to principles established by international law.

3 Economic Issues

Economic aspects of SAR operations related to aviation have been on the agenda of ICAO for a considerable time. At ICAO Conference on the Economics of Airports and Air Navigation Services (ANSConf 2000) held in Montreal from 19 to 28 June 2000, the Conference considered that, in 1996 a recommendation had been made by an ICAO Air Navigation Services Economics Panel, that existing policy be amended to allow for costs of SAR services performed by establishments other than permanent civil establishments such as military, to be included in the cost basis for air navigation services charges. The ICAO Council had not approved the Panel recommendations pending a Secretariat Study of the implications concerned. A subsequent survey carried out by the ICAO Secretariat of Contracting States had resulted in only a limited number of responses, precluding a conclusion as to the wishes of States on this issue. The Conference therefore agreed that there was a need for follow-up of the Secretariat Study, as well as information from many States that had not responded to the survey in the first instance.[26]

The Secretariat drew attention to the humanitarian aspects of SAR operations where States did not wish to charge for services rendered spontaneously and on an emergency basis. The Conference noted that under the *International Convention on Maritime Search and Rescue*, States were obligated to render gratuitous assistance to any person in distress and that there was no attendant cost-recovery mechanism in SAR in the maritime field. Based on the above deliberations the Conference

[25]Christol (1991) at 231.

[26]See Report of the Conference on the Economics of Airports and Air Navigation Services, Montreal, 19-28 June 2000, Doc 9764 ANSConf 2000 at p. 37.

recommended that ICAO undertake further study as to the position of States and the implications of amending ICAO policy with regard to recovery of costs for civil aviation related to SAR services presided by other than permanent civil establishments.[27] As for further work on the subject, the Conference recommended that ICAO develop guidance on the establishment of organizations at the regional level for SAR activities and conduct a study on the establishment of regional or sub-regional SAR mechanisms and how they might be funded as regards civil aviation.[28]

4 Humanitarian Issues

Search and rescue operations conducted gratuitously and with intent to save human lives and property are what legal commentators call "humanitarian intervention", which is considered to be a basic moral response of one human being to another, to save the latter's life. One definition identifies:

> humanitarian intervention as the proportionate transboundary help, including forcible help, provided by governments to individuals in another [S]tate who are being denied basic human rights and who themselves would be rationally willing to revolt against their oppressive government.[29]

The general principle of intervention for the provision of relief on moral grounds has been subject to a great degree of intellectual polarization. One view is that if human are dying, one has got to help that at all costs.[30] The other is that the mere act of treating humanitarian intervention as an extant legal doctrine would be to erode the applicable provision of the United Nations Charter on recourse to force.

The latter view, which discourages humanitarian intervention is substantiated by the following arguments:

- The good Samaritan must fight for the right to perform his act of humanitarian intervention and may end up causing more injury than he averts;
- The authorization for forceful and unilateral humanitarian assistance may be abused; and,
- Unilateral recourse to force even for genuinely humanitarian purposes may heighten expectations of violence within the international system and concomitantly erode the psychological constraints on the use of force for other purposes.[31]

---

[27] *Id.* Recommendation 23 at p. 38.

[28] *Id.* Recommendation 24 at p. 38.

[29] Teson (1956) at 5.

[30] See letter to the Editor by Professor Leff, Yale Law School, *New York Times* October 4 1968 at 46 Column 3, cited in Lillich (1973) at 151.

[31] The principle of non-intervention has been strongly espoused in order that sovereignty of a State be retained as sacrosanct. See Vattel, *le driot des geus*, Bk II, Chapter V (Scott ed. 1916) at 135. Also, Hall, *International Law* (Higgins 8th ed. 1924) at 343. Lawrence, *Principles of International Law* (Winfield 7th ed. 1923) at 126.

The essence of intervention is compulsion. Compulsion could either take place through the use of force, armed or otherwise. The legal question, with regard to the inviolability of the sovereignty of a State is not whether the intervention concerned was an armed or unarmed one, but whether it was effected unilaterally under compulsion or threat by the intervening State.[32]

Some authorities in international law also believe that intervention should, if absolutely necessary, be effected when there is cogent evidence of a breakdown in the minimum guarantees of humanity.[33] Accordingly, it may be argued that any act of intervention aimed at saving the lives of human beings which are in danger, would be legally and morally justifiable. Fernando Teson[34] argues that since the ultimate justification for the existence of States is the protection and enforcement of the natural rights of the citizens, a government that engages in substantial violations of human rights betrays the very purpose for which it exists and so forfeits not only its domestic legitimacy, but also its international legitimacy as well. He goes on to say:

> I suggest that from an ethical standpoint, the rights of States under international law are properly derived from individual rights. I therefore reject the notion that States have any autonomous moral standing - that they hold international rights that are independent from the rights of individuals who populate the State.[35]

Schwarzenberger analyses the concept somewhat clinically and concludes that in the absence of an international *jus cogens* which corresponds to municipal *jus cogens* of advanced communities, where the latter prevents the worst excesses of inequality of power, the supremacy of the rule of force would prevail.[36]

There is also a contrasting view that humanitarian intervention is generally resorted to by States only in instances of serious abuses of human rights by one State upon its people or others. Dr Michael Akehurst argues that if a State intervenes forcibly on the territory of another in order to protect the local population from serious human violations, such an armed intervention could inevitably constitute a temporary violation *de facto* of the territorial integrity of the latter State, and to an extent of its political independence, if carried out against its wishes.[37] Dr. Akehurst goes on to assert:

> Any humanitarian intervention, however limited, constitutes a temporary violation of the target State's political independence and territorial integrity if it is carried out against the State's wishes.[38]

---

[32]de Lima (1971) at 16.

[33]Hall *International Law* (Higgins 8th ed. 1924) at 343. Lawrence, *Principles of International Law* (Winfield 7th ed. 1923) at 126/ *op. cit.* at 347. Lawrence, *op.cit.* at 132. Hyde, *International Law*, Volume 1. (2nd Review:1945) at 253. *Stowell's Intervention at International Law* 1921 at 126 and 350. Also, Wehberg, *La Guerre Civil et le Droit International* 63 Hague Recueil, 1938 at 115.

[34]*Supra* at note 29.

[35]*Id.* at 16.

[36]Schwarzenberger (1971) at 63.

[37]Akehurst (1977) 3, at 16.

[38]Akehurst (1984) 95 at 105.

The doctrine of humanitarian intervention is thought of by some commentators as an invention of strategy to circumvent the strong *jus cogens* nature of the principle of sovereignty and inviolability of States which Dr. Akehurst refers to. Professor Brownlie is of the view that States have generally invoked the doctrine to give support to their commercial and strategic considerations.[39] The United Kingdom legislature recently considered the view of the British Minister of State who was of the view:

> When members of the United Nations act in a forcible manner either they should do so within and under the authority of the United Nations or that which they do should be authorised by the principles of international law.[40]

Clearly, this statement establishes the view that international law in the context of intervention is *jus cogens*. The British Foreign Office has supported this position in the following language:

> the best case that can be made in support of humanitarian intervention is that it cannot be said to be unambiguously illegal. . .but the overwhelming majority of contemporary legal opinion comes down against the existence of a right of humanitarian intervention.[41]

Despite this strong alignment towards *anti* humanitarian intervention, it is believed that there is a school of thought within the British legislature that is prepared to accept unilateral intervention as justifiable under customary international law in cases of "extreme humanitarian need."[42]

The author supports the view that despite these divergent views, the non-intervention principle remains sacrosanct as a contemporary postulate of international law and deviations from the principle, although recognized as ethical and moral in certain instances by scholars, would be justified only in extreme cases.[43]

The essence of search and rescue operations in aviation is cooperation, which is embodied as a fundamental principle in the Preamble to the Chicago Convention which states, *inter alia*, that it is desirable to avoid friction and to promote that cooperation between nations and peoples upon which the peace of the world depends. At the root of international cooperation is the element of assistance, and in this sense the maritime regulations which admit of gratuitous help are both significant and laudable. Although it is not the intention of this paper to recommend that all search and rescue operations be gratuitous, it certainly behoves the community of States to encourage all States who are in a position to give assistance without charge, to do so. Humanitarian assistance is an integral element of diplomatic unity and co-existence.

[39]Brownlie (1963) at 338–340.

[40]*Hansard* H.C, Vol 219, col. 784 (23 February 1993).

[41]UKMIL (1986) 57 *B.Y.I.L.* 619.

[42]See Current Developments: International Law, A.V. Lowe and Colin Warbrick ed., *I.C.L.Q.* Vol 42 October 1993 Part 4, 938 at 944.

[43]See Vincent (1974), at 313.

References

Akehurst M (1977) The use of force to protect nationals abroad. Int Rel 5(3):16

Akehurst M (1984) Humanitarian intervention. In: Bull H (ed) Intervention in world politics. Clarendon, Oxford

Brownlie I (1963) International law and the use of force by States. Clarendon, Oxford

Christol CQ (1991) Space law past, present and future. Kluwer Law and Taxation Publishers, Deventer

de Lima FX (1971) Intervention in international law. Vitgeverij Pax Nederland, Den Haag

Joyner CC (ed) (1997) The United Nations and international law. Cambridge University Press, Cambridge

Lillich RB (ed) (1973) Humanitarian intervention and the United Nations. University Press of Virginia, Charlottesville

Schwarzenberger G (1971) International law and order. Stevens and Sons, London

Teson FR (1956) Humanitarian intervention: an inquiry into law and morality. Transnational Publishers Inc., Dobbs Ferry

Vincent RJ (1974) Non-intervention and international order. Princeton University Press, New Jersey

Article 26
Investigation of Accidents

In the event of an accident to an aircraft of a contracting State occurring in the territory of another contracting State, and involving death or serious injury, or indicating serious technical defect in the aircraft or air navigation facilities, the State in which the accident occurs will institute an inquiry into the circumstances of the accident, in accordance, so far as its laws permit, with the procedure which may be recommended by the International Civil Aviation Organization. The State in which the aircraft is registered shall be given the opportunity to appoint observers to be present at the inquiry and the State holding the inquiry shall communicate the report and findings in the matter to that State.

Contents

1 Intricacies of Accident Investigations

This provision has several conditions which have to be satisfied: the accident must involve death or injury; or indicate serious technical defect in the aircraft or air navigation facilities; the State in which the accident occurs will hold the inquiry; the State of Registry of the aircraft involved in the accident may be an observer in that inquiry; and ICAO may recommend the procedure for the accident investigation.

The ICAO Assembly, at its 4th Session (Montreal, 30 May–20 June 1950) adopted Resolution A4-9 (Activities and Programme of ICAO in the Field of Accident Investigation) whereby it was acknowledged that uniformity of interpretation and application of the provisions of Article 26 is necessary to the effective execution by the Organization of its obligations under Article 37 (which is to adopt Annexes to the Chicago Convention) relating to accident investigation. The Assembly therefore requested that Council to undertake a study of the interpretation, application and limitations of Article 26 of the Convention.

A significant Assembly Resolution in this regard is Resolution A4-14 (Examination of Article 26: Privileges and obligations of Contracting States other than the State of Registry or the State of Occurrence with respect to Accident Investigation) by which the Assembly resolved that it was of great importance for the general improvement of the safety of air navigation that, to the greatest possible extent, a Contracting State other than the State of Registry or the State

DOI 10.1007/978-3-319-00068-8_27,

of Occurrence, whose facilities or services were used by an aircraft prior to an accident, submit to the State which conducting an inquiry any pertinent evidence in its possession for consideration at the inquiry. The Resolution also resolved that such State, which provides information should be accorded observer status for purposes of the inquiry. Finally the Assembly resolved that responsibility devolved upon the State that a State in which an accident occurred involved an aircraft not manufactured in that State should communicate to the State of manufacture any pertinent information which results from the inquiry and which may reflect on the airworthiness of the particular type of aircraft.

The Assembly at its 14th Session (Rome, 21 August–15 September 1962) adopted Resolution A14-22 (Reports of Aircraft Accident Investigations and Inquiries) which directed the Council to study the possibility of initiating a uniform procedure to be used by States to make available promptly the reports of aircraft accident investigations and inquiries, particularly when related to large modern transport aircraft, so that the dissemination of such reports by all Contracting States may be improved. The Resolution also directed the Council to study whether it was practicable to establish procedures by which the State of Manufacture or the State that first certified the aircraft type would, in appropriate cases and upon invitation, make available competent experts for advice or consultation in the investigation of accidents.

An important consideration of the Assembly in this regard was that the Council should determine the most practicable means of ensuring that the fullest possible advantage will be taken of the specialized knowledge of such experts and notify all Contracting States accordingly. The Contracting States were requested to cooperate in the use of such experts so as to contribute to the safety of air navigation.

At its 15th Session (Montreal, 22 June–16 July 1965) the Assembly adopted Resolution A15-8 (Consolidated Statement of Continuing ICAO Policies Related Specifically to Air Navigation—Accident Investigation) Appendix P of which stated that that it was of great importance for the general improvement of the safety of air navigation that a Contracting State in which an accident occurred involving aircraft (especially large aircraft) other than of its manufacture communicate to the State of Manufacture o the State which certified the aircraft type any pertinent information resulting from the inquiry that may affect the airworthiness of the aircraft type so that such information may be used for the improvement of safety.

At its 21st Session (Montreal, 24 September–15 October 1974) the Assembly adopted Resolution A21-20 (Co-operation among Contracting States in Investigations of Certain Aircraft Accidents) where the Assembly recognized that it was becoming increasingly evident that owing to the growing sophistication and complexity of modern aircraft, the conduct of an accident investigation requires participation of experts from many specialized technical fields and access to specially equipped facilities for testing. In this context the Assembly resolved that Contracting States cooperate in the investigation of major aircraft accidents or accidents in which the investigation requires highly specialized experts and facilities, and to this end Contracting States, to the extent possible, provide on request by other Contracting States, expert assistance and facilities for the investigation of major aircraft accidents and also afford opportunity to Contracting States seeking investigation experience to attend investigations of major aircraft accidents in the interest of developing and furthering investigation experts.

2 Annex 13 to the Chicago Convention

It is a platitude to say that the causes of an aircraft accident[1] or serious incident[2] must be identified in order to prevent repeated occurrences. Article 26 of the Chicago Convention provides that, in the event of an accident to an aircraft of a Contracting State occurring in the territory of another Contracting State, and involving death or serious injury, or indicating serious technical defect in the aircraft or air navigation facilities, the State in which the accident occurs will institute an inquiry into the circumstances of the accident, in accordance, so far as its laws permit, with the procedure which may be recommended by ICAO. Article 26 goes on to say that the State in which the aircraft is registered shall be given the opportunity to appoint observers to be present at the inquiry and the State holding the inquiry shall communicate the report and findings in the matter to that State.

The identification of causal factors is best accomplished through a properly conducted investigation.[3] To emphasise this point, Annex 13 to the Chicago

---

[1]The word "accident" is defined in Annex 13 as "an occurrence associated with the operation of an aircraft which takes place between the time any person boards the aircraft with the intention of flight until such time as all such persons have disembarked, in which: (a) a person is fatally or seriously injured as a result of: being in the aircraft, or direct contact with any part of the aircraft, including parts which have become detached from the aircraft, or direct exposure to jet blast, *except* when the injuries are from natural causes, self-inflicted or inflicted by other persons, or when the injuries are to stowaways hiding outside the areas normally available to the passengers and crew; or (b) the aircraft sustains damage or structural failure which: adversely affects the structural strength, performance or flight characteristics of the aircraft, and would normally require major repair or replacement of the affected component, *except* for engine failure or damage, when the damage is limited to the engine, its cowlings or accessories; or for damage limited to propellers, wing tips, antennas, tires, brakes, fairings, small dents or puncture holes in the aircraft skin; or (c) the aircraft is missing or is completely inaccessible".

[2]An "incident" has been defined in Annex 13 as "an occurrence, other than an accident, associated with the operation of an aircraft which affects or could affect the safety of operation". The types of incidents which are of main interest to the International Civil Aviation Organization for accident prevention studies are listed in the Accident/Incident Reporting Manual (Doc 9156).

[3]An "investigation" has been defined in Annex 13 as "a process conducted for the purpose of accident prevention which includes the gathering and analysis of information, the drawing of conclusions, including the determination of causes and, when appropriate, the making of safety recommendations".

Convention on accident and incident investigation states that the objective of the investigation of an accident or incident is prevention.[4] Annex 13 provides the international requirements for the investigation of aircraft accidents and incidents. It has been written in a way that can be understood by all participants in an investigation. As such, it serves as a reference document for people around the world who may be called on, often without any lead time, to deal with the many aspects involved in the investigation of an aircraft accident or serious incident. As an example, the Annex spells out which States may participate in an investigation, such as the States of Occurrence, Registry, Operator, Design and Manufacture. It also defines the rights and responsibilities of such States.

The inter-relationship between Article 26 and Annex 13 is reflected in a resolution of the ICAO Council. The Council, at the 20th meeting of its Twelfth Session on 13 April 1951, adopted the following additional resolution:

> "*Whereas* Article 26 of the Convention provides that a State in which an accident to an aircraft occurs within the terms of the Article, 'will institute an inquiry into the circumstances of the accident in accordance, in so far as its laws permit, with the procedure which may be recommended by the International Civil Aviation Organization'; and
>
> "*Whereas* the Council, at the 18th meeting of its Twelfth Session on 11 April 1951, adopted Annex 13 on Aircraft Accident Inquiry;
>
> "*The Council recommends* the Standards and Recommended Practices for Aircraft Accident Inquiry contained in Annex 13 to the Convention, as the procedure to be followed by Contracting States for inquiries into accidents involving death or serious injury and instituted in accordance with the provisions of Article 26;
>
> "*It being understood:*
>
> "1) that States may in accordance with Article 38 of the Convention, deviate from any provision of Annex 13, except that, with respect to accidents covered by terms of Article 26 of the Convention and pursuant to this Article, 'the State in which the accident occurs will institute an inquiry', 'the State in which the aircraft is registered shall be given the opportunity to appoint observers to be present at the inquiry' and 'the State holding the inquiry shall communicate the report and findings in the matter to that State'; and
>
> "2) that the procedure here recommended is not applicable when an accident to an aircraft not involving death or serious injury 'indicates serious technical defect in the aircraft or air navigation facilities', in which cases and until ICAO recommends a procedure to this effect, the inquiry shall be conducted in accordance with the national procedure of the State concerned, subject to the obligations deriving from the provisions of Article 26."

The accredited representative and the advisers referred to in the Annex together comprise the observers that are given the right to be present at an inquiry under Article 26.

The ninth edition of Annex 13 consists of eight chapters, an appendix and four attachments. The first three chapters over definitions, applicability and general

---

[4]Standards and Recommended Practices for Aircraft Accident Inquiries were first adopted by the Council on 11 April 1951 pursuant to Article 37 of the Convention on International Civil Aviation (Chicago, 1944) and were designated as Annex 13 to the Convention. The Standards and Recommended Practices were based on recommendations of the Accident Investigation Division at its First Session in February 1946 which were further developed at the Second Session of the Division in February 1947.

information. Chapter 3 includes the protection of evidence and the responsibility of the State of Occurrence for the custody and removal of the aircraft. It also defines how that State must handle requests for participation in the investigation from other States. All States that may be involved in an investigation must be promptly notified of the occurrence. Procedures for this notification process are contained in Chapter 4. The same chapter outlines the responsibilities for conducting an investigation depending on the location of the occurrence, e.g. in the territory of an ICAO Contracting State, in the territory of a non-Contracting State, or outside the territory of any ICAO State. Following the formal notification of the investigation to the appropriate authorities, Chapter 5 addresses the investigation process.

As discussed earlier, responsibility for an investigation belongs to the State in which the accident or incident occurred. That State usually conducts the investigation, but it may delegate all or part of the investigation to another State. If the occurrence takes place outside the territory of any State, the State of Registry has the responsibility to conduct the investigation. States of Registry, Operator, Design and Manufacture who participate in an investigation are entitled to appoint an accredited representative to take part in the investigation. Advisers may also be appointed to assist accredited representatives. The State conducting the investigation may call on the best technical expertise available from any source to assist with the investigation.

The investigation process includes the gathering, recording and analysis of all relevant information; the determination of the causes; formulating appropriate safety recommendations and the completion of the final report. Chapter 5 also includes provisions regarding: the investigator-in-charge, flight recorders, autopsy examinations, coordination with judicial authorities, informing aviation security authorities, disclosure of records, and re-opening of an investigation. States whose citizens have suffered fatalities in an accident are also entitled to appoint an expert to participate in the investigation.

Chapter 6 contains the Standards and recommended practices dealing with the development and publication of the final report of an investigation. The recommended format for the final report is contained in an Appendix to the Annex. Computerized databases greatly facilitate the storing and analysing of information on accidents and incidents. The sharing of such safety information is regarded as vital to accident prevention. ICAO operates a computerized database known as the Accident/Incident Data Reporting (ADREP) system, which facilitates the exchange of safety information among Contracting States. Chapter 7 of Annex 13 addresses the reporting requirements of the ADREP system which is by means of Preliminary and Accident/Incident Data Reports.

Chapter 8 of Annex 13 deals with accident prevention measures. The provisions in this chapter cover incident reporting systems, both mandatory and voluntary, and the necessity for a non-punitive environment for the voluntary reporting of safety hazards. This chapter then addresses database systems and a means to analyse the safety data contained in such databases in order to determine any preventive actions required. Finally, it recommends that States promote the establishment of safety information sharing networks to facilitate the free exchange of information on actual and potential safety deficiencies. The processes outlined in this chapter form part of a safety management system aimed at reducing the number of accidents and serious incidents worldwide.

3 Criminalization of the Accident Process

A critical provision in Annex 13 is Standard 3.1 which states that the sole objective of the investigation of an accident or incident shall be the prevention of accidents and incidents and that it is not the purpose of that activity to apportion blame or liability. Roderick Van Dam, Head of Legal Service, EUROCONTROL states:

> Recent years have shown a growing concern on the part of aviation professionals and the aviation industry about the interpretation of the general public, as well as the criminal judiciary, of flight safety and aviation accidents…the fear of legal proceedings and involvement of judicial authorities can have an impact on the level of reporting of safety incidents. With respect to aviation, failure to gather all available safety data may have potentially serious consequences. The ability to learn from mistakes and prevent new ones is one of the most valuable tools for improvement of aviation safety.[5]

As Van Dam states, there are two protagonists involved in this equation: one who has the objective of preserving justice by investigating and prosecuting possible perpetrators and the other with the aim of enhancing aviation safety through independent investigation and reporting.[6] Corporate entities are apprehensive of developments in common law, particularly in the England, Wales, Northern Ireland and Scotland that would prompt them to be reluctant to divulge information that may be helpful in an accident or incident investigation process. *The Corporate Manslaughter and Corporate Homicide Act* of 2007,[7] provides that an organization[8] is guilty of an offence if the way in which its activities are managed or organized causes a person's death, and amounts to a gross breach of a relevant duty of care owed by the organization to the deceased.[9] The Act applies *inter alia* to a corporation. The offence is termed "corporate manslaughter", in so far as it is an offence under the law of England, Wales or Northern Ireland; and "corporate homicide", in so far as it is an offence under the law of Scotland. An organization that is guilty of corporate manslaughter or corporate homicide is liable on conviction to a fine and the offence of corporate homicide is indictable only in the High Court of Justiciary.[10]

---

[5]Michaelides-Mateou and Mateou (2010), Foreword at xxi.

[6]*Ibid.*

[7]http://www.opsi.gov.uk/acts/acts2007/ukpga_20070019_en_1#pb1-l1g1.

[8]An organization that is a servant or agent of the Crown is not immune from prosecution. *Id.* Section 11.

[9]*Id.* Section 1.

[10]*Id.* Section 1.5.

The Act provides that the concept of "relevant duty of care", in relation to an organization, means: a duty owed to its employees or to other persons working for the organization or performing services for it; a duty owed as occupier of premises; a duty owed in connection with the supply by the Organization of goods or services (whether for consideration or not); and the carrying on by the Organization of any construction or maintenance operations, the carrying on by the Organization of any other activity on a commercial basis, or the use or keeping by the Organization of any plant, vehicle or other thing.[11] Section 8 of the Act addresses the issue of "gross breach" and provides that where it is established that an Organization owed a relevant duty of care to a person, and it falls to the jury to decide whether there was a gross breach of that duty, the jury must consider whether the evidence shows that the Organization failed to comply with any health and safety legislation that relates to the alleged breach, and if so how serious that failure was; how much of a risk of death it posed. The jury may also consider the extent to which the evidence shows that there were attitudes, policies, systems or accepted practices within the Organization that were likely to have encouraged any such failure or to have produced tolerance of it; taking into consideration any health and safety guidance that relates to the alleged breach. The provision does not prevent a jury from having regard to any other matters they consider relevant. For purposes of this provision, "health and safety guidance" means any code, guidance, manual or similar publication that is concerned with health and safety matters and is made or issued (under a statutory provision or otherwise) by an authority responsible for the enforcement of any health and safety legislation.

The possible application of this piece of legislation to air transport is a reality, given the nature of the air transport product and the operation of aircraft. The profession of aeronautics, particularly relating to the piloting of aircraft, remains one of the most responsible, particularly in the context of the many lives that are entrusted to the airline pilot at any given time. Commercial airline pilots operate in a highly complex environment, particularly in single pilot operations. The difficulties faced by pilots in the work environment are compounded by the fact that often inadequate information aggravates the problem. Pilots rely heavily on their visual and auditory senses while flying, and it is of paramount importance that accurate information be available to the pilot at all times. Most importantly, pilots have usually the predilection to complete their given schedule no matter what, such as competing a flight as planned, meeting schedules, impressing their employees and pleasing the people they carry. Therefore negligent issues concerning the professional conduct of a pilot form quintessential elements for a highly esoteric legal debate, there being several recorded instances of criminal and civil prosecution of aviation professionals following from aircraft accidents.[12]

[11]*Id.* section 2.1. (a) to (c).

[12]See Mateou and Mateou (2010), The Chicago Convention, Preamble *supra* note 1, defines, in Article 2, "territory of a State" as the land areas and territorial waters adjacent to the State under the sovereignty, suzerainty, protection and mandate of such State, Chapter 9, pp. 163–184.

Act, which introduces a new offence in England, Wales, Northern Ireland and Scotland, may have some relevance to and bearing on the *Helios trial* which opened on 26 February 2009 in Cyprus. The trial pertains to the island's worst air tragedy, when 121 people perished on a charter plane that slammed into a Greek hillside nearly 4 years ago. According to reports[13] Helios Airways and four airline officials faced charges of manslaughter and reckless endangerment in one of the most complex and high-profile cases in the eastern Mediterranean island's legal history. Plaintiffs, who are relatives of the dead, have called for criminal action against those deemed responsible when the Helios Airways Boeing 737–300 ran out of oxygen and crashed outside Athens in August 2005. It has also been reported[14] that, although the authorities have not named those to be charged, the accused are known to be officials who held top management positions in the airline at the time of the crash.

A look at current tends brings to bear the fact that lessors and lessees as business entities could be heading for toward a time where they could face both tortuous and criminal liability for their negligent acts. In most globalized economies, tort law,[15] which is the branch of law that provides compensation for injuries to persons and property caused by the act of another, is a constantly evolving area of the law. This continuous evolution is caused by new and emerging social and economic activities brought about by technological advancement and increasing and varied commercial activity. For example, in the field of environmental law, concerning a $333 million class action[16] which was successfully argued in court against a California utility for polluting the water supply of a local community, is a good indicator of corporate liability. In the area of hospitality there is the "hot coffee" paradigm of Macdonald's fame where two States in the United States went on for law reform after learning of the true facts of the case.[17]

In so far as criminalizing corporate activity is concerned, this could set a dangerous trend against efforts by the air transport industry which are calculated to ensure safety. Negligent entrustment, unless in extreme cases of criminal negligence, in its classical sense is a tort and must be treated as such, with the damage being calculated in terms of monetary compensation. Any extension of the principles under this head of liability to criminal law would cast an undue burden on those involved in providing services that are usually given out by experts such as pilots and surgeons and their employers. They would be forced to concentrate on covering their tracks rather than ensuring the protection of those under their charge.

---

[13] *Kathimerini*, Thursday February 2009. http://www.ekathimerini.com/4dcgi/_w_articles_world_1_26/02/2009_105057.

[14] *Ibid.*

[15] A 'tort'is simply the Norman word for 'wrong' but 'torts' have typically been distinguished from wrongs identified with contractual relations. Tort law is concerned with civil wrongs not arising from contracts. See G.E. White, *Tort Law in America* (1980) XI.

[16] Cohen et al. (2000) at 22.

[17] See *State ex. Rel Ohio Academy of Trial Lawyers* v. *Sheward*, 86 Ohio St.3d. 451 (Ohio 1999) and *Best* v. *Taylor Machine Works* 689 N.E. 2d. 1057.

Two significant provisions in Annex 13 are a Standard and Recommended Practice. Standard 5.4 provides that the accident investigation authority shall have independence in the conduct of the investigation and have unrestricted authority over its conduct, consistent with the provisions of this Annex. The investigation would include:

(a) the gathering, recording and analysis of all available information on that accident or incident; (b) if appropriate, the issuance of safety recommendations; (c) if possible, the determination of the causes; and (d) the completion of the final report. When possible, the scene of the accident shall be visited, the wreckage examined and statements taken from witnesses. Recommendation 5.4.1 suggests that any judicial or administrative proceedings to apportion blame or liability should be separate from any investigation conducted under the provisions of this Annex.[18]

References

Cohen A et al (2000) Are lawyers running America? Their lawsuits are setting policy on guns, tobacco and now HMOs, who elected them? Time (17 July 2000)

Michaelides-Mateou S, Mateou A (2010) Flying in the face of criminalization. Ashgate, Surrey

[18]5 Standard 5.12 of Annex 13 provides that the State conducting the investigation of an accident or incident shall not make the following records available for purposes other than accident or incident investigation, unless the appropriate authority for the administration of justice in that State determines that their disclosure outweighs the adverse domestic and international impact such action may have on that or any future investigations: (a) all statements taken from persons by the investigation authorities in the course of their investigation; (b) all communications between persons having been involved in the operation of the aircraft; (c) medical or private information regarding persons involved in the accident or incident; (d) cockpit voice recordings and transcripts from such recordings; and (e) opinions expressed in the analysis of information, including flight recorder information.

Article 27

Exemption from Seizure on Patent Claims

(a) While engaged in international air navigation, any authorized entry of aircraft of a contracting State into the territory of another contracting State or authorized transit across the territory of such State with or without landings shall not entail any seizure or detention of the aircraft or any claim against the owner or operator thereof or any other interference therewith by or on behalf of such State or any person therein, on the ground that the construction, mechanism, parts, accessories or operation of the aircraft is an infringement of any patent, design, or model duly granted or registered in the State whose territory is entered by the aircraft, it being agreed that no deposit of security in connection with the foregoing exemption from seizure or detention of the aircraft shall in any case be required in the State entered by such aircraft.

(b) The provisions of paragraph (a) of the Article shall also be applicable to the storage of spare parts and spare equipment for the aircraft and the right to use and install the same in the repair of an aircraft of a contracting State, provided that any patented part or equipment so stored shall not be sold or distributed internally in or exported commercially from the contracting State entered by the aircraft.

(c) The benefits of this Article shall apply only to such States, parties to this Convention, as either 1) are parties to the International Convention for the Protection of Industrial Property and to any amendments thereof; or 2) have enacted patent laws which recognize and give adequate protection to this Convention.

R. Abeyratne, *Convention on International Civil Aviation*,
DOI 10.1007/978-3-319-00068-8_28,

Article 28
Air Navigation Facilities and Standard Systems

Each contracting State undertakes, so far as it may find practicable, to:

(a) Provide, in its territory, airports, radio services, meteorological services and other air navigation facilities to facilitate international air navigation, in accordance with the standards and practices recommended or established from time to time, pursuant to this Convention;

(b) Adopt and put into operation the appropriate standard systems of communications procedure, codes, markings, signals, lighting and other operational practices and rules which may be recommended or established from time to time, pursuant to this Convention;

(c) Collaborate in international measures to secure the publication of aeronautical maps and charts in accordance with standards which may be recommended or established from time to time, pursuant to this Convention:

Contents

1 Provision of Services That Assist Air Navigation

The Assembly, at its 14th Session (Rome, 21 August–15 September 1962) adopted Resolution A14-20 (Structure of Air Navigation Regions and the Regional Planning Process) by which the Assembly directed the Council to study the then prevailing structure of air navigation regions and the regional planning processes, with a view to their possible modification in the light of current and future requirements, including the need for improved coordination in respect of the overlap areas, and also to consider the areas not within the existing ICAO regions. At its 15th Session (Montreal, 22 June–16 July 1965) the Assembly adopted Resolution A15-11 (Study of Regional Structure and Planning Processes) the Assembly directed the Council to continue the study.

At its 21st Session (Montreal, 24 September–15 October 1974) the Assembly adopted Resolution A21-21 (Consolidated Statement of Continuing and Associated Practices Related Specifically to Air Navigation) in Appendix K (Formulation of Regional Plans including Regional Supplementary Procedures) the Assembly, while acknowledging that the Council had established regional plans setting forth the facilities and services and regional supplementary procedures to be provided or

R. Abeyratne, *Convention on International Civil Aviation*, DOI 10.1007/978-3-319-00068-8_29, © Springer International Publishing Switzerland 2014

employed by Contracting States pursuant to Article 28, called for the revision of these plans when it became apparent that they were no longer consistent with current and foreseen requirements of international civil aviation, and when a change to such plans was warranted such will be effected through correspondence between ICAO and the Contracting States concerned. The same Resolution in Appendix L (Regional Air Navigation (RAN) Meetings) acknowledged that RAN meetings are critical instruments in the determination of the facilities and services the Contracting States are expected to provide pursuant to Article 28 and resolved that RAN meetings, convened by the Council, shall be the principle means of conducting comprehensive reviews and revisions of the regional plans as necessary to keep them abreast of changings requirements.

Resolution A21-21 also resolved that RAN meetings would be convened taking into consideration the need for such as called for by any shortcomings in the region. At its 37th Session (September–October 2010) the Assembly adopted Resolution A35-15 (Consolidated Statement of continuing ICAO policies and associated practices related specifically to air navigation) where in Appendix B it is stated:

> *Whereas* the holding of worldwide air navigation meetings is an important function of ICAO and entails substantial expenditures of effort and money by the Contracting States and the Organization; and
>
> *Whereas* it is necessary to ensure that maximum benefit is obtained from these meetings without imposing any undue burden upon the Contracting States or the Organization;
>
> The Assembly resolves that:
>
> 1. meetings, convened by the Council, in which all Contracting States may participate on an equal basis shall be
>
> the principal means of progressing the resolution of problems of worldwide import, including the development of amendments to the Annexes and other basic documents in the air navigation field;
>
> 2. such meetings shall be convened only when justified by the number and importance of the problems to be dealt with and when there is the likelihood of constructive action on them; meetings convened on this basis may also be requested to conduct exploratory discussions on matters not mature for definite action;
>
> 3. the organization of such meetings shall be arranged so that they are best suited to carry out the assigned task and to provide proper coordination among the technical specialities involved; and
>
> 4. unless necessitated by extraordinary circumstances, not more than two such meetings shall be convened in a calendar year, and successive meetings dealing extensively with the same technical specialty shall be separated by at least twelve months.

The Assembly also recognized that before deciding to refer a matter to a worldwide meeting, the Council should consider whether correspondence with States or use of machinery such as panels or air navigation study groups could dispose of it or facilitate subsequent action on it by a future meeting and that the agenda should be sufficiently explicit to define the task to be performed and to indicate the types of specialized expertise that will be needed at the meeting. In an agenda including more than one technical specialty the types of expertise called for should be kept to the minimum compatible with efficiency.

2 Regional Aspects

Another practice associated with this resolution was the facilitation of the participation of all Contracting States, the Council should so plan the meeting programme as to keep to the minimum, consistent with efficiency, the demands upon the time of States' technical officials. It was also recognized that the planned duration of a meeting should allow adequate time for completion of the agenda, study of the report as drafted in the working languages of the meeting and approval of the report. Following the meeting, the Secretariat should make any necessary minor editorial amendments and typographical corrections to the meeting report.

Following Appendices J, K and L also contains resolving clauses relating to Article 28. Whereas the Council establishes Regional Plans setting forth the facilities, services and Regional Supplementary Procedures to be provided or employed by Contracting States pursuant to Article 28 of the Convention. The principles enunciated in Appendix J are that since Regional Plans require amendment from time to time to reflect the changing needs of international civil aviation and since ICAO has established an approach to planning of facilities and services that centres on the Global ATM Operational Concept and the Global Air Navigation Plan which should be based on a performance-based approach to planning. Regional Plans will be revised when it becomes apparent that they are no longer consistent with current and foreseen requirements of international civil aviation. They also require that when the nature of a required change permits, the associated amendment of the Regional Plan should be undertaken by correspondence between the Organization and Contracting States and International Organizations concerned; and when amendment proposals are associated with the services and facilities provided by States and such amendment proposals:

- Do not represent changes to the requirements set by the Council in the Regional Plans;
- Do not conflict with established ICAO policy; and
- Do not involve issues which cannot be resolved at the regional level;

the Council may delegate authority for processing and promulgating such amendments to the regional level.

Associated practices to these requirements are that the Council should ensure that the structure and format of regional plans is aligned with the Global Air Navigation Plan and in support of a performance-based approach to planning and that, the Council, taking into account the requirement to improve still further existing safety levels, should keep under review the effect of changing requirements on the Regional Plans to ascertain in good time any need for their revision.

In assessing the urgency of any revision of the Regional Plans the Council should take into account the time needed by Contracting States to arrange for the provision of any necessary additional facilities and services. It is also required that the Council ensure that implementation dates in Regional Plans involving the procurement of new types of equipment are realistically related to the ready availability of suitable equipment. Furthermore, the Council is required to ensure that an electronic database of regional plans is developed, with supporting planning tools, in order to improve efficiency and expedite the amendment cycle and to use the planning groups it has established throughout the regions to assist in keeping up to date the Regional Plans and any complementary documents.

APPENDIX K (Regional air navigation (RAN) meetings) recognizes that whereas RAN meetings are important instruments in the determination of the facilities and services the Contracting States are expected to provide pursuant to Article 28 of the Convention; and whereas these meetings entail substantial expenditures of effort and money by Contracting States and the Organization; whereas it is necessary to ensure that maximum benefit is obtained from these meetings without imposing any undue burden on Contracting States or the Organization; and considering that regional air navigation planning is normally accomplished by Planning and Implementation Regional Groups (PIRGs);

The Assembly resolved that:

- RAN meetings shall be convened only to address issues which cannot be adequately addressed through PIRGs;
- The convening of such meetings and their agenda shall be based on the existence or expectation of specific shortcomings in the Regional Plans of the respective areas;
- The geographical area to be considered, account being taken of the existing and planned international air transport and international general aviation operations, the technical fields to be dealt with and the languages to be used shall be decided for each such meeting;
- The organization best suited to deal with the agenda and to ensure effective coordination among the components of the meeting shall be used for each such meeting; and
- Meetings of limited technical and/or geographical scope shall be convened when specific problems, particularly those requiring urgent solution, need to be dealt with or when convening them will reduce the frequency with which full scale RAN meetings must be held.

The associated practices to this Appendix are that the Council should endeavour to hold RAN meetings at sites within the areas concerned and should encourage the Contracting States within those areas to serve as host, either individually or jointly and that the approved agenda and the main supporting documentation should be made available, by electronic means, not less than 10 months in advance of the convening date in the case of the agenda and not less than 3 months in the case of the main supporting documentation.

The Council is also required to ensure that adequate guidance is made available to RAN meetings on operational and technical matters relevant to their agenda. Additionally, each participating Contracting State should inform itself, in advance of a meeting, on the plans of its air transport operators and its international general aviation for future operations and, similarly, on the expected traffic by other aircraft on its registry and on the overall requirements of these various categories of aviation for facilities and services.

The Council, taking into account the requirement to improve still further existing safety levels, is expected to foster the establishment, for and by RAN meetings, of up-to-date planning criteria which would aim to ensure that Regional Plans satisfy the operational requirements and are economically justified.

APPENDIX L (Implementation of Regional Plans) acknowledged that whereas in accordance with Article 28 of the Convention Contracting States undertake, insofar as they may find practicable, to provide air navigation facilities and services necessary to facilitate international air navigation; and whereas the Regional Plans set forth the requirements for facilities and services for international civil aviation

The Assembly resolved that:

priority shall be given in the implementation programmes of Contracting States to the provision, including continuing operation, of those facilities and services the lack of which would likely have a serious effect on international air operations;

- The identification and investigation of and action by the Organization on serious deficiencies in the implementation of Regional Plans shall be carried out in the minimum practicable time; and
- Regional planning and implementation groups shall identify problems and shortcomings in Regional Plans and in the implementation thereof, along with suggested remedial measures.

The associated practices to this Appendix are that the Council, taking into account the requirement to improve still further existing safety levels, should inform fully and promptly each Contracting State of the recommendations for the provision of air navigation facilities and services that are applicable to that State under the Regional Plans and that Contracting States should prepare and keep up to date suitable plans, including the requirements for personnel, for the orderly implementation of the parts of Regional Plans applicable to them. The Council is also required to arrange for the monitoring of the status of implementation of the Regional Plans and for the issue of periodic progress reports which should include information on serious shortcomings in implementation of the Regional Plans. The users of air navigation facilities and services should report any serious problems encountered due to the lack of implementation of air navigation facilities or services required by Regional Plans. The reports should be addressed to the Contracting States responsible for implementation. These States should act on such reports to resolve the problems, but when remedial action is not taken users should inform ICAO, through the medium of an international organization where appropriate.

3 Standardization

Article 28 requires a State to adopt and put into operation the appropriate standard systems of communications procedure, codes, markings, signals, lighting and other operational practices and rules which may be recommended or established from time to time, pursuant to this Convention. Obviously, the Convention, through an inarticulate premise requires in Article 28 that States provide *functional* airport services among other services prescribed in the provision. This fundamental concept of State responsibility has to be viewed from the perspective of modern exigencies of the supply and demand curve of air traffic services where such services are currently being provided both regionally and on a flight information region (FIR) basis. The need for a shift of focus of the modern air navigational system is determined by two factors: the growing air traffic demand and the need for enhanced and more efficient air traffic services; and the transition into a seamless air traffic management system calculated to obviate inconsistencies caused by boundaries.[1] The goals of a global seamless air traffic management system are: to provide greater flexibility and efficiency by accommodating user-preferred flight profiles; to improve existing levels of safety; to accommodate the full range of aircraft types and airborne capabilities; to improve the provision of information to users, including weather conditions, the traffic situation and the availability of facilities; to organize air space in accordance with air traffic management (ATM) provisions and procedures; to increase user involvement in ATM decision making, including air-ground computer dialogue for flight plan negotiation; to create, to the extent possible, a single continuum of airspace where boundaries are transparent to users; and to increase capacity to meet future traffic demand.[2]

Reference

Sudharshan HV (2003) Seamless sky. Ashgate, Aldershot

[1]Sudharshan (2003) at 2.

[2]*Global Air Navigation Plan for CNS/ATM Systems*, Second Edition: 2002, ICAO Doc 9750, AN/963, p. 1-4-3 at paragraph 4.12.

Article 29
Documents Carried in Aircraft

Documents carried in aircraft Every aircraft of a contracting State, engaged in international navigation, shall carry the following documents in conformity with the conditions prescribed in this Convention:

(a) Its certificate of registration;

(b) Its certificate of airworthiness;

(c) The appropriate licenses for each member of the crew;

(d) Its journey log book;

(e) If it is equipped with radio apparatus, the aircraft radio station license;

(f) If it carries passengers, a list of their names and places of embarkation and destination;

(g) If it carries cargo, a manifest and detailed declarations of the cargo.

DOI 10.1007/978-3-319-00068-8_30,

Article 30
Aircraft Radio Equipment

(a) Aircraft of each contracting State may, in or over the territory of other contracting States, carry radio transmitting apparatus only if a license to install and operate such apparatus has been issued by the appropriate authorities of the State in which the aircraft is registered. The use of radio transmitting apparatus in the territory of the contracting State whose territory is flown over shall be in accordance with the regulations prescribed by that State.

(b) Radio transmitting apparatus may be used only by members of the flight crew who are provided with a special license for the purpose, issued by the appropriate authorities of the State in which the aircraft is registered.

R. Abeyratne, *Convention on International Civil Aviation*,

DOI 10.1007/978-3-319-00068-8_31,

Article 31
Certificates of Airworthiness

Every aircraft engaged in international navigation shall be provided with a certificate of airworthiness issued or rendered valid by the State in which it is registered.

Contents

1 Mandatory Documentation

Articles 30 and 31 refer to the need for aircraft to carry radio equipment as required and be issued with certificates of airworthiness by the State of Registry, respectively. Article 32 (a) requires the pilot and other members of the crew to be provided with certificates of competency and licenses issued by the State of Registry. The Council, on 8 February 1949 at the sixth meeting of its sixth session, adopted standards for aircraft nationality and Registration Marks and designated them as Annex 7 to the Chicago Convention. One of the evolving features of documents carried in aircraft under Article 29 is the list of passengers' names and their places of embarkation and disembarkation. An extension to this provision, in terms of modern exigencies of security is advance passenger information and the nuances of extra territoriality it brings to bear that may veer Article 28 to a whole new tangent.

One of the most dramatic events pertaining to aviation security occurred in July 2005 when United States air traffic controllers turned back a KLM flight en route to Mexico City from Amsterdam, which was flying over US airspace. The action was

R. Abeyratne, *Convention on International Civil Aviation*,

DOI 10.1007/978-3-319-00068-8_32,

grounded on the basis that two of the passengers in the passenger list earlier provided to the US authorities were on a "no fly" list. The importance of this drama to modern day aviation is that the aircraft was merely over-flying the territory of a State. Even more important is the fact that at the time of the incident, there was no US legislation covering the act of refusal to grant over-flying permission to an aircraft in that situation.[1] However, within days, The US Transportation Security Administration (TSA) announced that rules will be adopted to require that passengers on all flights landing in and overflying US territory will be screened against a "no fly" list.[2]

The Passenger Name Record (PNR) is a subject that has been under intense scrutiny by the Council of ICAO, which has developed PNR Data Guidelines that have been transmitted to Contracting States for their comments[3] This exercise was carried out on the understanding that, in the present context of the compelling need for the enhancement of aviation security, the global aviation community has shown an increased interest[4] in adding the PNR data as a security measure in addition to the already existing Advanced Passenger Information (API)[5] and the Machine

---

[1]Consequent upon the events of 2001, President George Bush signed a new *American Transportation & Security Act* on November 25th 2002 making mandatory API transmission and the provision of PNR data pertaining to all passengers arriving in the United States. Such information, required prior to departure and arrival in the United States should include in the passenger and crew manifest for each flight, in accordance with , Section 115 of the *Transportation & Security Act* is:

a. The full name of each passenger and crew member;
b. The date of birth and citizenship of each passenger and crew member;
c. The sex of each passenger and crew member;
d. The passport number and country of issuance of each passenger and crew member if required for travel;
e. The United States visa number or resident alien card number of each passenger and crew member, as applicable;
f. Such other information as the under Secretary, in consultation with the Commissioner of Customs, determines is reasonably necessary to ensure aviation safety.

[2]Crossing the Line, *Airline Business*, August 2005, at 9.

[3]See Attachment to State Letter EC 6/2-05/70, Passenger Name Record (PNR) data, 9 June 2005.

[4]The advantage of collection by States of PNR Data was first discussed by the global aviation community at the Twelfth Session of the ICAO Facilitation Division that was held in Cairo, Egypt from 22 March to 1 April 2004. Consequently, the Division adopted Recommendation B/5, that reads as follows:

> It is recommended that ICAO develop guidance material for those States that may require access to Passenger Name Record (PNR) data to supplement identification data received through an API system, including guidelines for distribution, use and storage of data and a composite list of data elements [that] may be transferred between the operator and the receiving State.

Pursuant to this recommendation, In June 2004, the Air Transport Committee of the ICAO Council requested the Secretary General to establish a Secretariat Study Group to develop Guidelines on PNR data transfer. The Council, in endorsing Recommendation B/5, directed that these Guidelines were to be submitted early in 2005.

[5]See, Abeyratne (2002a). Also by Abeyratne (2001): pp. 153–162, and also by Abeyratne (2003).

Readable Travel Document (MRTD), which, although primarily are facilitation tools, greatly assist States authorities in ensuring border security.

One of the issues that emerge from PNR data collection is extraterritoriality and the question as to whether at law a State can require information held by other States relating to flights that originate and end in the latter States. An example is Canada, which may be required by the US to divulge information pertaining to passengers on domestic flights operating within the territorial limits of Canada but over-fly United States' territory for reasons of expediency and fuel efficiency. While there is no room for doubt that usually, requirements for safety and security of a State are based on sound legal justification with a view to protecting A State's integrity and internal security, a requirement for information by a particular State of those that do not enter the territory of that State might open itself to question, as to whether such would impinge upon another sovereign State's right to privacy[6] and dignity.

2 Passenger Name Record

A new Recommended Practice concerning the PNR data has been included in Annex 9 to the Chicago Convention (Facilitation) after being adopted by the ICAO Council in March 2005.[7] This Recommended Practice, which supplements an already existing Recommended Practice[8] provides that Contracting States requiring Passenger Name Record (PNR) access should conform their data requirements and their handling of such data to guidelines developed by ICAO. It is worthy of note that Article 13 of the Chicago Convention provides that the laws and regulations of a Contracting State as to the admission to or departure from its territory of passengers, crew or cargo of aircraft, such as regulations relating to entry, clearance, immigration, passports, customs, and quarantine shall be complied with, by or on behalf of such passengers, crew or cargo upon entrance into or departure from, or while within the territory of that State. This provision gives a State the discretion to specify the information it requires relating to persons wishing to gain entry into its territory. Accordingly, a State may require aircraft operators operating flights to, from or in transit through airports within its territory to provide its public authorities, upon request, with information on passengers such as PNR data.

---

[6]See Abeyratne, The Exchange of Airline Passenger Information—Issues of Privacy, *supra*, note 7.

[7]Recommended Practice 3.48 which provides: "Contracting States requiring Passenger Name Record (PNR) access should conform their data requirements and their handling of such data to guidelines developed by ICAO".

[8]Recommended Practice 3.47, which provides *inter alia* that Contracting States should, where appropriate, should introduce a system of advance passenger information which capture certain passport and visa information prior to departure, for onward transmission to relevant public authorities by electronic means.

The philosophy underlying the importance of PNR data and their efficient use by States for enhanced expediency in border crossing by persons is embodied in the General Principles set out in Chapter 1 of Annex 9 which require Contracting States to take necessary measures to ensure that: the time required for the accomplishment of border controls in respect of persons is kept to the minimum[9]; the application of administrative and control requirements causes minimum inconvenience; exchange of relevant information between Contracting States, operators and airports is fostered and promoted to the greatest extent possible; and, optimal levels of security, and compliance with the law, are attained.

Contracting States are also required to develop effective information technology to increase the efficiency and effectiveness of their procedures at airports.[10]

3 Definition and Application of PNR

The air transport industry regards a *Passenger Name Record* (PNR), as a generic term applicable to records created by aircraft operators or their authorized agents for each journey booked by or on behalf of any passenger. The data is used by operators for their own commercial and operational purposes in providing air transportation services.[11] The definition applicable in the United States identifies a PNR as a repository of information that air carriers would need to make available upon request under existing regulations and refers to reservation information contained in a carrier's electronic computer reservation system.[12]

The above definitions and identifiers go to show that a PNR is developed and constructed from data that has been provided by or on behalf of the passenger concerning all the flight segments of a journey.[13] This data may be added to by the

---

[9]There is an abiding symbiosis between security and facilitation in the field of air transport. While security is of paramount interest to the global aviation community, it must not unduly disrupt or in any adversely affect the expediency of air transport. To this end, Recommended Practice 2.2 of Annex 9—Facilitation—to the Chicago Convention suggests that Each Contracting State should whenever possible arrange for security controls and procedures to cause a minimum of interference with, or delay to the activities of civil aviation provided the effectiveness of these controls and procedures is not compromised. See McMunn (1996) at 7.

[10]It must be noted that Annex 9 specifies that the provisions of the Annex shall not preclude the application of national legislation with regard to aviation security measures or other necessary controls.

[11]The Industry Standards related to PNR creation are detailed in IATA's *Passenger Services Conference Resolutions* and in the ATA/IATA *Reservations Interline Message Procedures (AIRIMP) Manual*.

[12]Passenger Name Record Information Required for Passengers on Flight in Foreign Air Transportation to or from the United States of 2001, 66 *Fed. Reg*. 67482 (2002).

[13]There are two possible methods of PNR data transfer currently available: (a) the "pull" method, under which the public authorities from the State requiring the data can reach into the aircraft operator's system and extract ("pull") a copy of the required data into their database; and (b) the "push" method, under which aircraft operators transmit ("push") the required PNR data elements into the database of the authority requesting them.

operator or his authorized agent, for example, in the form of changes to requested seating, special meals, additional services requested, etc. PNR data could be obtained in many ways. For instance, information captured through reservations created by international sales organizations (global distribution systems "GDS" or computer reservation systems "CRS") with pertinent details of the PNR could be transmitted to the operating carrier(s). When reservations are made directly by the aircraft operator and the complete PNR is stored within the operator's automated reservations systems, the information therein could be a useful repository of PNR data. Information contained in records of some operators who may hold sub-sets of the PNR data within their own automated departure control systems (DCS), for their information or for onward transmittal to contracted ground handling service providers, calculated to support airport check-in functions would be another way in which PNR data could be provided. However, it must be noted that in each case, operators (or their authorized agents) will have access to, and be able to amend only that data that has been provided to their system(s). An important consideration in this regard is that some DCS systems are programmed such that details emerging from check-in (i.e. seat and/or baggage information) can be overlaid into the existing PNR for each passenger. However, that capability is limited—covering less than 50 % of operating systems today.

The time element, with regard to the capture and relevance of PNR data, is relevant to the use of such data. For instance, Data could be entered into a reservation system many days or weeks in advance of a flight. This could extend to as long as 345 days in advance of departure. Under such circumstances, both the provider and the receiver of PNR data must bear in mind that Information in reservation systems is dynamic and may change continuously from the time when the flight is open for booking. On the other hand, passenger and flight information in the DCS, becomes available only from the time the flight is "open" for check-in (up to 48 h prior to departure). In such an instance, departure control information for a flight will be finalized only upon flight closure, and may remain available 12–24 h after arrival of a flight at its final destination.

Aircraft operators specializing in charter air services, who often do not hold PNR data in an electronic form, but still use a DCS which will only enable them to have a limited PNR record after the flight has closed, would still be required to provide any captured data to States requesting it regardless of the process by which they receive PNR data. States could also require supplemental or "requested service" information which may be contained in the PNR, such as information relating to special dietary and medical requirements, "unaccompanied minor" information, requests for assistance etc.

Operators should take particular care in refraining from incorporating in PNR data any information that is not essential to facilitate the passenger's travel. Such information would include, but not be necessarily restricted to details of the passenger's racial or ethnic origin, political opinions, religious or political beliefs, trade-union membership, marital status or data relating to a person's sexual orientation. The ICAO guidelines make specific mention of the fact that Contracting States should not require aircraft operators to collect such data in their PNRs.

The above notwithstanding, any information which would legitimately facilitate the carriage of the passenger, such as details of meal preferences and health issues as well as free text and general remarks, could comprise the PNR. Sensitive data contained in the PNR and is submitted in compliance with a regulation of a State should not be used as the primary source for assessment of risk that the passenger might present to the State concerned.

4 The Importance of PNR Data to States

From a regulatory perspective, the two main areas to which PNR data make a contribution are expedition of customs and immigration processing at airports; and facilitation of passenger traffic and the safeguard of the legitimate rights of the passenger. The Chicago Convention provides a sound basis for States to require PNR data in the current context. The Convention, in Article 22, recognizes the importance of facilitating the passage of a person through borders by requiring each Contracting State to adopt all practicable measures, through the issuance of special regulations or otherwise, to facilitate and expedite navigation by aircraft between the territories of Contracting States, and to prevent unnecessary delays to aircraft, crews, passengers and cargo, especially the administration of the laws relating to immigration, quarantine, customs and clearance.

The main reason for States to require the advance submission of PNR data is that such data could prove to be a valuable tool in ensuring aviation security. PNR data are critically important for the threat assessment value that can be derived from the analysis of such data, not only in possible instances of unlawful interference with civil aviation but also in relation to the fight against terrorism. This critical value of PNR data has prompted some States to enact legislation or develop draft legislation for approval by their Legislatures requiring that aircraft operators provide their public authorities with PNR data.

PNR data primarily enable States, through the identification of potentially high-risk passengers through PNR data analysis, to improve aviation security; enhance national and border security; prevent and combat terrorist acts and related crimes and other serious crimes that are transnational in nature, including organized crime; and to enforce warrants and prevent flight from custody for such crimes. Such data could also protect the vital interests of passengers and the general public, including their health.

States are aware that, if the guidelines are implemented in a uniform manner, they would provide a global framework enabling all States to benefit from the value-added analysis of PNR data for shared security/safety purposes. Air carriers would also benefit from having to comply with only one set of common requirements for PNR data transfer. As for the consumer of air transport, all passengers would benefit from basic protection afforded to them by the exchange of PNR data between air carriers and State authorities.

The above notwithstanding, there are certain fundamental obligations that the State receiving the data has to fulfill. Firstly, States should require PNR data only of those passengers on flights that are scheduled to enter, depart or transit through airports situated in their territories. Secondly, a State obtaining PNR information should, as a minimum, limit the use of data to the purpose for which it collects it. States must restrict access to such data, ensure that the data is adequately protected, and limit the period of data storage, consistent with the purposes for which data is transferred. States must also ensure that individuals are able to request disclosure of the data that is held on them, consistent with the guidelines, in order to request corrections or notations, if necessary. More importantly, they must ensure that individuals aggrieved by the PNR data collection and usage process have an opportunity for redress.

The responsibility of ensuring that their public authorities have the appropriate legal authority to process PNR data requested from aircraft operators, in a manner that observes the guidelines, devolves entirely upon the States. They have been requested by ICAO to forward the full texts of legislation pertaining to PNR data dissemination and use to ICAO for online dissemination to other States, for information. The State concerned will be responsible for responding to any queries arising from such legislation.

5 Advantages of Unified Guidelines

Through the PNR Data Guidelines ICAO has introduced uniform measures for PNR data transfer and the subsequent handling of that data by the States concerned. The guidelines are both durable and easy to follow, making them cost effective for the parties concerned. They would ensure accuracy of information, while at the same time protecting the data subject against encroachment of his privacy. The Guidelines call for completeness of data and the need for timely submissions and effective collection of data. They also ensure that data management will be efficient and efficacious. From a practical perspective, the guidelines also provide useful directions assisting States in designing data requirements and procedures, in order to minimize technical difficulties that might prove too onerous and may impair the implementation of the uniform measures suggested. The Guidelines also contain detailed instructions with a view to assisting both air carriers and States on PNR data transfer from an operator's system to a State and the management of the data including arrangements for storage and protection.

States are enabled, by the guidelines, to design systems and establish arrangements that are compatible with the guidelines while not impairing their ability to implement their laws and enforce them. The guidelines do not interfere with the preservation of national security and public safety of a State. Arguably, one of the most important features of the unified PNR data guidelines is that, by their very nature, they would effectively obviate the complexities that aircraft operators could face with regard to legal, technical and financial issues if they were to be required to respond to multiple, unilaterally imposed or bilaterally agreed PNR data transfer requirements that differ substantially from one another.

It must be noted that States also have the responsibility of enacting explicit legal provisions concerning data transfer. Such legislation should clearly elaborate on the reasons for requiring PNR data, or provide explanatory material accompanying such laws or regulations, as appropriate. Since an aircraft operator is obliged to comply with the laws of both the State from which it transports passengers (State of departure) and the State to which these passengers are transported (State of destination), when a destination State legislates with regard to its PNR data transfer requirements, it should do so cognizant of the fact that *existing* laws of other States may affect operators' ability to comply with these requirements. Therefore, where there could be an inconsistency between two legal regimes of the departure State and the destination State, or where a conflict arises between any two States, or where an operator advises of a conflict, The ICAO guidelines suggest that the States involved should consult each other to determine what might be done to enable affected operators to continue to operate within the bounds of the laws in both States.

6 Extra Territoriality

Strictly interpreted, extra-territoriality at international law means the attempt of one State to apply its laws outside its territory[14] and there is a general presumption against the application of extra-territoriality.[15] In the 1979 case of *Mannington Mills* v. *Congoleum Corporation*[16] the United States Supreme Court extended the concept of extra territorialty by introducing a test of balance that ensured consideration by one State for the interests of another State.

The above principle of extra-territoriality might not sit comfortably in the instance of a State requiring PNR data from a flight over-flying its territory as there is no *stricto sensu* application of a requirement in a foreign territory. The most fundamental principle of public international law, that of State sovereignty, is embodied in Article 1 of the Chicago Convention, thus importing the principle into the tenets of air law. This Article provides that Contracting States recognize that every State has complete and exclusive sovereignty over the air space above its territory. The territory of a State, for the purposes of the Convention, cover the land areas and territorial waters adjacent to and under the sovereign, suzerainty, protection and mandate of the State concerned. Arguably, these provisions would give the United States the right *in limine* to prescribe requirements on aircraft flying over its

[14]Shaw (2003), at 611–612.

[15]*Holmes* v. *Bangladesh Biman Corporation*, [1989] 1 AC 1112 at 1126. Also, *Air India* v. *Wiggins* [1980] 1 WLR 815 at 819. In the 1991 case of *EEOC* v. *Arabian American Oil Company and ARAMCO Services* 113 L E 2d 274, the US Supreme Court held that the practice of extra territoriality by one State against another cannot in any way be justified under the principles of public international law.

[16]595 F.2d 1287; 66 ILR at 487. See also *Timberlane Lumber Company* v. *Bank of America*, 549 F 2d 597 (1976); 66 ILR at 270.

territory. Article 12 of the Chicago Convention provides, *inter alia*, that each Contracting State undertakes to adopt measures to insure that every aircraft flying over or maneuvering within its territory and that every aircraft carrying its nationality mark, wherever that aircraft may be, shall comply with the rules and regulations relating to the flight and maneuver of air raft there in force. This rule can apply to a foreign carrier who is over-flying the territory of any State having a regulation that certain data pertaining to a flight that over-flies its territory has to be submitted to that State. Also important is Article 9 of the Convention, which allows a Contracting State to restrict or prohibit an aircraft from flying over its territory for reasons of military necessity or public safety. The provision goes on to say that each Contracting State could also reserve the right, in exceptional circumstances or during a period of emergency, or in the interest of public safety and with immediate effect, temporarily to restrict or prohibit flying over the whole or part of its territory, provided such action would apply without distinction of nationality to aircraft of all States.[17]

At the 28th Session of the International Law Association held in Madrid in 1913, the meeting drew up text which stated that it was the right of every State to enact prohibitions, restrictions and regulations as it may think proper in regard to passage of aircraft through the airspace above its territory and territorial waters.[18] However, the text contained a caveat that such restrictions should be subject to the rights of subjacent States and the liberty of passage of aircraft of every nation.[19] The balance advocated at the Madrid meeting of the ILA goes to show that even as early as the beginning of the last century, the thinking was that a State ought to allow other States free passage for their aircraft through the airspace above its territory. There is no doubt that the same position prevails even now, particularly through the currently applicable International Air Services Transit Agreement (IASTA) which was concluded at the same time as the Chicago Convention in December 1944 and has been ratified by as many as122 ICAO Contracting States. IASTA[20] allows aircraft of foreign States freedom of peaceful transit (over the airspace of a State) and freedom of making non-traffic (non-revenue) stops for such purposes as refueling and repair. It has been acknowledged that without these two freedoms, the air transport industry could not survive.[21]

---

[17]*Id.* Article 9 (b).

[18]International Law Association, 28th Report, Madrid, 1913, 533–545 at 540.

[19]Madrid Report, *Id.*, at 538.

[20]The United States ratified IASTA in 1945. With regard to Canada, it is interesting to note that Canada signed the Agreement on 10 February 1945 and deposited an instrument of acceptance thereto on the same date with the Government of the United States of America. On 12 November 1986, a notice of denunciation of the Agreement by the Government of Canada was received by the Government of the United States of America which was to have taken effect on 12 November 1987. However, this notice was revoked by a note dated 10 November 1987. By a second note dated 10 November 1987, the Government of Canada gave a new notice of withdrawal from the Agreement, which took effect on 10 November 1988.

[21]Honig (1956), at 29.

The above discussion brings one to the inexorable conclusion that there are two major issues at stake. The first is whether the PNR is an acceptable tool which helps in enhancing facilitation and security measures in air transport. The answer to this question, as provided by the 12th ICAO Facilitation Division in March/April 2004 and subsequently by the ICAO Council[22] is a resounding "yes". This affirmation brings to bear the need to consider whether the PNR should be used strictly as intended, firstly to facilitate customs and immigration procedures regarding persons, and secondly to advise States in advance of persons on board an aircraft approaching their territory for purposes of landing there, thus enabling States to determine appropriate security clearance measures. The security angle of the PNR brings one to the second issue, as to whether a State can use information contained in the PNR to disallow the right of passage to an aircraft flying over its territory, thereby denying that aircraft a fundamental right acknowledged by States through IASTA.

The second issue raises the question of extra territoriality, which can be answered by invoking Articles 9 and 12 of the Chicago Convention, as earlier discussed. These provisions clearly give a State the right to prohibit an aircraft from over-flying its territory if it believes that such over-flying could be a security hazard. The final issue would be to determine the extent to which a State could exercise its right without touching the sensitivities and dignity of a State in an instance where an aircraft plying domestic services within two points in its territory but passes through the airspace of the prohibiting State is disallowed from using the right of passage.

The entire issue of diversion of an aircraft which is exercising its fundamental right of passage, and the justification of a State for disallowing that aircraft from using that fundamental right hinges on the circumstances prevailing at the time. As was mentioned earlier, this is no legal issue as the question of extra-territoriality does not arise with regard to action taken by a State within its territory. The fundamental postulate in the debate is that sovereignty should no longer mean the mere exercise by one State of rights over its territory but should mean also the right of that State to ensure the safety and security of its citizens as well as to protect the integrity of the State.

Public international law is increasingly becoming different from what it was a few decades ago. It can be said with some justification that international law is the thread which runs through the fabric of international politics and provides the latter with its abiding moral and ethical flavour. Without principles and practices of international law, foreign policy would be rendered destitute of its sense of cooperation and become dependent on a nation's self interest. As President Woodrow Wilson once claimed:

It is a very perilous thing to determine the foreign policy of a nation in the terms of material interests ... we dare not turn from the principle that morality and not expediency is the thing that must guide us, and that we will never condone equity because it is convenient to do so.[23]

---

[22]*Ibid.*

[23]Quoted in Morganthau and Thompson (1950) at p. 24.

This statement, made in 1950, has great relevance today, when continued progress is being made in technological and economic development and policy decisions of States have far reaching consequences on a trans-boundary basis. Nation States are becoming more interdependent, making decisions made by a particular State in its own interest have a significant negative impact on the interests of other States. Therefore ethics in foreign policy has largely become a construct which combines cultural, psychological and ideological value structures. Within this somewhat complex web of interests, decisions have to be made, which, as recent events in history have shown, require a certain spontaneity from the international community. For example, when Iraq invaded Kuwait in 1990, the members of the United Nations chose economic sanctions against Iraq, claiming that war was the last resort to be embarked upon against Iraq if economic sanctions did not prove to have any effect. In hindsight, one could argue one way or another, firstly, as did the United States, that the use of force bore quick results and, on the other hand, as did many officials in Paris, Moscow, Ottawa and Washington, that the decision to wage war against Iraq was too precipitous as not enough time had been given to economic sanctions to compel Iraq to retreat from Kuwait. The precipitous but quick action taken in going to war with Iraq might be justified by some with the analogy of Britain appeasing Hitler in the 1930s without adopting a more aggressive and perhaps belligerent attitude toward German atrocities. This action, which was later labeled as folly by most political scientists, was applauded and endorsed at that time in the British Parliament.

In the absence of extra territoriality the only balancing factor in favor of a State which orders the diversion of an aircraft over-flying its territory, on the basis that persons therein are unacceptable is that the State must have sound justification for doing so in the interests of security and safety. It is very much the call of the State which is enforcing the action, and its evaluation of the pros and cons of the action from a diplomatic perspective as weighed against its own compelling security interests.

7 Advance Passenger Information

Advance passenger information and other methods of data processing of air travel find their fundamental legal roots within the *Convention on International Civil Aviation* of 1944[24] by promoting safety of flight in international air navigation and by the promotion of the developments of all aspects of civil aeronautics. This Convention was signed in Chicago and created the International Civil Aviation Organization. The objectives of this organization are set forth in article 44 of the Convention which includes the more specifics aspects of facilitation and aviation security:

[24]*Convention on international Civil Aviation*, 7th of December 1944, ICAO Doc. 7300/8(entered into force 4th April 1947) [hereinafter: Chicago Convention].

The aim and objectives of the Organization are to develop the principles and techniques of international air navigation and to foster the planning and development of international air transport so as to:

- Insure the safe and orderly growth of international civil aviation throughout the world;
- Encourage the arts of aircraft design and operation for peaceful purposes;
- Encourage the developments of airways, airports, and air navigation facilities for international civil aviation;
- Meet the needs of the peoples of the world for safe, regular, efficient and economical air transport;
- Prevent economic waste caused by unreasonable competition;
- Insure that the rights of Contracting States are fully respected and that every Contracting State has a fair opportunity to operate international airlines;
- Avoid discrimination between Contracting States;
- Promote safety of flight in international air navigation; and
- Promote generally the development of all aspects of international aeronautics.[25]

In addition to ICAO's objectives, the modern traveler requires rapid processing through the different stages of air transport, whether it implicates the air carrier's processes or those set forth by the border control agencies. In article 22, the Chicago Convention recognizes the importance of facilitating its formalities with respect to each passenger:

> Each Contracting State agrees to adopt all practicable measures, through the issuance of special regulations or otherwise, to facilitate and expedite navigation by aircraft between the territories of Contracting States, and to prevent unnecessary delays to aircraft, crews, passengers and cargo, especially the administration of the laws relating to immigration, quarantine, customs and clearance.[26]

To achieve such conditions of travel, States have adopted ICAO's recommendations into national laws but have as well implemented some extraterritorial applications to the existing legislations of legality. For example, the United States by *Enhanced Border Security and Visa Reform Act*,[27] which objectives are to reform the *US Patriot Act* by different new technologies of data capturing such as biometrics. In fact, the government acknowledges the need for expedited clearance of passengers at airports using machine readable travel documents (MRTDs) technology. This should be used in a secure environment in order to ensure that the proper border control authorities remain the only agency in possession of such

[25]Chicago Convention-Art. 44.

[26]Ibid-Art. 22.

[27]*Enhanced Border And Visa Entry Reform Act of 2002*; 107 established by the Congress of the United States of America at the second session, 22nd January 2002.

data. Furthermore, these procedures have been standardized on a worldwide basis and are currently being developed with the initiative of ICAO and ISO with new biometric procedures[28]:

With modern methods comes the inextricable discussion on privacy rights and their possible violations. Most prominent authors recognize four rights of privacy relating to the storage and use of personal data, which can be classified into four sections:

- The right to determine what personal information to share with others, and to control the disclosure of such data;
- The right to know what personal data is disclosed, collected and stored;
- The right to dispute incomplete or inaccurate data;
- The right of those who possess legitimate reasons for information on data pertaining to health and safety of society.[29]

As a counterpart, air carriers are stricken with the possible burden of financially assuming new technologies of data processing. According to the *Simplifying Travel Organization*,[30] the technology implemented will entail capturing of the passengers information by additional manpower at check-in wit purchase of hardware in order to comply with Annex 9's provision of machine-readable documents.[31]

Furthermore, the usage of advance passenger information not only can be considered as a facilitation aspect, but it is also one of aviation security. The Chicago Convention stipulates at its article 44-subsection d the necessity of safe and efficient air transport. ICAO has recognized the fact that security and facilitation must act at a joint venture.

---

[28] Heitmeyer, R., "Biometric ID and Airport Facilitation" Airport World (ACI) 5:1 (February–March 2000) 18–20.

[29] Abeyratne (2002b).

[30] Refer to the SPT Brochure 2002. The Simplifying Travel Group is a joint venture with IATA in order to develop new technologies in biometrics for the screening of passengers: "The SPT Program is a joint initiative amongst a number of organizations, representing passengers, airlines, airports, control authorities, travel agents and broad government interests, *to measurably improve the passenger experience and enable security enhancement by:*

- Implementing biometrics and other new technologies;
- Sharing information amongst service providers;
- Enabling controls and services to be effected more efficiently.

[31] "6.5.1 The principal costs for carriers are associated with system development/integration and capture of passenger details for transmission to the destination country of a flight. Costs will likely be incurred in other areas as well; e.g. additional check-in staff to cope with the extended period of time required to complete check-in formalities, additional check-in desks, hardware acquisitions, etc. Various techniques can be used to offset these costs to some degree; e.g. agreements with governments, as is the case in Australia, machine-readable passports, "up-stream" capture of passenger data at the time of booking, etc.[. . .]" World Customs Organization, "Advance Passenger Information : Guidelines for Customs and Air Carriers" (2003) WCO Annex I to Doc PW0072E1 11.

A recent organizational change at ICAO, in which the administration of the security and facilitation programs was merged, recognizes formally the importance of establishing a good balance between the need for effective aviation security and the need to facilitate air travel.[32]

In fact, by transmitting data in advance to a border control agency, it becomes more and more probable to control inadmissible passengers, such as potentially high-risk passengers who have been banned into entering the State.[33] The information shared consists of identifying these individuals that could cause a potential threat to national security. As the Fourth Panel Meeting Facilitation Panel stated:

Moreover, the events of 11 September 2001 and afterwards have demonstrated that national programmes of travel document issuance and security, and the efficacy of inspection systems in controlling smuggling and illegal migration, can have a significant effect on the security of civil aviation.[34]

In addition, due to the fact that security emphasized at article 2.2 of Annex 9 on Facilitation, Annex 17 on Security also stipulates the importance of its collateral concept at the recommended practice 2.2:

> Each Contracting State should whenever possible arrange for security controls and procedures to cause a minimum of interference with, or delay to the activities of civil aviation provided the effectiveness of these controls and procedures is not compromised,[35]

which corresponds to the obligation by States for proper control set forth within the Chicago Convention at its article 13.[36] Each State can therefore exercise an effective control on the individuals crossing the border. However, the fundamental right of privacy of mankind is governed on principles of the right to be informed as to which public agency should be entitled to dispose of such information as well as the content of such data tracing versus the public's recognized right to justify under national security such a process.[37] It is therefore tantamount to conduct proper automated procedures rather than collect manually data by ground staff.

---

[32]McMunn (1996), 7.

[33]"The Facilitation programme has taken a proactive stance against law enforcement problems, particularly narcotics trafficking and travel by inadmissible passengers.[. . .] At its first meeting in 1997, the ICAO Facilitation Panel will review all of the Annex 9 provisions related to inadmissible passengers and will attempt to devise some means to implement them more effectively." McMunn (1996), at 9.

[34]ICAO Secretariat, "Facilitation Panel Fourth Meeting Information Paper" (Montreal, 2–5 April 2002), ICAO Doc FAL/4-IP/3. This paper was first introduced to the High-Level Ministerial Conference of February 2002 (1P/1).

[35]Abeyratne (1998) at 78.

[36]Supra note 1 at article 13: "The laws and regulations of a contracting State as to the admission to or departure of its territory of passengers, crew or cargo of aircraft, such as regulations relating to entry, clearance, immigration, passports, customs, and quarantine shall be complied with by or on behalf of such passengers, crew or cargo upon entrance into or departure from, or while within the territory of that State."

[37]"One of the issues as important in the API process is that the data required must be collectable by machine or already contained in the airline's system. Manual collection and data entry at the

This API summary will serve as an introduction to the concept of advance passenger information and biometrics procedures, a comparative study of its applications both by public entities as well as the financial and legal implications of such transmission of data on air carriers and border control agencies as well as these new methods of identification.

The conclusion will be to demonstrate that security and facilitation are both complimentary concepts. In fact, as Annex 9 and Annex 17 of the Chicago Convention explicitly state, the concept of aviation security and facilitation are inter-related and form an intertwined relationship: "[...] the relationship between facilitation and security at airports should not be seen as a "trade-off" or a "balance" between adversarial programmes. Rather, the enhancement of one results in enlargement of the success of the other.[38]" In addition, the composite opinion objective by ICAO's Facilitation Section in regards to security and facilitation annexes is: "[...] that legitimate-and only legitimate-traffic be boarded on aircraft, carried by air and cleared to cross international boundaries, safely and in good time."[39]

8 Definition

The concept of advance passenger information involves the capture of passport details by the carrier prior to departure and the transmission of the details by electronic means to the authorities at destination. With this capture, the authorities can screen the passengers by their databases in order to identify potentially high-risk individuals. The positive aspect is to reduce congestion at airports and consequently decrease delays in border control processing.[40]

API has begun to be for certain States a compulsory method for public authorities to manage risk prior to arrival in order to expedite clearance.[41] The

check-in desk for a scheduled flight is time-consuming and prone to errors, and or life. The foundations of "information privacy", whereby the individuals would determine when, how, and to what extent information about themselves would be communicated to others, inextricably drawing the right of control of information about oneself, is a cornerstone of privacy" Abeyratne (2001).

[38]McMunn, M.K. for ICAO Secretariat, "Facilitation And Security-Not A Zero-Sum Game" (March 1999) ICAO Doc AFCAC/ATC/4-IP at point 9.

[39]Unofficial statement given by Mary K. McMunn, Chief of the Facilitation Section at ICAO.

[40]"[...] This technique is beginning to be used by Border Control Agencies and it has the potential to reduce considerably the inconvenience and delay experienced by some travellers due to border controls." Facilitation Division-Eleventh Session, (1995) ICAO Doc FAL/11-IP/2.

[41]Refer to Recommendation Practice 3.34 of Annex 9: "Where appropriate Contracting States should introduce a system of advanced passenger information which involves the capture of certain passport or visa details prior to departure, the transmission of the details by electronic means to public authorities, and the analysis of such data for risk management purposes prior to arrival in order to expedite clearance. To minimize handling time during check-in, document reading devices should be used to capture the information in machine readable travel documents. When specifying the identifying information on passengers to be transmitted, Contracting States should only require information that is found in the machine readable zones of passports and visas that comply with the specifications contained in Doc 9303 (series), Machine Readable Travel

implementation of such a system requires a great deal of regulation as it involves data capturing and processing.

9 History

Since 1948, ICAO's Facilitation Division invoked the presence of reducing exit visas and granting a time constraint to any visa that has been given to a traveler[42] with criterias in order for public authorities to manage the number of entries of a passenger, for example, and to standardize the required information on each visas. In 1959, during the Fifth Session,[43] Rome determined additional criteria, which is not to apply different procedures that would be less favorable to the airline industry in comparison to any other means of transport.

API was first brought to life by the recommendations of adopting Annex 9 in 1963 in Mexico.[44] Following the United Nations Conference on International Travel and Tourism, a recommendation[45] was set forth in order for the UN to

---

Documents. All information required should conform to specifications for UN/EDIFACT PAXLST message formats.

[42]"8.1 (RP) In order to facilitate the unilateral and bilateral elimination of entrance visas for non-immigrants, but at the same time to provide a simplified form of control with respect to the movement of non-immigrants where such control is deemed necessary, the following uniform system should be adopted [. . .] 8.4 (RP) Each State should abolish exit visas, and reduce any other emergency exit formalities to an absolute minimum." Facilitation Division, "Final Report Of The Second Session" (Geneva, June 1948) ICAO Doc 5464-FAL/535.

[43]"3.1(ST) Governmental regulations and procedures applied to persons travelling by air shall be no less favorable than those applied to persons travelling by other means of transport.

3.2(ST) Contracting States shall make provisions whereby the procedures for clearance of persons travelling by air will be applied and carried out in such a manner to retain advantage of speed inherent in air transport.

3.3(ST) No documents other than those provided for in this Chapter shall be required by Contracting States for the entry into and departure from their territories of tourists. And other temporary visitors." Facilitation Division, "Report of The Fifth Session" (Rome, December 1959) ICAO Doc 8043-FAL/562 Recommendation A-17.

[44]Facilitation Division, "Report Of The Sixth Session" (Mexico, March–April 1963) ICAO Doc 8324-FAL/563.

[45]"Recommendation B-6: WHEREAS the UN Conference on International Travel and Tourism, to be held later in the year 1963, will consider the question of formalities to be complied with by tourists on entry and departure;

WHEREAS the provisions of Annex 9 relating to the movement of persons have been carefully developed throughout the years and have been thoroughly reviewed at the Sixth Session of the Facilitation Division, the conclusions of which will be communicated to the Secretary General of the UN Conference; and

WHEREAS it is essential that any action taken by the UN Conference should not be inconsistent with the pertinent International Standards and Recommended Practices contained in Annex 9 to the Convention on International Civil Aviation and should actually encourage States to implement that Annex;

THE DIVISION RECOMMENDS that the Council request the UN Conference to preface any recommendations it will ultimately adopt by a reference, in a preamble or otherwise, to the

introduce the implementation of the different member States to Annex 9. This conference took place in Rome and began to define concepts such as visitors, tourist and other facilitation aspects that eventually lead to the implementation of Annex 9.[46]

Furthermore, we can notice the willingness to incorporate security awareness within the concept of facilitation where a recommendation[47] stated the obligation for the air carrier to return an individual, which has been refused by the said State. With this new proposition, it is foreseeable to notice that airlines will need to be much more vigilant when verifying if a passenger can travel.[48]

The concept of API was discussed during the Tenth Session of the Facilitation Division in Montreal in 1988. It suggested a recommendation[49] that was only introduced during the Eleventh Session following a conference in Djerba in 1997, which gave background information on API and comments from Member States.

---

continuing obligations of the Contracting States of the International Civil Aviation Organization to implement the provisions of Annex 9." *Ibid* at 32.

[46]UN Conference on International Travel and Tourism, online: <http://www.oas.org/TOURISM/docnet/Iatc2en.htm> (date accessed: 15 January 2003).

[47]"3.25 (ST) Upon refusal of admission and transfer back of any person, the operator shall be responsible for promptly returning him to the point where he commenced the use of the operator's aircraft or to any other place where the person is admissible." Ibid at 40.

[48]Refer to Article 3.58 of Annex 9: "The public authorities shall without delay inform the operator when a person is found inadmissible and consult the operator regarding the possibilities for removal.

Note 1.-A person found inadmissible shall be transferred back into the custody of the operator who transported that person directly to the final destination or, where appropriate, into the custody of one of the operators who carried the person to one of the transit destinations.[. . .]"

[49]Facilitation Division, "report Of The Tenth Session" (Montreal, September 1988) ICAO Doc 9527, FAL/10 at 54 : Recommendation B-11 : "IT IS RECOMMENDED THAT:

a. Contracting States, where possible, undertake projects to examine the effects of various advance passenger information programmes (including as appropriate various manual and electronic collection and transmission methods) in facilitating the clearance of arriving passengers through the inspection processes at major international airports;
b. Where data are transmitted by Electronic Data Interchange, procedures should conform to international message standards and formats;
c. ICAO would undertake a study of Contracting States' experiences from the projects undertaken under a) above in the advance passenger information privacy issues and the facilitation and other benefits and costs, by types of programmes, for passengers, air carriers and Contracting States; ICAO should liaise with the Customs Co-operation Council and other appropriate international bodies to ensure proper co-ordination in this area, and to safeguard the interests of immigration authorities;
d. ICAO would keep Contracting States fully informed of developments; and
e. ICAO would, no later than 1992, report on the study to the Council, which would decide whether the findings and recommendations should be recommended to Contracting States."

The recommendation on the format of API was implemented within the 10th edition of Annex 9.[50]

In its report, it was stated that the Members of ICAO were concerned about privacy issues that could arise from the usage of electronic information provided by the API system.[51] It is also noted in this report that any electronic messaging should be processed under the Electronic Data Interchange [E.D.I.] format, and become international practice, therefore being common between Contracting States.

One of the WCO mission, through the Permanent Technical Committee was to develop a convention in order to adapt the changing structure of international trade and the evolution of Customs techniques and therefore facilitate States adopting national legislation. In 1973, the Council of the WCO adopted in Tokyo the Convention on the Simplification and Harmonization of Customs Procedures.[52]

The WCO's main objective is to simplify travel and create effective border control for the rapid clearance of passengers. It is stipulated in its recommended practice in the Kyoto Convention as well as in the associated benefit[53]:

> The benefit to Customs is the receipt, in advance of the arrivals of travelers, of information that will aid risk management with the objective of more precise targeting of Customs control. A benefit to travelers is that, on the basis of Customs analysis and evaluation of API, their risk status can be determined prior to arrival in the country concerned. Greater

---

[50]ICAO Secretariat, " Informal Facilitation Area Meeting in Consultation with ACI on Advance Passenger Information" ICAO Doc INF/FAL/DJE WP/11 (2 July 1997):

"2.1 Article 29 of the Chicago Convention requires every aircraft engaged in international navigation to carry certain documents, including, for passengers, "a list of their names and places of embarkation and destination". Annex 9 specifies, in Standard 2.7, the presentation of a passenger manifest document shall not normally be required, and notes that if the information is required it should be limited to the data elements included in the prescribed format, i.e. names, places of embarkation and destination, and flight details.

2.2 It should be noted that the opinion of this Standard contemplated the passenger manifest as a paper document which would have to be typed or written and delivered by hand.[. . .] It is widely recognized that in any system involving the exchange of information (automated or not), it is the collection of data which is the major expense. Increases in data collection requirements should result in benefits which exceed the additional costs. This principle was a central issue during the debate over API in the Tenth Session of the Facilitation Division (FAL/10) and the eventual adoption by FAL/11 of API systems as a Recommended Practice." (refer to Article 3.14.2 of the 10th Edition of Annex 9)."

[51]"There was, however, considerable support for both B-type Recommendations although several delegates pointed out that there would be a need for the programmes concerned to take into account the importance of the privacy of the individuals reflected in the data protection laws already adopted in many States."*Ibid* at 53.

[52]*Convention On the Simplification And Harmonization Of Customs Procedures*, [hereinafter referred to as Kyoto Convention], online: <http://www.unece.org/trade/kyoto/ky-01-e1.htm#Historica> (date accessed: 3 January 2003).

[53]Refer to the Kyoto Convention, ibid. at Annex J at article 5.5: "Recommended Practice 8: The Customs, in co-operation with other agencies and the trade, should seek to use internationally standardized advance passenger information, where available, in order to facilitate the Customs control of travellers and the clearance of goods carried by them."

precisions in Customs targeting should result in the vast majority of travelers being assessed as presenting negligible or no risk and thus subject to minimal or no Customs control on their arrival.[54]

It is also noted in the general field of applications that the Convention is aimed at developing a system of pre-clearance to utilize wanting time prior to the departure of an aircraft in order to carry out formalities, which might otherwise delay passengers upon arrival of that aircraft at destination.

10 Advance Passenger Information Guidelines

During the Eleventh Session of Facilitation Division held in Montreal in 1995, the position of the WCO, formerly CCC in 1992, was stated into guidelines for API mainly due to:

- Information Technology
- Greater co-operation between Border Control Agencies domestically;
- Greater international co-operation between Customs administrations and with other Border Control Agencies;
- Greater co-operation between Border Control Agencies and carriers.[55]

In order to fulfil the roles of the CCC, the system of API can facilitate such an information system by:

4.1.4[...] (a) Providing its Members with information on the technique of API benefits it can bring;

(b) Providing a forum in which the constraints on API can be discussed and hopefully resolved; and

(c) Seeking to jointly agreed standards with the airline industry so that API does not develop and proliferate in an inconsistent or unstructured way.[56]

In April of 2002, during a Facilitation Panel in Montreal on API, it recommended:

The usage of API for immigration, quarantine and aviation security (AVSEC) applications to customs;

The internet or other PC-based systems and wireless technologies should be considered for the exchange of data rather than specify UN/EDIFACT syntax for data interchange;

API should be part of a border system management, machine readable passports with electronic visas, automated entry/exit records instead of embarkation or disembarkation cards and as well as interoperability of API systems with other States;

[54]*Ibid.*

[55]Facilitation Division, "Eleventh Session Information Paper on Advance Passenger Information (API) Guidelines adopted by the WCO" (Montreal, April 1995) ICAO Doc FAL/11-IP/2 at point 3.

[56]Facilitation Division, "Eleventh Session Information Paper on Advanced Passenger Information (API) Guidelines adopted by the WCO" (Montreal, April 1995) ICAO Doc FAL/11-IP/2 at point 1.3.

Applicable Standards and Recommended Practices (SARPs) should leave the possibility of including biometrics into article 3.34 of Annex 9 (11th Edition);

ICAO should measure the programme's success in operational efficiency and reduction of airport congestion.[57]

The API topic became more and more a priority during the 1993/1994/1995 triennium and IATA comprehended in a greater capacity the necessity for API implementation.[58] IATA and the WCO formally introduced the formal WCO/IATA guidelines in 1993 following the Working Paper presented by the ICAO Secretariat during the Eleventh Session.[59] In the preamble of the guideline,[60] it stipulates that because of the increase of passenger traffic, Customs are strained to process much more additional data when it clears border control. Furthermore, in order to prevent increase in delays, the need for efficient automated processing has become a necessity. This position has also been supported by IATA.[61] Where API should be considered uniform electronic text capturing by the UN/Edifact PAXLST Messaging system. In fact:

---

[57]*Ibid* at point 4.1.4.

[58]Facilitation panel presented by the Secretariat, "Advance Passenger Information Further Development of ICAO Doctrine" ICAO Working Paper FALP/4-WP/2 (Montreal, April 2–5 2002).

[59]"4.2.4 Furthermore, given the practical and cost constraints of data capture and transmission, limiting the required information to that which can be captured by machine reading passports and visas, augmented by basic flight details, is a prerequisite. To this end, IATA sees particular benefit in co-operating with the CCC to define the data and message sets for API within the UN/EDIFACT PAXLST development, and in establishing jointly agreed principles which can expand the benefits of automating and integrating all elements of the passenger process from origin to destination." See supra note 9 at point 4.2.4.

[60]*Ibid* at clause no. 4: The Customs Co-operation Council recommended a standardization for API interoperability and an objective to control costs to airlines. It also : "[. . .] requests Members of the United Nations Organization or its specialized agencies, and Customs or Economic Union which accept this Recommendation to notify the Secretary General of the Council of the date from which they will apply the Recommendation and of the conditions of its application. The Secretary General will transmit this information to the Customs administrations of all Members of the United Nations Organization or its specialized agencies and to Customs or Economic Unions which have accepted this Recommendation."

[61]Ibid. at attachment clause no.5: "IATA has constantly sought to eliminate unnecessary forms and procedures min international air transport and the abolition of the passenger manifest has been an important policy objective for the Association. Recent opportunities to automate government control processes have, however, let to a close look at the concept of API and its potential for facilitation improvements.

Collection of passenger details at departure presents a problem of additional workload for airlines at point in the system where staff and facilities are frequently already stretched to maximum capacity and beyond. Consequently, carrier support of API depends heavily on there being truly realizable benefits for airline passengers on arrival at destination.

Furthermore, given the practical cost constraints of data capture and transmission, limiting the required information to that which can be captured by machine reading passports and visas, augmented by basic flight details, is a prerequisite. To this end, IATA sees particular benefit in co-operating with the CCC to define the data and message sets for API within UN/EDIFACT PAXLST development, and in establishing jointly agreed principles which can expand the benefits of automating and integrating all elements of the passenger process from origin to destination."

API permits a very thorough and rigorous screening of inbound passengers to be carried out, targeting those that present the highest risk and allowing for the faster throughput of low risk[62]

IATA also notes the necessity to create a limitation of standardization to identify data would prevent abuse in the transfer of data. As a suggestion, the data pertaining to the flight should consist of :

- Flight Identification;
- Scheduled departure date;
- Last place/port of call of aircraft; and
- Place/port of aircraft initial arrival.

11 Contracting States' Positions

11.1 The United States Legislation pertaining to API and PNR

Due to the most recent events of 2001, President George Bush signed a new *American Transportation & Security Act* on November 25th 2002 making mandatory API transmission and PNR access to all passengers arriving in the United States. The Department of Homeland Security will therefore ensure that the air carriers, airobia and other governmental agencies comply with this new bill.

According to the WCO/IATA guideline,[63] there is a stipulation that API transmissions should originate from the last port before entering into the port of arrival. However, the US concluded agreements between different States that seem to violate the general guidelines of IATA and the WCO[64] in the sense that under these guidelines no data from an API transmission would only be provided to the port of entry. Under this Act, API data submissions to the US now have been made mandatory on flights bound for another State. These agreements find extra-territorial applications of American legislation where it imposes to another State submitting API as well as PNR passenger information. According to the US Customs Service, it implemented a Canada Smart Border/30 Point Action Plan, better known as the Manley Ridge Agreement. According to this plan adopted in

[62] *Ibid.* at attachment clause 9.

[63] *Ibid.* at attachment clause 8.1.5: "It should be noted that API transmissions will contain data for passengers carried into a country (initial place/port of arrival) from the last place/port of call of that aircraft abroad. API transmissions will not provide information of passengers' previous flights or ports of call before joining the flight at the last foreign port of call. Neither will API transmissions provide information on onward flights to other countries. Put simply, the API transmission contains only details of passengers carried from last port of call to the first port of call in the country of arrival without regards for the passengers´ initial point of departure or their ultimate destination."

[64] Refer to the "US-Mexico Border Partnership Action Plan", online: <http://www.whitehouse.gov/infocus/usmxborder/22points.html> (date accessed: 17 December 2002).

December 2001, the United States and Canada agreed to share API and passenger name records as of spring 2003.[65]

The new American Transportation & Security Act stipulates as section 115 the required information from each flight prior to departure and arrival in the United States:

> A passenger and crew manifest for a flight required under paragraph (1) shall contain the following information:
> The full name of each passenger and crew member;
> The date of birth and citizenship of each passenger and crew member;
> The sex of each passenger and crew member;
> The passport number and country of issuance of each passenger and crew member if required for travel;
> The United States visa number or resident alien card number of each passenger and crew member, as applicable;
> Such other information as the under Secretary, in consultation with the Commissioner of Customs, determines is reasonably necessary to ensure aviation safety.[66]

Furthermore, according to sub-section three and four of the same section on Passenger Manifests, the Customs service also can prescribe the time frame it can expect to receive electronic messaging from air carriers as well as passengers name records and all pertinent identification necessary for screening.

As a response to this new API/PNR data transmission, American Airlines and Continental Airlines have agreed to comply with the new legislations but have requested the US Customs Service to review its penalty procedures if issued erroneously.[67]

11.2 The Canadian Position

The *Canadian Immigration and Refugee Act*[68] came into force as of June 2002 dealing with required documentations and obligations on air carriers in conjunction with Part 17 of the Regulations issued by Citizenship and Immigration Canada.[69]

As first point of interest, in section 148 of the IRPA, air carriers are required not to carry any person that is not in possession of required documents of travel. In the

[65]Refer to the "US–Canada Smart Border/30 Point Action Plan", online: <http://wwww.whitehouse.gov/news/2002/12/20021206-1.html> (date accessed: 17 December 2002): "The United States and Canada have agreed to share Advanced Passenger Information.

[66]One Hundred Seventh Congress of the United States, Aviation and Transportation Security Act", HR 5005 EAS, Chapter 1 of title 49 S. 1447 at section 115 sub-section 2.

[67]Unofficial letter by American Airlines dated February 28th, 2002 and unofficial letter by Continental Airlines dated February 28th, 2002.

[68]*Immigration And Refugee Protection Act*, L.c. 2001, c.27 [hereinafter referred to as: IRPA].

[69]*Citizenship and Immigration Canada* [hereinafter referred to as: CIC].

event that such obligations are not fulfilled, section 278 describes the different penalties, which will be imposed to the transportation companies.

Within Part 17 of the implemented regulations by the CIC, section 269 contains relevant advance passenger information legislation including:

> 269: Details data elements that will be required under the Canadian Advance Passenger Information programs, including;
> Surname, first name and initial(s) of any middle names;
> Date of birth;
> Country that issued a passport or travel document, the citizenship or nationality of the airobia;
> Gender;
> Passport number or, if a passport is not required, the number on the travel document that identifies them; and,
> Reservation record locator or file number.
> This part also provides for government access to airline reservation systems at 269(2), and seemingly indicates that the government shall have access to any record at any following its creation.[70]

Furthermore, it is important to note that paragraph 2 of the same legislation includes a disposition where any electronic messaging follows the existing UN EDIFACT PAXLST format.

According to the Privacy Commissioner of Canada, complying with the United States new interim rule would infringe upon fundamental privacy rights and could possibly be used for other purposes, such as verifying income tax and other criminal information. It views this as a comparison as a "Big Brother" database.[71] However, it should be mentioned that when a passenger travels, he or she implicitly gives up a certain amount of privacy in order to receive clearance at different border control authorities.

As for the air carrier, a leading charter Montreal based airline, Air Transat A.T. requested from the US Customs a delay until December 15th, 2003 in order to fully comply with the new Interim Rule.[72] The airline's representative in government affairs indicated that the airline does not possess at this time any central reservation system as most of its bookings are done through tour operators and other travel agencies. The costs relating to changing to a fully electronic method would represent an investment of 1.3 million dollars. It further criticized the deadlines imposed by the United States:

> We trust that such best efforts to date will be properly considered and that the Final Rule will not unduly penalize or burden smaller or less sophisticated air carriers such as Air Transat, in terms of passenger reservation and seat inventory management, with an unreasonably expeditious effective date.[73]

[70]*IRPA*, *Part 17 Transportation*, supra note 55 at section 269.

[71]"Privacy Commissioner of Canada : News Release", online: <http://www.privcom.gc.ca/media/nr-c/02_05_b_020926_2_e.as> (date accessed: 8 November 2002).

[72]Unofficial letter dated January 24th 2003 [not published].

[73]*Ibid*. at page 4.

As a response to the United States Customs Directive on API, the legal director of ICAO gave an opinion and confirmed that although punitive recourses are at this time being imposed to different carriers, it appears to follow the guidelines set forth by the Chicago Convention:

> Another essential feature of API as an effective facilitation measure is the accuracy of the information provided. The accuracy of the data contained in a Passenger Manifest is an essential requirement of this document, whether it is transmitted in advance (API) or not, and such requirement should be equally enforceable. Based on this principle, the requirement that the data provided by airlines must comply with an increasing percentage of accuracy only means that the US authorities intend to reduce their degree of tolerance of errors, possibly aiming at tolerance zero which is consistent with such principle. Punitive measures against airlines failing to comply with the required accuracy, the level of severity of such measures and the empowerment of the authorities to apply them are matters of national policy and law, provided that the applicable measures are enforceable within the territory of the State concerned, which appears to be the case [. . .].
>
> In view of the foregoing, it is concluded that the US Customs Directive on API, in spite of worsening the airline's burden, is consistent with the relevant provisions of the Chicago Convention and its Annex 9, and therefore does not raise extraterritoriality issue*s*."[74]

11.3 The United Kingdom Position

A standing committee on the *Draft Immigration* (*Leave to Enter Remain*) *Order 2000*[75] was first introduced by Mrs. Barbara Roche, Minister of State, Home Office. During this parliamentary discussion, it was made clear that API systems were to have a dual positive impact: not only would it permit a rapid clearance process within an airport for the possibility for officials to detect the presence of potentially high-risk individuals. Furthermore, according to Roche, this will in no way diminish the role of customs officers who will have the possibility of examining each passenger as well as their baggage and other belongings.[76] She also stated that the immigration officer could still at any point of border control monitor the passenger traffic and inspect the traveller.[77]

---

[74]Weber, L., "Inter-Office Memorandum on United States Customs Directive on Advanced Passenger Information"., 7 June 2002 [not published].

[75]House of Commons Standing Committee on Delegated Legislation, *Draft Immigration* (*Leave to Enter and Remain*) *Order 2000*, online: <http://www.hmso.gov.uk> (date accessed: 4 March 2003).

[76]*Ibid.* at p.3: " The power to grant or refuse leave to enter before a person arrives in the UK has two benefits. Advance passenger information could pre-clear certain low-risk school groups and recognized reputable tour groups, thereby speeding their progress through immigration control and removing the need for detailed, individual examination on arrivals. Alternatively, we might send immigration officers overseas, with the agreement of the Government concerned, to address particular pressure points. It also allows us to take advantage of future technological developments such as biometrics. Such measures will benefit the travelling public, carriers and the immigration service."

[77]*Ibid.* at page 3: "As I said, the role of the immigration officer is not diminished, as he or she can still examine a person with continuing leave."

The UK, under this new legislation has enabled many different enforcement agencies in order to collect an intelligence map of potential high-risk individuals and prevent entry.[78] Furthermore, under the assurance of such agencies, both carriers and government agencies determined that such API legislation would be applied through a very rigorous and fair process[79] and compliance would have to be effective within the next 6 months period of time.

We can as well study the position of the major UK carrier, British Airways, towards the new American legislation pertaining to API and PNR access. In a letter in early March 2002,[80] it informed the Office of Regulations and Rulings of the United States of its support towards the API system and its compliance by a Memorandum of Understanding [hereinafter referred to as : MOU] with the US Customs of 1998. This MOU consisted of voluntary release of passenger information to the US. According to this new *Aviation and Transportation and Security Act*, British Airways believes that an automated version should be considered in order for the collection of data such as PNR information so long as it is not stored manually and complies with the machine readable information.

British Airways also considers this as the best method of transferring passenger information from the airline system to the government. It considered the automatic PNR release and ruled that it would not affect the *UK Data Protection Act*.[81] BA's pretension is that the carrier should not be obliged to change any reservation system, which could encounter additional costs or in the least reduce them to a strict minimum.[82] The WCO also considered this position to be viable as it announced in its recommendation that information should be kept to a strict minimum or it otherwise, such operations become time and cost consuming:

> 8.2.1 Perhaps the most critical aspect of API is the means by which the data to be transmitted to the Border Control Agencies in the destination country is captured. Data capture can be costly, time consuming, labour intensive and error prone. The capture of data concerning departing passengers at the airport of departure introduces a delay in the check in process that could, if not managed properly, offset the potential advantage to passengers provided by efficient API applications. If the check-in process in unduly prolonged, then API will simply shift much of the delays and congestion away from the arrival area to the

---

[78]Regulatory Impact Assessment : Introduction to Extended Powers of Information Collection On Passenger and Goods, Schedule 7 to the Terrorism Act 2000 (Information) Order 2002, online: <http://www.homeoffice.gov.uk/atoz/pax_and_goods.pdf> (date accessed: 8 November 2002) at point 12: "The measure will enable the police to build an intelligence picture which will allow them to target and track terrorists in a way that has become essential in the aftermath of September 11 and the subsequent ongoing campaign against the threat of global terrorism."

[79]*Ibid.* at point 39: "We are confident that the enforcement agencies would apply the legislation fairly, proportionately and appropriately requesting the information and the police utilizing it. This approach has been confirmed by representatives of the police at meetings with the carriers."

[80]British Airways letter dated March 1rst 2002 [not published].

[81]*UK Data Protection Act*, online: <http://www.legislation.hmso.gov.uk/acts1998/19980029.htm> (date accessed: 10 January 2003).

[82]Refer to supra note 9 at clause 6.5.1 and 6.5.2.

departure area. It is vital therefore that the effect of API on the check-in process is kept to the absolute minimum.[83]

However, the WCO also claims that API can also reduce staff costs because of this automated process that can therefore bring some form of saving for the air carrier.[84]

Furthermore, such information should not permit access to any other passenger that is not on a flight bound to the US.[85]

This air carrier also had specified its concerns to the US Customs in a previous correspondence whereas the American legislation gave no assurance that other information pertaining to that were not bound to the US would be transmitted.[86]

Virgin Atlantic also expressed grave concerns over PNR transmission as some private information relating to passenger's file are private and would contravene the UK Data Protection Act, according to local management.[87] According to authorities, it is now necessary under the UK Data Protection Act that recording of personal data of passengers do not leave the territory of the European Economic Area unless enough protection can be assured. In order to achieve this, the US Customs Service Agency would have to adopt the *Safe Harbour Principles*, set forth by the European Commission, under EU Directive (94/46/EC).[88] In this Directive, European companies can only send out information to any foreign

[83]*Ibid.* at clause 8.2.1.

[84]*Ibid.* at clause 6.9.3. and 6.3.

[85]Refer to supra note 59 at attachment A:

> "(3) A general request to oblige the carrier to give access only to passenger name record information relating to passengers whose itineraries include at least one flight operated to or from or within the United States. In the event that carrier's systems are not designed or configured so as to allow such access without also giving access to information about other passengers, the Customs Service shall adopt procedures or take other appropriate measures to ensure that its officers do not access information relating to such other passengers. In addition, prior to implementing any online processes, the Customs Service will agree to appropriate security protocols with the carrier.
>
> (4) No carrier shall be obliged to change or modify its computer systems (hardware or software) in order to comply with a general or specific request, unless the changes or modifications and the allocation of the cost of making them are agreed in advance between the carrier and the Customs Service."

[86]British Airways letter date August 26th, 2002[not published]: "There appears to be nothing in the Interim Rule to protect the security and integrity of the carrier's systems. This is essential for British Airways to have confidence that cooperation will protect the integrity of its departure control systems and the legitimate rights and interests of its passengers. The Rule should provide such protection and British Airways respectfully requests the Customs Service agree to a security protocol prior to any direct systems access [...] British Airways requests that the agreements be finalized before access is activated."

[87]Unofficial letter by Virgin Atlantic Airways Ltd. dated August 30th, 2002.

[88]Frashfields Bruckhaus Deringer. "Data Protection", online: <http://www/freshfields.com/practice/ipit/publications/22367.pdf> (date accessed: 6 February 2003).

country outside of the community if it should so correspond to a reasonable protection of sharing of information.

11.4 Safe Harbour Principles

In order to fully comply with the EU Directive (95/46EC), it introduced seven principles, otherwise referred to as Safe Harbour Principles:

- Notice must be given to individuals informing them of the purposes for which their data has been collected and how it will be used;
- Choice must be offered to individuals, allowing them to choose (opt out) whether and how their personal information is disclosed to third parties or used for purposes which differ from the ones which were originally notified;
- Onward transfer of personal data by organizations to third parties must be consistent with the principles of notice and choice;
- Security of personal data must be maintained using reasonable precautions;
- Data integrity must be ensured so that personal data is relevant for the purposes for which it used, not processed in ways which are incompatible with the purposes for which it has been collected and steps taken to ensure that it remains accurate;
- Access to personal data must be maintained so that individuals can ensure that it is corrected or deleted where inaccurate;
- Enforcement should be available through independent recourse mechanisms to deal with complaints, disputes and remedies, and provide sufficiently rigorous sanctions to ensure compliance.[89]

BA's charter counterpart, Britannia, had mentioned to the American authorities that it was not able to comply with the interim rule on passenger name records requirements because it did not process the necessary computerized reservation system.[90] Furthermore, in regards to API, it urged the US Customs Agency to waive applicability of data transmission on flights that are not bound for the US. It should also consider reducing the penalties imposed on air carriers if compliance cannot be performed on time. The time frame allotted to airlines in regards to changing their reservation systems is also an aspect that should be considered for API transmission.

11.5 The Australian Position

The Australia Immigration and Customs have already implemented API systems in order to accelerate the process and enhance border control.[91] In order to achieve these goals, Australia implemented the Advance Passenger Processing, hereinafter referred to as APP. This system provides a rapid clearance by the participating carriers. In essence, at foreign check-in points, the airobia's passport is read and a

---

[89] *Ibid.* at page 1.

[90] Unofficial letter by Air 2000 Limited (August 26, 2002).

[91] Manning, J. (Australian Delegate), "Facilitation Panel Fourth Meeting Information Paper", (Montreal, 2–5 April 2002) ICAO Doc FAL/4-IP/8.

magnetic card is then given if authorization to travel is granted with the individual's details enabling him or her to use the "Express Lane" upon arrival in Australia.[92] The government has also amended their national legislation in order to permit capture of data of API and PNR without infringing on privacy rights. This APP system works in cooperation with the Electronic Travel Authority (ETA), which is a communications network. When data is captured, it is sent through the ETA system that verifies the validity of the visa for those passengers who require such travel documents as well as the status of Australian and New Zealand passports.

The APP/ETA system has been accepted by air carriers as it meets the individual need of each and every one of them. As some carriers have voluntarily participated to this plan, it is of the government wish to implement mandatory procedures for APP.[93] According to Australian Customs Service, APP gives a quantifiable reduction in undocumented travels and therefore reducing the possibility of being imposed fines by other Contracting States.

Qantas Airways, Australia's leading air carrier, expressed a similar concern to the one of British Airways but asked the US Customs Agency for further precisions on the interim rule pertaining to PNR information. It has concerns pertaining to the legislation when it permits sharing of all relevant information to the different Federal Agencies.[94] Furthermore, Qantas has asked the authorities to sign an agreement in order to prevent the US Customs Service[95] to send such data to an undetermined amount of agencies.

11.6 The German and Swiss Positions

According to present privacy laws, certain data is protected by the *Federal Data Protection Act* ("*Bundesdatenschutzgesetz*", *BDSG*)[96] and requires special permission from each airobia before permitting access to such information by other

[92]*Ibid.* at clause 3.2.5 and 3.2.6: "At check-in, the airline prints the passenger's bio data and flight number on a special Australian Incoming Passenger Card with the word "EXPRESS" indicated. The card also has a magnetic strip that is coded with an identifier to retrieve that data on arrival in Australia.

On arrival in Australia, the passenger will be directed to the appropriate processing lanes by use of dynamic signage and Customs marshals who are on-hand. APP passengers using the Express lanes are expected to be cleared in about half the time of other passengers who are not APP."

[93]Permanent Technical Committee, "Review of the WCO/IATA Guidelines on Advance Passenger Information" WCO Doc PW0045E1 (Brussels, August 20th, 2001).

[94]*Ibid.* at clause 4.2.

[95]Qantas Airways letter dated August 22, 2002 [not published]: "Prima facie, Qantas has not identified any incompatibility between USCS Passenger Name Record (PNR) requirements and Australia's national protection laws. However the statement in the CFR that "PNR information that is made available to Customs electronically may, upon request, be shared with other Federal Agencies", requires further clarification. Specifically, whether or not carriers will be notified when and with whom this information is being shared and how the integrity of the data will be maintained during this process."

[96]*Bundesdatenschutz*, online: < http://www.datenschutz-berlin.de/recht/de/bdsg/bdsg1.htm#abschl > (date accessed: 17 January 2003).

States. Furthermore, all data must be deleted from any banks after a certain amount of time. In response to this legislation, Germany's leading air carrier, die Deutsche Lufthansa Aktiengesellschaft, had informed the US Customs Service of these legislative impediments and was awaiting assistance in order to comply. According to the in-house legal counsel department, it appears that this could be implemented throughout the year of 2003.[97] Unless clear amendments can be made, violations of the Federal Data Protection Act can be of substantial financial and legal consequence.[98]

As for the Swiss Government, personal data is regulated by the *Data Protection Act* (*DSG*),[99] all transmissions must be transferred in good faith and must be done in a secure manner. As the API transmission is currently used is relevant to current border inspections, it therefore does not go against these legal previsions.

As for Swiss International, Switzerland's main air carrier, API does not cause any infringement on the DSG, however the compulsory PNR unless certain conditions are meet could cause legal consequences to the carrier:

> The unlimited access by a third party in a foreign jurisdiction to the entire PNR data of a Swiss air carrier, without legal safeguards described above, turns out to cause major legal problems for the carrier concerned.
>
> However, provided that data can be restricted to PNR data on in–/–outbound US-flights, SWISS might be able to comply with national data protection laws when providing PNR access to US Customs. Compliance with Swiss and European Data Protection law could be achieved, if (a) the air carrier receives permission from the Swiss National Data Protection Officer and (b) obtains the required guarantees from the US authorities (see Point 2.2 above), eventually by applying the "Safe Harbour" principles. Furthermore (c), the air carrier would have to change its booking procedures by asking the passenger for additional data and an explicit consent to make this data available to U.S. Customs and other explicitly named U.S. authorities.[100]

Swiss International AirLines also raised the matter of implementing a filter system in order to protect a leakage of information to other authorities in the US. This filter concern was also brought up by the Bundeskriminalamt but is currently being resolved by the creation of new biometric procedures slowly being introduced in Frankfurt's airport that facilitates the creation of such a filtering data processing that would not infringe on any federal data protection legislation.[101]

---

[97]Unofficial letter dated August 30th 2002 by the Deutsche Lufthansa Aktiengesellschaft: "Implementation by Lufthansa in the first quarter of 2003 appears feasible, provided that the present legal issues can be resolved."

[98]*Ibid.* at page 2: "Administrative offences are applicable and punishable by fines up to Euros 250,000.00 to anyone who, whether intentionally or through negligence, collects or processes personal data which are not generally accessible without authorization (Section 43 BDSG); additionally, certain violations of this law can also carry criminal penalties of up to 2 years imprisonment and/or fines up to Euros 250,000.00 per offense (Section 44 BDSG).

[99]*Swiss Federal law On Data Protection*, online: <http://www.datenschutz-berlin.de/recht/de/bdsg/bdsg1.htm#abschl> (Date accessed: 5 March 2003).

[100]Unofficial letter by Swiss International Air Lines dated August 26th, 2002 [not published].

[101]Unofficial interview with Dr. Edgar Friedrich, Bundeskriminalamt, Wiesbaden Germany in February of 2003.

11.7 The Mexican Approach

As of the present time, Mexico submits all API information on passengers and crew on all international flights and intended as of July 2002 to submit to the United States to a minimum of 95 % sufficiency all information in the UN-EDIFACT messaging format. As a counterpart, Mexico also plans to fully request API information and penalize air carriers that are either late or not submitting such information. Contrary to the US, it does not plan to request any passenger name record other than the Record Locator reference to be used upon request.

Mexico also signed an agreement with the United States, the *Smart Border 22 Point Agreement*, stipulating that on a voluntary basis, Mexico would exchange some information with the United States on a mutual level in order to prevent illegal migration and detection of high potential risk passengers. According to the US Customs Service, there is presently exchange of information on international flights bound for Mexico even though the port of arrival is not the Unites States. For example, at this time, a flight from Frankfurt to Mexico City non-stop may have to submit to the US API and PNR information on its passengers.

The Instituto Nacional De Migracion proposed an electronic database collecting information of passengers when making a reservation and together with ICAO/ IATA and the Simplifying Travel Procedures established a data processing that will be later discussed regarding the understanding of biometric procedures.[102]

Other airlines have as well expresses grave concerns at the new Final Rule RIN 1515-A06 on Passenger Name Record Information Required for Passengers on Flights in Foreign Air Transportation To or From the United States.[103] In fact, according to VARIG, Brazil's leading air carrier, PNR violates the Brazilian Constitution where unless express authorization is given by the competent authority, it cannot comply with this new legislation.[104]

IATA, the International Air Transport Association, which represents 274 member airlines noted that considerable discussions should continue to be held with the US Bureau of Customs and Border Protection in order to assure its carriers that privacy laws are being complied with. It founded its remarks on the *EC Directive (95/46EC)*[105] that regulates the processing of personal data for all

---

[102]Secretaria De Gobernacion, Instituto Nacional De Migracion, "Technical Specifications INM Fast-Track" Confidential INM Presentation [not published].

[103]Passenger Name Record Information Required For Passengers On Flights In Foreign Air Transportation To Or From The United States, 67 Fed. Reg. 42710 (June 25, 2002).

[104]Unofficial letter by Varig's legal counsel, Mrs. Constance O'Keefe dated September 18th 2002 [not published]: "Due to Constitutional provision, information contained in air travel reservations, which is of a confidential nature, can only be disclosed upon written request by competent public authorities, by public administrative agencies, by an individual passenger—with proper identification—or by a legal representative duly authorized by the passenger."

[105]*EC Data Protection Directive (95/46EC), Protection of the individuals in relation to the processing of personal data*, online <http://wwwdb.europarl.eu.int/oeil/oeil4.Res213> (date accessed: 5 march 2003).

countries falling under the European Union. According to IATA, with the EC Directive, if the United States would adopt the Safe Harbour Principles, it would give sufficient protection for other States and their air carriers to comply without being held liable for data transmission only if all agencies of the US receiving such data also adopt such principles. Furthermore, the US Customs Service should:

- Self-certify under the Department of Commerce "Safe Harbour" Principles or develop and implement self-regulatory data privacy policies that conform to those Principles;
- Communicate that self-certification or privacy policy development to all governments having data privacy legislation adopted in accordance with the EU Directive;
- Provide guarantees that limit sharing of data obtained through access to airline systems only to those agencies that have self-certified under, or fully adopted the "Safe Harbour" principles;
- Limit its access to "read only" capability and provides assistance in blocking illegal outside access; and,
- Provide assurances to governments and to carriers alike that it will limit access to information pertaining only to those flights touching U.S. territory."[106]

In conclusion, although many air carriers deem that potential liability could be foreseeable, it is to be noted that under the Chicago Convention, a State has the right to request information in order for proper border control to be established.[107] Therefore, as national policy of a member State has the right to infringe upon others requesting clearance into their sovereign State, it can request or infringe upon another its principles for proper border control.[108]

References

Abeyratne R (1998) Aviation and regulatory issues. Ashgate, London

Abeyratne R (2001) The exchange of airline passenger information-issues of privacy. Comm Law J 6(5):153

Abeyratne RIR (2002a) Intellectual property rights and privacy issues: the aviation experience in API and biometric identification. J World Intellect Property 5(4):631–650

Abeyratne R (2002b) Attacks on America-privacy implications of heightened security measures in the United States, Europe and Canada. J Air Law Comm 67:86

[106]Unofficial letter by IATA dated August 26th, 2002 [not published].

[107]Chicago Convention-Art.13: "The laws and regulations of a contracting State as to the admission to or departure from its territory of passengers, crew or cargo of aircraft, such as the regulations relating to entry, clearance, immigration, passports, customs, and quarantine shall be complied with by or on behalf of such passengers, crew or cargo upon entrance into or departure from, or while within the territory of that State."

[108]Chicago Convention-Art.1: "The contracting States recognize that every State has complete and exclusive sovereignty over the airspace above its territory."

Abeyratne (2003) Profiling of passengers at airports - imperatives and discretions. Eur Transp Law XXXVIII(3):297–311
Honig JP (1956) The legal status of aircraft. Martinus Nijhoff, The Hague
McMunn MK (1996) Aviation security and facilitation programmes are distinct but closely intertwined. ICAO J 51:9
Morganthau H, Thompson KW (1950) Principles and problems of international politics. Knopf, New York
Shaw MN (2003) International law, 5th edn. Cambridge University Press, London

Article 32
Licenses of Personnel

(a) **The pilot of every aircraft and the other members of the operating crew of every aircraft engaged in international navigation shall be provided with certificates of competency and licenses issued or rendered valid by the State in which the aircraft is registered.**
(b) **Each contracting State reserves the right to refuse to recognize, for the purpose of flight above its own territory, certificates of competency and licenses granted to any of its nationals by another contracting State.**

Contents

1 Ensuring Competence

At its 21st Session the Assembly (Montreal, 24 September to 15 October 1974) adopted Resolution A21-21 (Consolidated Statement of Continuing Policies and Associated Practices Related Specifically to Air Navigation), Appendix G of which stated that certificates of airworthiness and certificates of competency and licences of the crew of an aircraft issued or rendered valid by the Contracting State in which the aircraft is registered shall be recognized as valid by the other Contracting States for the purposes of flight over their territories, including landings and take offs, subject to the provisions of Articles 33 and 32(b) of the Chicago Convention.

Appendix G of Assembly resolution A37-15 on Certificates of airworthiness, certificates of competency and licences of flight crews *states:*

> *Whereas* Article 33 of the Convention does not explicitly define the purposes for which recognition is to be accorded to certificates and licences;
>
> *Whereas* several interpretations exist as to whether or not there is any obligation on Contracting States to recognize certificates and licences issued or rendered valid by other Contracting States pending the coming into force of SARPs applicable to the aircraft or airmen involved; and
>
> *Whereas* with respect to certain categories of aircraft or classes of airmen, it may be many years before SARPs come into force or it may be found most practicable not to adopt SARPs for some categories or classes;

DOI 10.1007/978-3-319-00068-8_33,

The Assembly resolves that:

1. certificates of airworthiness and certificates of competency and licences of the crew of an aircraft issued or rendered valid by the Contracting State in which the aircraft is registered shall be recognized as valid by other Contracting States for the purpose of flight over their territories, including landings and take-offs, subject to the provisions of Articles 32 (b) and 33 of the Convention; and

2. pending the coming into force of international Standards respecting particular categories, classes or types of aircraft or classes of airmen, certificates and licences issued or rendered valid, under national regulations, by the Contracting State in which the aircraft is registered shall be recognized by other Contracting States for the purpose of flight over their territories, including landings and take-offs.

Standards and Recommended Practices for Personnel Licensing were first adopted by the Council on 14 April 1948 pursuant to the provisions of Article 37 of the Convention on International Civil Aviation (Chicago 1944) and designated as Annex 1 to the Convention. They became effective on 15 September 1948. The most recent amendment to Annex 1, was Amendment 168 (Annex 1, Tenth Edition), adopted by the Council on 23 February 2007. The amendment concerns: the replacement of the approach and area radar control ratings by approach and area control surveillance ratings to reflect the fact that surveillance systems are not limited to radar; the harmonization of the Human Factors knowledge requirements for air traffic controllers with those recently adopted as part of Amendment 167 to Annex 1 for flight crew; the applicability of the existing Standards on approved training for flight crew (Annex 1, 1.2.8 and Appendix 2) to the approved training required for the air traffic controller licence and ratings; and new provisions for student air traffic controllers receiving instruction in an operational environment.

Annex 1 contains Standards and Recommended Practices adopted by the International Civil Aviation Organization as the minimum standards for personnel licensing. The Annex is applicable to all applicants for and, on renewal, to all holders of the licences and ratings specified herein. The ICAO Council has decided that, in principle, amendments affecting existing licensing specifications are applicable to all applicants for, and holders of, licences but, in considering their application to existing holders of licences, the assessment, if necessary, by re-examination of the knowledge, experience and proficiency of individual licence holders is left to the discretion of Contracting States.

As long as air travel cannot do without pilots and other air and ground personnel, their competence, skills and training will remain the essential guarantee for efficient and safe operations. Adequate personnel training and licensing also instil confidence between States, leading to international recognition and acceptance of personnel qualifications and licences and greater trust in aviation on the part of the traveller. Standards and Recommended Practices for the licensing of flight crew members (pilots, flight engineers and flight navigators), air traffic controllers, aeronautical station operators, maintenance technicians and flight dispatchers, are provided by Annex 1 to the Chicago Convention.

Related training manuals provide guidance to States for the scope and depth of training curricula which will ensure that the confidence in safe air navigation, as intended by the Convention and Annex 1, is maintained. These training manuals also provide guidance for the training of other aviation personnel such as aerodrome emergency crews, flight operations officers, radio operators and individuals involved in other related disciplines.

Today's aircraft operations are so diverse and complex that protection must be provided against the possibility, however remote, of total system breakdown due to either human error or failure of a system component. The human being is the vital link in the chain of aircraft operations but is also by nature the most flexible and variable. Proper training is necessary so as to minimize human error and provide able, skilful, proficient and competent personnel. Annex 1 and ICAO training manuals describe the skills necessary to build proficiency at various jobs, thereby contributing to occupational competency. The medical standards of the Annex, in requiring periodic health examinations, serve as an early warning for possible incapacitating medical conditions and contribute to the general health of flight crews and controllers.

The Human Factors programme addresses known human capabilities and limitations, providing States with basic information on this vital subject as well as the material necessary to design proper training programmes. ICAO's objective is to improve safety in aviation by making States more aware of, and responsive to, the importance of human factors in civil aviation operations. Licensing is the act of authorizing defined activities which should otherwise be prohibited due to the potentially serious results of such activities being performed improperly. An applicant for a licence must meet certain stated requirements proportional to the complexities of the task to be performed. The licensing examination serves as a regular test of physical fitness and performance ensuring independent control. As such, training and licensing together are critical for the achievement of overall competency.

One of ICAO's main tasks in the field of personnel licensing is to foster the resolution of differences in licensing requirements and to ensure that international licensing standards are kept in line with current practices and probable future developments. This is ever more crucial as the flight crew will be exposed to increasing traffic density and airspace congestion, highly complicated terminal area patterns and more sophisticated equipment. To accomplish this task, Annex I is regularly amended to reflect the rapidly changing environment.

2 Cultural Diversity of Crews

Licences of personnel merely attest to the technical qualifications of crew members. They do not speak to the ability of crew members to communicate with each other and with air traffic control. Traditionally, industrial accidents have gone through three phases of causation. The first phase involved regular instances of technical malfunctions or system failures. This phase was followed, after technological development had reached a reasonable level of sophistication, by the second phase dominated by human error. The third phase was known as the sociotechnical period which involved human interaction with technology, which in turn led to the phase we are in at the present time which one commentator[1] calls the "organizational culture period" wherein operators perform as a coordinated team according to a particular safety culture instilled in them by their employers. This evolutionary pattern also applies to the air transport industry.

The need for operators to perform as a coordinated team has become increasingly apparent with globalization, where pilots of differing ethnic backgrounds are thrown together to fly an aircraft. Members of different cultures have not only been found to differ in their styles and attitudes towards leadership but also to exhibit distinct conversational norms.[2] *A fortiori*, culture influences how juniors relate to their seniors. Therefore, a multinational, multicultural crew has to mesh their communications coherently, and, above all, transcend ethnic and cultural inhibitions in the cockpit.

Human error has been ascribed to 70–80 % of civil and military aviation accidents.[3] Of these many have been due to improper communications either between cockpit crew or between cockpit crew and air traffic control on the ground. Malcolm Gladwell, in his book *Outliers—The Story of Success*[4] has a chapter which he has entitled *The Ethnic Theory of Plane Crashes*.[5] Gladwell suggests that the ethnicity of the pilots and first officers in the cockpit of an aircraft may have an impact on the safety of flight and cites the famous example (among others) of the crash of the Colombian airliner Avianca flight 052 in January 1990. Here the first officer, in his communications with the air traffic controller had indulged in what was called "mitigated speech" which downplayed or sugar coated critical information that was needed both by the pilot and by the air traffic controller. Gladwell quotes an expert, Earl Weener, a former chief engineer for safety at Boeing who said:

> The whole flight deck design is intended to be operated by two people, and that operation works best when you have one person checking the other, or both people willing to participate. Airplanes are very unforgiving if you don't do things right. And for a long time it's been clear that if you have two people operating the airplane cooperatively, you will have a safer operation than if you have a single person who is simply there to take over if the pilot is incapacitated.[6]

Let us analyse what happened to the Avianca flight. The aircraft was dangerously low on fuel and needed immediate landing. The Captain instructed the first officer to tell air traffic control "We are in an emergency". The First Officer relayed the following message: "That's right to one-eight-zero on the heading and, ah, we'll

[1]Haris Amin, Promoting a Safety Culture in Aviation, http://harisamin.hubpages.com/hub/Safety-Culture-in-Aviation.

[2]Merritt and Helmreich (1996) at 5–24.

[3]O'Hare et al. (1994) at 1855–1869. See also generally Taneja (2002).

[4]Gladwell (2008).

[5]*Id.* Chapter Seven, at 177–223.

[6]*Id.*, 185.

try once again. We are running out of fuel".[7] This does not, by any means, tell ground control that the aircraft had an emergency. The phrase "running out of fuel" does not convey the grave emergency at all. After a period of silence in the cockpit (which was extremely unusual for an emergency) the aircraft slammed into an estate in Long Island town of Oyster Bay, New York. Seventy three passengers perished.

The first officer mitigated his communication to the air traffic controller because he wanted to be polite, presumably because he held the controller in high esteem. Another aspect to the disaster was that Klotz (the first officer) who was Colombian, expected his pilot to take the decisions Gladwell says:

> Klotz sees himself as a subordinate. It's not his job to solve the crisis. It's the captain's. Then there's the domineering air traffic controllers at Kennedy Airport ordering planes around. Klotz is trying to tell him he's in trouble. But he is using his own cultural language, speaking as a subordinate would be, to a superior. The controllers though, aren't Colombian. They are low-power distance New Yorkers. They don't see any hierarchical gap between themselves and the pilots in the air. To them, mitigated speech from a pilot doesn't mean the speaker is being appropriately deferential to a superior. *It means the pilot doesn't have a problem.*[8]

The first measure which should be taken in ensuring team work is to create a positive work environment where each employee feels valued. Listening to one another and honouring the other's point of view is the most effective way to have a common base of equality in a team. In the Avianca instance the first officer, due to his ethnic and cultural background, did not expect his captain to seek his point of view. There did not exist in the cockpit environment a feeling of respect for every individual.

The second measure is to build mutual trust. Inasmuch as trust is the basic tenet for all relationships, it should not be on a one-way basis, as was in the case of Klotz who trusted his captain to take the decisions, thereby relinquishing his own responsibilities of saving the aircraft and passengers. Trust is about doing what you say you are going to do and being who you say you are. It's about showing one's peers and team mates in everything one does that one is reliable, responsible and accountable, and that they can rely on that person for consistency. That was certainly not what Klotz did with his captain.

As for the captain of the Avianca flight, what he did not do was to make his first officer feel that the latter played an important role. The captain did not encourage an environment of cooperation. Rather, he created an environment of competition where Klotz felt he was the inferior and weaker member of crew. The effective leader lets each member of staff know he is a valued part of the team, and that will create a work environment where staff members will respect each other for their unique contributions. This essentially requires the leader to create team spirit by demonstrating that he is open to communication from everyone.

---

[7] *Id.* 193.

[8] *Id.* 207.

The Avianca example resonates with the aviation community a similar trend that existed in civil aviation in Korea until recently. The strongly hierarchical nature of Korean culture[9] was often instrumental in blocking a second in command in the cockpit of a Korean airliner from advising the captain when he knew that the captain was taking wrong action during the flight. This resulted in a high crash rate, until Korean Air trained its flying crew not to behave hierarchically in the cockpit.[10]

Another example that has been cited is that of China Air, the national carrier of Taiwan, which had one of the worst accident records from 1986 to 1998 which reported 561 fatalities.[11] Donald Davis, an expert who carried out a three year study commissioned by the National Aeronautics and Space Administration (NASA) of the United States on how differing styles impede or otherwise adversely affect cockpit communication and coordination has reportedly said of Air China during the period in question:

> a reasonable inference is that the autocratic decision-making style preferred by Air China pilots, most of whom received their training in the Taiwanese Air Force, can be counter-productive during times of crisis. Compared to pilots of other nations, the Taiwanese place the greatest emphasis on rules, order, strict time limits and a preference for finding a single correct answer to any problem. Chinese subordinates are unlikely to question or challenge their superiors, even if they are aware of situation-critical information that senior pilots are not.[12]

In discussing an accident involving an aircraft which was operated by Filipino crew one commentator has observed that, during the final moments before the impact the pilot's foremost thoughts, as reflected in his expostulations recorded in the black box, focussed on his family and creator, not on battling the defects that were bringing his aircraft down. The commentator mentions that Filipinos bend towards spirituality and messianic self sacrifice as a response to stress.[13] In contrast, American and Australian pilots are lone men braving the elements and tend to fix problems in the cockpit by themselves.[14]

One of the reasons ascribed to the breakdown of communications between the captain and his subordinate crew is that the former is generally considered an elite, confident and competent professional in the aircraft who is often unwilling to admit impairment of judgement or failure. The captain is a self sufficient hero who does

---

[9]Culture is the behaviour, customs, values, language and beliefs of a social group. See Tam and Duley (2005). www.sjhfes.org/miniconference/PDFs/01-Tam.pdf.

[10]Korean Airline Pilots, Arrogant Physicians, and Life-or-Death Decision Making, December 13, 2008 at http://physioprof.wordpress.com/2008/12/13/korean-airline-pilots-arrogant-physicians-and-life-or-death-decisionmaking/.

[11]James Schultz, Hear What They Are Saying: the Influence of Culture on Cockpit Communication, http://www.odu.edu/ao/instadv/quest/CockpitCommun.html.

[12]*Ibid.*

[13]Lima (2000) at 86.

[14]*Ibid.*

not make a mistake.[15] This brings to bear the gender issue and the question as to whether a male pilot generally conducts himself and functions differently from his female counterpart. Although a study has concluded that there is no difference between pilot error accident rates attributed to male and female pilots,[16] male pilots have been reported to be more task-oriented and exude more confidence than women pilots, whereas women pilots naturally are reported to be more sensitive towards the needs of the passengers and be better at communicating.[17]

A significant aspect of good communication and leadership in the work place is approachability, where the boss demonstrates that he is available to hear any point of view and give credit for success of staff. Above all, the boss should always take responsibility. For example, it could be argued that in the case of the Avianca disaster, the captain should have taken over if he was not satisfied that the gravity of the emergency was not being relayed appropriately by the first officer to ground control.

At the end of the day there is no established magic formula for effective communication except for best practices as discussed. However, it is prudent to shed inhibitive cultural nuances if crew members are to perform at their optimum, particularly in the face of potential disasters.

3 Cockpit Communications

Inadequate communications between technical crew and air traffic control and between members of technical crew have led to serious accidents. In 1977, the worst aviation disaster occurred at Tenerife in the Canary Islands, caused by heavy accents and improper terminology among a Dutch KLM crew, an American Pan Am crew and a Spanish air traffic controller. 583 passengers perished. In 1980, another Spanish air traffic controller at Tenerife gave a holding pattern clearance to a Dan Air flight by saying "turn to the left" when he should have said "turns to the left"—resulting in the aircraft making a single left turn rather than making circles using left turns. The jet hit a mountain killing 146 people. As already mentioned, In 1990, Colombian Avianca pilots in a holding pattern over Kennedy Airport told controllers that their 707 was low on fuel. The crew should have stated they had a "fuel emergency," which would have given them immediate clearance to land. Instead, the crew declared a "minimum fuel" condition and the plane ran out of fuel, crashing and killing 72 people. Three years later Chinese pilots flying a U.S.-made MD-80 were attempting to land in northwest China. They just did not understand a simple instruction in English which caused them to crash their aircraft. The pilots were baffled by an audio alarm from the plane's ground proximity warning system. A cockpit recorder picked up the pilot's last words: "What does 'pull up' mean?"

---

[15] *Supra* note 1.

[16] Mcfadden (1996) at 443–450.

[17] Turney (1995) at 262–268.

In 1995, an American Airlines jet crashed into a mountain in Colombia after the captain instructed the autopilot to steer towards the wrong beacon. A controller later stated that he suspected from the pilot's communications that the jet was in trouble, but that the controller's English was not sufficient for him to understand and articulate the problem. Another glaring example of failed communications occurred in New Delhi, India on November 13, 1996, when a Saudi Arabian airliner and a Kazakhstan plane collided in mid-air. All indications were that the Kazak pilot may not have been sufficiently fluent in English and was consequently unable to understand an Indian controller giving instructions in English.

On 22 May 2010 a Boeing 737-888 operated by Air India Express between Mangalore and India, carrying 160 passengers on board, four of whom were infants, approached the runway too high, crossing the runway threshold at almost 200 ft whereas the normal approach would be 50 ft. The aircraft broke through the boundary fence, plunged into the steeply wooded gorge beyond it, where it broke into three and caught fire. Eight passengers escaped through gaps in the fuselage. All others on board died. The Captain was Serbian and the First Officer and four other cabin crew were Indian. It has been reported that:

> Apart from the captain's numerous procedural violations during approach, cultural factors played a role in the development of this accident. The captain and the first officer did not communicate adequately with each. When the captain continued the approach in an unstabilized condition, despite the fact that it was not in accordance with standard procedures, and then failed to take corrective action, the first officer did not assert himself.[18]

The accidents cited above are symptomatic of a lack of aircrew coordination or crew resource management (CRM)[19] which were the most common determinants and causal factors.[20] Applebaum and Fewster hold the following view:

> Research has long shown that accidents and poor service quality are rooted primarily in sociotechnical human factors, not technology per se. Sub-optimisation, or poor quality with regard to management, decision-making, teamwork, employee motivation or communication, can translate into loss of customers, loss of market share, loss of organization assets and, above all, loss of life.[21]

---

[18]Macarthur Job, Falling Off the Mountain, *Flightsafety*, September–October 2011 Issue 82, 62–65.

[19]Crew Resource Management is the species of the genus Human Resource Management which has been defined as: "a set of processes, which – through the recruitment, training, motivation, appraisal, reward, and development of individuals, and through the effective handling of industrial relations - translates strategy into action". See Holloway (1998), cited in Steven H. Applebaum and Brenda M. Fewster, Human Resource Management Strategy in the Global Airline Industry—A Focus on Organizational development, *Business Briefing: Aviation Strtegies: Challenges and Opportunities of Liberalization* at 70. See also *infra*, note 33 at Article 1.

[20]Yavacone (1993).

[21]Steven H. Applebaum and Brenda M. Fewster, Human Resource Management Strategy in the Global Airline Industry—A Focus on Organizational development, *Business Briefing: Aviation Strtegies: Challenges and Opportunities of Liberalization* at 70.

The Tenerife disaster personifies this statement. The crash involved two Boeing 747 aircraft, one belonging to PAN-AM which was taxing on the runway on which the other aircraft, operated by KLM was taking off in thick fog where visibility was minimal. The KLM captain was the most experienced in the airline' fleet, with 1,100 flying hours' experience. The Accident Investigation Report reflects:

> Conclusions from all of this it may be ascertained that the KLM 4805 captain, as soon as he heard the ATC (altitude) clearance, decided to takeoff.

The fundamental cause of this accident was the fact that the KLM captain: (1) Took off without clearance (to take off). (2) Did not obey the "stand by for take-off" from the tower. (3) Did not interrupt take-off when Pan Am reported that they were still on the runway. (4) In reply to the flight engineer's query as to whether the Pan Am airplane had already left the runway, replied emphatically in the affirmative...

The report records the following additional causative factor *inter alia*:

> Inadequate language. When the KLM co-pilot repeated the ATC clearance, he ended with the words, "we are now at take-off". The controller, who had not been asked for take-off clearance, and who consequently had not granted it, did not understand that they were taking off. The "O.K." from the tower, which preceded the "stand by for take-off" was likewise incorrect - although irrelevant in this case because take-off had already started about six and a half seconds before.[22]

The Report made just three recommendations: placing of great emphasis on the importance of exact compliance with instructions and clearances; use of standard, concise and unequivocal aeronautical language; avoidance of the word "take off" in the ATC clearance; and adequate time separation between the air traffic control clearance and the take off clearance.[23]

4 Language Proficiency and Aeronautical Communications

An integral part of a person's culture is his language and the expressions he uses. This is particularly relevant in many countries where English is the second language, the usage of which has changed over the years and been mixed with cultural expressions and nuances of those countries. The use of standardized phraseology in aeronautical communications has been a critical issue addressed

[22]Secretary of Aviation Report On Tenerife Crash: KLM, B-747, PH-BUF and Pan Am B-747 N736 collision at Tenerife Airport Spain on 27 March 1977. Report dated October 1978 released by the Secretary of Civil Aviation,. *Aircraft Accident Digest* (ICAO Circular 153-AN/56) page 22–68.

[23]*Ibid.*

by the international aviation community for several years. The surging growth in aviation involving numerous pilots of various nationalities whose first language would differ from those countries they fly over, brings to bear the need for a common level of English to allow for safe and efficient air traffic management. Global enforcement of language proficiency regulation is an essential pre requisite in this endeavour, since variations in teaching methods around the world could only result in confusion in the skies. Arguably the most dangerous factor in this equation could be the presumption of a pilot who is conversant in the English language as required by international standards, that the personnel in the tower directing him have the same degree of proficiency in language as he does.

Exponential growth in air traffic, involving longer flights operated by ultra long range aircraft crossing several territorial boundaries in a given flight could cause crew fatigue, leading the pilot to misunderstand instructions, particularly if he is unfamiliar with the language of communication with the tower.

Such failures in communication bring to bear the compelling need for a standardized aviation language that all those who are involved in the technical operation of a flight could speak and understand. The unique challenges posed by communications in air transport are also due to the fact that the protagonists are in two different places. In this regard, there has been a suggestion from a seasoned air traffic controller that controllers should have "familiarization" privileges of riding in the cockpit up to eight times a year, so that they could learn what goes on in the flight deck during landing and take off.[24] Conversely, pilots would have the opportunity to ask questions as to how the tower works.

Ironically, new developments and the use of electronic communications may be adding to the problem. Prior to the introduction of paperless cockpits and MFD[25] and FMS[26] displays, which replaced the hard copy paper manuals, the crew could highlight and annotate on the paper documents, which enabled them to familiarize themselves with unfamiliar terminology. These two display systems are essentially visual and, without the appropriate language background, a pilot could find it difficult to decipher the visuals.

---

[24]This practice was, in fact in place before the events of 11 September 2001 in the United States. See Richards (2007) at 343.

[25]The MFD (Multi Function Device) is a big, multicolour GPS moving map on the screen of the dashboard. The MFD is a small screen in an aircraft surrounded by multiple buttons that can be used to display information to the pilot in numerous configurable ways. Often an MFD will be used in concert with a PFD (Primary Flight Display). MFDs are part of the digital era of modern planes or helicopter. The first MFDs were introduced by air forces. The advantage of an MFD over analog display is that an MFD does not consume much space in the cockpit.

[26]A flight management system or FMS is a computerized avionics component installed in most commercial and business aircraft to assist pilots in navigation, flight planning, and aircraft control functions. FMS is composed of three major components: FMC—Flight Management Computer; AFS—Auto Flight System, and Navigation System including IRS—Internal Reference System and the Global Positioning System—GPS.

5 Regulatory Developments

The International Civil Aviation Organization (ICAO[27] has, throughout its many years of serving the international aviation community, carried out sustained efforts at harmonizing language requirements for aeronautical communications. These efforts to address language proficiency for pilots and air traffic controllers is long standing and was first made by the 32nd Session of the Assembly[28] in September 1998. At that session, ICAO member States adopted Resolution A 32-16 on the subject of proficiency in the English language for radio telephony communications which was adopted as a direct response to an accident that cost the lives of 349 persons, as well as previous fatal accidents where the lack of proficiency in English was a causal factor. This resolution acknowledged the fact that recent major accident investigations had indicated lack of proficiency and comprehension of the English language by flight crews and air traffic controllers as a contributing factor and that in order to prevent such accidents, it was essential that ICAO devise ways and means to ensure that all member States take steps to ensure that air traffic control personnel and flight crew involved in flight operations in air space where the use of the English language is required, are proficient in conducting and comprehending radiotelephony communications in the English language.

---

[27]The International Civil Aviation Organization is the United Nations specialized agency dealing with international civil aviation. ICAO was established by the Convention on International Civil Aviation (Chicago Convention), signed at Chicago on 7 December 1944. Fifty two States signed the Chicago Convention on 7 December 1944. The Convention came into force on 4 April 1947, on the thirtieth day after deposit with the Government of the United States. Article 43 of the Convention states that an Organization to be named the International Civil Aviation Organization is formed by the Convention. ICAO is made up of an Assembly, which is the sovereign body of the Organization composed of the entirety of ICAO member (Contracting) States, and a Council which elects its own president. The Assembly, which meets at least once every three years, is convened by the Council. The Council is a permanent organ responsible to the Assembly, composed of 36 Contracting States. These 36 Contracting States are selected for representation in the Council in three categories: States of chief importance to air transport; States not otherwise included which make the largest contribution to the provision of facilities for international air navigation; and States not otherwise included whose designation will insure that all the major geographic areas of the world are represented on the Council. Article 47 of the Chicago Convention provides that ICAO enjoys "such legal capacity as may be necessary for the performance of its functions" and goes on to say that "full juridical personality shall be granted to the Organization wherever compatible with the constitution of the laws of the State concerned." The Council has two main subordinate governing bodies, the *Air Navigation Commission* and the *Air Transport Committee*. *The Air Navigation Commission* is serviced by The Air Navigation Bureau and is responsible for the examination, coordination and planning of all of ICAO's work in the air navigation field. This includes the development and modification of SARPS) contained in the ICAO Annexes (all except Annexes 9 and 17), subject to the final adoption by the ICAO Council. At the time of writing, ICAO had 191 member States.

[28]The ICAO Assembly, comprised of the Organization's 191 Member States, meets once every three years. An extraordinary meeting of the Assembly may be convened by the Council at any time. The powers and duties of the Assembly are stated in Article 49 of the Convention on International Civil Aviation.

Accordingly, Resolution A32-16 urged the ICAO Council[29] to direct the Air Navigation Commission—a subordinate body of the Council- to consider the matter with a high level of priority. Subsequently, and in response to action taken by the Council, the Air Navigation Commission initiated the development of language provisions in the following Annexes to the Chicago Convention: Annex 1—*Personnel Licensing*, Annex 6—*Operation of Aircraft*, Annex 10—*Aeronautical Telecommunications*, and Annex 11—*Air Traffic Services*. On 5 March 2003, the Council adopted Amendment 164 to Annex 1.[30] As of 5 March 2008, the ability to speak and understand the language used for radiotelephony that is currently required for pilots and air traffic controllers will have to be demonstrated based on the ICAO holistic descriptors and language proficiency rating scale (at Level 4 or above). Additionally, since November 2003, Annex 10 to the Convention on International Civil Aviation has required the availability of English language at all stations on the ground serving designated airports and routes used by international air services.

Since 2003, several steps have been taken to assist States with the implementation of these requirements. The first edition of the *Manual on the Implementation of ICAO Language Proficiency Requirements*[31] and the training aid entitled *ICAO Language Proficiency Requirements—Rated Speech Samples* were produced. The second edition of Document 9835 is presently under development. To date, eleven regional seminars have been conducted. Two ICAO Aviation Language Symposia were conducted in September 2004 and in May 2007 respectively.

These symposia have been popular and well attended. The Second ICAO Aviation Language Symposium was attended by 221 participants from sixty-two States and eight international organizations. While some participants were from State authorities, many of the participants were from air operators, air navigation service providers and language training and testing entities. During the Symposium, concerns were expressed to suggest that some Member States were encountering difficulties in implementing the language proficiency requirements including the establishment of language training and testing capabilities. Some support was expressed for ICAO to establish a system for the endorsement of language testing

---

[29]The ICAO Council is a permanent body responsible to the Assembly. It is composed of 36 Member States elected by the Assembly. In electing the members of the Council, the Assembly gives adequate representation to States of chief importance to air transport; States not otherwise included which make the largest contribution to the provision of facilities for international air navigation; and States not otherwise included whose designation will ensure that all the major geographic areas of the world are represented on the Council. The mandatory and permissive functions of the Council are stipulated in Articles 54 and 55 of the Convention on International Civil Aviation respectively. The Council has its genesis in the Interim Council of the Provisional International Civil Aviation Organization (PICAO). PICAO occupied such legal capacity as may have been necessary for the performance of its functions and was recognised as having full juridical personality wherever compatible with the Constitution and the laws of the State concerned. See Interim Agreement on International Civil Aviation, opened for signature at Chicago, December 7 1944, Article 3. Also in Hudson, *International Legislation*, Vol 1X, New York: 1942–1945, at 159.

[30]See C-DEC 168/9.

[31]Doc 9835, AN/453, First Edition 2004.

as a means to identify testing services that meet harmonized ICAO criteria. Several participants also requested clarification on the steps States should take if they did not implement the requirements by 5 March 2008.

During its deliberations in June 2007, the Council recognized that a single, universally applicable aviation language proficiency test, although desirable, would be inappropriate. However, the Council supported the development of globally harmonized language testing criteria. The implementation of such criteria could effectively be achieved through the establishment of an ICAO endorsement mechanism for aviation language testing. The Council recognized, however, that budgetary resources would be required to establish an ICAO endorsement mechanism for aviation language testing.

ICAO advised its member States at the 36th Session of its Assembly (Montreal: September 18–28, 2007) that it was widely recognized that implementation of the language provisions is resource intensive. Since the language provisions have become effective, several States have invested considerable resources and efforts to comply with the provisions by 5 March 2008. While some States may not be compliant by March 2008, the applicability date established a milestone that would help to retain the focus required to implement the safety Standards related to language proficiency as soon as practicable.

The Assembly was also advised that understanding the consequence of non-compliance was important in order to take appropriate action. A negative impact on safety would be considered the most serious consequence of non-compliance. In addition, the multilateral recognition of pilots' licences provided for under Article 33[32] of the Chicago Convention could also be impacted when a State is unable to meet the minimum Standards prescribed in *Annex 1* to the Convention. Transparency and regular communications among Member States would be the best means of mitigating the potential impact. It should be noted that the Convention provides for the means to deal with such situations and to ensure the continuity of international civil aviation. In the case of flight crew licences, and in application of Articles 33, 39[33]

[32]Article 33 is on the subject of Recognition of certificates and licenses. It provides that certificates of airworthiness and certificates of competency and licenses issued or rendered valid by the member State in which the aircraft is registered, shall be recognized as valid by the other member States, provided that the requirements under which such certificates or licences were issued or rendered valid are equal to or above the minimum standards which may be established from time to time pursuant to this Convention.

[33]Article 39 is on the subject of endorsement of certificates and licenses. It provides: (*a*) that any aircraft or part thereof with respect to which there exists an international standard of airworthiness or performance, and which failed in any respect to satisfy that standard at the time of its certification, shall have endorsed on or attached to its airworthiness certificate a complete enumeration of the details in respect of which it so failed; and (*b*) that any person holding a license who does not satisfy in full the conditions laid down in the international standard relating to the class of license or certificate which he holds shall have endorsed on or attached to his license a complete enumeration of the particulars in which he does not satisfy such conditions.

and 40[34] of the Convention, those pilots that do not meet the operational Level 4 requirements would require permission from other States to operate in the airspace under their jurisdiction.

ICAO also recommended that, in all cases, including those of States where air traffic controllers are not yet compliant, States should notify ICAO of the differences in accordance with Article 38[35] of the Convention and ensure that these differences are indicated in their Aeronautical Information Publication (AIP). They will also have to endorse the personnel licences in accordance with Article 39 of the Convention.

States that may not be in a position to comply with the language proficiency requirement by the applicability date should also provide information on their implementation plans and interim measures taken to mitigate risk. It is important for safety purposes that each State has sufficient information to make a proper risk analysis. This analysis will be required in order to allow an aircraft with pilots who may not meet the language proficiency requirement to fly in the airspace under the jurisdiction of another State. This analysis will also be required for States to authorize their operators to fly in the airspace under the jurisdiction or responsibility of another State that may not be compliant. The purpose of the risk analysis is to ensure that the lack of language proficiency is minimized as a potential causal factor of incidents and accidents. This step will not only help to eliminate or mitigate risk, but to actually strengthen a Standard that could otherwise be ignored by some States.

To this end ICAO planned to provide guidance on the development of implementation plans by the end of October 2007 and to conduct seminars in each ICAO Region as soon as practicable. At the time this article was being written, work on the guidance material was being finalized by ICAO.

---

[34]Article 40 concerns the *validity of endorsed certificates and licenses* and provides that No aircraft or personnel having certificates or licenses so endorsed shall participate in international navigation, except with the permission of the State or States whose territory is entered. The registration or use of any such aircraft, or of any certificated aircraft part, in any State other than that in which it was originally certificated shall be at the discretion of the State into which the aircraft or part is imported.

[35]Article 38 concerns departures from international standards and procedures and provides that any State which finds it impracticable to comply in all respects with any such international standard or procedure, or to bring its own regulations or practices into full accord with any international standard or procedure after amendment of the latter, or which deems it necessary to adopt regulations or practices differing in any particular respect from those established by an international standard, shall give immediate notification to the International Civil Aviation Organization of the differences between its own practice and that established by the international standard. In the case of amendments to international standards, any State which does not make the appropriate amendments to its own regulations or practices shall give notice to the Council within sixty days of the adoption of the amendment to the international standard, or indicate the action which it proposes to take. In any such case, the Council shall make immediate notification to all other states of the difference which exists between one or more features of an international standard and the corresponding national practice of that State.

In response to the recommendations of the ICAO Council, ICAO member States, at the 36th Session of the Assembly convened in September 2007, adopted Assembly Resolution A 36-11[36] (Proficiency in the English Language used for radiotelephony) which superseded the earlier resolution A32-16 on the same subject. This resolution, while recognizing ICAO's work in introducing language provisions to ensure that air traffic personnel and pilots are proficient in conducting and comprehending radiotelephony communications in the English language, including requirements that the English language shall be available on request at all stations on the ground serving designated airports and routes used by international air services; acknowledged that the language provisions reinforce the requirement to use ICAO standardized phraseology in all situations for which it has been specified. The Assembly gave due recognition to the fact that ICAO member States had made substantial efforts to comply with the language proficiency requirements by 5 March 2008. However, some States still encounter considerable difficulties in implementing the language proficiency requirements including the establishment of language training and testing capabilities. Furthermore, some member States will require additional time to implement the language proficiency provisions beyond the applicability date.

There was also the reality according to Resolution A36-11 that some States were finding it impracticable to comply in all respects with any international standard or procedure but were obliged to give immediate notification of a deviation from a Standard in an Annex to ICAO in accordance with Article 38 of the Convention. The Resolution makes mention of the requirements in Articles 39 and 40 which have been already referred to in this article and urges the member States to use ICAO standardized phraseology in all situations for which it has been specified. It also directs the Council to support Member States in their implementation of the language proficiency requirements by establishing globally harmonized language testing criteria and urges member States that are not in a position to comply with the language proficiency A-2 requirement by the applicability date to post their language proficiency implementation plans including their interim measures to mitigate risk, as required, for pilots, air traffic controllers and aeronautical station operators involved in international operations on the ICAO website as outlined in accordance with the practices recommended by ICAO together with ICAO guidance material. Finally the Resolution directs the Council to provide guidelines to States on the development of implementation plans, including an explanation of the risk mitigation measures so as to enable Member States to post their plans as soon as practicable, but prior to 5 March 2008.

The Resolution also calls upon the member States to waive the permission requirement under Article 40 of the Convention, in the airspace under their jurisdiction for pilots who do not yet meet the ICAO language proficiency requirements, for a period not exceeding three years after the applicability date of 5 March 2008,

---

[36]This Resolution has been transmitted to all the 190 ICAO member States by the Secretary General of ICAO per State Letter *AN 12/44.6-07/68* dated 26 October 2007.

provided that the States which issued or rendered valid the licences have made their implementation plans available to all other Member States. States are also requested not to restrict their operators, conducting commercial or general aviation operations, from entering the airspace under the jurisdiction or responsibility of other States where air traffic controllers or radio station operators do not yet meet the language proficiency requirements for a period not exceeding three years after the applicability date of 5 March 2008, provided that those States have made their implementation plans available to all other Member States.

Member States were required, in accordance with Resolution A36-11, provide data concerning their level of implementation of the Language Proficiency Requirements when requested by ICAO.

At the 37th Assembly (Montreal, 18 September–8 October 2010) Resolution A 36-11 was superseded by Resolution A 37-10 (proficiency in the English language used for radiotelephony communications) which recognized *inter alia* that language provisions introduced ICAO helps in preventing accidents and ensuring that air traffic personnel and pilots are proficient in conducting and comprehending radiotelephony communications in the English language, including requirements that the English language shall be available on request at all stations on the ground serving designated airports and routes used by international air services. The resolution also recognized that the language provisions reinforce the requirement to use ICAO standardized phraseology in all situations for which it has been specified and that some Contracting States encountered considerable difficulties in implementing the language proficiency requirements including the establishment of language training and testing capabilities. It was also noted by the Assembly that some Contracting States required additional time to implement the language proficiency provisions beyond the applicability date.

Resolution A 37-10 urged the Contracting States to use ICAO standardized phraseology in all situations for which it has been specified to assist each other in their implementation of the language proficiency requirements. It also urged ICAO member States that have not complied with the language proficiency requirement by the applicability date to post their language proficiency implementation plans including their interim measures to mitigate risk, as required, for pilots, air traffic controllers and aeronautical station operators involved in international operations on the ICAO website as outlined in accordance with the associated practices below and ICAO guidance material. The Resolution directed the ICAO Council to continue to support Contracting States in their implementation of the language proficiency requirements.

It also called upon Contracting States to waive the permission requirement under Article 40 of the Chicago Convention,[37] in the airspace under their jurisdiction for

[37]Article 40 provides that no aircraft or personnel having certificates or licenses so endorsed shall participate in international navigation, except with the permission of the State or States whose territory is entered. The registration or use of any such aircraft, or of any certificated aircraft part, in any State other than that in which it was originally certificated shall be at the discretion of the State into which the aircraft or part is imported.

pilots who do not yet meet the ICAO Language Proficiency Requirements, for a period not exceeding three years after the applicability date of 5 March 2008, provided that the States which issued or rendered valid the licences have made their implementation plans available to all other Contracting States and have notified ICAO of the differences pertaining to language provisions. The Resolution furthermore urged Contracting States to take a flexible approach after 5 March 2011 towards States that do not yet meet the Language Proficiency Requirements, yet are making progress as evidenced in their implementation plans. It is a requirement under the Resolution that decisions concerning operations should be made on a non-discriminatory basis and not be made for the purpose of gaining economic advantage and directed the Council to monitor the status of implementation of the Language Proficiency Requirements and take necessary actions to advance safety and maintain the regularity of international civil aviation.

6 Crew Resource Management (CRM)

CRM as a system of management[38] which encompasses a wide range of knowledge, skills and attitudes including communications, situational awareness, problem solving, decision making, and teamwork; together with all the attendant subdisciplines which each of these areas entails.[39] The essence of good airmanship is situational awareness and good decision making particularly in stressful situations.[40] Communications act as an integral link between these two essentials. The paper goes on to say:

[38]Crew Resource Management is the species of the genus Human Resource Management which has been defined as: "a set of processes, which – through the recruitment, training, motivation, appraisal, reward, and development of individuals, and through the effective handling of industrial relations - translates strategy into action". See Holloway (1998), cited in Steven H. Applebaum and Brenda M. Fewster, Human Resource Management Strategy in the Global Airline Industry – A Focus on Organizational development, *Business Briefing: Aviation Strategies: Challenges and Opportunities of Liberalization* at 70.

[39]http://www.raes-hfg.com/reports/crm-now.htm.

[40]Stress is a factor which can quickly undermine the emotional climate in which the crew is operating is stress—defined as a state of highly unpleasant emotional arousal associated variously with overload, fear, anxiety, anger and hostility—all of which threaten both individual performance and teamwork. Stress often arises as a result of a perceived gap between the demands of a situation and an individual's ability to cope with these demands. As stress involves the processes of perception and evaluation, it impinges directly on the cognitive and interpersonal skills which form the basis of good CRM. Both arousal and alertness are necessary to enable each individual to achieve optimum performance in CRM-related skills, but too much or too little arousal will have a significantly adverse impact on the ability of the crew to function effectively as a team. It is therefore important for crew members not only to be aware of the symptoms of stress in themselves and others, but also to understand the effects which stress can have on CRM, and to mitigate these effects where possible by taking measures to counter them.

> it is evident that effective communication between crew members is an essential prerequisite for good CRM. Research has shown that in addition to its most widely perceived function of transferring information, the communication process in an aircraft fulfils several other important functions as well. It not only helps the crew to develop a shared mental model of the problems which need to be resolved in the course of the flight, thereby enhancing situational awareness, but it also allows problem solving to be shared amongst crew members by enabling individual crew members to contribute appropriately and effectively to the decision-making process. Most importantly, it establishes the interpersonal climate between crew members and is therefore a key element in setting the tone for the management of the flight.[41]

The starting point for CRM is incontrovertibly Annex 6 to the Chicago Convention which addresses aeronautical aspects of the operations of aircraft. Prudent operation of aircraft is vital in avoiding accidents and incidents. The essence of Annex 6, simply put, is that the operation of aircraft engaged in international air transport must be as standardized as possible to ensure the highest levels of safety and efficiency. In 1948 the Council first adopted Standards and Recommended Practices for the operation of aircraft engaged in international commercial air transport. They were based on recommendations of States attending the first session of the Operations Divisional Meeting held in 1946, and are the basis of Part I of Annex 6.

In order to keep pace with a new and vital industry, the original provisions have been and are being constantly reviewed. For instance, a second part to Annex 6, dealing exclusively with international general aviation, became applicable in September 1969. Similarly, a third part to Annex 6, dealing with all international helicopter operations, became applicable in November 1986. Part III originally addressed only helicopter flight recorders, but an amendment completing the coverage of helicopter operations in the same comprehensive manner as aeroplane operations covered in Parts I and II was adopted for applicability in November 1990.

It would be impractical to provide one international set of operational rules and regulations for the wide variety of aircraft which exist today. Aircraft range from commercial airliners to the one-seat glider, all of which cross national boundaries into adjacent States. In the course of a single operation, a long-range jet may fly over many international borders. Each aircraft has unique handling characteristics relative to its type and, under varying environmental conditions, may have specific operational limitations. The very international nature of commercial aviation, and of general aviation to a lesser degree, requires pilots and operators to conform to a wide variety of national rules and regulations.

However, it is certainly possible to provide one international set of rules on CRM. In 1994, the Air Navigation Commission (ANC) of ICAO adopted a proposal to include a Standard in Annex 6, pursuant to a review of the Annex, pertaining to

[41]See Holloway (1998), cited in Steven H. Applebaum and Brenda M. Fewster, Human Resource Management Strategy in the Global Airline Industry – A Focus on Organizational development, *Business Briefing: Aviation Strategies: Challenges and Opportunities of Liberalization* at 70.

the initial and recurrent human performance training for flight crews. The Standard, promulgated through Amendment 21 to the Annex became applicable in November 1995. The new Standard, which appears as 9.3.1 in Chapter 9 Part 1 of the Annex provides that performance training for flight crews shall also include training in knowledge and skills related to human performance and that such training will be given on a recurrent basis as determined by the State of the operator of the aircraft. A further review by the ANC of Annex 6 resulted in Amendment 23 to the Annex which brought out in Chapter 8, Standard 8.7.5.4 which provides that the training programme established by the maintenance organization shall include training in knowledge and skills[42] related to human performance, including coordination with other maintenance and flight crew. This provision established the clear requirement to mesh coordination among crew members through knowledge and skills. In Chapter 10 this requirement is extended to the flight operations officer and flight dispatcher.

Some of the more important CRM skills are;

- Communication and interpersonal skills

Communication skills essentially involve the need to be conversant with and apply as polite assertiveness and participation, active listening and feedback. Communication skills also ensure clear and effective use of language and responsiveness to feedback.[43] Ambiguities in communication are the antithesis of good communication skills. Communications are improved by taking into account cultural influences as well as factors such as rank, age and crew position, all of which can create barriers to communication in the cockpit. Polite assertiveness is a skill frequently ignored in communications training but vital to a healthy cockpit. A pilot-in-command may be open to communication but temporarily unable to receive and comprehend. As the disasters of Avianca and Teneriffe showed, other crewmembers must be aware of the importance of the information they hold and have a strong feeling of self-value and a single hesitant attempt to communicate important information and data constitutes a failure to discharge individual responsibility. Pilots-in-command must constantly strive to emphasize this responsibility in their team-building efforts. There must be created in the cockpit environment a legitimate window of dissent and disagreement so that a clear line of unequivocal communication exists between the captain and other flight crew members.

---

[42]A CRM skill is a goal-directed, well-organized behaviour that is acquired through practice and performed with economy of effort. See Proctor and Dutta (1995) at 18.

[43]Colloquialisms should be avoided at all cost. There is a recorded instance of an air traffic controller in an airport in the United States deciding to ignore the standard rules governing Federal Aviation Administration (FAA) radio phraseology in communications to cockpits. He strongly exhorted more hustle on the part of pilots taxiing in a line of planes waiting to take off in the narrow gaps of airspace between planes landing on the same runway. As a result, the airport's hourly rate of flights improved. The controller's "hurry-up" attitude led to what is known as the "Pete Rose departure," a reference to the former Cincinnati Reds baseball star nicknamed "Charlie Hustle." Richards (2007) at 136–137.

- Situational awareness

Situational awareness is complete knowledge and understanding of a situation and environment through one's ability to accurately perceive what is going on in the cockpit and outside the aircraft. It also involves the instant planning and the ability to execute one or more of several responses to any emergency situation that could occur in the immediate future. Awareness of one's situation could be a complex process that is driven by the appreciation that one's perception of reality may differ from reality itself. Therefore, to remain within the realm of reality it is necessary to indulge in ongoing questioning, cross-checking and refinement of one's perception. A tenaciously constant and conscious monitoring of the situation is also required.

- Problem solving, decision making and judgment

These three areas are integrally linked and interrelated. The accepted fact among accident investigators is that an accident is an ultimate cause of a concatenation of errors and that the resolving of one error may break the chain and prevent the accident. Therefore problem solving is necessary to tackle an overall cycle of events beginning with information input and ending with pilot judgment in making a final decision. A danger in the information receiving phase of problem solving is that several interpretations of information and conflicting points of view may be represented. Skills in resolving conflict are therefore especially appropriate at this time. The pilot in command is in charge of making a decision[44] with the agreement of others. At least there must be a clear understanding among the crew that no crew member has a radically different view from that of the captain.

- Leadership and "followership"

To exercise leadership is to use appropriate authority to ensure focus on task and crew member concerns. Therefore the role of the captain carries a special responsibility, where although other members of the crew are engaged in carrying out their own responsibilities, it is for the captain to be responsible for supervising the overall management of the flight. His authority and command must be acknowledged at all times. Recognition and acceptance of the captain's cannot be assumed by position alone. The credibility of a leader is built over time and must be accomplished through conscious effort. Similarly, every non-command crewmember is responsible for actively contributing to the team effort, for monitoring changes in the situation and for being assertive when necessary. This last aspect is extremely relevant, as was shown in the instance of the First officer's reticence to differ with the Captain in the Avianca instance.

---

[44]The responsibility of the pilot in command under rules of the air is both critical and grave. Standard 2.4 of Annex 2 to the Chicago Convention (Rules of the Air) states that the pilot-in-command of an aircraft shall have final authority as to the disposition of the aircraft while in command. It is therefore quite obvious that irresponsible conduct that would lead to an accident or incident would bring to bear negligence of the part of the pilot in command.

- Vigilance

The crew must be vigilant at all times and consciously avoid complacency during the flight. In some cultures, there is no emphasis on readiness for anticipated risk. CRM training must inculcate in the crew member the need for him to keep watch over system and environment changes and inform other crew members of potential threats and errors.

- Team-building

Crew members must be trained to establish prioritization of tasks and appropriate utilization of resources within the flight deck.

The purpose of Annex 6 is to contribute to the safety of international air navigation by providing criteria for safe operating practices, and to contribute to the efficiency and regularity of international air navigation by encouraging ICAO's Contracting States to facilitate the passage over their territories of commercial aircraft belonging to other countries that operate in conformity with these criteria. ICAO Standards do not preclude the development of national standards which may be more stringent than those contained in the Annex. In this context it becomes essential that States establish a safety culture that espouses a non-punitive response to operational error so that they could identify errors that lead to incidents and accidents.[45]

The organizational culture period which was referred to in the first paragraph of this article, and in which we are at present, calls for a particular mind-set in the safety culture of States and of airlines. It is a common feature of globalization that we tend to interpret the world through our individual cultural lens or worldview. However, it must be remembered that aviation is a multicultural domain and crew members should go through a rigid CRM system that would enable them to shed their cultural roots in the flight deck and adapt themselves to a common culture of communicating clearly and assertively, putting their point across to both fellow crew members and air traffic and ground traffic controllers, irrespective of rank.

The most important is for pilots to have the comfort, irrespective of their misjudgement or inhibition brought about by their cultural origin, of knowing that they are operating under a non-punitive corporate culture which would not criminalize their actions unless they are guilty of negligence. A critical provision in Annex 13 to the Chicago Convention (Accident Investigation) is Standard 3.1 which provides that the sole objective of the investigation of an accident or incident shall be the prevention of accidents and incidents and that it is not the purpose of that activity to apportion blame or liability.

Two significant provisions in Annex 13 in this context are a Standard and Recommended Practice. Standard 5.4 provides that the accident investigation authority shall have independence in the conduct of the investigation and have unrestricted authority over its conduct, consistent with the provisions of the Annex. The investigation would include: (a) the gathering, recording and analysis of all available information on that accident or incident; (b) if appropriate, the issuance of

[45] *Human Factors Training Manual,* ICAO Doc 9683-AN/950, ICAO: Montreal First Edition—1998, at 2-2-7.

safety recommendations; (c) if possible, the determination of the causes; and (d) the completion of the final report. When possible, it required that the scene of the accident be visited, the wreckage examined and statements taken from witnesses. Recommendation 5.4.1 suggests that any judicial or administrative proceedings to apportion blame or liability should be separate from any investigation conducted under the provisions of this Annex.[46]

It is a serious issue that in accident investigations persons with knowledge of the circumstances that lead to an accident are reluctant to come forward to give evidence in fear of criminal prosecutions against them. This anomaly has serious connotations in instances where cultural differences may have caused miscommunication among the crew or between the crew and air traffic control which leads to an accident. In the absence of clear evidence given by the crew involved investigators could find it difficult to obtain valuable information, particularly when judicial proceedings are launched at the same time as the safety investigation. EUROCONTROL suggests a "just culture" which has been defined as

> a culture in which front line operators or others are not punished for actions, omissions or decisions taken by them that are commensurate with their experience and training, but where gross negligence, wilful violations and destructive acts are not tolerated. This is important in aviation, because we know we can learn a lot from the so-called honest mistakes[47]

The just culture as suggested by EUROCONTROL draws a delicate balance between protection of the public from errors of such persons as pilots and air traffic controllers and the protection of such professionals from being arbitrarily prosecuted for mistakes that are not grounded by negligence or gross negligence. The idea is for a just culture to protect people against being blamed for honest mistakes, but also to lay them open for prosecution for their reprehensible conduct. The idea is neither to create a list of offences nor is it to label certain conduct as culpable. The idea is to determine the extent of accountability and responsibility exhibited by the person concerned and to draw the line in each case. The role of the judiciary in determining these factors becomes vital. Judicial action as a driver of this determination has a dichotomy of a balance between two fundamental societal interests that go to the importance of serving the interest of the public through the maximising of safety (through incident and accident investigation and reporting) and the

[46]5 Standard 5.12 of Annex 13 provides that the State conducting the investigation of an accident or incident shall not make the following records available for purposes other than accident or incident investigation, unless the appropriate authority for the administration of justice in that State determines that their disclosure outweighs the adverse domestic and international impact such action may have on that or any future investigations: (a) all statements taken from persons by the investigation authorities in the course of their investigation; (b) all communications between persons having been involved in the operation of the aircraft; (c) medical or private information regarding persons involved in the accident or incident; (d) cockpit voice recordings and transcripts from such recordings; and (e) opinions expressed in the analysis of information, including flight recorder information.

[47]Just Culture Guidance Material for Interfacing with the Judicial System, Edition date: 11.02.2008 Reference nr: 08/02/06-07.

maximising of justice (through the application of laws). The two can well conflict with each other and the prudent approach would be to establish a reasonable balance between the two while not prejudicing the interest of the professional who acts within the parameters of good employment.

Finally, there must be a sustained environment of trust between the flight crew and their employer. The critical issue is whether crew members would usually volunteer information regarding their decisions and actions taken in the flight deck that could be ascribed to the cause of an incident or accident. The answer is, as long as the crew member suspects or has reason to believe that such disclosure would result in punitive measures or loss of pay (which could even lead to loss of career) he/she would not usually volunteer information even if such decisions or actions may have jeopardized the safety of the flight. Such is human nature. One approach to this question would lie in the triumvirate of regulator, operator and crew member reaching a cohesive relationship of trust that would obviate jeopardising the interests of the pilot's career and more importantly the interests of safety. Attitudes should change, resulting in a culture change. The regulator must ensure that there are regulations in place that would protect both the operator and crew member. The following elements are critical to this process: shared responsibility; acknowledgment of the complexity of stress and cultural diversity in the flight deck as a risk factor that can never be totally eliminated; multiple solutions for multiple problems; scientific progress; and continuous evaluation of CRM with a view to enhancing guidance.

References

Gladwell M (2008) Outliers – the story of success. Little Brown and Company, New York

Holloway S (1998) Changing planes: a strategic management perspective on industry in transition. Volume 1: Situation analysis. Ashgate, Aldershot

Lima EP (2000) Paradigm shift in the cockpit. Air Transp World 37(11):85–89

Mcfadden KL (1996) Comparing pilot-error accident rate of male and female airline pilots. Omega Int J Manag Sci 24:443–450

Merritt AC, Helmreich RL (1996) Human factors on the flight deck: the influence of national culture. J Cross-Cult Psychol 27:5–24

O'Hare D, Wiggins M, Batt R, Morrison D (1994) Cognitive failure analysis for aircraft accident investigation. Ergonomics 37:1855–1869

Proctor RW, Dutta A (1995) Skill acquisition and human performance. Sage, Thousand Oaks

Richards B (2007) Secrets from the tower. Ithaca, New York

Tam L, Duley J (2005) Beyond the west: cultural gaps in aviation human factors research. In: Proceedings of the mini-conference on human factors in complex sociotechnical systems, 1–5 January 2005

Taneja N (2002) Human factors in aircraft accidents: a holistic approach to intervention strategies. In: Proceedings of the 46th annual meeting of the human factors and ergonomics society, Santa Monica

Turney MA (1995) Womens' learning and leasdrship styles: impact on CRM, applications of psychology to the aviation system. In: McDonald N, Johnston N, Fuller R (eds) Proceedings of the 21st conference of the European Association for Aviation Psychology (EAAP), vol 1. Ashgate, Aldershot

Yavacone D (1993) Mishap trends and cause factors in naval aviation: a review of naval safety centre data, 1986-1990. Aviat Space Environ Med 64:392–395

Article 33

Recognition of Certificates and Licenses

Certificates of airworthiness and certificates of competency and licenses issued or rendered valid by the contracting State in which the aircraft is registered, shall be recognized as valid by the other contracting States, provided that the requirements under which such certificates or licences were issued or rendered valid are equal to or above the minimum standards which may be established from time to time pursuant to this Convention.

R. Abeyratne, *Convention on International Civil Aviation*,
DOI 10.1007/978-3-319-00068-8_34,

Article 34
Journey Log Books

There shall be maintained in respect of every aircraft engaged in international navigation a journey log book in which shall be entered particulars of the aircraft, its crew and of each journey, in such form as may be prescribed from time to time pursuant to this Convention.

R. Abeyratne, *Convention on International Civil Aviation*,
DOI 10.1007/978-3-319-00068-8_35, © Springer International Publishing Switzerland 2014

Article 35
Cargo Restrictions

(a) No munitions of war or implements of war may be carried in or above the territory of a State in aircraft engaged in international navigation, except by permission of such State. Each State shall determine by regulations what constitutes munitions of war or implements of war for the purposes of this Article, giving due consideration, for the purposes of uniformity, to such recommendations as the International Civil Aviation Organization may from time to time make.

(b) Each contracting State reserves the right, for reasons of public order and safety, to regulate or prohibit the carriage in or above its territory of articles other than those enumerated in paragraph (a): provided that no distinction is made in this respect between its national aircraft engaged in international navigation and the aircraft of the other States so engaged; and provided further that no restriction shall be imposed which may interfere with the carriage and use on aircraft of apparatus necessary for the operation or navigation of the aircraft or the safety of the personnel or passengers.

R. Abeyratne, *Convention on International Civil Aviation*,
DOI 10.1007/978-3-319-00068-8_36,

Article 36
Photographic Apparatus

Each contracting State may prohibit or regulate the use of photographic apparatus in aircraft over its territory.

R. Abeyratne, *Convention on International Civil Aviation*,
DOI 10.1007/978-3-319-00068-8_37, © Springer International Publishing Switzerland 2014

Article 37

Adoption of International Standards and Procedures

Each contracting State undertakes to collaborate in securing the highest practicable degree of uniformity in regulations, standards, procedures, and organization in relation to aircraft, personnel, airways and auxiliary services in all matters in which such uniformity will facilitate and improve air navigation.

To this end the International Civil Aviation Organization shall adopt and amend from time to time, as may be necessary, international standards and recommended practices and procedures dealing with:

(a) Communications systems and air navigation aids, including ground marking;

(b) Characteristics of airports and landing areas;

(c) Rules of the air and air traffic control practices;

(d) Licensing of operating and mechanical personnel;

(e) Airworthiness of aircraft;

(f) Registration and identification of aircraft;

(g) Collection and exchange of meteorological information;

(h) Log books;

(i) Aeronautical maps and charts;

(j) Customs and immigration procedures;

(k) Aircraft in distress and investigation of accidents; and such other matters concerned with the safety, regularity, and efficiency of air navigation as may from time to time appear appropriate.

R. Abeyratne, *Convention on International Civil Aviation*,

DOI 10.1007/978-3-319-00068-8_38,

Article 38

Departures from International Standards and Procedures

Any State which finds it impracticable to comply in all respects with any such international standard or procedure, or to bring its own regulations or practices into full accord with any international standard or procedure after amendment of the latter, or which deems it necessary to adopt regulations or practices differing in any particular respect from those established by an international standard, shall give immediate notification to the International Civil Aviation Organization of the differences between its own practice and that established by the international standard. In the case of amendments to international standards, any State which does not make the appropriate amendments to its own regulations or practices shall give notice to the Council within sixty days of the adoption of the amendment to the international standard, or indicate the action which it proposes to take.

In any such case, the Council shall make immediate notification to all other states of the difference which exists between one or more features of an international standard and the corresponding national practice of that State.

Contents

R. Abeyratne, *Convention on International Civil Aviation*,
DOI 10.1007/978-3-319-00068-8_39,

1 ICAO's Role in Regulation

Pursuant to Article 37 the Council has developed and under the authority granted under Article 54 (l)[1] it has adopted18 Annexes (at the time of writing Annex 19 on Safety Management was before the council for adoption). The laws and regulations pertaining to the Annexes have been discussed comprehensively elsewhere[2] and therefore will not be discussed in this book. There have been numerous Assembly Resolutions governing the development of Standards and Recommended Practices (SARPs) in the Annexes.[3] Resolution A21-21 (Consolidated Statement of Continuing Policies and Associated Practices Related Specifically to Air Navigation), which has already been discussed, in Appendix A (Formulation of Standards and Recommended Practices and Procedures for Air Navigation Services) Resolves that a *Standard* is

> any specification for physical characteristics, configuration, material, performance, personnel or procedure, the uniform application of which is recognized as necessary for the safety of international air navigation and to which Contracting States will conform in accordance with the Convention; in the event of impossibility of compliance, notification to the Council is compulsory under Article 38 of the Convention.

The same Resolution defines a *Recommended Practice* as:

> any specification for physical characteristics, configuration, material, performance, personnel or procedure, the uniform application of which is recognized as desirable in the interest of safety, regularity or efficiency of international air navigation and to which Contracting States will endeavour to conform in accordance with the Convention.

These definitions are repeated verbatim in Assembly Resolution A37-15 (already discussed) in Appendix A (Formulation of Standards and Recommended Practices and Procedures for Air Navigation Services).

Appendix A of Assembly Resolution A37-15 resolved that:

> SARPs and PANS shall be amended as necessary to reflect changing requirements and techniques and thus, inter alia, to provide a sound basis for regional planning and the provision of facilities and services;
>
> Subject to the foregoing clause, a high degree of stability in SARPs shall be maintained to enable the

[1]Article 54 of the Chicago Convention contains the mandatory functions of the Council. Article 54 (l) states that the Council shall adopt, in accordance with the provisions of Chapter VI of the Convention, international standards and recommended practices; for convenience, designate them as Annexes to this Convention; and notify all contracting States of the action taken.

[2]See Abeyratne (2012). Also by the same author, *Aviation Security Law*, Springer: Heidelberg, 2010.

[3]Beginning with Resolution A1-13(Implementation of Standards and Recommended Practices); A1-25 (Future Work of the Technical Divisions; A1-31 (Definition of International Standards and Recommended Practices); A2-42 (Submission of Comments for Proposals on Standards and Recommended Practices); A4-8 (Annexes Prescribing Standards Related to Facilities); A7-9 Notification of Differences From International Standards) and A10-16 (Allocation of Effort Between the Formulation of Standards and Regional Plans and Their Implementation).

Contracting States will maintain stability in their national regulations. To this end amendments shall be limited to those significant to safety, regularity and efficiency, and editorial amendments shall be made only if essential;

SARPs and PANS shall be drafted in clear, simple and concise language. SARPs shall consist of broad,

mature and stable provisions specifying functional and performance requirements that provide for the requisite levels of safety, efficiency and interoperability. Supporting technical specifications, when developed by ICAO, shall be placed in separate documents to the extent possible;

In the development of SARPs, procedures and guidance material, ICAO should utilize, to the maximum extent appropriate and subject to the adequacy of a verification and validation process, the work of other recognized standards making organizations. Material developed by these other standards-making organizations may be deemed appropriate by the Council as meeting ICAO requirements; in this case such material should be referenced in ICAO documentation;

To the extent consistent with the requirements of safety and regularity, Standards specifying the provision of facilities and services shall reflect a proper balance between the operational requirements for such facilities and services and the economic implications of providing them;

Contracting States shall be consulted on proposals for the amendment of SARPs and PANS before the Council acts on them, except when the Council may deem urgent action to be necessary. Furthermore, subject to the adequacy of the verification and validation process, technical specifications may be acted upon by the Council without consultation with States. Such material shall however be made available to States upon request;

The applicability dates of amendments to SARPs and PANS shall be so established as to allow the Contracting States sufficient time for their implementation;

No Annex or PANS document shall be amended more frequently than once per calendar year.

By way of Associated Practices Associated practices The Resolution recognized that the Council should ensure that provisions of SARPs and PANS are completely consistent with each other and that it should endeavour to improve the processing, presentation and usefulness of ICAO documents containing SARPs, PANS and other related provisions, especially for complex systems and their associated applications. To that end the Council was required to promote the development and upkeep of broad system-level, functional and performance requirements. The Council was also requested to continue seeking the most appropriate means of development, translation, processing and dissemination of technical specifications. Contracting States were requested to comment fully and in detail on the proposals for amendment of SARPs and PANS or at least should express their agreement or disagreement on their substance. They should be allowed at least 3 months for this purpose. Furthermore, Contracting States should receive at least 30 days of notification of the intended approval or adoption of detailed material on which they are not consulted. Contracting States were to be allowed a full 3 months for notifying disapproval of adopted SARPs amendments; in establishing a date for notifying disapproval the Council should take into account the time needed for transmission of the adopted amendments and for receipt of notifications from States.

Finally Appendix A required the Council to ensure that, whenever practicable, the interval between successive common applicability dates of amendments to Annexes and PANS is at least 6 months.

In terms of the Implementation of SARPS Appendix D (Implementation of Standards and Recommended Practices and Procedures for Air Navigation Services) of Resolution A37-15 states:

"*Whereas* Article 37 of the Convention on International Civil Aviation requires each Contracting State to collaborate in securing the highest practicable degree of uniformity in regulations and practices in all matters in which such uniformity will facilitate and improve air navigation;

Whereas in accordance with Article 38 of the Convention any Contracting State which finds it impractical to comply in all respects with any international standard or procedure and deems it necessary to adopt regulations or practices differing therefrom is obliged to give immediate notification to ICAO; and

Whereas it is important that all available means of the Organization be employed in encouraging and assisting

Contracting States in overcoming their difficulties in implementation of SARPs and PANs;

The Assembly resolves that:

1. Contracting States shall be encouraged and assisted in the implementation of SARPs and PANs by available means;

2. the differences between the regulations and the practices of Contracting States and the SARPs and PANS shall be monitored by the Council with the aim of encouraging the elimination of those differences that are important for the safety and regularity of international air navigation or are inconsistent with the objectives of the international Standards; and

3. the Council shall analyse the root cause for non-implementation and take appropriate action.

The Appendix has the following Associated Practices:

1. In encouraging and assisting Contracting States in the application of SARPs and PANs the Council should make use of all existing means including the resources of Headquarters, the ICAO Regional Offices and the United Nations Development Programme;

2. Contracting States should continue, and where necessary should intensify, their efforts to apply at theiroperating installations practices and procedures that are in accordance with the current SARPs and PANS. In this regard, Contracting States should consider the practicability of modifying the internal processes by which they give effect to the provisions of SARPs and PANs, if such modifications would expedite or simplify the processes or make them more effective;

3. The Council should urge Contracting States to notify the Organization of any differences that exist between their national regulations and practices and the provisions of SARPs as well as the date or dates by which they will comply with the SARPs. If a Contracting State finds itself unable to comply with any SARPs, it should inform ICAO of the reason for non-implementation, including any applicable national regulations and practices which are different in character or in principle. The notifications of differences from SARPs received should be promptly issued in supplements to the relevant Annexes. Contracting States should also be requested to publish in their AIPs any significant differences from the SARPs and PANs

4. In the monitoring of the differences from SARPs and PANS, the Council should request reports from Contracting States that have not or have incompletely reported to the Organization the implementation of SARPs;

Furthermore, the Council should also request Contracting States that have not published in their AIPs information on the implementation of SARPs and PANs to publish it.

The ICAO Assembly, at its 4th Session (Montreal, 30 May–20 June 1950) adopted Resolution A4-7 (Relation of Procedures for Air Navigation Services to Annexes) Resolved that the Council ensure that Procedures for Air Navigation Services are incorporated in the appropriate Annexes to the Convention as soon as these have become sufficiently stable and that the Council continue to urge Contracting States to notify the Organization of national differences from relevant procedures for air navigation services when such differences affected the safety or regularity of international air navigation. At a later Session during its 7th Assembly (Brighton, 16 June–6 July 1953) it was resolved in Resolution A7-9 (Notification of differences from International Standards) that the Council initiate a more effective and simplified programme with regard to the reporting by States of differences, pursuant to Article 38 of the Convention, to the end that the Organization (ICAO) may be better informed of the actual state of implementation among the Contracting States. The Resolution also requested that more emphasis be given to a programme of monitoring outstanding differences. With the objective of encouraging the elimination of those differences which are important for the safety of air navigation or which are inconsistent with the objective of the international standards.

This last statement of Resolution A7-9 indubitably was the precursor to the current ICAO Continuous Monitoring Approach (CMA). It is encouraging that a later Assembly continued this progressive trend of monitoring adhering by States of SARPs and PANs.[4]

2 What Is an Annex to the Chicago Convention?

In dealing with this question one immediately hits a snag. As stated earlier, Article 54 (l) states that the Council shall adopt, international standards and recommended practices, and, *for convenience* (my emphasis), designate them as Annexes to this Convention. If the Annexes are called as such for convenience, would they form an integral part of the Chicago Convention? Aust is of the View:

> When a treaty has an annex it is normal to provide, though not necessarily in a separate article, that the annex is an integral part of the treaty. Since there are often other documents produced at the time the treaty is adopted, such as agreed minutes, declarations and interpretative exchanges of notes, it is important to know whether they are an integral part of the treaty or merely associated with it[5]

Therefore is accepted legal practice that if a treaty has attachments it will usually say in the text of the treaty that they are integral to the instrument. Such integral portions of a treaty are Annexes, Protocols, Appendices or Schedules. In the context

[4]Resolution A21-21 in Appendix D merely reiterated the significance of Articles 37 and 38 and stated that States should be encouraged to adhere to SARPs and PANs and that their ability to comply with Articles 37 and 38 should be monitored.

[5]Aust (2000) at 348.

of the 18 Annexes to the Chicago Convention one could say that such long documents could form a series of separate instruments which use terminology different from the Treaty provisions. For instance, if a treaty uses the word "shall" the Annex should usually not use the word. However, both the Chicago Convention and the SARPs of its Annexes use the word "shall" which deviates from this trend. Therefore from a strictly legal perspective, it is arguable that the Annexes to the Chicago Convention are not integral parts of the Convention.

The above notwithstanding, and from a practical perspective, there seems to be no way out other than to consider an Annex in this context as part of the Convention. If one does not do so, there will be no work for ICAO and the international community in following a structured regulatory practice in the safety and security of air navigation and assisting States with deficiencies.

Another difficulty is perceived in Article 38 of the Convention which clearly does not obligate States to adhere to Standards in an Annex. In this context, one wonders what the status of the Annexes are and one could be perplexed as to what ICAO's role in pursuing the Standards of the Annexes with Contracting States is. This ambivalence prevailed until ICAO initiated the proactive move of its safety and security audits whereby ICAO could follow progress of States in complying with SARPs (in particular the Standards) and assist States with overcoming any deficiencies with regard to adhering the the Standards of the Annexes.

The question arises as to whether a Contracting State is formally bound by Standards contained in an Annex to the Chicago Convention, particularly when such a State has no convincing argument that it is impracticable to implement such Standards or when it has not notified the ICAO Council of differences as required. This is a vexed debate, particularly in the face of two blatant facts. The first is that the *travaux preparatoires* to the Convention contains a statement that "the Annexes are given no compulsory force".[6] The second is that, as stated earlier, in Article 54 of the Convention, which lays down the mandatory functions of the Council, it is provided that one of the mandatory functions is to

> Adopt, in accordance with the provisions of this Convention, international standards and recommended practices; **for convenience** (emphasis added) designate them as Annexes to this Convention; and notify all Contracting States of action taken.[7]

One could argue therefore that the Annexes are not an integral part of the Convention by virtue of the statement in Article 54 and therefore do not form binding law.

There have been numerous views of legal scholars who have cautioned against this approach and advocated that the words of the Convention should not be taken literarily. One commentator is of the view that:

> The debate is largely academic. Whether or not ICAO standards are formally binding in the treaty law sense, they are highly authoritative in practice. This reflects their recognized

---

[6]See Whiteman (1968) at p. 404

[7]Chicago Convention, Preamble *supra* note 1, Article 54 (l).

importance for the safety and efficiency of civil air travel and the thorough process by which they are promulgated.[8]

It is therefore arguable that all Standards contained in the ICAO Annexes are formally binding on Contracting States, except when a State opts out of under the procedure set forth in the Convention in Article 38. ICAO's international Standards are identified by the words "Contracting States shall" and have a mandatory flavour (infused by the word "shall") while Recommended Practices identified by the words "Contracting States may" have only an advisory and recommendatory connotation (infused by the word "may"). It is interesting that at least one ICAO document[9] requires States under Article 38 of the Convention, to notify ICAO of all significant differences from both Standards and Recommended Practices, thus making all SARPS regulatory in nature.

The legal effect and the compulsive nature of the Standards and Recommended Practices of the ICAO Annexes have to be recognized if a meaningful implementation of the Annexes to the Chicago Convention were to be carried out by Contracting States. Milde states:

> The Chicago Convention, as any other legal instrument, provides only a general legal framework which is given true life only in the practical implementation of its provisions. Thus, for example, Article 37 of the Convention relating to the adoption of international standards and recommended procedures would be a very hollow and meaningless provision without active involvement of all Contracting States, Panels, Regional and Divisional Meetings, deliberations in the Air Navigation Commission and final adoption of the standards by the Council. Similarly, provisions of Article 12 relating to the rules of the air applicable over the high seas, Articles 17 to 20 on the nationality of aircraft, Article 22 on facilitation, Article 26 on the investigation of accidents, etc., would be meaningless without appropriate implementation in the respective Annexes. On the same level is the provision of the last sentence of Article 77 relating to the determination by the Council in what manner the provisions of the Convention relating to nationality of aircraft shall apply to aircraft operated by international operating agencies.[10]

Milde goes on to say that the manner in which the International Civil Aviation organization has exercised its regulatory functions in matters relating to the safety of international air navigation and the facilitation of international air transport provides a fascinating example of international law making. He further observes that the Organization has consequently not had to contend with any of the post war

[8]Kirgis (1995), p. 109 at 126. There is a similar process in operation under the World Meteorological Organization, whereby a certain amount of decision making authority is given to the WMO Congress. Article 9 (a) of the WMO Convention provides that all members shall do their utmost to implement the decisions of the Congress. Article 9 (b) allows any member to opt out by notifying the Secretary General, with reasons if it finds it impracticable to give effect to the technical requirement in question. WMO Convention, reprinted in *International Organization and Integration*, (P Kapteyn et al. eds) 2nd Revised Edition, 1981, pt. I.B.1.9 a. Also in *WMO Basic Documents*, No. 1. WMO Doc. No. 15 at 9 1987.

[9]*Aeronautical Information Services Manual*, ICAO Doc 8126-0 AN/872/3.

[10]Milde (1989) at 208. See also Schenkman (1955), at p. 163.

ideological differences that have impeded international law making on politically sensitive issues.[11]

Policy decisions of ICAO are usually adopted and transmitted through the Council and in this respect the Council is a powerful and visible body in international aviation.[12] Furthermore, a decision taken by the Council is juridically dignified by Article 86 of the Convention, where the Article provides that unless the Council decides otherwise, any decision by the Council on whether an international airline is operating in conformity with the provisions of the Convention shall remain in effect unless reversed in appeal. The Council also has powers of sanction granted by the Convention, if its decision is not adhered to.[13] Schenkman states:

> The power of sanctions in this field is an entirely new phenomenon, attributed to an aeronautical body...none of the pre-war instruments in the field of aviation had the power of sanctions as a means of enforcement of its decisions.[14]

A most interesting aspect of the ICAO Council remains to be that one of its mandatory functions is to consider any subject referred to it by a Contracting State for its consideration.[15] or any subject which the President of the Council or the Secretary General of the ICAO Secretariat desires to bring before the Council. Although the Council is bound to consider a matter submitted to it by a Contracting State it can refrain from giving a decision as the Council is only obligated to consider a matter before it.

3 Law Making Powers of the ICAO Council

The words "legislative power" have been legally defined as "power to prescribe rules of civil conduct",[16] while identifying law as a "rule of civil conduct". The word "quasi" is essentially a term that makes a resemblance to another and classifies it. It is suggestive of comparative analogy and is accepted as:

> the conception to which it serves as an index and its connection with the conception with which the comparison is instituted by strong superficial analogy or resemblance.[17]

The question *stricto sensu* according to the above definition is therefore whether the ICAO Council now has power to prescribe rules of civil conduct (legislative power) or in the least a power that resembles by analogy the ability to prescribe rules of conduct (quasi-legislative power). Since legislative power is usually

[11] *Id.* p. 210.

[12] Sochor (1991), at 58.

[13] Chicago Convention, Preamble *supra* note 1, Article 87.

[14] Schenkman (1955), at p. 162.

[15] Chicago Convention, Preamble *supra* note 1, Article 54 (n).

[16] *Schaake* v. *Dolly* 85 Kan. 590, 118 Pac. 80.

[17] *People* v. *Bradley* 60 Ill. 402, at 405. Also, *Bouviers Law Dictionary and Concise Encyclopedia* 3 ed. Vol 11, Vernon Law Book Co., New York 1914.

attributed to a State, it would be prudent to inquire, on a general basis, whether the ICAO Council has law making powers (in a quasi-legislative sense). Therefore, all references hereafter that may refer to legislative powers would be reflective of the Council's law making powers in a quasi-legislative sense.

Article 54 (l) of the Chicago Convention prescribes the adoption of international Standards and Recommended Practices (hereafter, SARPS) and their designation in Annexes to the Convention, while notifying all Contracting States of the action taken. The adoption of SARPS was considered a priority by the ICAO Council in its Second Session (2 September–12 December 1947)[18] which attempted to obviate any delays to the adoption of SARPS on air navigation as required by the First Assembly of ICAO.[19] SARPS inevitably take two forms: a negative form e.g. that States shall not impose more than certain maximum requirements; and a positive form e.g. that States shall take certain steps as prescribed by the ICAO Annexes.[20]

Article 37 of the Convention obtains the undertaking of each Contracting State to collaborate in securing the highest practical degree of uniformity in regulations, standards, procedures and organization in relation to international civil aviation in all matters in which such uniformity will facilitate and improve air navigation. Article 38 obligates all Contracting States to the Convention to inform ICAO immediately if they are unable to comply with any such international standard or procedure and notify differences between their own practices and those prescribed by ICAO. In the case of amendments to international Standards, any State which does not make the appropriate amendment to its own regulations or practices shall give notice to the Council of ICAO within 60 days of the adoption of the said amendment to the international Standard or indicate the action which it proposes to take.

The element of compulsion that has been infused by the drafters of the Convention is compatible with the "power to prescribe rules of civil conduct" on a *stricto sensu* legal definition of the words "legislative power" as discussed above. There is no room for doubt that the 18 Annexes to the Convention or parts thereof lay down rules of conduct both directly and analogically. In fact, although there is a conception based on a foundation of practicality that ICAO's international Standards that are identified by the words "Contracting States shall" have a mandatory flavour (infused by the word "shall") while Recommended Practices identified by the words "Contracting States may have only an advisory and" recommendatory connotation (infused by the word "may"), it is interesting that at least one ICAO document requires States under Article 38 of the Convention, to notify ICAO of all significant differences from both Standards and Recommended Practices, thus making all SARPS regulatory in nature.[21]

---

[18]Proceedings of the Council 2nd Session 2 September–12 December 1947, Doc 7248-C/839 at 44–45.

[19]ICAO Resolutions A-13 and A-33 which resolved that SARPS relating to the efficient and safe regulation of international air navigation be adopted.

[20]ICAO Annex 9, Facilitation, Foreword.

[21]*Aeronautical Information Services Manual*, ICAO Doc 8126-0 AN/872/3. ICAO Resolution A1-31 defines a Standard as any specification for physical characteristics. . .the uniform application of which is recognised as *necessary*... and one that States *will conform to*. The same

Another strong factor that reflects the overall ability and power of the Council to prescribe civil rules of conduct (and therefore legislate) on a strict interpretation of the word is that in Article 22 of the Convention each Contracting State agrees to adopt all practical measures through the issuance of special regulations or otherwise, to facilitate and expedite air navigation... It is clear that this provision can be regarded as an incontrovertible rule of conduct that responds to the requirement in Article 54 (l) of the Convention. Furthermore, the mandatory nature of Article 90 of the Convention—that an Annex or amendment thereto shall become effective within 3 months after it is submitted by the ICAO Council to Contracting States is yet another pronouncement on the power of the Council to prescribe rules of State conduct in matters of international civil aviation. *A fortiori*, it is arguable that the Council is seen not only to possess the attribute of the term "jurisfaction" (the power to make rules of conduct) but also the term "jurisaction" (the power to enforce its own rules of conduct). The latter attribute can be seen where the the Convention obtains the undertaking of Contracting States not to allow airlines to operate through their air space if the Council decides that the airline concerned is not conforming to a final decision rendered by the Council on a matter that concerns the operation of an international airline.[22] This is particularly applicable when such airline is found not to conform to the provisions of Annex 2 to the Convention that derives its validity from Article 12 of the Convention relating to rules of the air.[23] In fact, it is very relevant that Annex 2, the responsibility for the promulgation of which devolves upon the Council by virtue of Article 54 (l), sets mandatory rules of the air, making the existence of the legislative powers of the Council an unequivocal and irrefutable fact.

Academic and professional opinion also favours the view that in a practical sense, the ICAO Council does have legislative powers.

Professor Milde concludes that ICAO has regulatory and quasi-legislative functions in the technical field and plays a consultative and advisory role in the economic sphere.[24] A similar view had earlier been expressed by Buergenthal who states:

> the manner in which the International Civil Aviation organization has exercised its regulatory functions in matters relating to the safety of international air navigation and the facilitation of international air transport provides a fascinating example of international law making... the Organization has consequently not had to contend with any of the post war ideological differences that have impeded international law making on politically sensitive issues.[25]

---

resolution describes a Recommended Practice as any specification for physical characteristics... which is recognised as *desirable* ...and one that member States *will endeavour to conform to*... Buergenthal (1969), at 10 also cites the definitions given in ICAO's Annex 9 of SARPS.

[22]Article 86 of the Convention.

[23]Article 12 stipulates that over the high seas, the rules in force shall be those established under the Convention, and each contracting State undertakes to insure the prosecution of all persons violating the applicable regulations.

[24]Milde (1989), *op. cit.* 122.

[25]Buergenthal (1969), at 9.

Paul Stephen Dempsey endorses in a somewhat conservative manner, the view that ICAO has the ability to make regulations when he states:

> In addition to the comprehensive, but largely dormant adjudicative enforcement held by ICAO under Articles 84–88 of the Chicago Convention, the Agency also has a solid foundation for enhanced participation in economic regulatory aspects of international aviation in Article 44, as well as the Convention's Preamble.[26]

A significant attribute of the legislative capabilities of the ICAO Council is its ability to adopt technical standards as Annexes to the Convention without going through a lengthy process of ratification.[27] Eugene Sochor refers to the Council as a powerful and visible body in international aviation.[28] It is interesting however to note that although by definition, the ICAO Council has been considered by some as unable to deal with strictly legal matters, since other important matters come within its purview,[29] this does not derogate the compelling facts that reflect the distinct law making abilities of ICAO. Should this not be true, the functions that the Convention assigns to ICAO in Article 44—that ICAO's aims and objectives are to "develop the principles and techniques of international air navigation and to foster the planning and development of international air transport"—would be rendered destitute of effect.

Under the Interim Agreement[30] the PICAO Council was required to act as an arbitral body on any differences arising among member States relating to matters of international civil aviation which may be submitted to it, wherein the Interim Council of PICAO was empowered to render an advisory report or if the parties involved so wished, give a decision on the matter before it.[31] The Interim Council, which was the precursor to the ICAO Council, set the stage therefore for providing the Council with unusual arbitral powers which are not attributed to similar organs of the specialised agencies of the United Nations system.[32] *A fortiori*, since the ICAO Council is permanent and is almost in constant session, Contracting States could expect any matter of dispute brought by them before the Council to be dealt with, without unreasonable delay.[33]

Most Contracting States have, on their own initiative, enacted dispute-settlement clauses in their bilateral air services agreements wherein provision is usually made to refer inter-State disputes relating to international civil aviation to the ICAO Council, in accordance with Chapter XV111 of the Convention. In this context, it is

---

[26]Dempsey (1987), at 302.

[27]Sochor (1991), at 58.

[28]*Ibid.*

[29]Tobolewski (1979) at 359.

[30]See note 41.

[31]Interim Agreement, Article 111, Section 6(i).

[32]Schenkman (1955), 160.

[33]See statement of R. Kidron, Israeli Head Delegate, Statement of the Second Plenary Meeting of the Seventh Assembly on June 17, 1953, reported in *ICAO Monthly Bulletin*, August–October 1953, at 8.

also relevant to note that the President of the Council is empowered by the Convention to appoint an arbitrator and an umpire in certain circumstances leading to an appeal from a decision of the Council.[34]

A most interesting aspect of the ICAO Council remains to be that one of its mandatory functions is to consider any subject referred to it by a Contracting State for its consideration.[35] or any subject which the President of the Council or the Secretary General of the ICAO Secretariat desires to bring before the Council.[36] Although the Council is bound to consider a matter submitted to it by a Contracting State it can refrain from giving a decision as the Council is only obligated to consider a matter before it.

4 Safety Oversight

Paragraph 2.1.1 of the *Safety Oversight Manual* of ICAO defines safety oversight as:

> a function by means of which States ensure effective implementation of the safety related Standards and Recommended Practices (SARPs) and associated procedures in the Annexes to the Convention on International Civil Aviation and related ICAO Documents.[37]

The ICAO Assembly at its 29th Session (1992), adopted Resolution A29-13 which reaffirmed that each individual State's responsibility for safety oversight is one of the tenets of the Convention on International Civil Aviation (Chicago Convention).

The Assembly, at its 37th session adopted Resolution A37-5 (The Universal Safety Oversight Audit Programme USOAP continuous monitoring approach) which recognized that safety oversight and the safety of international civil aviation in general, is the responsibility of the Contracting States, both collectively and individually.

According to the Chicago Convention, safety oversight and a State's responsibility pertaining thereto can be subsumed into two broad areas: nationality and registration of the aircraft and responsibilities that flow therefrom; and territorial jurisdiction of that State. Accordingly, there are certain mandatory responsibilities of a State in ensuring safety oversight.

---

[34]Article 85.

[35]Rules of Procedure for the Council. *op. cit.* Section 1V, Rule 24 (e). Also, Article 54 (n) stipulates that one of the mandatory functions of the Council is to consider any matter relating to the Convention which any contracting State refers to it.

[36]Rules of Procedure for the Council, *op. cit.* Section 1V Rule 24 (f). The two additional multilateral agreements stemming from the Convention and providing for the exchange of traffic rights—the Air Services Transit Agreement and the Air Transport Agreement, also contain provisions that empower the ICAO Council to hear disputes and " make appropriate findings and recommendations... see Air Services Transit Agreement Article 11 Section 1, and the Air Transport Agreement Article 1V Section 2.

[37]ICAO Doc 9734, AN/959, 2nd ed. (2006) Part A.

Article 17 of the Convention provides that aircraft have the nationality of the State in which they are registered. Article 18 prohibits the registration of aircraft in more than one State, which effectively precludes dual nationality of aircraft and responsibility of more than one State.

The registration of an aircraft in a particular State devolves upon that State certain safety related obligations. For example Article 12 of the Chicago Convention states:

> Each Contracting State undertakes to adopt measures to insure that every aircraft flying over or manoeuvring within its territory and that every aircraft carrying its nationality mark, wherever such aircraft may be, shall comply with the rules and regulations relating to the flight and manoeuvre of aircraft there in force. Each Contracting State undertakes to keep its own regulations in these respects uniform, to the greatest possible extent, with those established from time to time under this Convention. Over the high seas, the rules in force shall be those established under this Convention. Each Contracting State undertakes to insure the prosecution of all persons violating the regulations applicable.

Article 31 of the Convention provides:

> Every aircraft engaged in international navigation shall be provided with a certificate of airworthiness issued or rendered valid by the State in which it is registered.

Along the same lines, Article 32 provides:

> *a*) The pilot of every aircraft and the other members of the operating crew of every aircraft engaged in international navigation shall be provided with certificates of competency and licenses issued or rendered valid by the State in which the aircraft is registered.
> *b*) Each Contracting State reserves the right to refuse to recognize, for the purpose of flight above its own territory, certificates of competency and licenses granted to any of its nationals by another Contracting State.

In the context of equipment on board aircraft, Article 30 of the Convention requires that over the territory of States other than their State of registration, aircraft shall carry radio transmitting apparatus only if a license to install and operate such apparatus has been issued by the appropriate authorities of the State in which the aircraft is registered. The use of radio transmitting apparatus in the territory of the Contracting State whose territory is flown over shall be in accordance with the regulations prescribed by that State.

The only exception to these mandatory principles is found in Article 83 *bis*, where, for purposes of expediency, when an aircraft registered in a Contracting State is operated pursuant to an agreement for the lease, charter or interchange of the aircraft or any similar arrangement by an operator who has his principal place of business or, if he has no such place of business, his permanent residence in another Contracting State, the State of registry may, by agreement with such other State, transfer to it all or part of its functions and duties as State of registry in respect of that aircraft under Articles 12, 30, 31 and 32 (a). In such an instance the State of registry is relieved of responsibility in respect of the functions and duties transferred.

In pursuance of the above obligations, one could validly conclude that Article 37 of the Chicago Convention imposes an obligation on each Contracting State to undertake to collaborate in securing the highest practicable degree of uniformity in regulations, standards, procedures, and organization in relation to aircraft, personnel, airways and auxiliary services in all matters in which such uniformity will facilitate and improve air navigation. These Standards and Recommended Practices are contained in the 18 Annexes to the Chicago Convention, all of which except Annexes 9 (Facilitation) and 17 (Security) are applicable to safety oversight either directly or indirectly.

5 Regional Safety Oversight

Regional governance under State responsibility is an undeniable necessity in ensuring safety in air transport. However, since international air transport crosses boundaries and all regions of the world, a certain stringent global regulatory system is required, and this objective cannot be reached unless the safety oversight system is de-centralized into safety agencies in each region. This is because although air transport is still considered the safest mode of travel, constant and diligent vigilance in ensuring the safety of aviation is required. The International Civil Aviation Organization (ICAO), which is the global forum for aviation, provides this guidance and direction and encourages States to form regional civil aviation bodies in regions where they do not exist. Since almost all regional civil aviation bodies have their own Constitutions, these Constitutions should recognize the need for increased cooperation with ICAO and ICAO should actively seek such an inclusion. Cooperation could include assignment of tasks related to oversight, accident investigation, sharing of information and the conduct of training in collaboration with ICAO.

Some of the measures that are suggested in this article include, the initiation by ICAO where appropriate, of periodic meetings with regional civil aviation bodies and the setting up of panels that include members from ICAO and each regional body on a selective basis. This panel could recommend and monitor issues of cooperation, including cooperation between the various regional organizations.

There is also the compelling need to define the role that regional organizations and regional civil aviation bodies would play in working closely with ICAO; and the adoption of a policy with regard to cooperation with regional organizations and regional civil aviation bodies. ICAO should also ascertain the role to be played by the regional offices in coordinating ICAO cooperation with such bodies. Consideration should be given to providing appropriate resources in regional offices where necessary.

This article also discusses in some depth the various regional safety oversight organizations agencies (RSOOs) and examines the legal and regulatory basis of ICAO's role with regard to the establishment and functions of RSOOs with reference to selected regional oversight organizations around the world.

The role played by Regional Safety Oversight Organizations (RSOOs)[38] has sustained serious discussions in ICAO and in particular in the Council[39] of that

[38]Sometimes referred to as Regional Safety Oversight Agencies (RASAs).

[39]3. The ICAO Council is a permanent body responsible to the Assembly. It is composed of 36 Member States elected by the Assembly. In electing the members of the Council, the Assembly

Organization over the past year. At its 187th Session on 10 June 2009, the Council accepted several principles with regard to the role of ICAO in regional governance[40] which were placed before it by a multidisciplinary group comprising some Representatives on the Council and members of the ICAO Secretariat, which had been tasked with producing a report on ICAO and Regional Governance, by the Council.

At its meeting, the Council agreed ICAO should enhance its cooperation with regional organizations and regional civil aviation bodies and vice-versa, both in the technical and economic fields and should ensure that the interests of States which do not belong to regional civil aviation bodies should not be jeopardized or compromised in the above context. The Council also recognized that while ICAO encourages the activities of States and regional civil aviation bodies in facilitating the implementation of Standards and Recommended Practices of the Annexes to the Convention on International Civil Aviation (Chicago Convention) (SARPs), States ultimately remained responsible for their obligations under the Convention on International Civil Aviation, notwithstanding whatever arrangements States may conclude with their regional civil aviation bodies. It was also agreed that ICAO should encourage States to form regional civil aviation bodies in regions where they do not exist, and that ICAO should define the role that regional organizations would play in working closely with ICAO.

Article 55 (a) of the Chicago Convention states that the Council may, where appropriate and as experience may show to be desirable, create subordinate air transport commissions on a regional or other basis and define groups of States with or through which it may deal to facilitate the carrying out of aims of the Convention. Article 65 of the Convention allows the Council to enter into agreements with other international bodies for the maintenance of common services with and for common arrangements concerning personnel and, with the approval of the Assembly, allows the Council to enter into such other arrangements as may facilitate the work of the Organization.

This provision prompted the ICAO Assembly, at its First Assembly held on 6–27 May 1947 to adopt Resolution A1-10, which recognized that there were a number

---

gives adequate representation to States of chief importance to air transport; States not otherwise included which make the largest contribution to the provision of facilities for international air navigation; and States not otherwise included whose designation will ensure that all the major geographic areas of the world are represented on the Council. The mandatory and permissive functions of the Council are stipulated in Articles 54 and 55 of the Convention on International Civil Aviation respectively. The Council has its genesis in the Interim Council of the Provisional International Civil Aviation Organization (PICAO). PICAO occupied such legal capacity as may have been necessary for the performance of its functions and was recognized as having full juridical personality wherever compatible with the Constitution and the laws of the State concerned. See Interim Agreement on International Civil Aviation, opened for signature at Chicago, December 7 1944, Article 3. Also in Hudson, *International Legislation*, Vol 1X, New York: 1942–1945, at 159. For a detailed discussion on the functions of the Council See Abeyratne (1992).

[40]C-WP/13339, 12/05/09.

of public organizations whose activities affect or are affected by ICAO and that the work of ICAO and the advancement of international civil aviation will be enhanced by close cooperation with such organizations. The Resolution authorized the Council to make appropriate arrangements with public international organizations whose activities affect international civil aviation and suggested that such arrangements be established through informal working arrangements rather than formal arrangements, wherever practicable.

At its Tenth Session conducted in Caracas from 19 June to 16 July in 1956, the ICAO Assembly adopted Resolution A10-5, which established a policy framework for ICAO and the European Civil aviation Conference (ECAC) which had its focus on regional cooperation. An agreement effective 12 July 1969 was signed by ICAO and ECAC which allowed the appointment of ICAO staff to the Secretariat of ECAC, and the provision of qualified and expert staff to the latter. A similar agreement effective 1 January 1978 was signed between ICAO and the African Civil Aviation Commission (AFCAC0 which provided *inter alia* that ICAO will provide, to the extent possible, secretariat services to AFCAC. There is also an agreement, which became effective on 1 January 1979 between ICAO and the Lating Civil Aviation Commission (LACAC) which called for close cooperation between the two Organizations and the provision of secretariat services to LACAC by ICAO.

There are numerous other instances of such regional cooperation which followed such as the signing of a memorandum of understanding by ICAO and LACAC on 1 October 1999 followed by an MoU between ICAO and ECAC on 11 November 1999. These were to be followed by a management service agreement between ICAO and AFCAC which came into effect on 1 January 2007.

These regional bodies are by themselves not RSOOs, although they could, and indeed should play an integral role in assisting States in their oversight functions and the RSOOs in their work. In this regard, it is critical to recognize that regional governance and regional safety oversight are prerogatives which devolve consequent responsibility on the State which cannot be delegated. The RSOO acts as an agent[41] of the State (which is the Principal) with delegated functions from the Principal. ICAO, to the extent of its empowerment by its member States, can assist in the oversight process, offer expertise, advice, technical support and even manage an RSOO if necessary.

Therefore, ICAO's position with regard to RSOOs is that it would give its full cooperation to regional governance as determined by States through carefully formulated policy which would in no way abdicate or delegate ICAO's responsibilities under the Chicago Convention. In the agency equation the State remains the Principal and the RSOO remains the Agent and ICAO remains the organization empowered by the States to carry out audits and report results thereof.

[41]An agency is a consensual relationship created by contract or law where the Principal grants authority to agents to act on behalf of or under the control of the Principal to deal with a third party.

6 The Regional Oversight Manual

ICAO's perspectives on regional safety oversight are reflected in its *Regional Safety Oversight Manual*[42] which provides guidance for States that wish to form a regional safety oversight system. It formulates a regional strategy which attenuates and brings together the efforts, experiences and resources of Contracting States. From ICAO's perspective the reasons for adopting a strategy to establish RSOOs are to: eliminate a duplication of effort by standardizing regulatory and enforcement provisions over a large area of aviation activities; achieve economies of scale leading to efficiency and effectiveness; pool human and financial resources; institute effective regional programmes through the joint action of States; address external factors and constraints more effectively; develop and implement a safety management system[43] that would allow for the implementation of similar standards and procedures to measure the safety performance of civil aviation organizations in the region; supplement shortfalls in the scope of domestic and bilateral interventions; prove organizational ability by testing activities before making important commitments under national programmes; meet industry expectations; demonstrate improved regional solidarity; improve the objectivity and independence of inspectors; and develop a capability for drafting and amending regulations and procedures.[44]

ICAO believes that RSOOs should have goals which reflect national priorities, and recommends that States consider providing adequate and efficient resources and ensure coordination between high level government officials who are responsible for aviation safety. The most incontrovertible principle ICAO believes in is, as mentioned earlier, that the responsibility for safety oversight within the State remains with the State and cannot be delegated.[45] The critical activities of a State are identified in the *Manual* as: the licensing of operational personnel; the certification of aircraft, air operators, aerodromes, and maintenance organizations; the control and supervision of licensed personnel, certified products and approved organizations; the provision of air navigation services[46]; and aircraft accident and incident investigation.[47]

ICAO's role, according to the *Manual*, is to provide assistance to States that are willing to enter into a cooperative agreement for the establishment and management of an RSOO on the basis of an agreement between ICAO and the interested parties. ICAO may also manage an RSOO until such time as the State concerned develops

---

[42]ICAO Doc 9734 AN/959 Part B (The Establishment and Management of a Regional Safety Oversight System).

[43]For a detailed discussion on safety management systems see Abeyratne (2007).

[44]*Id.* Para 2.2.4 at pp. 2–3.

[45]*Id.* Para. 2.2.5 at pp. 2–4.

[46]Including meteorological information, aeronautical telecommunications, search and rescue services, charts and the distribution of information.

[47]*Id.* Para. 2.4.5 at pp. 2–6.

an ability to perform the task. The Organization may also provide technical and logistical support as well as necessary information and documents that may be needed by an RSOO and also, upon a request of a State monitor the effectiveness of an RSOO. Where necessary and as requested, ICAO could, according to the Manual, provide advice or propose action required in the proper running of an RSOO.[48]

In this regard the Symposium was also advised that ICAO has conducted a series of safety audits throughout the world and information of these audits were available.

7 The ICAO AFI Comprehensive Implementation Programme (ACIP)

The ICAO Assembly, at its 36th Session held in September–October 2007, adopted Resolution A36-1 which recognized that it was essential that there be increased coordinated efforts under ICAO leadership to reduce serious deficiencies in the Africa-Indian Ocean (AFI) Region which are detrimental to the functioning an further development of international civil aviation. The resolution also noted that the Council of ICAO had already taken steps to address safety issues through the development of a Comprehensive Regional Implementation Plan for Aviation Safety in Africa (the AFI Plan)[49] and that many Contracting States in the AFI Region may not have the technical or financial resources to comply with the requirements of the Chicago Convention and its Annexes. It also recognized the need to coordinate, under the ICAO umbrella, activities of all stakeholders providing assistance to States in the AFI Region. Resolution A36-1 therefore urged Contracting States of the AFI Region to commit to the achievement of the goals and objectives of the AFI Plan and to ongoing transparency with regard to the progress accomplished. It instructed the Council to coordinate the contributions towards the implementation of the AFI Plan and to ensure a stronger ICAO leadership role in coordinating activities, initiatives and implementation strategies aimed specifically at meeting the goals and objectives of the Plan, in order to achieve sustained improvement of flight safety in the AFI Region and to allocate resources to the relevant Regional Offices accordingly; Furthermore the Council was instructed to ensure the continued development of new working relationships integrating the capabilities of the bureaux at Headquarters with the resources of Regional Offices.

The Council started with the initial premise that the problems facing the States in the AFI Region were not dissimilar to what other States around the world were facing. The one unique feature of the AFI region, however, was that the acute

[48]*Id.* Para 3.5 at pp. 3–5.

[49]The Comprehensive Regional Implementation Plan for Aviation Safety in Africa (AFI Plan) was developed by ICAO with a view to addressing the various concerns expressed by the ICAO Council on the status of safety of aircraft operating in the African and Indian Ocean Region.

economic and political issues facing States of the AFI region were indeed complex in nature and therefore posed a challenge to aviation safety in the region which clearly required a new approach. The AFI Plan was a response to this particular exigency and was developed with a view to addressing aviation safety concerns and supporting African States in their endeavour to meet their international obligations for safety oversight under the applicable Annexes to the Chicago Convention.

The AFI Plan was considered by a high-level conference convened in Montreal on September 17, 2007, which unreservedly endorsed it. . Consequently, the AFI Plan was presented to the 36th Session of the ICAO Assembly, held from 18 to 28 September 2007 and the Assembly requested both the Council and the Secretary General to implement the AFI Plan within the shortest possible period. The Assembly's AFI Plan Resolution, A36-1, also emphasized a heightened leadership role and accountability by ICAO for the Plan's effective implementation, supported by strong programme management and coordination activities. Thus, under ICAO's leadership, the AFI Plan calls for optimal collaboration between regulatory agencies and industry in the implementation of initiatives aimed at rectifying safety deficiencies.

To this end, specific objectives have been developed requiring ICAO to: (a) Increase compliance with ICAO SARPs and industry best practices; (b) Increase the number of qualified personnel at the industry and oversight levels. (c) Improve the quality of inspectors and other civil aviation staff through training. (d) Ensure impartial and unimpeachable investigation and reporting of serious accidents and incidents. (e) Enhance regional cooperation. (f) Enhance capacity of regional and sub-regional safety oversight systems. (g) Improve assistance in oversight to least developed States. (h) Provide expert aviation knowledge within the reach of the targeted States via the web.

The Secretary General established the AFI Comprehensive Implementation Programme (ACIP) with a view to ensuring that the aforesaid objectives are met. For this purpose, he nominated members to the Steering Committee of ACIP to oversee its work. ACIP, which is aimed at implanting the AFI Plan recognizes the basic requirements of a safety oversight programme to be a regulatory framework; responsibility and accountability; accident and incident investigation; and enforcement policy, all of which were necessary for a State safety programme. He identified the essential elements for a safety risk management programme as an adequate oversight organization with trained personnel, equipment, tools, guidance material and processes and procedures for exchanging safety critical information.

The basic requirements for an effective safety culture are management commitment and safety accountability; appointment of key safety personnel; coordination of emergency response planning and safety management system documentation.

8 Safety Oversight in Africa

It is heartening that safety levels in Africa were improving although the accident rates in Africa were still above world average levels. The main instances of accidents in African States were mostly seen in conflict ridden States, particularly in the Democratic Republic of Congo, Angola and Sudan. The main reasons for accidents were ageing aircraft, lack of financing for fleet modernization and the lack of adequate skilled personnel. The reduction of accident rates in Africa could be attributed to the adoption of IOSA,[50] the ICAO Global Safety Roadmap[51] and the proactive approach taken towards safety by the key stakeholders. He concluded that RSOOs would enhance safety oversight and that toward this end there was a need for African States to have autonomous and regionalized civil aviation authorities. He was also of the view that there was a need to pool scarce resources in training and to harmonize standards on conditions of service and the operation of regional airlines. One of the key features of RSOOs was their ability to be better able to provide attractive conditions of service for their personnel and to mobilize resources for infrastructure.

9 The Banjul Accord Group (BAG)

The BAG States (Cape Verde, Gambia, Ghana, Guinea Conakry, Liberia, Nigeria and Sierra Leone) play a significant role in leadership in aviation safety oversight. The 10th Plenary Session of the BAG States held on 30–31 October 2008 reached the conclusion that a regional safety oversight agency and a regional accident investigation agency must be established. In this regard BAG States had sent a letter seeking assistance to ICAO. ICAO was requested for assistance in developing a framework for the establishment of the BAG Regional Safety Oversight Organization (BAGASOO) and the development of a framework for the establishment of a Regional Accident Investigation Agency (BAGAIA). ICAO was also requested to assist in developing operational regulations and guidance material for the implementation of BAGASOO and BAGAIA and the development of a guidance manual for the implementation of a regional safety programme.

It is also noteworthy that the Directors General of Civil Aviation of the BAG States met with ACIP management in Accra on 15–16 December 2008 and developed an action plan and determined the scope of cooperation, resulting in a letter of understanding.

---

[50]The IATA Operational Safety Audit (IOSA) programme is an internationally recognised and accepted evaluation system designed to assess the operational management and control systems of an airline. IOSA's quality audit principles are designed to conduct audits in a standardised manner. With the implementation and international acceptance of IOSA, airlines and regulators achieve the following benefits: a reduction of costs and audit resource requirements for airlines and regulators; continuous updating of standards to reflect regulatory revisions and the evolution of best practices within the industry; a quality audit programme under the continuing stewardship of IATA; accredited audit organisations with formally trained and qualified auditors; accredited training organisations with structured auditor training courses; a structured audit methodology, including standardised checklists; elimination of audit redundancy through mutual acceptance of audit reports; and development of auditor training courses for the airline industry.

[51]For a discussion on the Roadmap, see Abeyratne (2009a), pp. 29–36.

Three events which had already been held and were significant in the evolution of BAGASOO were: The Global Aviation Roadmap Workshop (Abuja, April 2008); GAP Analysis with GASRs in All BAG States (May–June 2008); and the 10th Plenary Session of the BAG States (Banjul, October 2008). Priority action of BAGASOO, as already presented to the ICAO Council included: development of a framework for the accelerated establishment of BAGASOO; Development of a framework for the establishment of BAGAIA; development of operational regulations and guidance material for the implementation of BAGASOO and BAGAIA and the development of guidance material for the implementation of the regional programme.

The ICAO Council, in November 2008, had requested Contracting States, industry and donors to assist the BAG States in implementing priority projects and actions as determined during the 10th Plenary Session of the BAG States, and that ACIP take necessary measures to assist BAG States in the implementation of such projects.

9.1 Safety Oversight in Central America (ACSA)

ACSA, the Central American Agency for Aeronautical Safety, created in 2001, is committed to the regional capability of safety oversight, through reduced bureaucracy and the avoidance of duplication of efforts. ACSA concentrates on sharing expertise and resources among member States and was ISO 9001 compliant in terms of quality management. ACSA is an agency of COCESNA, which was tasked with implementing plans pertaining to COCESNAs Regional Information Aeronautical System (SIAR) on safety issues, continuous surveillance, licensing and qualified personnel.

9.2 Safety Oversight in the Caribbean

The Civil Aviation Authorities of Barbados, Belize, Guyana, Haiti, Jamaica, OECS (Anguilla, Antigua and Barbuda, Dominica, Grenada, Montserrat, St. Lucia, St. Kitts and Nevis, St. Vincent and the Grenadines), Suriname, and Trinidad and Tobago, Member States of the Caribbean Community, signed an agreement in late 2001, formalizing their participation in and support for a cooperative approach to aviation safety oversight. The Agreement provided for the establishment of an "Association of Civil Aviation Authorities of the Caribbean" (ACAAC) under the umbrella of the Caribbean Community (CARICOM) Secretariat and to form its operating arm, the Regional Aviation Safety Oversight System (RASOS). Each RASOS member authority was mandated to implement the provisions of all ICAO Annexes. It was the RASOS mandate to assist them with specific regard to Annexes 1, 6, and 8 of the Chicago Convention. The mandate also involved aiding, facilitating, harmonizing and sharing resources for the provision of aviation safety oversight services in thirteen small nations in the Caribbean region. Although all participating authorities belong to States that are members of CARICOM, membership in the CARICOM was never a pre-requisite for membership in the ACAAC.

RASOS Member States developed a formal agreement signed by Heads of State in 2008 to widen the regional organization's mandate to include all ICAO Annexes. This marked a major step forward in elevating the RASOS status by establishing it as a new entity, and renaming RASOS as the Caribbean Aviation Safety and Security Oversight System (CASSOS) and having it designated as an Institution of the Community by the Conference of Heads of Government pursuant to Article 21 of the *Revised Treaty of Chaguaramas Establishing the Caribbean Community Including the CARICOM Single Market and Economy*. The agreement subsumed RASOS into CASSOS and ACAAC no longer exists. CASSOS has full juridical personality, and its Board of Directors report to the Ministers who constitute the CARICOM Council for Trade and Economic Development (COTED).

Originally, the Members of ACAAC implemented the Regional Aviation Safety Oversight System (RASOS), to share resources and reduce the cost of providing the required airworthiness and flight operations oversight services to individual member states. The RASOS concentrated on the optimization of the use of the region's technical resources. Its focus was to strengthen the civil aviation authorities, promote the upgrading and harmonization of regulations, standards, procedural guidance material, inspector training and to enhance the ability of the individual states to fully discharge their safety oversight responsibilities. The sub-regional approach chosen is consistent with the global strategy promoted by ICAO to address safety oversight problems of Contracting states. RASOS core operations were funded by equal annual member CAA contributions and it was self sufficient during its existence, and by its frugality was able to commence CASSOS with no additional expenditures.

Assistance from the FAA between the years 2003–2008 provided numerous training courses for RASOS Member CAA's inspectors and other technical staff, ICAO courses were delivered in the region with regard to aerodrome certification and dangerous goods and PEL. Transport Canada assisted with some compliance/enforcement training, and medical examiner training. The FAA provides extensive in-country assistance by providing technical experts in an effort that was aimed at advising and assisting Members to achieve IASA category one and compliance with Annexes 1, 6 and 8. This assistance included mentoring of inspectors and technical advice provided during re-certification of air operators. Other ongoing FAA technical assistance pursuant to Technical Assistance Agreements was aimed at full implementation in the first half of 2008 of a common and ICAO compliant computer based written knowledge testing system. A regionally developed, harmonized, common license format and production system has been installed in all member authorities and could be made be available commercially to any other authorities that are interested in such a system. Transport Canada continues to assist with training of civil aviation medical examiners and cabin safety inspectors and is offering ongoing training support in SMS and aviation compliance and enforcement. Common qualifications and training standards for inspectors have been enunciated to facilitate resource sharing, that is, the transnational use of inspectors, and procedures for designation and delegation of authority and for requesting, tasking and deploying transnational inspectors have been developed and approved. A Policy and Procedures Manual was developed to guide the management and operation of RASOS and in its latest version will now be used to guide and direct CASSOS activities. Inspector guidance material is shared freely between Members and is well harmonized already. It is anticipated that development of unitary common guidance material will follow the development of common "regional" civil aviation regulations while at present, all regulations are based on adaptations of the ICAO MCAR and are virtually identical. Other initiatives include harmonized enforcement procedures and inspection procedures.

In 2009 four of the original seven RASOS Member CAAs continued to meet the IASA Category One standards. CASSOS, in a manner similar to that used by the European Aviation Safety Agency (EASA), performs reviews of its Members using experienced inspectors from the region and checklists derived from IASA and ICAO USOAP checklists. Reports developed for the Member authorities are reviewed and the results are used to determine, prioritize and respond to region-wide needs. The reviews assist with harmonization activities and have also assisted members to prepare for IASA and ICAO audits some Members find them useful to develop compliance action plans. There remains an ongoing need for on-site mentoring and training of technical personnel and for technical assistance in all Member CAAs, particularly as the CASSOS mandate has been significantly widened and new expectations arise resulting from changes to the Annexes and technology.

Using needs assessment methodology the regional body has identified the need for professional training and recurrent qualification training of airport operators' personnel. It has from its own resources and assisted by a member of the FAA airport standards staff, delivered a 3 day seminar on aerodrome manual preparation to some 33 aerodrome specialists from the region.

The RASOS web site contains public information and members' only sections. The inspectors' section contains downloadable inspection forms, some common guidance material, flight test forms and other data required by the region's technical staff. The site also hosts a safety newsletter, links to Member CAA sites, links contact to RASOS and provides a secure 128 bit encrypted e-mail service for the Directors, RASOS staff and all technical safety inspectors in the RASOS group and other selected officials who have been working with RASOS. It is a very strong tool for communication, information and data sharing and for providing a public identity for the organization as well as serving as a virtual office for RASOS personnel. The website is being updated and changed to reflect CASSOS operations and that should be completed by October, 2009.

CASSOS has adopted the European Center for Civil Aviation Incident Reporting System (ECCAIRS) for incident and accident reporting and, in a regional project, CASSOS Members use a common, harmonized regional inspection planning, tracking and reporting system. While this system respects national security, confidentiality and sovereignty as required, it provides a valuable tool for analysis and tracking of trends and allow development of appropriate safety and regulatory interventions. CASSOS will share inspection data as in the European Safety Audit of Foreign Aircraft (SAFA) system. Seminars in ALAR CFIT accident reduction have been delivered and this will continue under CASSOS with a much widened safety promotion mandate. The regional organization has assisted members with accident and incident investigation and it is envisaged that this will grow into a truly regional service as the benefits of a centralized investigating office are beyond question. The foregoing summarizes the major efforts of the past 7 years toward safety oversight harmonization within the CARICOM CSME framework.

All of the above initiatives have been aimed at building a strong regional regulatory and Safety Management System to enhance civil aviation safety in Member states and throughout the region and are continuing under CASSOS. Funding at this moment in time is limited to provision of two technical experts and one administration person. Future development and strengthening of the regional safety oversight capability may require increasing member contributions or new sources of funding or assistance.

The direct beneficiaries of the regional CASSOS institution activities are the participating States of CARICOM and will include any other States or Territories in the region that might become part of the regional aviation safety oversight mechanism during the next few years. Other direct beneficiaries of CASSOS activities are the owners and operators of aircraft and all who use the aviation industry infrastructure and services in the CASSOS States. One must not overlook the indirect beneficiaries of the air transport, aviation services and infrastructure that includes the tourism and business sectors of the economies. External benefits flow to the States from the improved aviation safety environment resulting from the upgraded aviation infrastructure and the increased surveillance and enforcement of the safety standards established by ICAO.

CASSOS has matured from its fairly humble beginnings during 7 years of hard work by all persons involved and now has 7 years of successful operating experience in coordinated, cooperative, harmonized, self-sufficient group efforts aimed at providing safety oversight services to the high economic value air transportation system in all participating States as well as to other States whose airlines operate into the region. This high level of achievement will continue as CASSOS continues its growth into a truly regional institution.

The immediate benefits of regional cooperation are evident from the constantly improving track record of results of the ICAO and FAA safety oversight audits of member CAAs. Benefits are also accruing to members from the mutual technical cooperation, mutual technical assistance, attainment of greater numbers of trained and qualified technical inspectors, and the valuable technical expertise contributions made by all Members in their efforts to achieve and sustain compliance with international aviation safety oversight standards at affordable costs. A strong regional safety oversight partnership has been forged. Future activities are aimed at establishing a permanent Headquarters, undertaking new regional projects such as a single upper airspace control system, introduction of new ATM surveillance technology and air navigation technology, and managing safety initiatives and interventions to keep the aviation system loss rates as low as possible.

9.3 Safety Oversight in South Asia

South Asian Regional Initiative (SARI) comprises a group of authorities from South Asia including those of Bangladesh, Bhutan, India, Maldives, Nepal, Pakistan and Sri Lanka. SARI was created during the EU-South Asia Civil Aviation Cooperation Programme in 2006. SARIs objective is to create a forum for civil aviation authorities in South Asia to foster regulatory convergence. It was mentioned that EASA and the European aerospace industry support the activities of SARI with a view to reviving the Cooperation Programme.

SARI was created to update insufficient and outdated legislation, regulations and guidance on aviation safety and provide a legal foundation for regional cooperation in South Asia. It is also aimed at assisting airlines and regulators in interpreting rules. One of the main goals of SARI is to ensure that the region retains qualified technical personnel and to eliminate duplication of efforts and confusion of functions within civil aviation authorities.

9.4 Safety Oversight in Europe

Arguably the most developed regional implementation system of safety oversight lies in Europe, through the European Aviation Safety Agency (EASA). EASA has evolved through a gradual process of regulation in Europe where the roots of institutionalized regional regulation of aviation safety goes back to 1954 with the creating of the European Civil Aviation Conference (ECAC). ECAC is a consultative body whose resolutions and other recommendations for Europe are subject to approval of the European Union (EU) member States. The predecessors of EASA were the Joint Airworthiness Authorities (JAA) who began their work in 1970 with the initial mandate given by the EU to produce common certification codes for large aircraft and engines. The Authorities were intrinsically associated with ECAC and subsequently came to be renamed as the Joint Aviation Authorities, retaining the earlier acronym. The new JAA had an extended mandate reaching into operations, maintenance, licensing, certification and design standards for all types of aircraft. One of the products of JAA was JARs (Joint Aviation Requirements) which carried regulatory thrust within the EU.[52]

EASA came into being on 15 July 2002,[53] replacing JAA in 2009. It is an agency of the European Union and is governed by EU public law. It stands separate from the EC institutions such as the European Council, European Parliament and the European Commission, and has its own legal *persona*. Its main areas of focus, in terms of ensuring the highest standards within Europe, are aviation safety and environmental protection (from an aviation perspective) in which it develops

[52]As per European Regulation 3911/11.

[53]EASA was created by Regulation EC No. 1592/2002 by the European Parliament and the Council of the EU.

rules within Europe. EASA employs inspectors to monitor the implementation within EU States of ICAO SARPs and cooperates with national authorities on the issuance of certificates of airworthiness and the licensing of technical crew.

EASA's responsibilities include: giving expert advice to the EU for drafting new legislation; implementing and monitoring safety rules, including inspections in the Member States; type-certification of aircraft and components, as well as the approval of organisations involved in the design, manufacture and maintenance of aeronautical products; authorization of third-country (non EU) operators; and safety analysis and research. These responsibilities are growing to meet the challenges of the fast-developing aviation sector. It is envisioned that, in a few years, the Agency will also be responsible for safety regulations regarding airports and air traffic management systems.

On 28 September 2003, EASA took over responsibility for the airworthiness and environmental certification of all aeronautical products, parts, and appliances designed, manufactured, maintained or used by persons under the regulatory oversight of EU Member States.

The Agency's certification work also includes all post-certification activities, such as the approval of changes to, and repairs of, aeronautical products and their components, as well as the issuing of airworthiness directives to correct any potentially unsafe situation. All type-certificates are therefore now issued by the European Aviation Safety Agency and are valid throughout the European Union.

On the same date the Agency became the competent authority to approve and oversee the organisations involved in the design of aeronautical products, parts and appliances. It also carries out the same role for foreign organisations involved in the manufacture or maintenance of such products.

To execute its tasks within the present period of building up its resources, the Agency relies on national aviation authorities who have historically filled this role and concludes contractual arrangements to this effect.

If one were to compare the sharing of responsibilities between national and regional authorities involving ACSA, BAGASOO, CASSOA, CASSOS and EASA, on initial airworthiness, EASA, ACSA and BAGASOO share responsibility with national authorities on functions (not accountability) while CASSOA and CASSOS do not show such sharing. On continuing airworthiness all agencies show shared responsibility at national and regional levels. This is also the case in operations and licensing. On air traffic management and airports there is consistency in all agencies in sharing responsibility of functions.

On regulatory aspects, all agencies in the comparison exercise share responsibilities on the development of regulations while none have a mandate to develop hard law (which is as should be, as legislation is the sole purview of sovereign States). CASSOS and EASA have the power to issue certificates while the other three do not have this power. In terms of judicial control, ACSA is under the jurisdiction of the American Court of Justice, while CASSOA is under the East African Court of Justice. CASSOS comes under the Caribbean Court of Justice while EASA falls under the jurisdiction of the European Court of Justice.

So where does this leave ICAO? To start with, it is incontrovertible that ICAO's involvement in RSOOs is both inevitable and pervasive. Whatever the regional or national initiatives may be in this area, since air transport involves the operation of foreign built aircraft in States, their certification and licensing has to be carried out according to standards set by ICAO in the relevant Annexes to the Chicago Convention.[54] This is particularly recognized by Article 33 of the Convention which provides for the acceptance by one State of certificates of airworthiness/ competency and licenses issued by another, provided such document conform to the specifications set by the Convention. Therefore, whether it be the International Safety Assessment Programme (IASA) of the United States or EASA of the European Union, It is likely that any prudent State will refuse to recognize documents if they do not meet the Standards of the ICAO Annexes. In terms of empowerment and enforcement with regard to implementation, Article 87 of the Convention provides that the ICAO Council can ban the operation of an airline worldwide if such operation does not conform to a final decision of the Council. Therefore, ICAO is indispensable for the global operation of aircraft within the safety oversight equation.[55]

However, regional initiatives such as those discussed above can give ICAO support and an impetus in its work. In this regard it must be stated that there are many measures that ICAO could take in making its work easier and more effective. Firstly, ICAO should enhance its cooperation with regional organizations and regional civil aviation bodies and *vice-versa*, both in the technical and economic fields. In doing so, ICAO should ensure that the interests of States which do not belong to regional civil aviation bodies should not be jeopardized or compromised. It should be noted that, while ICAO encourages the activities of States and regional civil aviation bodies in facilitating the implementation of SARPs, States ultimately remain responsible for their obligations under the Convention on International Civil Aviation, notwithstanding whatever arrangements States may conclude with their regional civil aviation bodies;

ICAO should also encourage States to form regional civil aviation bodies in regions where they do not exist. Since almost all regional civil aviation bodies have their own Constitutions, these Constitutions should recognize the need for increased cooperation with ICAO and ICAO should actively seek such an inclusion. Cooperation could include assignment of tasks related to oversight, accident investigation, sharing of information and the conduct of training in collaboration with ICAO.

ICAO should, where appropriate, initiate periodic meetings with regional civil aviation bodies and set up a panel that includes members from ICAO and each regional body on a selective basis. This panel could recommend and monitor issues of cooperation, including cooperation between the various regional organizations.

---

[54]Regulation EC No. 2111/2005 requires that certificates and licences issued in the EC member States have to conform to international safety standards contained in the Chicago Convention and its Annexes.

[55]For a more in depth discussion on the powers of the ICAO Council in the field of Safety, See Abeyratne (2009b) at 196–206.

ICAO should define the role that regional organizations and regional civil aviation bodies would play in working closely with ICAO; and should adopt a policy with regard to cooperation with regional organizations and regional civil aviation bodies. ICAO should also ascertain the role to be played by the regional offices in coordinating ICAO cooperation with such bodies. Consideration should be given to providing appropriate resources in regional offices where necessary.

Regional governance under State responsibility is an undeniable necessity and ICAO has been unreserved in supporting it. ICAO's policy would be one driven by its ability to provide assistance and to implement this policy ICAO needs close cooperation with the regional civil aviation bodies and the political will of States. Any ICAO policy, however enthusiastically adopted would be destitute of effect if there is no endorsement and support of its member States.

9.5 The ICAO Safety Roadmap

In the field of aviation safety, there are three incontrovertible truths. The first is that the attainment of safety is the highest priority in aviation. The second is that safety cannot be de-regulated. The third is that safety is a global concern and therefore States cannot have their own individual plans for ensuring safety. The last brings to bear a compelling need for an action plan of global dimension that identifies the roles to be played by the regulatory and industry partners in ensuring safety.

Significant progress was made throughout 2007 in transforming ICAO into a performance-based and results oriented Organization, in keeping with the Organization's Business Plan. The most significant improvements are highlighted in this Annual Report which, in its more accessible format and with links to the ICAO website, is a graphic illustration of this new way of doing business. Responding to the outcome of the 36th Session of the ICAO Assembly and to facilitate the transition to results-based planning and results-based budgeting, the Council, in November, reviewed the Business Plan for the next triennium (2008–2010) to support the implementation of the Strategic Objectives. The task involved identifying and implementing additional ways and means of further increasing ICAO's efficiency as an ongoing process throughout the Organization.

One of ICAO's Strategic Objectives is to enhance aviation safety globally, and to this extent the ICAO Safety Roadmap serves as an effective tool to assist member States of the Organization in addressing their safety issues efficiently. This article addresses the main issues involved in ICAO's role in this regard.

The Safety Roadmap (hereinafter referred to as the Roadmap), which is applicable to all these statements and responds to needs arising therefrom, was put together by a group of industry partners called the Industry Safety Strategy Group (ISSG), the establishment of which was inspired by the International Civil Aviation Organization (ICAO) at its Seventh Air Navigation Commission Industry Meeting in May 2005. Although the title of this article ascribes ownership of the Safety Roadmap to ICAO, it is a joint effort of the ISSG which comprises Airports Council International, Airbus Industrie, Boeing, the Civil Air Navigation Services

Organization, Flight Safety Foundation, International Air Transport Association and the International Federation of Airline Pilots Associations.

The Roadmap presents a phased approach that would ensure safety in aviation through a proactive modality using ICAO as the main protagonist. Part 1 of the Roadmap—A Strategic Action Plan for Future Aviation Safety—provides the framework for action by Contracting States of ICAO, regions and the industry to correct inconsistencies and weaknesses in 12 main focus areas, including implementation of international standards, regulatory oversight, incident and accident investigation, Safety Management Systems (SMS) and sufficient qualified personnel. SMS are processes which proactively manage the projected increase in aircraft incidents and accidents brought about by the increase in air traffic movements. SMS require vigilance in the liberalization of air transport and the correspondent increase in capacity. At the Directors General of Civil Aviation Conference on a Global Strategy for Aviation Safety, convened by ICAO in Montreal from 20 to 22 March 2006, Canada defined a Safety Management System as a business-like approach to safety. An SMS is a systematic, explicit and comprehensive process for the management of safety risks that integrates operations and technical systems with financial and human resource management, for all activities related to an air operator as an approved maintenance organization's certificate holder.

The Roadmap sets one or more short-term and medium-term objectives for each focus area over the next 10 years. Part 2—Implementing the Global Aviation Safety Roadmap—describes and prioritizes specific coordinated actions by industry to reduce risk and improve safety worldwide. For each objective identified in Part 1, the Roadmap proposes best practices with related industry references and compliance metrics. Part 2 also includes Annexes containing recommendations on existing and proven technologies (and associated training programmes) to further enhance safety in flight operations, airport operations and air traffic control domains, as well as regional implementation through a knowledge-based regional assessment and deployment strategy.

The Roadmap provides a common frame of reference to all stakeholders focussing on States' responsibilities in addressing inconsistent implementation of international standards; inconsistent regulatory oversight; impediments to reporting of errors and incidents and ineffective incident and accident investigation. The key focus areas for the industry are: impediments to reporting and analysing errors and incidents; inconsistent use of safety management systems; inconsistent compliance with regulatory requirements; inconsistent adoption of industry best practice; non-alignment of industry safety strategies; insufficient numbers of qualified personnel; and gaps in the use of technology to enhance safety.

ICAO has a grave responsibility and important role to play in implementing the plans contained in the Roadmap and making sure that the Roadmap proves to be an implementing tool. This responsibility can be attenuated from the fundamental aim of ICAO, enshrined in Article 44 of the Convention on International Civil Aviation to develop the principles and techniques of international air navigation. The Air Navigation Commission of ICAO will keep the Roadmap in review with a view to ensuring that it assists in implementing Strategic Objective A of ICAO's Business Plan, which is the enhancement of safety.

9.6 The Role of ICAO

The Roadmap calls for leadership, particularly from ICAO as the global forum for international civil aviation. The hallmark of leadership is action. The action required of ICAO in this regard is to create a paradigm shift that would take key stakeholders away from the reactive role of implementing safety measures to a proactive role of actively reducing safety related accidents and incidents in aviation. ICAO as a true leader must possess the two recognized values of leadership—actual leadership, in giving guidance or direction, and potential leadership that carries the capacity or ability to lead. Therefore, ICAO's role should be that of a leader who influences people to strive willingly for the group objective of ensuring safety in aviation.

The simplest way to measure the effectiveness of leadership involves evaluating the size of the following that the leader can muster. To measure leadership more specifically, one may assess the extent of influence on the followers, that may involve testing the results of leadership activities against the goal, vision, or objective.

It is fair to say that ICAO is at the crossroads of re- defining its continuing role towards achieving its aims and objectives as set out in the Convention on International Civil Aviation. With a view to setting its course in line with rapidly evolving trends of globalization and regionalization, the Organization has embarked on implementing an aggressive business plan that calls for a cultural transition and change of mind-set that rids ICAO from the shackles of the Convention. New leadership and new thinking have been catalysts in this process, and, through a fog of rhetoric which sometimes accused the Organization of being a bureaucracy that was rapidly headed towards obsolescence, a flight path seems to have cleared that enables the Organization to steer towards a more relevant role in the new century. The Roadmap is a crucial test for ICAO in playing that role.

Today, ICAO is a results-based, performance and values driven Organization guided by its own business plan. The Business Plan translates its Strategic Objectives into action plans and establishes a link between planned activities, organizational cost and performance assessment. A vital dimension of this approach is the integration of programmes and activities of Bureaus and Regional Offices for optimum allocation of resources based on agreed priorities.

Together, the Business Plan and the related budget provide the basis for a reporting framework that unites strategies, activities, funds, human resources and time frames into a coherent and effective means of monitoring and evaluating outcomes. By engaging staff at all levels in the performance improvement process, highlighting responsibilities, and by holding managers accountable for their performance and regularly measuring, monitoring and evaluating results, the Organization will strengthen accountability, demonstrate value for money and improve overall performance at the operational and strategic levels. Moving from concept to action and results also involves a set of Supporting Implementation Strategies and the successful Technical Co-operation Programme which has a long tradition of supporting Contracting States in the implementation of ICAO regulations, procedures and policies.

Many of ICAO's 191 member States are already facing problems with respect to safety oversight. A glaring fact emerging from safety audits conducted by ICAO on States is that the findings of the initial safety oversight audit conducted by ICAO relating to the three Annexes to the Chicago Convention—Annex 1—Personnel Licensing, Annex 6—Operation of Aircraft and Annex 8—Airworthiness of Aircraft, indicated that of the 181 Contracting States that were audited between March 1999 and July 2004, considerable numbers of States had deficiencies in respect of a number of requirements under these Annexes. Furthermore, audit follow-up missions have revealed that in many cases, significant deficiencies identified during the initial audits remain.

Under ICAO's Universal Safety Oversight Audit Programme (USOAP), States are required to establish a safety programme where aircraft operators, maintenance organizations and services providers implement appropriate safety management systems.

The first ICAO safety audit cycle was conducted between 1999 and 2004 where almost all Contracting States (except a few who could not be accessed due to adverse conditions in their territories) were audited. These audits were conducted on the basis of Annexes 1, 6 and 8 of the Chicago Convention. From January 2005, ICAO started its 6 year cycle of audits based on sixteen of the eighteen ICAO Annexes (all except Annex 9 and Annex 17) and by the end of April 2008 had audited 90 States.

The responsibilities of ICAO in the area of safety are grave and compelling. For the past 60 years or so, ICAO has been active in its standardization role, which has been blended in recent years with a burgeoning implementation role that is gradually blurring the former. In a world that is becoming largely globalized and regionalized, ICAO has vastly to focus on not so much what it does but how it does its work. In this context, ICAO has a dual role to play. The first is to act as a global forum for aviation, which is primarily the role expected of ICAO by the developed nations which are largely self reliant in regulatory matters. However, they need ICAO to set global standards that could apply to all ICAO's 190 member States. On the other hand, ICAO has to be both a global forum and a mentor to the developing world which expects ICAO to assist and guide them.

In order to serve its 191 member States, irrespective of whether they are in the developed or developing category, ICAO has to justify its performance and values based stature. In other words, ICAO needs to undergo a whole system change. For the Business Plan to be implemented and results to be produced, firstly ICAO's leaders (its Council and the senior managers of the Secretariat) have to drive the process of transition from service to performance. They need to be the ambassadors of the Organization's mission and vision statements and set values and behaviours. They must "talk the talk" and "walk the walk".

The next step is to ensure that the mission and vision statement influences all decision making. This should permeate right to the bottom of the ICAO Secretariat. Thirdly, the new culture and its results must be measured by causal performance indicators. In other words, ICAO's new culture should be constantly monitored. The final measure would be to ensure that the values of the Organization's culture pervade and drive every aspect of decision making and be seen in every system and process.

In a way, ICAO is already undergoing a cultural transformation. It has come a long way in developing a mission and vision statement and a business plan driven by strategic objectives. There is a leadership that is committed to its work. There is also every indication that the leadership is ready and willing to involve the entire Organization in defining the mission, vision and values of the Organization. However, in order to achieve this successfully, strong tools and aggressive goals have to be in place through a robust and energized operational plan that is not disaggregated among the Organization's constituent bureaux and other offices.

Such an operational plan must have objectives and key performance indicators, as in the end it is measurement that matters. For this there must be targets set, not just improvement of performance. The Roadmap has to be part of target setting. In this regard it must be noted that ICAO's Business plan is on the right track, as it has all three types of indicators: causal indicators—which relate to values and behaviours (which are known in other words as core competencies); output indicators—which measure performance in terms if efficiency and productivity; and outcome indicators, which relate to the result or effect on clients and stakeholders.

Of these indicators, ICAO's concentration should be mainly on output indicators that measure productivity, efficiency, quality, innovation, creativity of the Organization as a whole and ensuing customer satisfaction. Innovation and creativity are key factors that serve to promote ICAO's contribution to its member States. Just as an example, since many States do not have the volume and scope of aviation activities which generate the resources and the base-line activity necessary to support a workable safety oversight system, ICAO's role must be to take the leadership in providing States with templates of different models of safety oversight and recommend what is best suited for them. ICAO could also further the involvement of regional safety oversight organizations that are successful; and provide guidance to States as to the modalities of the transfer of responsibilities or tasks, depending on the model used, from participating State to regional safety oversight organizations.

Leadership is the key to ICAO's role in the Twenty First Century and the Roadmap is a tangible example of how this leadership can be demonstrated. The first step in driving the Roadmap is for ICAO to make a philosophical adjustment and ensure that it keep abreast of the new world order where States are increasingly being disaggregated into components which act in global networks, linking the world together in a manner that enables global trends to permeate the local environment . In other words, ICAO should facilitate interaction between States and their components that interact in matters of civil aviation. For example, in many member States, aviation has numerous players in different areas such as customs and immigration, medical and quarantine, tourism, police, airports and air navigation service providers. In most instances these players do not act in accord, thus resulting in disharmony in the ultimate delivery of an efficient air transport product. ICAO's Mission and Vision Statement exhorts ICAO to do just what is needed—to act as the global forum in the key areas of concern to international civil aviation through cooperation between its member States.

While promoting fluid dialogue and cooperation among its member States, ICAO should take the initiative to assist States both in technical issues. This assistance is not confined to providing technical assistance through projects administered by the Technical Cooperation Bureau but should also extend to providing guidance, mainly to States which still look up to ICAO as the global forum of aviation experts.

A critical area that has not been addressed in the Roadmap equation is the funding of safety. There is seemingly no conscious awareness on the part of States that a roadmap would be useless if there are no resources to correct the various deficiencies addressed by the safety audits and the overall application of the Roadmap. Neither ICAO nor its member States have aggressively addressed this issue.

In implementing the Roadmap, ICAO's leadership role hinges on two key factors: an aggressive operational plan with key performance indicators and targets; and the realization that organizational culture, which is an intangible asset, is the new frontier of competitive advantage. The latter is particularly important under the current circumstances of ICAO where human resources and expertise are in short supply. Cultural transformation starts with the leadership and individual and leadership values. When one looks at ICAO's current leadership structure, there is no room for doubt that this is not in short supply. However the trick is to motivate the staff sufficiently so that they would be impelled to follow their leaders in the transformation and forge ICAO's leadership forward in its various areas of work.

All this leads one to the bottom line, which is the need for change in the mindset of the Organization, from its service role to a role of implementation and assistance. The human factor is an essential consideration in this metamorphosis. The key and the starting point, however, is to recognize the need for the transition, which ICAO has already done. The next step is to recognize that ICAO needs its peoples' best efforts, both individually and collectively. ICAO's image and the perception of the outside world of ICAO as an effective Organization is anchored on the extent to which its workers represent themselves as good stewards of ICAO's business. They should therefore work together in the overarching interest of the Organization. When all these factors are considered together, there is nothing to suggest that ICAO is headed in the wrong direction. However, what seems to be badly needed is funding.

9.7 Security Oversight

One of the significant results of the 36th Session of the ICAO Assembly, held in September 2007, was the adoption of a Resolution calling for the sharing of information through the ICAO Council pertaining to security audits conducted by ICAO. This brings to bear a certain shift of focus from the original confidentiality of the audits to one of limited transparency. It also raises the more compelling issue as to what the legal principles applicable are that would attribute to the Council the ability to divulge information and the limitations if any, on carrying out the instructions of the Assembly, which is one of the mandatory functions of the Council. The question also arises as to whether such a function could be sustained in the face of other overriding factors, one of which is the extent to which ICAO stands empowered by its constituent member States to divulge information pertaining to aviation activities in their territories.

In the discussion to follow is an analysis of the developments in the ICAO security audit process and critically analyses the role of ICAO and its Council in the practical application of Article 54 (j) of the Convention on International Civil Aviation[56] which deals with the issue of disclosure of information—a provision which has hitherto not been invoked in ICAO's 60 year history.

10 The High Level Ministerial Conference

At the 33rd Session of the Assembly, held from 25 September to 5 October 2001, ICAO adopted Resolution A33-1 entitled "*Declaration on misuse of civil aircraft as weapons of destruction and other terrorist acts involving civil aviation*".[57] This Resolution, while singling out for consideration the terrorist acts which occurred in the United States on 11 September 2001, and, *inter alia*, recognizing that the new type of threat posed by terrorist organizations requires new concerted efforts and policies of cooperation on the part of States, urged all Contracting States to intensify their efforts in order to achieve the full implementation and enforcement of the multilateral conventions on aviation security, as well as of the ICAO Standards and Recommended Practices and Procedures (SARPs) relating to aviation security. The Resolution also called upon States to monitor such implementation, and to take within their territories appropriate additional security measures commensurate to the level of threat in order to prevent and eradicate terrorist acts involving civil aviation. Furthermore, the Resolution urged all Contracting States to make contributions in the form of financial or human resources to ICAO's aviation security mechanism to support and strengthen the combat against terrorism and unlawful interference in civil aviation; called on Contracting States to agree on special funding for urgent action by ICAO in the field of aviation security; and directed the Council to develop proposals and take appropriate decisions for a more stable funding of ICAO action in the field of aviation security, including appropriate remedial action.

---

[56] *Ibid.*

[57] *Assembly Resolutions in Force* (as of 5 October 2001), ICAO Doc 9790, at p. VII-1. Also of general interest is UN General Assembly Resolution 56/88, *Measures to Eliminate International Terrorism*, adopted at the Fifty Sixth Session of the United Nations which calls upon States to take every possible measure in eliminating international terrorism. See A/RES/56/88, 24 January 2002.

Resolution A33-1 also directed the Council to convene, at the earliest date, an international high-level ministerial conference on aviation security in Montreal with the objectives of preventing, combating and eradicating acts of terrorism involving civil aviation; of strengthening ICAO's role in the adoption of SARPs in the field of security and the audit of their implementation; and of ensuring the necessary financial means to strengthen ICAO's AVSEC Mechanism, while providing special funding for urgent action by ICAO in the field of aviation security.

On 19 and 20 February 2002, in keeping with the requirement of Assembly Resolution A33-1a high level ministerial conference on aviation security was held in the Headquarters of the International Civil Aviation Organization, Montreal. In the words of Dr. Assad Kotaite, then President of the ICAO Council who opened the Conference (and later served as the Chairman of the Conference), the Conference was being held

> . . .at a critical juncture for civil aviation and for society at large. . . and would review and develop global strategy for strengthening aviation security with the aim of protecting lives both in the air and on the ground, restoring public confidence in air travel and promoting the health of air transport in order that it can renew its vital contribution to the world economy. . .[58]

Dr. Kotaite stated that this was a historic moment in the evolution of civil aviation.[59]

At this Conference, attended by Member States of the International Civil Aviation Organization, Some 714 participants from 154 Contracting States and observers from 24 international civil aviation organizations endorsed a global strategy for strengthening aviation security worldwide and issued a public declaration at the conclusion of their 2 day meeting.

The High Level Ministerial Conference came to several conclusions and adopted numerous recommendations containing guidance for follow up action. The Conference concluded that the events of 11 September 2001 have had a major negative impact on world economies and an impact on air transport which is unparalleled in history and restoration of consumer confidence in air transport and assurance of the long-term health of the air transport industry are both vital, and many States have already initiated a range of measures to this effect. It was also the view of the Conference that the effective application of enhanced uniform security measures, commensurate with the threat, will help to restore confidence in air transport, but these measures will need to be passenger and cargo user-friendly and not overly costly for the industry and its consumers if traffic growth is to be regenerated. Accordingly, the Conference recommended that, consistent with Assembly Resolution A33-1, States should intensify their efforts to achieve the full implementation and enforcement of the multilateral conventions on aviation security as well as of the ICAO SARPs relating to aviation security and take within their territories

[58] ICAO News Release *PIO 02/2002*.

[59] *Ibid*.

appropriate additional security measures which are commensurate with the level of threat and are cost effective. Since restoration of confidence in air transport is a collective responsibility, the Conference called upon States to enhance international cooperation in aviation security and assist developing countries to the extent possible.

With regard to the compelling need to strengthen aviation security worldwide, the Conference concluded that a strong and viable aviation security (AVSEC) programme was indispensable and that a global uniform approach to the implementation of the international aviation security standards is essential, while leaving room for operational flexibility. It was also considered useful to establish regional and sub-regional approaches which could make a significant contribution to ICAO's aviation security activities. The Conference concluded that aviation security was a responsibility of Contracting States, and States which outsource aviation security programmes should therefore ensure that adequate governmental control and supervision are in place. The Conference also observed that, since gaps and inadequacies appear to exist in international aviation security instruments with regard to new and emerging threats to civil aviation, further study was needed in this regard. There was a need for a comprehensive ICAO *Aviation Security Plan of Action* for strengthening aviation security, through a reinforced AVSEC mechanism, an ICAO aviation security audit programme, technical cooperation projects, promotion of aviation security quality control functions and appropriate performance indicators.

Based on the above conclusions the Conference recommended that States take immediate action to lock flight deck doors for aircraft operated internationally, while maintaining measures on the ground to provide the highest level of aviation security. States were also requested to actively share threat information in accordance with Standards in Annex 17 and employ suitable threat assessment and risk management methodologies appropriate to their circumstances, based on a template to be developed by ICAO and ensure that aviation security measures are implemented in an objective and non-discriminatory manner.

As for ICAO's role in this process, the Conference recommended that the Organization develop, as a matter of high priority, amendments to the appropriate Annexes to require protection of the flight deck door from forcible intrusion; continue its efforts to identify and analyze the new and emerging threats to civil aviation with the purpose of assisting in the development of security measures and to actively collaborate with other associated agencies; carry out a detailed study of the adequacy of the existing aviation security conventions and other aviation security-related documentation with a view to proposing and developing measures to close the existing gaps and remove the inadequacies, including amendment where required, so as to deal effectively with the existing, as well as the new and emerging threats to international civil aviation; develop and take action to deal with the problem of aviation war risk insurance; and develop and implement a comprehensive *Aviation Security Plan of Action* and any additional actions approved by the Council, including a clear identification of priorities.

One of the key conclusions of the Conference was that, in order to further enhance safety and security and to ensure the systematic implementation of the critical elements of a State's aviation security system, there was an urgent need for a comprehensive ICAO programme of aviation security audits and that such a programme should audit national level and airport level compliance with Annex 17 and with aviation security related provisions of other Annexes on a regular, mandatory, systematic and harmonized basis. It was the view of the Conference that the ability to determine whether an airport or State is in compliance will require that auditors have a solid aviation security background and be sufficiently trained and certified by ICAO to ensure that auditing is conducted in a consistent and objective manner. The Conference was strongly convinced that such an audit programme should be undertaken under the auspices of ICAO's AVSEC Mechanism which could be guided by proven and successful concepts used in viable programmes already developed by the European Civil Aviation Conference (ECAC), the United States and other States in the development of the framework for a security audit programme.

11 Security Oversight

ICAO has a security oversight programme called the Universal Security Audit Programme (UASP). The ICAO Universal Security Audit Programme (USAP), launched in June 2002, represents an important initiative in ICAO's strategy for strengthening aviation security worldwide and for attaining commitment from States in a collaborative effort to establish a global aviation security system.

The programme, which is part of the Aviation Security Plan of Action, provides for the conduct of universal, mandatory and regular audits of the aviation security systems in all ICAO member States. The objective of the USAP is to promote global aviation security through the auditing of States on a regular basis to assist States in their efforts to fulfil their aviation security responsibilities. The audits identify deficiencies in each State's aviation security system, and provide recommendations for their mitigation or resolution.

Implementation of the programme commenced with the first aviation security audit taking place in November 2002 and between three and four audits continue to be conducted around the world each month. The 35th Session of the Assembly held from 28 September to 8 October 2004 mandated ICAO to maintain strict confidentiality of all State-specific information derived from audits conducted under the Universal Security Audit Programme (USAP). However, in order to promote mutual confidence in the level of aviation security between States, the Assembly urged all Contracting States to "share, as appropriate and consistent with their sovereignty, the results of the audit carried out by ICAO and the corrective actions taken by the audited State, if requested by another State".[60]

[60] A35-9, Appendix E, Resolving Clause 4; and Recommended Practice 2.4.5 of Annex 17—*Security*).

While noting the importance of continuing bilateral exchanges of information between States, the 36th Session of the Assembly, held from 18 to 28 September 2007, also recognized the value of proposals presented by the Council and Contracting States for the introduction of a limited level of transparency with respect to ICAO aviation security audit results.[61] The Assembly directed the Council to consider such an introduction of a limited level of transparency, balancing the need for States to be aware of unresolved security concerns with the need to keep sensitive security information out of the public realm. In doing so, the Assembly emphasized that it was essential that any methodology developed to provide for increased transparency also ensure the appropriate safeguarding of a State's security information in order to prevent specific information that could be used to exploit existing vulnerabilities from being divulged.

The 36th Session of the ICAO Assembly adopted Resolution A36-20,[62] *Appendix E* of which addresses the USAP (this resolution has since been superceded by Resolution A37-17). As mentioned earlier, it must be emphasized that the Resolution *inter alia* directed the Council to consider the introduction of a limited level of transparency with respect to ICAO aviation security audits, balancing the need for States to be aware of unresolved security concerns with the need to keep sensitive security information out of the public realm and requested the Council to report to the next ordinary session of the Assembly (in 2010) on the overall implementation of the USAP.

Since the launch of the USAP in 2002, 169 aviation security audits and 77 follow-up missions have been conducted.[63] The audits have proven to be instrumental in the ongoing identification and resolution of aviation security concerns, and analysis reveals that the average implementation rate of Annex 17 Standards in most States has increased markedly between the period of the initial audit and the follow-up mission.

A critical part of the audit process is the requirement that all audited States submit a corrective action plan to address deficiencies identified during an audit. As directed by the Council, all States are notified (by State letter and on the USAP secure website) of those states that are more than 60 days late in submitting a corrective action plan. As of 31 July 2007, there were seven States that were more than 60 days late. In the case of late corrective action plans, repeated reminders are sent to States, including at the level of the Secretary General and with the involvement of the applicable Regional Office, and ICAO assistance is offered should the

---

[61]Resolution A36-20, A36-WP/336 and Plenary Action Sheet No. 3.

[62]Resolution A36-20, Consolidated statement on the continuing CA policies related to the safeguarding of international civil aviation against acts of unlawful interference, Report of the Executive Committee (Report Folder) Assembly, Thirty-sixth Session, A36-WP/336, p/46, at 16–2.

[63]The 36th Session of the ICAO Assembly was informed that there are some 150 certified auditors on the USAP roster, from 59 States in all ICAO regions. The participation of certified national experts in the audits under the guidance of an ICAO team leader has permitted the programme to be implemented in a cost-effective manner while allowing for a valuable interchange of expertise.

State require advice or support in the preparation of its action plan. Extensive feedback is provided to each audited State on the adequacy of its corrective action plan, and an ongoing dialogue is maintained where necessary to provide support in the implementation of proposed actions.

ICAO performs comprehensive analyses of audit results on levels of compliance with Annex 17—*Security* Standards on an on-going basis (globally, by region and by subject matter). This statistical data is made available to authorized users on the USAP secure website and is shared with other relevant ICAO offices as a basis for prioritizing training and remedial assistance projects. USAP audits are regular, mandatory, systematic and harmonized. They cover States' aviation security oversight capabilities as well as auditing security measures at selected airports. In addition, the USAP is based on the following nine principles: sovereignty of States; universality; transparency; objectivity; fairness; quality; timeliness; all-inclusiveness; and confidentiality.

The results of the audit follow-up visits have demonstrated that, overall, States had made improvements in their obligations to meet Annex 17 Standards. However, varying levels of improvement were identified between regions, and, in many cases, between States within a region. States having difficulties in addressing deficiencies identified during an audit were offered the opportunity to request assistance from ICAO through the Implementation Support and Development (ISD) Programme—Security in coordination with the Technical Co-operation Programme.

In recognizing that the USAP had proven to be instrumental in identifying aviation security concerns and providing recommendations for their resolution, the 36th Session of the ICAO Assembly, in Resolution A36-20: Consolidated statement of continuing ICAO policies related to the safeguarding of international civil aviation against acts of unlawful interference, since superseded by Assembly Resolution A37-17, requested the ICAO Council to ensure the continuation of the USAP following the initial cycle of audits at the end of 2007.

Aviation security audits under the ongoing ICAO USAP second cycle commenced in January 2008. The primary objectives of the second-cycle audits are to:

- Determine the State's capability for aviation security oversight by assessing whether the critical elements of an aviation security oversight system have been implemented effectively;
- Determine the State's degree of compliance with Annex 17 Standards and the security related provisions of Annex 9;
- Assess the State's adherence to security procedures, guidance material and security related practices associated with the relevant ICAO SARPs; and
- Provide recommendations to the audited State on how to improve its aviation security system and security oversight capabilities. These missions are normally conducted by the applicable Regional Office, with close coordination through Headquarters. The results of the follow-up visits indicate that the majority of States have made significant progress in the implementation of their corrective action plans.

A high-level ICAO Secretariat Monitoring and Review Board (MARB) has been established as part of an overall coordinated strategy for working with States that are found to have significant compliance shortcomings with respect to ICAO Standards and Recommended Practices (SARPs). The purpose of the MARB will be to continue the work of the former Audit Results Review Board (ARRB) by focusing on developing and implementing broad, high-level assistance strategies. It will also focus on coordinating ICAO assistance and monitoring activities (the term monitoring in this document refers to both auditing and continuous monitoring activities), and on States with Significant Safety and/or Security Concerns.

The MARB is an ICAO Secretariat senior management forum responsible for:

- Coordinating and evaluating the effectiveness and efficiency of monitoring activities to ensure that opportunities for continuous improvement are identified and acted upon; and
- Identifying, coordinating and validating assistance strategies and other courses of action to improve the responsiveness of States to ICAO monitoring activities and to address identified needs.

The MARB examines both the safety and security histories of specific States and provides an internal advisory forum for coordination among ICAO's safety, security and assistance programmes.

12 The Role of the ICAO Council

As future measures in the audit programme of ICAO, the ICAO Council approved in 2007, the practice that not all States need to be audited at the same frequency, although the USAP should always preserve the principle of universality. The Council was of the view that, with a solid baseline of audit results established for all States by the end of 2007, a more effective use of resources can be achieved by developing an appropriate scheduling/frequency model to determine the priority of future audits and frequency of visits to States. It remains as a requirement however that the principle of universality will be maintained with all States audited at least once within a 6-year period.

Another decision of the Council was that future audits under the USAP should be expanded to include relevant security-related provisions of Annex 9—*Facilitation.* With the recent expansion of the Universal Safety Oversight Audit P to a comprehensive systems approach covering all safety-related Annexes, Annex 9 is currently the only Annex which is not included in either of ICAO's two audit programmes. There are a number of security-related provisions contained in Annex 9, particularly as related to the security and integrity of travel documentation, which can be audited under the USAP along with the related Standards of Annex 17.

The Council also decided that wherever possible, ICAO aviation security audits should be focused on a State's capability to provide appropriate national oversight of its aviation security activities. Using the results of the initial audits and follow-up visits, the scope of future ICAO audits should be adjusted to the prevailing situation in each audited State. Those States that have demonstrated the requisite national infrastructure necessary to oversee security activities at their airports may undergo a targeted oversight audit to verify adequate implementation of the State's national quality control programme. Such oversight audits would continue to include a verification of the implementation of ICAO provisions through spot checks at the airport level.

The decision of the 36th Session of the Assembly—that a "limited level of transparency" with respect to ICAO aviation security audit results be maintained, ensures a balance between the need to divulge certain information while protecting the interests of States. As such the Council has to draw a fine line between potentially conflicting interests. As for safety, the 35th Session of the Assembly, when it addressed the issue of expanding the audits from a limited Annex basis to a comprehensive systems approach, instructed the Secretary General to make the final safety audit reports available to all Contracting States and also to provide access to all relevant information derived from the Audit Findings and Differences Database (AFDD) maintained by ICAO.[64] Furthermore, in Resolution A36-2 (Unified Strategy to Resolve Safety Related Deficiencies) the Assembly, in operative Clause 6 of the Resolution, has directed the Council to apply and review, as necessary, the procedures to inform Contracting States, within the scope of Article 54 (j) of the Chicago Convention, in the case of a State having significant shortcomings with respect to ICAO safety related SARPs in order for other Contracting States to take action in an adequate and timely manner.

Article 54 (j) makes it a mandatory function of the Council to report to any Contracting State any infraction of the Chicago Convention as well as any failure to carry out recommendations and determinations of the Council. There are various dimensions to this provision in the context of Resolution A36-2. Firstly, it is surprising that the Assembly Resolution does not also request the Council to perform its mandatory function in Article 54 (k), which is to report to the Assembly any infraction of the Convention where a Contracting State fails to take appropriate action within a reasonable time after notice of the infraction. This would have arguably been a more coercive and effective tool than the measure prescribed in Article 54 (j) in that States would be quite concerned if their shortcomings were to be aired out in front of 190 Contracting States at an ICAO Assembly.

The second dimension to the Resolution is that it the function of the Council in this case, to use the words of operative clause 6 of Resolution A36-2 to

> apply and review... the procedure to inform Contracting States within the scope of Article 54j) of the Chicago Convention, in the case of a State having significant shortcomings with respect to ICAO safety related SARPs in order for other Contracting States to take action in an adequate and timely manner.

Surprisingly the Council is asked by the Assembly to restrict itself to determining the adherence to SARPs and report its findings thereof, which is already a

[64]Resolution A35-6, Operative Clause 7.

function handed down in the Convention to the Council in Article 38.[65] Again, it is not clear as to why the Assembly refrained from applying the rest of Article 54 (j) to its Resolution, which makes it incumbent upon the Council to report the failure to carry out recommendations or determinations of the Council. This application would have served the purpose of the Assembly better than the mere restriction to the SARPs in the Annexes.

The third dimension is that the Council, under the Convention, has only functions (which are in essence duties) and no powers.[66] On the other hand the Assembly has powers and duties accorded to it in the Chicago Convention,[67] one of which is to delegate to the Council the powers and authority necessary or desirable for the discharge of the duties of the Organization and revoke or modify the delegations of authority at any time.[68] However, in this instance there is no indication that the Assembly exercised its powers to delegate its authority or power to the Council to apply and review the procedure in Article 54 (j). If this had been the case, the Council would have had the same right and the authority of the Assembly to take appropriate action as deemed necessary in the manner in which the information derived from safety and security audits would be disseminated and reported to other States.

A power is the capacity to direct the decisions and actions of others. A function on the other hand is to perform, execute or administer.[69] A power is also defined as an ability on the part of a person to produce a change in a given legal relation by doing or not doing a certain act.[70] In this context the Council only has a duty or function to report to States shortcomings of other States detected during the course of safety and security audits with regard to adherence by the ICAO member States of SARPs. It is therefore incontrovertible that Assembly Resolution A36-2 merely hands over to the Council the function to report an infraction of the Chicago Convention as well as shortcomings with regard to SARPs and recommendations and determinations of the Council in that regard.

The Chicago Convention bestows neither the ability nor the power on the Council to investigate and determine on its own initiative whether there has been an infraction of the Convention. There is also no specific provision which entitles the Council to notify the State concerned that an infraction has taken place. However, Article 54 (n) provides that the Council can consider any matter relating

[65]Article 38 provides: *inter alia* that any Contracting State can file a difference to a standard and notify the Council which in turn is required to make immediate notification to all other States of the difference which exists between one or more features of an international standard and the corresponding national practice of that State.

[66]Although Jacob Schenkman, in his well documented and logically reasoned treatise on ICAO states that "The Council has been entrusted with duties, powers and functions. . ." he does not give a single example of such a power. See Capt. Schenkman (1955) at 158.

[67]Article 49 of the Convention.

[68]Article 49 (h).

[69]Deluxe Black's Law Dictionary, Sixth Edition, St. Paul. Minn: 1009, at 673.

[70]*Id.* at 1189.

to the Convention which any Contracting State refers to it, giving the Council the capacity to make its own determination and recommendations pertaining to a matter referred to it. It is also noteworthy that both Article 15 of the Convention, which allows the Council to report and make recommendations resulting from a review by the Council of charges imposed for the airports and other facilities, and Article 69, which gives the Council competency to make recommendations to member States for the improvement of air navigation facilities, are two instances of specific provision being made within the Convention where the Council can make recommendations for the consideration of ICAO member States.

Clearly, non compliance with SARPs and shortcomings or deficiencies in security cannot be classified as infractions of the Convention. An infraction is a violation and arguably applicable to the Chicago Convention itself and not to the Annexes which only contain SARPs that are not strictly legally binding so as to constitute a violation if not followed. Therefore, the Assembly, in A36-2 quite clearly meant the reportage of failure to carry out recommendations and determinations of the Council with regard to SARPs. This is clearly an administrative function and not a judicial function, since an administrative act is usually referred to as similar or related activities regarding the handling and processing of information.

Another important dimension to the Council's role as per A36-2 in divulging security information is that ICAO has already entered into memoranda of understanding with the States audited that audit reports will be confidential and made available to the State audited and relevant ICAO staff on a need-to-know basis. These agreements also require that, concurrently with the preparation of the report, a non-confidential audit activity report limited to the name of the audited State, the identity of airports visited during the audit, and the completion date of the audit will be developed for release to all Contracting States. Reports to the Council are required to be in a form that maintains the confidentiality of the audit report in relation to the State concerned. Accordingly, ICAO has restricted itself for purposes of confidentiality to giving only limited and non specific details of audits to its member States. This raises a legal issue as to ICAO's right to contravene its agreement with member States in deference to an Assembly Resolution. This issue also seemingly goes to the root of ICAO's empowerment by its member States and ICAO's accreditation to such States.

International organizations can generally only work on the basis of legal powers that are attributed to them. Presumably, these powers emanate from the sovereign States that form the membership of such organizations.[71] Therefore, the logical conclusion is that if international organizations were to act beyond the powers accorded to them, they would be presumed to act *ultra vires*.[72] It should be noted that ICAO does not only derive implied authority from its Contracting States based on universality but it also has attribution from States to exercise certain powers. The doctrine of attribution of powers comes directly from the will of the founders, and

[71]See de Witte (1998) at pp. 277–304.

[72]Klabbers (2002) at p. 60.

in ICAO's case, powers were attributed to ICAO when it was established as an international technical organization and a permanent civil aviation agency to administer the provisions of the Chicago Convention. In addition, ICAO could lay claims to what are now called "inherent powers" which give ICAO power to perform all acts that the Organization needs to perform to attain its aims not due to any specific source of organizational power but simply because ICAO inheres in organizationhood. Therefore, as long as acts are not prohibited in ICAO's constituent document (the Chicago Convention), they must be considered legally valid.[73]

Over the past two decades the inherent powers doctrine has been attributed to the United Nations Organization and its specialized agencies on the basis that such organizations could be stultified if they were to be bogged down in a quagmire of interpretation and judicial determination in the exercise of their duties. The advantages of the inherent powers doctrine is twofold. Firstly, inherent powers are functional and help the organization concerned to reach its aims without being tied by legal niceties. Secondly, it relieves the organization of legal controls that might otherwise effectively preclude that organization from achieving its aims and objectives. The ability to exercise its inherent powers has enabled ICAO to address issues on aviation insurance and establish an insurance mechanism; perform mandatory audits on States in the fields of aviation safety and security; and establish a funding mechanism to finance aviation safety projects, all of which are not provided for in the Chicago Convention but are not expressly prohibited.

With regard to the conferral of powers by States to ICAO, States have followed the classic approach of doing so through an international treaty. However, neither is there explicit mention of such a conferral on ICAO in the Chicago Convention nor is there any description of ICAO's powers, except for an exposition of ICAO's aims and objectives. The Council of ICAO is designated both mandatory and permissive "functions", although the Council could impose certain measures when provisions of the Convention are not followed. Therefore States have not followed the usual style of conferral of powers in the case of ICAO, which, along the lines of the decision of the International Court of Justice in the 1996 *WHO Advisory Opinion* case[74] was that the powers conferred on international organizations are normally the subject of express statement in their constituent instruments.[75] This notwithstanding, it cannot be disputed that ICAO Contracting States have conferred certain powers on ICAO to perform its functions independently. For example, ICAO is a legal entity having the power to enter into legal agreements with legal entities including other international organizations with regard to the performance of its functions.

Conversely, an international organization must accept conferred powers on the basis of Article 34 of the Vienna Convention on the Law of Treaties which stipulates that a treaty does not create rights or obligations of a third State without

[73]Seyersted (1963), at p. 28.

[74]*Legality of the Threat or Use of Nuclear Weapons, Advisory Opinion*, ICJ Reports, 1996, p. 64.

[75]*Id.* p. 79.

its consent. This principle can be applied *mutatis mutandis* to an international organization such as ICAO. The conferral of powers on an international organization does not *ipso facto* curtail the powers of a State to act outside the purview of that organization unless a State has willingly limited its powers in that respect. This principle was recognized in the *Lotus* Case[76] where the Provisional International Court of Justice held that a State can exercise powers on a unilateral basis even while the conferral to the Organization remains in force. The Court held that restrictions upon the independence of States cannot be presumed.[77]

ICAO's conferred powers enable the Organization to adopt binding regulations by majority decision (which is usually unnecessary as most of ICAO policy is adopted through consensus). However, States could opt out of these policies or make reservations thereto, usually before such policy enters into force. This is because States have delegated power to ICAO to make decisions on the basis that they accept such decisions on the international plane. In such cases States could contract out and enter into binding agreements outside the purview of ICAO even on subjects on which ICAO has adopted policy. The only exception to this rule lies in the adoption of Standards in Annex 2 to the Chicago Convention on Rules of the Air, in particular navigation over the high seas and other overflight areas where freedom of flight prevails which all Contracting States are bound to follow in order to maintain global safety.

The 35th Session of the Assembly, when it addressed the issue of expanding the audits from a limited Annex basis to a comprehensive systems approach, instructed the Secretary General to make the final safety audit reports available to all Contracting States and also to provide access to all relevant information derived from the Audit Findings and Differences Database (AFDD) maintained by ICAO.[78] Furthermore, in Resolution A36-2 (Unified Strategy to Resolve Safety Related Deficiencies) the Assembly, in operative Clause 6 of the Resolution, directs the Council to apply and review, as necessary, the procedures to inform Contracting States, within the scope of Article 54 (j) of the Chicago Convention, in the case of a State having significant shortcomings with respect to ICAO safety related SARPs in order for other Contracting States to take action in an adequate and timely manner.

From the above discussion it becomes clear that, while on the one hand the ICAO Assembly, which in essence is the representative voice of the 190 member States comprising ICAO, has directed the Council to apply and review procedures to inform member States within the scope of Article 54 (j) of shortcomings, on the other hand, the overriding separate and individual memoranda signed by ICAO with its member States in the area of security would have to be revised in terms of the confidentiality clause. Additionally, the Council would have to set in place an understanding with States and appropriate mutually agreed guidelines on the content of such information and the manner in which it is to be divulged.

---

[76] *PCIJ Reports* Series A, No. 10, p. 4.

[77] *Id.* p. 18.

[78] Resolution A35-6, Operative Clause 7.

States retain the powers to act unilaterally and they are not bound to comply with obligations flowing from the Organization's exercise of conferred powers. States which have delegated powers on ICAO have the legal right under public international law to take measures against a particular exercise by ICAO of conferred powers which is considered to be *detournement de pouvoir*, *ultra vires* or an internationally wrongful act with which the objecting States do not wish to be associated. A State could also distance itself from the State practice of other Contracting States within the Council if such activity is calculated to form customary international law that could in turn bind the objecting State if it does not persist in its objections.[79]

The above notwithstanding, a significant issue in the determination of ICAO's effectiveness as an international organization is the overriding principle of universality and global participation of all its 190 Contracting States in the implementation of ICAO policy. This principle, which has its genesis in the Chicago Conference of 1944, has flowed on gaining express recognition of legal scholars. This is what makes ICAO unique as a specialized agency of the United Nations and establishes without any doubt that ICAO is not just a tool of cooperation among States.

The second cycle of the USAP was launched in January 2008 and is on course to be completed by the end of 2013. Recognizing the need to determine the future nature and direction of the USAP, the ICAO Council, during its 187th Session (C-DEC 187/8), directed the Secretary General to undertake a study to assess the feasibility of applying a Continuous Monitoring Approach (CMA) to the USAP after the conclusion of the current audit cycle in 2013. Any application of a CMA approach to security audits would need to take into consideration the principle of confidentiality and the appropriate level of transparency associated with data collection and reporting under the USAP.

13 The Continuous Monitoring Approach (CMA)

It was in January 199 that the ICAO Universal Safety Oversight Audit Programme (USOAP) was launched, in response to a resolution adopted by the ICAO Assembly. The objective of the USOAP is to promote global aviation safety by regularly auditing ICAO Member States to determine their capability for effective safety oversight. The USOAP is managed by the Continuous Monitoring and Oversight Section of the Air Navigation Bureau.

The current cycle of USOAP Comprehensive Systems Approach (CSA) audits, which assess the level of effective implementation by States of the critical elements of a safety oversight system, began in 2005 and was completed at the end of 2010 ICAO is now looking ahead to the implementation of a USOAP Continuous

[79]See Sarooshi (2005) at p. 110.

Monitoring Approach (CMA) which was adopted by the Council of ICAO as a more proactive approach which will incorporate the analysis of safety risk factors.

The USOAP embarked on 2-year transition period to the CMA beginning in 2011, with the launch of this new approach planned for 2013.

The CMA is designed to be long-term, cost-effective, flexible and sustainable, generating valuable data and contributing to the improvement of global aviation safety. This will be accomplished by using a harmonized and consistent approach to monitoring the safety oversight capabilities of Member States on an on-going basis.

The CMA will identify safety deficiencies, assess associated safety risks, develop assistance strategies, and enable the prioritization of assistance. CSA audits will continue to be carried out by ICAO and will be tailored to the level and complexity of aviation activities in the State and could be either full scale or of limited scope additional activities are envisaged; which will include, but not limited to, safety audits (CSA audits carried out at the request of States and on a cost recovery basis) and ICAO Coordinated Validation Missions (ICVM).

Under CMA, the objective of the USOAP is to promote global aviation safety through continuous monitoring of the Member States' safety oversight capabilities. The CMA enables ICAO to collect vast amounts of safety information, which is provided primarily by States. Safety information is also gathered from relevant external stakeholders, as well as through audits and other USOAP-CMA activities. Using the CMA, ICAO will be able to enhance States' safety oversight and safety management capabilities by:

- Identifying safety deficiencies.
- Assessing associated safety risks.
- Developing strategies for CMA activities and assistance.
- Prioritizing assistance.

Since CMA relies on multiple inputs, many of which may be received simultaneously, it is important when examining this new approach to look first at the big picture before breaking it down into component steps. While scheduled CMA activities will provide much important data and information, a vast amount of additional safety data will be collected and provided to the USOAP under the CMA by three types of stakeholders.

References

Abeyratne RIR (1992) Law making and decision making powers of the ICAO council - a critical analysis. Z Luft-und Weltraumrecht 41(4):387–394

Abeyratne RIR (2007) State responsibility for safety management systems. J Aviat Manag II:7–13

Abeyratne R (2009a) The role of ICAO in implementing the safety roadmap. Eur Transp Law XLIV(1):29–36

Abeyratne R (2009b) Aircraft registration, legal and regulatory issues. Ann Air Space Law XXXIV:173

Abeyratne R (2012) Air navigation law. Springer, Heidelberg

Aust A (2000) Modern treaty law and practice. Cambridge University Press, New York

Buergenthal T (1969) Law making in the international civil aviation organization. Syracuse University Press, Syracuse
de Witte B (1998) Sovereignty and European integration: the weight of tradition. In: Slaughter A-M et al (eds) The European court and national courts: doctrine and jurisprudence. Hart Publishing, Oxford
Dempsey PS (1987) Law and foreign policy in international aviation. Transnational Publishers Inc., Dobbs Ferry
Kirgis FL Jr (1995) Specialized law making processes. In: Schachter O, Joiner C (eds) United Nations legal order, vol 1. The American Society of International Law
Klabbers J (2002) An introduction to international institutional law. Cambridge University Press, Cambridge
Milde M (1989) The Chicago Convention - after forty years, IX. Ann Air Space Law IX:203
Sarooshi D (2005) International organizations and their exercise of sovereign powers. Oxford University Press, Oxford
Schenkman J (1955) International civil aviation organization. Librairie E. Droz, Geneve
Seyersted F (1963) Objective international personality of intergovernmental organizations: do their capacities really depend upon the conventions establishing them? Krohns Bogtnykker, Copenhagen
Sochor E (1991) The politics of international aviation. McMillan, London
Tobolewski A (1979) ICAO's legal syndrome. Ann Air Space Law 1V:349
Whiteman M (1968) Digest of international law. Kluwer, The Hague

Article 39
Endorsement of Certificates and Licenses

(a) Any aircraft or part thereof with respect to which there exists an international standard of worthiness or performance, and which failed in any respect to satisfy that standard at the time of its certification, shall have endorsed on or attached to its airworthiness certificate a complete enumeration of the details in respect of which it so failed.

(b) Any person holding a license who does not satisfy in full the conditions laid down in the international standard relating to the class of license or certificate which he holds shall have endorsed on or attached to his license a complete enumeration of the particulars in which he does not satisfy such conditions.

R. Abeyratne, *Convention on International Civil Aviation*,
DOI 10.1007/978-3-319-00068-8_40,

Article 40

Validity of Endorsed Certificates and Licenses

No aircraft or personnel having certificates or licenses so endorsed shall participate in international navigation, except with the permission of the State or States whose territory is entered. The registration or use of any such aircraft, or of any certificated aircraft part, in any State other than that in which it was originally certificated shall be at the discretion of the State into which the aircraft or part is imported.

R. Abeyratne, *Convention on International Civil Aviation*,
DOI 10.1007/978-3-319-00068-8_41,

Article 41

Recognition of Existing Standards of Airworthiness

The provisions of this chapter shall not apply to aircraft and aircraft equipment of types of which the prototype is submitted to the appropriate national authorities for certification prior to a date three years after the date of adoption of an international standard of airworthiness for such equipment.

R. Abeyratne, *Convention on International Civil Aviation*,
DOI 10.1007/978-3-319-00068-8_42,

Article 42

Recognition of Existing Standards of Competency of Personnel

The provisions of this chapter shall not apply to personnel whose licenses are originally issued prior to a date one year after initial adoption of an international standard of qualification for such personnel; but they shall in any case apply to all personnel whose licenses remain valid five years after the date of adoption of such standard.

R. Abeyratne, *Convention on International Civil Aviation*,
DOI 10.1007/978-3-319-00068-8_43,

Part II
The International Civil Aviation Organization

Chapter VII. The Organization

Chapter VIII. The Assembly

Chapter IX. The Council

Chapter X. The Air Navigation Commission

Article 56. Nomination and Appointment of Commission
Article 57. Duties of Commission

Chapter XI. Personnel

Article 58. Appointment of Personnel
Article 59. International Character of Personnel
Article 60. Immunities and Privileges of Personnel

Chapter XII. Finance

Article 61. Budget and Apportionment of Expenses
Article 62. Suspension of Voting Power
Article 63. Expenses of Delegations and Other Representatives

Chapter XIII. Other International Arrangements

Article 64. Security Arrangements
Article 65. Arrangements with Other International Bodies
Article 66. Functions Relating to Other Agreements

Article 43
Name and Composition

An organization to be named the International Civil Aviation Organization is formed by the Convention. It is made up of an Assembly, a Council, and such other bodies as may be necessary

Contents

1 The UN Specialized Agency for Civil Aviation

ICAO's formation and purpose is given in the Proceedings of the International Civil Aviation Conference (Chicago, Illinois, November 1–December 7, 1944) as follows:

> On November 1944, representatives of 52 nations came together at Chicago, to create a framework for the growth anticipated in world civil aviation. The Convention on International Civil Aviation, also known as the Chicago Convention, provided the establishment of the International Civil Aviation Organization (ICAO) - an international body to guide and regulate international civil aviation. ICAO came into existence on 4 April 1947, after 26 states had ratified the convention. Between 1944 and 1947 a provisional organization (PICAO) operated, the purpose of which was to be of a technical and advisory nature of sovereign States for the purpose of collaboration in the field of international civil aviation and to lay down the foundation for a new international organization to be headquartered in Montreal, Canada.

The International Civil Aviation Organization is the United Nations specialized agency dealing with international civil aviation. ICAO was established by the Chicago Convention, signed at Chicago on 7 December 1944. On 11 September

R. Abeyratne, *Convention on International Civil Aviation*,
DOI 10.1007/978-3-319-00068-8_44, © Springer International Publishing Switzerland 2014

1944, the Government of the United States of America, on its own initiative, sent a letter of invitation to 53 States and two dignitaries whose governments were in exile, inviting them to a conference that would lead to the development of international air transport as a post-war measure. The US Government suggested that the proposed international conference consider *inter alia*, the establishment of provisional world route arrangements by general agreement which would form the basis for the prompt establishment of international air transport services by the appropriate countries. There was also a suggestion to set up a permanent international aeronautical body, and a multilateral aviation convention dealing with the fields of air transport and air navigation including aviation technical subjects.

Fifty two States signed the Chicago Convention on 7 December 1944. The Convention came into force on 4 April 1947, on the 30th day after deposit with the Government of the United States. In its Preamble, the Chicago Convention records the fact that the signatory States agree on certain principles and arrangements in order that international civil aviation may be developed in a safe and orderly manner and that international air transport services may be established on the basis of equality of opportunity and operated soundly and economically. *Ex facie*, this pronouncement blends the need for there to be order in civil aviation with that of the need for equality of opportunity and the economical operation of air services.

Article 43 of the Convention states that an Organization to be named the International Civil Aviation Organization is formed by the Convention. ICAO is made up of an Assembly, which is the sovereign body of the Organization composed of all the ICAO member (Contracting) States, and a Council which elects its own president. The Assembly, which meets at least once every 3 years, is convened by the Council. The Council is a permanent organ responsible to the Assembly, composed of 36 Contracting States. These 36 Contracting States are selected for representation in the Council in three categories: States of chief importance to air transport; States not otherwise included which make the largest contribution to the provision of facilities for international air navigation; and States not otherwise included whose designation will insure that all the major geographic areas of the world are represented on the Council.

The seeds for the creation of ICAO were planted by the delegates to the International Civil Aviation Conference, held from 1 November to 7 December 1944 in Chicago, Illinois, where they worked out an interim agreement setting up the International Organization on Civil Aviation,[1] which was later known as the Provisional International Civil Aviation Organization (PICAO), until the Chicago Convention came into force on 4 April 1947 formally establishing ICAO. At the fourth meeting of the interim Assembly of PICAO, it was decided, in accordance with Article 45 of the Chicago Convention, that the Organization would have its seat in Montreal.

The purpose of the Organization is derived from the *Preamble* to the Chicago Convention, which, *inter alia*, states that civil aviation should be developed in a safe and orderly manner and that international air transport services may be established on

[1]*Proceedings of the International Civil Aviation Conference*, Chicago, Illinois, November 1–December 7, 1944, Vol. 1, The Department of State, at p. 111.

the basis of equality of opportunity and operated soundly and economically. To this end, the aims and objectives of ICAO are set out in Article 44 of the Convention, as being to develop the principles and techniques of international air navigation and to foster the planning and development of international air transport. These aims and objectives are multiple and embrace several categories of competences and functions which are technical, administrative, supervisory, legislative and juridical. The supervisory and juridical functions have been called quasi legislative and quasi judicial functions, in order to avoid confusion with the parallel functions of governments.[2]

Article 47 of the Chicago Convention provides that ICAO enjoys "such legal capacity as may be necessary for the performance of its functions" and goes on to say that "full juridical personality shall be granted to the Organization wherever compatible with the constitution of the laws of the State concerned."

The ICAO Secretariat is headed by a Secretary General who is appointed by the Council and is in charge of five separate bureaux, each with its own director. Of these the Air Navigation Bureau, which deals with technical issues and the Air Transport Bureau, which addresses economic issues, are the two main bureaux of the Secretariat and are responsible for work on the 18 Annexes to the Chicago Convention containing Standards and Recommended Practices (SARPS) on international civil aviation which States are obligated to follow. The Legal Bureau handles international treaties adopted under the auspices of ICAO and other legal issues related to air law. The other two bureau are the Technical Cooperation Bureau, which is in charge of ICAO's technical assistance programme, and the Administration Bureau which assists the Secretariat with its administration. The Secretariat carries out the functions laid down by the Convention itself, by the directives of the Assembly and the decisions of the Council.

The Council has two main subordinate governing bodies, the *Air Navigation Commission*[3] and the *Air Transport Committee.*[4] *The Air Navigation Commission* is serviced by The Air Navigation Bureau and is responsible for the examination,

---

[2]Schenkman (1955), at p. 121.

[3]The Air Navigation Commission considers and recommends, for approval by the ICAO Council, Standards and Recommended Practices (SARPs) and Procedures for Air Navigation Services (PANS) for the safety and efficiency of international civil aviation. The Commission is composed of nineteen persons who, as outlined in the Convention on International Civil Aviation, have "suitable qualifications and experience in the science and practice of aeronautics". Commission Members, who act in their personal expert capacity, are nominated by Contracting States and are appointed by the Council of ICAO. The duties of the Commission are set out in Article 57 of the Chicago Convention: The Air Navigation Commission shall: (a) Consider, and recommend to the Council for adoption, modifications of the Annexes to this Convention; (b) Establish technical subcommissions on which any contracting State may be represented, if it so desires; (c) Advise the Council concerning the collection and communication to the contracting States of all information which it considers necessary and useful for the advancement of air navigation.

[4]The first meeting of the new Air Transport Committee was held in 19 January 1949. The ATC created by PICAO in 1945 ceased to exist; the functions of the original ATC were transferred to the new ATC, the membership remaining unchanged except for the Chairman. The work of the FAL Division was continued and a second meeting of this Division was held in Geneva, Switzerland, in May–June 1948.

coordination and planning of all of ICAO's work in the air navigation field. This includes the development and modification of SARPS) contained in the ICAO Annexes (all except Annexes 9 and 17), subject to the final adoption by the ICAO Council. The *Air Transport Committee* is serviced by the Air Transport Bureau and is responsible for matters outside the field of air navigation. The Committee is responsible for economic and commercial matters and related financial issues. It is charged with developing SARPS of Annexes 9 (Facilitation) and 17 (Aviation Security) and submitting them for approval of the Council. In addition, both these bodies submit other policy (such as guidance material on international civil aviation including but not limited to taxation of air transport and air navigation services economics) for the approval of the Council, based on the work of the Secretariat.

There are various other Committees of the Council, such as the *Legal Committee*,[5] a permanent body advising the Council on matters referred to it by the Council concerning the interpretation and amendment of the Chicago Convention and conducting studies on public international law; the *Committee on Joint Support of Air Navigation Services*,[6] which is responsible for the administration of over-flight charges concerned with North Atlantic flights; and the *Committee on Unlawful Interference*, which is a standing committee reporting to the Council on aviation security issues. *The Technical Cooperation Committee* assists the Council in supervising work carried out under ICAO's technical assistance programmes which assist States in various aviation related projects. In addition, the *Personnel Committee* and the *Finance Committee* report to the Council on administrative and financial issues relating to the Organization.

The ICAO Council has played a signal role in dispute resolution in the nineties up to date over the past two decades. One of the best examples of ICAO's role in the international community was seen in The Iranair Incident—IR 655 (Iran, United States 1998). This concerned the shooting down of an Iran Air Airbus A300 (IR655) carrying commercial passengers on a scheduled flight from Bandar-Abbas (Iran) to Dubai. The aircraft was brought down by the *U.S.S. Vincennes* over the Persian Gulf, resulting in the death of all 290 persons on board the aircraft.

---

[5]The Assembly, at its 7th Session (Brighton, 16 June–6 July 1953) adopted Resolution A7-5 (Revised Constitution of the Legal Committee) which resolved that the Legal Committee shall be a permanent Committee of ICAO constituted by the Assembly and responsible to the Council except as otherwise specified. The duties of the Committee are t advise the Council on matters relating to the interpretation and amendment of the Chicago Convention referred to it by the Council; to study and make recommendations on such other matters relating to public international air law as may be referred to it by the Council or Assembly to study problems relating to private air law affecting international civil aviation as directed by the Council or Assembly of ICAO and to make recommendations on participation of non-Contracting States and other international Organizations at meetings of ICAO.

[6]The Committee is charged with the organization and supervision of agreements between groups of ICAO member nations providing for collective financing or collective operation of essential air navigation aids and facilities in areas of the world where these could otherwise not be supplied. Included *in* the Committee's responsibilities are the North Atlantic Ocean Weather Stations agreement and agreements which maintain Loran and other air navigation facilities in Greenland, Iceland and the Faroe Islands.

One of the emergent features of the ICAO Council which became clear at its deliberations was the Council's resolve to address its deliberations to purely technical issues pertaining to the incident, while stringently avoiding political issues and diplomatic pitfalls. This is certainly true of all incidents discussed above, where the Council restricted its scope to technical issues as applicable to the principles embodied in the Chicago Convention.

ICAO maintains regional offices in seven locations around the world: Asia Pacific Office in Bangkok, Thailand; Eastern and Southern African Office in Nairobi, Kenya; European and North Atlantic Office in Paris, France; Middle East office in Cairo, Egypt; North American, Central American and Caribbean Office in Mexico City, Mexico; and Western and Central African Office in Dakar, Senegal.

No express mention is made in the Chicago Convention of the status of the Organization. ICAO is primarily governed by international law, being recognized by the United Nations Charter as a specialized agency of the United Nations. It is also governed by two major agreements, one between the United Nations and ICAO and the other between the Government of Canada and ICAO. The Headquarters Agreement between ICAO and Canada,[7] in Article 2, explicitly provides that ICAO shall possess juridical personality and shall have the legal capacities of a body corporate including the capacity to contract; to acquire and dispose of movable and immovable property; and to institute legal proceedings in Canada.

ICAO is driven by its joint mission and vision statement which reads:

> The International Civil Aviation Organization, a UN specialized agency, is the global forum for civil aviation. ICAO works to achieve its vision of safe, secure and sustainable development of civil aviation through cooperation among its member States.

To realize this vision, the Organization has established six strategic objectives which are: to enhance global civil aviation safety; enhance global civil aviation security; minimize the adverse effect of global civil aviation on the environment; enhance the efficiency of aviation operations; maintain continuity of aviation operations; and strengthen the law governing international civil aviation. These Strategic Objectives drive ICAO's Business Plan which in turn is run by Key Activities and Critical Tasks that are required to ensure the realization of the Strategic Objectives through a stringent performance management process.

Membership of ICAO is open to all member States of the United Nations who deposit their instrument of ratification of the Chicago Convention. There are 191 Member States of ICAO, and the Organization works in six official languages: English, French, Spanish, Arabic, Chinese and Russian.

In determining ICAO's status as an international organization, the most basic question at issue for the courts would be, "what is an international organization? Unfortunately, there is no specific answer to this question as no overarching definition has been developed identifying what an international organization is in precise terms. One commentator is of the view that at best, we might recognize one if we see

---

[7]Headquarters Agreement Between the International Civil Aviation Organization and the Government of Canada, ICAO Doc 9591. For further information on the agreement see Milde (1992).

it. The main reason for the difficulty in reaching a precise definition or identifying all encompassing characteristic of an international organization is that it is in limine a social creation. Moreover, the creators of an international Organization do not set off to create it with a pre approved blueprint. Rather, they carve it to accord with their needs. ICAO was created by the International Civil Aviation Conference and given effect to by the Chicago Convention of 1944 which, in Article 44, recognizes that ICAO has aims and objectives—nothing more, nothing less. Courts would be somewhat constrained in determining ICAO's role within the parameters of these aims and objectives with regard to ICAO's capacity to be considered in the context of a contract entered into or tortious liability it might be accused of.

The two main characteristics of ICAO are: that it is created by States, more specifically, as States themselves are abstractions, by duly authorized representatives of States; and that they are created by treaty, which is a written agreement signed by the States' Parties to it and governed by international law. States can only act by and through their agents. Different government departments or Instrumentalities of State bear responsibility for different international organizations. In the case of ICAO, the most likely government department that would be responsible for the Organization within a State would be the Ministry or departments of transport or aviation as the case may be. The third characteristic that distinguishes an international organization as a "club" of States without just being the spokesperson or mouthpiece of those States is that it is expected to have a "will" of its own. As will be seen later in this article, ICAO's independent will, recognized by the Government of Canada for purposes of its activities within the country is encapsulated in a provision which states that ICAO has an identity of its own, capable of entering into contracts. This having been said, ICAO is by no means sovereign in its own right, although courts have on occasion referred to sovereign rights of an organization merely to seek a compromise between absolute acceptance of parity between a State and an organization and absolute refusal of an international organization's ability to perform *acta jure imperii* (governmental acts).

It is curious that, six decades after the establishment of the International Civil Aviation Organization (ICAO), some still refer to its powers and functions. There are some others who allude to ICAO's mandate. The fact is that ICAO has only aims and objectives, recognized by the Convention on International Civil Aviation which established the Organization. Broadly, those aims and objectives are to develop the principles and techniques of international air navigation and to foster the planning and development of international air transport. In effect, this bifurcation implicitly reflects the agreement of the international community of States which signed the Chicago Convention that ICAO could adopt Standards in the technical fields of air navigation and could only to offer guidelines in the economic field. This exclusive right in the technical field initially bestowed on gave ICAO the authority to set standards for equipment and procedures on international air routes in the first years (1947–1949) of ICAO. This key task of standardizing technical specifications in air navigation gave rise to the realization that, apart from States that were most advanced in technology and could implement ICAO's standards, there were numerous other States who could not implement the Standards, however, willing they would be, due to the lack of resources and know-how. This gave rise to offers of help by numerous member States of ICAO, and the technical assistance limb of ICAO was born.

One has, however, to note that the position of ICAO as an incipient regulator was quite different from the one it is placed in now. In 1949 ICAO was feeling the pulse of the world of civil aviation and formulating regulations. This was the year in which ICAO embarked upon a comprehensive study of the ground facilities operated by governments and the services that they provided. Traffic volumes were only beginning to pick up and passengers and cargo between North America and Europe were being carried by eleven carriers. Elsewhere, in South America direct east–west services went into operation for the first time.

Much water has flown under the bridge since then and ICAO now stands in a world of civil aviation bewildered by its own progress and exponential demand for capacity and services.

2 ICAO Today

Today, ICAO is a results-based, performance and values driven Organization guided by its own business plan. The Business Plan translates its six Strategic Objectives into action plans and establishes a link between planned activities, organizational cost and performance assessment. A vital dimension of this approach is the integration of programmes and activities of Bureaus and Regional Offices for optimum allocation of resources based on agreed priorities.

The Business Plan provides the basis for a reporting framework that unites strategies, activities, funds, human resources and time frames into a coherent and effective means of monitoring and evaluating outcomes. By engaging staff at all levels in the performance improvement process, highlighting responsibilities, and by holding managers accountable for their performance and regularly measuring, monitoring and evaluating results, the Organization will strengthen accountability, demonstrate value for money and improve overall performance at the operational and strategic levels. Moving from concept to action and results also involves a set of Supporting Implementation Strategies and the successful Technical Co-operation Programme which has a long tradition of supporting Contracting States in the implementation of ICAO regulations, procedures and policies.

The Business Plan addresses all areas of ICAO activity, the first of which, for purposes of this article is economic policy and infrastructure management. ICAO is active in the field of economic regulation and provides guidance to States in areas of economic policy and airports and air navigation services economics. ICAO's Manual on the Regulation of International Air Transport is designed to meet an ever-increasing need for a comprehensive and objective source of information about the many facets of this dynamic area. It has been prepared to meet the needs of all ICAO member States. The Airport Economics Manual provides practical guidance to States, airport managing and operating entities and designated charging authorities to assist in the efficient management of airports and in implementing ICAO's policies and charges for airports and air navigation services. Additionally, ICAO publishes the Manual on

Air Navigation Services Economics (Doc 9161) which provides guidance on the economic aspects of the provision of air navigation services.

Another important function of ICAO is to provide denominator data that are used to arrive at safety exposure data and safety indicators in lines with standardized defined taxonomy. It develops and provides to the Planning and Implementation Regional Groups (PIRGs) and States, passenger/freight traffic flows and aircraft movement forecasts and other planning parameters to optimize decision making as envisaged in the ICAO Global Air Navigation Plan. States are also provided guidance on development of business cases and financial analysis for planning and implementation of ANS.

ICAO has a busy work schedule both in the technical and economic fields. The current and future role of ICAO does not therefore have to be reinvented. However, for the past 60 years or so, ICAO has been active in its standardization role, which has been blended in recent years with a burgeoning implementation role that is gradually blurring the former. In a world that is becoming largely globalized and regionalized, ICAO has vastly to focus on not so much what it does but how it does its work. In this context, ICAO has a dual role to play. The first is to act as a global forum for aviation, which and this is primarily the role expected of ICAO by the developed nations world which are can largely be self reliant in regulatory matters in their own regions. However, they need ICAO to set global standards that could apply to all ICAO's 190 member States. On the other hand, ICAO has to be both a global forum and a mentor to the developing world which expects ICAO to assist and guide them.

In order to serve its 191 member States, irrespective of whether they are in the developed or developing category, ICAO has to justify its performance and values based stature. In other words, ICAO needs to should undergo a whole system change. For the Business Plan to be implemented and results to be produced, firstly ICAO's leaders (its Council and the senior managers of the Secretariat) have to drive the process of transition, from service to performance. They need to be the ambassadors of the Organization's mission and vision statements and set values and behaviours. They must "talk the talk" and "walk the walk".

he next step is to ensure that the mission and vision statement influences all decision making. This should permeate right to the bottom of the ICAO Secretariat. Thirdly, the new culture and its results must be measured by causal performance indicators. In other words, ICAO's new culture should be constantly monitored. The final measure would be to ensure that the values of the Organization's culture pervade and drive every aspect of decision making and be seen in every system and process.

In a way, ICAO is already undergoing a cultural transformation. It has comegone a long way in developing a mission and vision statement and a business plan driven by strategic objectives. There is a leadership that is committed to its work. There is also every indication that the leadership is ready and willing to involve the entire Organization in defining the mission, vision and values of the Organization. However, in order to achieve this successfully, strong tools and aggressive goals have to be in place through a robust and energized operational plan that is not disaggregated among the Organization's constituent bureaux and other offices.

Such an operational plan must have objectives and key performance indicators, as in. In the end, it is measurement that matters. For this there must be targets set, not just improvement of performance. In this regard it must be noted that ICAO's Business plan is on the right track, as it has all three types of indicators: causal indicators—which relate to values and behaviours (which are known in other words as core competencies); output indicators—which measure performance in terms if efficiency and productivity; and outcome indicators, which relate to the result or effect on clients and stakeholders.

Of these indicators, ICAO's concentration should be mainly on output indicators that measure productivity, efficiency, quality, innovation, creativity of the Organization as a whole and ensuing customer satisfaction. Innovation and creativity are key factors that serve to would promote ICAO's contribution to its member States. Just as an example, since in the technical field, many States do not have the volume and scope of aviation activities which generate the resources and the base-line activity necessary to support a workable safety oversight system, ICAO's role must be to take the leadership in providing States with templates of different models of safety oversight and recommend what is best suited for them. ICAO could also further the involvement of regional safety oversight organizations that are successful; and provide guidance to States as to the modalities of the transfer of responsibilities or tasks, depending on the model used, from participating State to regional safety oversight organizations.

Leadership is the key to ICAO's role in the twenty-first century, and nowhere is its need more pronounced than in the economic field. The inhibitive mind-set created by the Chicago Convention where Article 44 ascribes to ICAO only a watered down role to merely "foster the planning and development of international air transport" would militate against any cultural transformation that would demand a more greater leadership role from ICAO. The archaic and hopelessly obsolete premise that ICAO's aim should be "to meet the needs of the peoples of the world for safe, regular, efficient and economical air transport" is diametrically opposed to the new thinking in the ICAO leadership that, in a competitive world, ICAO has to perform 100 % in safety and security in helping develop an air transport system that is sustainable and efficient.

One particular area in which ICAO should rid itself of the shackles of the Chicago syndrome is liberalization of air transport. The expertise of the ICAO Council and the Secretariat would amply suffice to initiate policy without and not merely setting guidelines and conducting studies in this area. Since The Council has, over the years not hesitated to adopt resolutions, recommendations and policy statements on such areas as taxation of air transport, and it is somewhat puzzling as to why no attempt has so far been made to initiate some policy on liberalization of air transport.

ICAO should provides assistance in the liberalization process by carrying out studies and analyses of air transport situations globally, and should extend is in the process of extending this research to countries with a view to providing necessary guidance material, manuals and procedures along with recommendations as to how best to liberalize air transport in their regions. To achieve the abovementioned objective, ICAO should will endeavour to provide States with different models of liberalization packages and encourage States to use them as appropriate on a regional and/or multilateral basis. ICAO should also hopes to carry out studies that apprise States of the benefits of liberalization.

Furthermore, liberalization at regional level is a good way forward in this process and can contribute to a more flexible approach to market access. With regard to promoting regional cooperation among States in the field of liberalization, ICAO should study the potential that lies in facilitating consistency in the design and implementation of liberalization policies by different regions. Since liberalization can be achieved nationally, regionally, or multilaterally. ICAO could provide more active support to regional liberalization initiatives as a means to promote the sustainable economic development of air transport, including (but not limited to) developing States. ICAO should also encourage States to take a comprehensive approach whereby common rules and convergence between regulations is developed in parallel with the removal of market access barriers at regional level.

Another philosophical adjustment that ICAO should make is to ensure that it keep abreast of the new world order where States are increasingly being disaggregated into components which act in global networks, linking the world together in a manner that enables global trends to permeate the local environment. In other words, ICAO should facilitate interaction between States and their components that interact in matters of civil aviation. For example, in many member States, aviation has numerous players in different areas such as customs and immigration, medical and quarantine, tourism, police, airports and air navigation service providers. In most instances these players do not act in accord, thus resulting in disharmony in the ultimate delivery of an efficient air transport product. ICAO's Mission and Vision Statement exhorts ICAO to do just what is needed—to in acting as the global forum in the key areas of concern to international civil aviation through cooperation between its member States.

While promoting fluid dialogue and cooperation among its member States, ICAO should take the initiative to assist States both in technical and economic issues. This assistance is not confined to providing technical assistance through projects administered by the Technical Cooperation Bureau but should also extend to providing guidance, mainly to States which still look up to ICAO as the global forum of aviation experts.

ICAO's leadership role in the economic and technical fields of civil aviation hinges on two key factors: an aggressive operational plan structure with key performance indicators and targets; and the realization that organizational culture, which is an intangible asset, is the new frontier of competitive advantage. The latter is particularly important under the current circumstances of ICAO where human resources and expertise are in short supply. Cultural transformation starts with the leadership and, their individual and leadership values. When one looks at Looking at ICAO's current leadership structure, there is no room for doubt that this is not in short supply. However the trick is would be to motivate the staff sufficiently so that they would be impelled to follow their leaders in the transformation and forge ICAO's leadership forward in its various areas of work.

All this leads one to the bottom line, which is the need for a change in the mind-set of the Organization, from its service role to a role of implementation and assistance. The human factor is an essential consideration in this metamorphosis. The key, however, and the starting point, however, is to recognize the need for the transition, which ICAO has already done. The next step is to recognize that ICAO needs its peoples' best efforts, both individually and collectively. ICAO's image and the perception of the outside world of ICAO as an effective Organization is would be anchored on the extent to which its workers represent themselves as good stewards of ICAO's business. They should therefore work together in the overarching interest of the Organization. When all these factors are considered together, there is nothing to suggest that ICAO is headed in the wrong direction.

3 Technical Assistance

Apart from its work in safety, security, environment and economics, ICAO has an important and vibrant technical assistance programme. It is curious that, six decades after the establishment of ICAO, some still refer to its powers and functions.[8] There are some others who allude to ICAO's mandate. The fact is that ICAO has only aims and objectives, recognized by the Chicago Convention. Broadly, those aims and objectives are to develop the principles and techniques of international air navigation and to foster the planning and development of international air transport. In effect, this bifurcation implicitly reflects the agreement of the international community of States which signed the Chicago Convention that ICAO could adopt Standards in the technical fields of air navigation and could only offer guidelines in the economic field. This exclusive right in the technical field initially bestowed on ICAO the authority to set standards for equipment and procedures on international air routes in the first years (1947–1949) of ICAO.[9] This key task of standardizing technical specifications in air navigation gave rise to the realization that, apart from States that were most advanced in technology and could implement ICAO's standards, there were numerous other States who could not implement the Standards, however, willing they would be, due to the lack of resources and know-how.[10] This gave rise to offers of help by numerous member States of ICAO, and the technical assistance limb of ICAO was born.

[8]MacKenzie (2008), Preface at 1.

[9]Max Hymans, President of the Second ICAO Assembly observed in 1948 that the standardization of equipment and procedures had progressed well and that ICAO had an incontestable authority in this respect. See Hymens (1948) at 422.

[10]See. ICAO Doc 6968, A4-P/1 Report of the Council to the Assembly on the Activities of the Organization in 1949, 23 March 1950.

One has, however, to note that the position of ICAO as an incipient regulator was quite different from the one it is placed in now. In 1949 ICAO was feeling the pulse of the world of civil aviation and formulating regulations. This was the year in which ICAO embarked upon a comprehensive study of the ground facilities operated by governments and the services that they provided.[11] Traffic volumes were only beginning to pick up and passengers and cargo between North America and Europe were being carried by eleven carriers. Elsewhere, in South America direct east–west services went into operation for the first time.[12]

In accordance with this new Policy, ICAO will cooperate with States and other relevant entities to provide technical assistance taking into account the priorities of States and ICAO policy in implementing the Strategic Objectives[13] of ICAO. This cooperation will be forged primarily through projects, particularly where such projects are necessary for the provision of vital air transport infrastructure and/or the economic development of a State. In implementing this Policy, ICAO will optimally use its resources both at Headquarters and its Regional Offices and apply the principles enunciated in the relevant ICAO Assembly Resolutions, guidance and policy.

Overall responsibility for the implementation and continued evolution of this Policy devolves upon the Secretary General of ICAO, assisted by the Director, Technical Cooperation Bureau. As necessary and where relevant, this Policy will be incorporated into the ICAO workplace through the Organization's Business Plan.

It is an interesting fact that the Chicago Convention, from which ICAO derives its legal legitimacy, does not contain a single reference to technical assistance to be rendered by ICAO. Yet the Technical Cooperation Programme (TCP)[14] of ICAO which is executed by the Technical Cooperation Bureau (TCB) that was established in 1952, has maintained a sustained record of technical assistance provided to States over a span of 60 years. TCP is the major operational tool for reinforcing the Organization's technical cooperation mission objectives, including enhancing the capacity of developing countries to implement ICAO Standards and Recommended Practices (SARPs). Its continuing importance has been reaffirmed by the Assembly in several resolutions, *inter alia* in the Consolidated Statement of ICAO Policies on Technical Co-operation (Resolution A36-17), which stipulates that the Technical Co-operation Programme is a permanent priority activity of ICAO that complements the role of the Regular Programme in providing support to States in the effective implementation of

---

[11] *Id.* A4-P/1 23/3/50 at 1.

[12] *Id.* 2.

[13] ICAO's Strategic Objectives pertain to the enhancement of aviation safety and security; reducing the adverse effects of aviation on the environment and working towards sustaining air transport.

[14] ICAO's Technical Co-operation Programme provides advice and assistance in the development and implementation of projects across the full spectrum of civil aviation aimed at the safety, security, environmental protection and sustainable development of national and international civil aviation. The Programme is conducted under the broad policy guidance of the ICAO Assembly and of the Council. Subject to general guidance by the Secretary General, the Technical Co-operation Programme is executed by the Technical Co-operation Bureau (TCB).

SARPs and Air Navigation Plans (ANPs) as well as in the development of their civil aviation administration infrastructure and human resources; and is furthermore one of the main instruments of ICAO to assist States in remedying the safety and security deficiencies identified through ICAO's audit programmes.

Since its establishment, TCB has implemented civil aviation projects with an accumulated value in excess of US$2 billion. With an average annual programme size of over US$120 million, it is involved in approximately 250 projects each year with individual project budgets ranging from less than US$20,000 to over US$120 million. To date, TCB has provided assistance to over 115 countries, deploying annually approximately 1,200 international and national experts.[15]

The story of technical assistance commenced in 1948, when the UN General Assembly (GA) decided to appropriate funds under its regular budget to enable the UN Secretary-General to supply teams of experts, offer fellowships, and organize seminars to assist national development projects at the request of governments. Contemporaneously, many of the specialized agencies had begun, under their own TCPs, to undertake similar projects. However, no sooner had the Regular Programs of Technical Assistance, as they were then called, begun to operate than it became apparent that the funds that could be allocated from the regular budget towards technical assistance would not be sufficient to meet the demand from States for assistance. Therefore, in 1949, the General Assembly set up a separate account for voluntary contributions toward technical assistance and decided to make it a central account to finance the activities not only of the UN itself but also of the specialized agencies. Machinery was established for distributing financial resources and coordinating projects, and the whole enterprise was called the Expanded Programme of Technical Assistance (EPTA), to distinguish it from the UN's technical assistance financed under the regular budget. The venture proved remarkably successful. Ten years later, in 1959, EPTA was financing technical assistance in some 140 countries and territories. Between 1950 and 1960, the number of governments contributing funds had grown from 54 to 85, and the total annual contributions had risen from $10 to $33.8 million.

For several decades, the United Nations Development Programme (UNDP) served as the central funding agency for technical cooperation within the UN system and ICAO, as a specialized agency, was a natural beneficiary of this scheme.[16] However, in later years UNDP transcended into being a substantive

[15]In the past 20 years TCB has developed an important portfolio of projects funded by governments or service providers from developing countries, which have greatly contributed to the enhancement of aviation safety and the development of civil aviation infrastructure worldwide, particularly in the Latin America region. Despite the limited resources available to some other regions, progress steadily continues in this regard with TCB's renewed focus on regional cooperation in the Asia Pacific, Africa and Europe and the Middle East regions, as a means to harness efficiencies and reap the benefits of south–south cooperation. To date, TCB has provided assistance to over 100 countries, through the implementation of an average of 300 projects per year, and field deployment of approximately 4,000 international and national experts annually.

[16]As an indicator, in 1973, approximately 85 % of ICAO technical assistance was funded by UNDP in 47 countries involving 175 ICAO field experts. See Vivian (1973) at 10.

agency while retaining vestiges of its funding functions that are channelled into direct national execution, by-passing the specialized agencies concerned. ICAO now has to largely fund itself, particularly through administrative charges levied on States for technical assistance.

The ICAO Policy on Technical Assistance meshes well with the vision of TCB which is to enable developing countries to attain, in the field of civil aviation, a standard comparable to that of the developed countries and thereby to share in the economic and social benefits that air transport and related civil aviation industries generate at the local, regional and national level throughout the world. This vision—to enable States to continuously improve their aeronautical infrastructures and services leading to self-sufficiency in the aviation field is calculated to bring better human, social and economical conditions. The TCB vision goes on to say that TCB understands the hopes and problems of the developing world and from experience knows how to realize those hopes and overcome those problems.

In reaching towards its vision, TCB's priorities are to improve the operational safety, security, efficiency and regularity of national and international civil aviation and to contribute to the global and uniform implementation of ICAO's Standards and Recommended Practices (SARPs). With over six decades of experience, and drawing upon all of the technical expertise and knowledge available within ICAO, The Bureau's mission is to provide unrivalled in-depth technological assistance to States with their aviation projects.[17]

4 Historical Perspectives

In March 1949, the ICAO Council observed that ICAO's activities in the field of *expert advisory assistance* (as technical assistance was then called) was slight compared to such bodies as FAO and WHO and called for international exchange of technical knowledge through the United Nations umbrella. The Council was aware that member States would like to draw on ICAO for direct advice on how to improve the operation of their facilities and efficiency of their ground operations.[18] Special mention was made of assistance in air navigation, safety and efficiency of air transport. At that time, it was thought that visits by field experts of ICAO to States would be the best form of assistance.

The ICAO Council was of the view that the best principle was that where extraordinary expenses were involved, the first stages of assistance could be given by ICAO's regular staff as a normal part of the work by ICAO, and with regard to anything that went beyond that, the cost would be borne by the beneficiary State through a joint project under Chapter XV of the Chicago Convention. The Council considered the most important aspect of assistance to be the provision of advice on the establishment of training programmes or direct assistance in training.[19]

---

[17]See http://www.icao.int/Secretariat/TechnicalCooperation/Pages/Whoweare.aspx.

[18]*C-WP/212, 11/3/49*

[19]See *DOC 6684, C/766, 5/4/49.*

On 15 August 1949 the United Nations Economic and Social Council adopted Resolution 222 (IX) A upon approval by the United Nations General Assembly (by Resolution of 16 November 1949), and endorsement by ICAO Resolution A4-20. Resolution 222 (IX) entrusted all Organizations of the United Nations system to participate fully in the Expanded Programme of Technical Assistance for Economic Development, and ICAO, as the United Nations specialized agency for civil aviation, began the execution of technical cooperation projects in 1951 with funding from the United Nations Special Account for Technical Assistance, established pursuant to the Resolution. Some specialized agencies thought it best to provide assistance through a decentralized plan where each agency raised its own funds.[20]

On 23 October 1953 the United Nations General Assembly adopted Resolution 723/VIII which approved a revised United Nations Programme in public administration comprising the provision, at the request of governments, of technical assistance related to public administration, including training for public service through the advisory service of experts; fellowships and scholarships; training institutes, seminars, conferences, working groups and other means; and the provision of technical publications.

For a long time, the United Nations (UN) and its Specialized Agencies followed an approach which provided capital-oriented assistance for supporting efforts of developing countries to close the gap with the industrialized States. It was not until the 1990s that it had become obvious that this policy did not bring about the desired impact and longer-term sustainability. Emphasis therefore shifted from the use of expatriate technical cooperation personnel to the development of national professionals for the strengthening of local capacities and institutions. These attempts to transfer skills and knowledge systems were initially labelled technical assistance, a term that gave way subsequently to technical cooperation, in order to make clear the two-way road in the interaction between the United Nations and the developing countries and the importance of national ownership of the programmes. ICAO embraced this new approach and consequently changed the name of the Technical Assistance Bureau to Technical Co-operation Bureau. In this respect, both assistance and cooperation are provided within the Technical Co-operation Programme, depending on the specific needs of recipient States, donors, financial organizations and private entities funding TCB activities under the umbrella of Management Service (MSA) and Civil Aviation Purchasing Services (CAPS) Agreements.

The 37th Session of the ICAO Assembly (Montreal, 28 September–8 October 2010) was advised that the Technical Cooperation Programme (TCP) was entrusted with distinct but complimentary roles under Article 44 (c) and (d) of the Chicago Convention. These roles were:

- The traditional "technical cooperation" role, derived from the mandate by the United Nations, to provide assistance to civil aviation projects, particularly where these are necessary for the provision of the vital air transport infrastructure and/or the economic development of a State (A36-17, Appendix B, *Resolving Clause* 4).

[20] *C-WP/330, 31/5/49.*

ICAO is guided by the Resolutions of the United Nations General Assembly, which repeatedly emphasize that operational activities should be carried out for the benefit of the countries and in accordance with their own policies and priorities for development. These fundamental principles have been reconfirmed by Council since the establishment of the Technical Cooperation Programme (TCP); and

- The "technical assistance" role entrusted by the ICAO Assembly to assist States in remedying their deficiencies in the field of civil aviation (A36-17, Appendix A, *Resolving Clause* 4). The Assembly (A36-17, Appendix A, Resolving Clause 1) "*Recognizes* the importance of the TCP for promoting the achievement of the Strategic Objectives of the Organization" and mentions that "ICAO can assist States in advancing their civil aviation and at the same time promote the realization of its Strategic Objectives".

The Assembly was also advised that A36-17 reaffirmed that the TCP is a permanent priority activity of ICAO which complements the role of the Regular Programme in providing support to States in the effective implementation of SARPs and ANPs as well as in the development of their civil aviation administration infrastructure and human resources. More importantly, A36-17 affirmed that improved coordination of ICAO's technical cooperation activities should be achieved through clear delineation of each Bureau's mandate and activities, and enhanced cooperation, as well as closer coordination of the TCP and other ICAO assistance programmes for the avoidance of duplication and redundancy.

In view of the above, it is fair to conclude that *technical assistance* and *technical cooperation* are not mutually exclusive in the context of ICAO's TCP and that both assistance and cooperation are provided within the parameters of the Programme.

5 Technical Assistance from ICAO's Regional Offices

The ICAO regional offices were set up almost concurrently with the coming into being of ICAO. They were established for the purpose of ensuring safety of aviation in the regions including but not limited to maintaining liaison with States on the regional plans, conducting missions to States, preparation of special studies of interest to States in the region; identifying and addressing deficiencies, and maintaining diplomatic relations with States.

Over the past several years, the Regional Offices and their functions have enjoyed some exclusivity with regard to Headquarters, despite basic guidelines in the Regional Office Manual which admit of the involvement of Headquarters Directors in areas of specialization particularly related to air navigation and air transport.

Recent conversations with some ROs have revealed (corroborated by TCB) that during the past 2 or 3 years there have been almost no liaison or contact between ROs and TCB, leading to a total breakdown in communications on TCB projects and other work of a technical cooperation nature.

5.1 Europe and the Middle East

In the Europe and Middle East geographical area (i.e. Paris and Cairo), the Regional Offices participate in a limited way in technical cooperation activities. The small participation of the Paris Office is understood, considering the small number of technical cooperation projects in Europe. In the Middle East region, where there are a number of important technical cooperation projects, the participation of the Cairo Office is more significant. However, there is currently no Technical Cooperation Officer in the Office, thus hindering the full potential of collaboration between the Regional Office and TCB. Nevertheless, it should be noted that the Cairo Regional Office is kept well informed of technical cooperation projects in the region, particularly with regards to regional projects. In fact, the Regional Office in Cairo participates actively in the Board meetings of the regional MID-RMA project.

5.2 Americas

In the Americas geographical area (i.e. Mexico and Lima), the Regional Offices participate actively in a number of technical cooperation activities. With regards to regional projects, meetings, seminars or conferences concerning technical cooperation, the Regional Offices are always invited to participate and in many instances have a lead role in the organization of such activities. For example, the Regional Office in Mexico already administers, in close collaboration with TCB, regional projects such as the CAPSCA, as well as coordinating important meetings such as the NACC/DCA steering committee meetings in the CAR sub-region. Still, neither of the Regional Offices in the Americas benefit from the support of a Technical Cooperation Officer, which could enhance ICAO's mission in the region, particularly with respect to technical cooperation project development.

5.3 Africa

In Africa (i.e. Dakar and Nairobi), the participation of the Regional Offices in technical cooperation activities is inconsistent, most notably in Dakar, which does not have a Technical Cooperation Officer. In Nairobi however, a Technical Cooperation Officer was appointed to assist in the promotion and management of technical cooperation activities. As a result, communication between the Regional Office and TCB has been enhanced, and the on-going follow up and management of technical cooperation projects has improved. Nevertheless, Regional Office involvement in technical co-operation activities must be better defined to avoid the doubling of tasks in both the Regional Office and Headquarters.

5.4 Asia and Pacific

In the Asia and Pacific region, the Regional Office in Bangkok has a high degree of participation in technical cooperation projects. Presently, a Technical Cooperation Assistant supports in the management of many projects in the region. Furthermore, the Regional Director is informed and invited to participate in all pertinent meetings, seminars and conferences, especially those regarding regional collaborative projects such as the various COSCAP projects in Asia. However, the role of the Regional Office remains incomplete insofar as no Technical Cooperation Officer is available to further assist in the identification and management of regional projects.

6 The Way Forward

Recognizing the need to reinforce the role of the Regional Offices in the identification, promotion and participation of technical cooperation activities to further support ICAO's Strategic Objectives, the Secretary General decided, in his IOM of 29 January 2010, to strengthen the Regional Offices by appointing Technical Cooperation Officers to each Office. As can be observed from the current trends, the Regional Offices play a role in technical cooperation projects. It is expected that the addition of Technical Cooperation Officers will assist TCB in identifying new projects as well as assist in the management of regional projects. However, to accomplish this, due regard must be given to certain criteria required to achieve this objective. Such a proposal must be done gradually, to ensure that the continuity of the on-going projects is not disrupted. A phased approach is recommended to avoid any major interference with existing project implementation.

Firstly, clear objectives and job descriptions must be established for the posts of Technical Cooperation Officers in the Regional Offices. Otherwise, the risk of redundancy, with the corresponding increase in costs will occur. The advantages of the TC Officer in the Region are his proximity to the States, thereby reducing mission costs and his awareness of the needs of the States, therefore permitting him to identify new technical cooperation projects. It is important to optimize such advantages in the elaboration of the posts tasks and in the selection of the suitable candidates.

Secondly, there is a need to transfer the management of regional collaborative projects to the Regional Offices must be done with due diligence. It is imperative that the Regional Offices be well equipped to take on the added responsibilities associated with this job.

In 2011 the Secretary General of ICAO Raymond Benjamin adopted a proactive approach to ICAOs Technical Cooperation Programme by integrating closely the activities of the Technical Cooperation Bureau (TCB) within the Regional Offices. It also elaborates on the procedures identified by the Secretary General in his memorandum, with a view to clarifying issues appurtenant thereto.

It is clear from the Secretary General's memorandum (and procedures identified therein) that by "integration" he means a "process of bringing people, activities and issues together to perform effectively". Therefore, there is seemingly no doubt that the Secretary General does not intend to transfer lock stock and barrel the functions of TCB into the regional offices, but rather, that he intends to establish closer cooperation between all parties concerned and to bring TCB closer to the States through the Regional Offices.

The first premise that flows from the aforesaid principle is that TCB retains an overall critical management role of projects to be exercised at Headquarters. However, the Regional Director, to whom the RO will report functionally, will represent ICAO in the States that his office is accredited to in the management of projects and other areas of representation, while receiving technical advice from TCB as necessary. The RO will perform his duties under the supervision of the Regional Director.

The vision of this approach is that the integration of TCB project management with the Regional Offices (ROs) by assigning TCB officers will improve ICAO assistance to States to which the ROs are accredited and facilitate the implementation of the ICAO Business Plan. The goal is to address aviation challenges pertaining to deficiencies and needs faced by States and provide assistance expeditiously and competently. In executing their duties it was expected that the ROs will pool their best resources of expertise in identifying the technical needs of States and nurturing partner coordination and regular exchange of views to set-up, implement, and critically review strategic assistance through TCB projects. In facilitating this platform, the TCB officer, under the ultimate supervision of the Regional Director, will advocate for additional interest and expertise from local and international partners and mobilize funds to rapidly scale up assistance from ICAO through TCB. This partnership can effectively address key technical and operational challenges to further achieve ambitious impact goals set by the ICAO Business Plan.

By placing a TCB Officer in each regional office the Secretary General has ensured that the following efficiency gains will ensue: (a) there will be better ICAO contact with States with a single point of contact in ICAO; (b) the interests of States will be served better; (c) ROs are in the best position to identify opportunities and potential projects; (d) missed opportunities, which result from the delay in getting timely responses from Headquarters, would be greatly diminished if not obviated; (e) each region will be served by a TCB person who is familiar with the region; (f) the TCB officer in the Regional Office will have a better appreciation and knowledge of the needs of the States; (g) there will be better promotion of the presence and visibility of technical cooperation in the region; (h) the quality of processes related to the selection of and cooperation with regional experts will improve; (i) briefing and debriefing of experts would be easier; (j) reports can be reviewed in the Regional Offices; (k) activities, plans and projects can be managed by the Regional Offices; (l) scholarship management in the regions would be easier; (m) project design would improve; (n) budgets of TCB projects would be better administered; (o) the reengineering of projects by an officer *in situ* would result in efficiency gains; (p) there will be a more streamlined process if regional projects are managed at the regional level (e.g. improved quality of draft projects); (q) there will be more visibility of TCB projects; (r) there will be less paper work; (s) there will be no more going back and forth between the regional offices and headquarters; (t) the

TCB officer will be on first name basis with the DGCA of each State and his/her staff; (u) time zone difficulties will be eradicated in terms of communications between ROs and TCB at Headquarters; (v) "who does what" will be clearer; (w) the more "demanding" States in a region can be served better; (x) costs of sending officers from Headquarters to the regions will be reduced.

ICAO is a specialized agency of the United Nations, deriving its legal legitimacy from Article 57[21] of the Charter of the United Nations (UN). The UN has established that four key principles should apply in the field of technical cooperation emanating from the UN. They are: (a) Lead responsibility for a given issue or activity should rest with the entity best equipped substantively to assume it; (b) Entities in the lead on a given issue or activity should work in close collaboration with the rest of the United Nations rather than attempt to duplicate expertise available elsewhere in the Organization; (c) More systematic efforts should be made to draw on the vast reservoir of knowledge and expertise that exists outside the United Nations system; and (d) Technical cooperation should be delivered to the maximum extent, possible by the entities that have an established field presence and experience. Secretariat entities should provide policy guidance and expertise, as appropriate.[22]

ICAO's role in these four key areas are rooted with the Council which directs the Organization (with the ultimate agreement of the Assembly) to deal with issues of particular concern i.e. matters pertaining to safety, security, environmental protection and the sustainable development of air transport. The genesis of ICAO's technical assistance, as related to the United Nations umbrella goes back to the Sixth Session of the Council in 1949 when it recognized a Resolution adopted by the UN Economic and Social Council (ECOSOC) in March 1948 which requested the UN Secretary General, in consultation with the executive heads of the interested specialized agencies to prepare a report setting forth a comprehensive plan for an expanded cooperative programme for economic development through the UN and its specialized agencies paying due attention to questions of a social nature which

[21]Article 57 stipulates that the various specialized agencies, established by intergovernmental agreement and having wide international responsibilities, as defined in their basic instruments, in economic, social, cultural, educational, health, and related fields, shall be brought into relationship with the United Nations in accordance with the provisions of Article 63. It also states that such agencies thus brought into relationship with the United Nations are hereinafter referred to as specialized agencies. Article 63 stipulates that The Economic and Social Council may enter into agreements with any of the agencies referred to in Article 57, defining the terms on which the agency concerned shall be brought into relationship with the United Nations. Such agreements shall be subject to approval by the General Assembly. It further states that the General Assembly may co-ordinate the activities of the specialized agencies through consultation with and recommendations to such agencies and through recommendations to the General Assembly and to the Members of the United Nations.

[22]Review of technical cooperation in the United Nations Report of the Secretary-General, Fifty-eighth session Agenda item 59, Strengthening of the United Nations system, A/58/382, 19 September 2003.

directly condition economic development and explore methods of financing such a programme including special budgets.[23]

The Council, in citing this ECOSOC Resolution observed that:

> In the interests of civil aviation alone, the Organization (ICAO) has an obligation to its member States to stand ready to extend assistance to the under developed countries. Any improvement in air navigation services helps to raise the levels of safety and efficiency on the international airlines. . .a proper response to such inquiries cannot be made by correspondence alone. It needs an early visit from a field office representative to discuss the needs with the officials of the State concerned and to make his own survey of the possibilities.[24]

This observation, made in 1949 by the Council, resonates well with the recent decision of the ICAO Secretary General to appoint dedicated technical cooperation officers in each of the ICAO regional offices.

The interoperability of technical assistance and technical cooperation within ICAO, particularly in the context of their separate definitions as discussed above, brings to bear the issue of funding. The question is, which activities would be funded by the regular programme, if any, and which activities would be funded through externally acquired resources from States for the specific purpose of rendering technical assistance? The President of the ICAO Council, at the 14th meeting of the Council in 1949 indicated that the principle he had in mind was that where extraordinary expenses were incurred, the first stages of the investigation and such small-scale assistance as could be given within a short space of time by ICAO's regular staff would be considered as a normal part of the work of the Organization, but that for anything that went beyond that the cost would be borne by the State benefitting or arranged through a joint project under Chapter XV of the Chicago Convention.[25] The Council noted that requests from States in this regard for ICAO assistance might take different forms. For instance, States could seek assistance in preparing a general programme of civil aviation; and for the development of air navigation facilities and services; they could seek specific advice on the planning and location of a particular facility such as an aerodrome; they may seek assistance in preparing detailed plans and specifications for the development and construction of particular projects or in letting contracts; they may seek advice as to the keeping of aeronautical statistics and other matters related to the economic activities of the Organization, or, on the drafting of civil navigation laws and other legal problems. Most important of all, States could seek advice on the establishment of personnel training programmes or even direct assistance in giving training.[26]

---

[23]C-WP/212, 42 tit.24, pr 22 S 4 at Appendix B. The Council of ICAO recognized that other specialized agencies such as the Food and Agriculture Organization (FAO), the International Labour Organization (ILO), the World Health Organization (WHO) and the International Bank for Reconstruction and Development (IBRD better known as the World Bank) had already conducted technical assistance missions to member States. *Id.* Appendix C.

[24]C-WP/212 (1949) *id.* at 2.

[25]Doc 6684, C/766, 5/4/49 at 19.

[26]*Id.* Appendix B at 32.

7 The Technical Cooperation Bureau

The Technical Co-operation Bureau is self-financing and must therefore recover all its administrative costs, to maintain a balance between income and expenditure, while ensuring that overhead charges to projects were kept to a minimum. a performance audit of the Technical Co-operation Bureau was conducted by the External Auditor in March 2009, resulting in a number of recommendations aiming at improving its efficiency and effectiveness at the management level, which were approved by the Council and are being addressed by the Secretary General through an action plan implemented under the supervision of the Council, ensuring that timely follow-up action has been undertaken.

There is a compelling need for a balanced approach so that the Technical Co-operation Programme could address both the sovereignty of States to determine their development priorities as well as the responsibility of ICAO to promote the Strategic Objectives of the Organization. Considering, however, that the Technical Co-operation Programme is financed, almost in its entirety, by those States requesting assistance on the basis of their national priorities and requirements, while deficiencies may exist in other areas, there is a need to strike a balance between a legitimate request for ICAO assistance by a sovereign State on the one hand and the promotion of the Organization's priorities in accordance with its Strategic Objectives on the other. In other words, it is necessary to facilitate technical cooperation in response to specific requirements of States while at the same time encouraging States to focus their implementation activities on ICAO priorities and for the Organization to provide technical assistance to remedy deficiencies identified in ICAO audits to those States lacking the required financial and technical resources. Furthermore, consideration should be given to developing an Organization-wide emergency response strategy to coordinate ICAO's response to natural disaster or national calamities situations in States, bearing in mind the assistance role of the Technical Co-operation Bureau and the Regional Offices in the affected areas.

Under the new policy on technical assistance, ICAO needs to achieve the following: strengthen TCB management in the region with a view to ensuring that it is operational at all levels of the regional system; improve capacity of host country governments to obtain, use, and disseminate quality information pertaining to deficiencies and respond to such; support States in developing national strategic policies that comprehensively address deficiencies; aid countries in obtaining and keeping external funding; discuss with regional civil aviation bodies ways and means to assist States in the region; create, maintain, and disseminate best-practice guides and reference resource; asses existing capacity development efforts in States; implement programmes and work plans that are data-driven at all levels; in the field of training, support States in developing national workforce forecasting and frameworks to accommodate changing national policies e.g. shifting from regulation to implementation; review and update curriculum of existing training courses taking into account new challenges and recent elimination objectives; and conduct workshops/courses to encourage, exchange, and teach best practices and evaluate training courses and program follow up of participants.

8 ICAO's Legal Work

8.1 Treaties and Arrangements Pertaining to Air Transport

One of the key roles of ICAO in the legal field is to function as depository of air law treaties and to register arrangements and agreements entered into by States in the field of aviation. ICAO is the depository of 35 international treaties and related instruments entered into between its member States, both on a bilateral and multilateral basis. In addition, ICAO performs registration functions with regard to other aeronautical agreements between States *inter se* and international organizations. These functions are provided by the Legal Affairs and External Relations Bureau.[27] Many aeronautical treaties have emanated from initiatives of the ICAO Legal Committee.[28] ICAO has a significant role to play in the procedural aspects of treaty work which brings to bear the need to discuss the salient principles of law and practice relating to aeronautical treaties and agreements. Such a study becomes essential to the understanding of treaties related to aviation security, safety and future agreements concerning aviation and environmental protection.

It is incontrovertible that every aeronautical treaty and agreement adopted with the involvement of ICAO is globally beneficial towards meeting the needs of the people of the world for safe, efficient, regular and economical air transport. However, not every treaty has attracted the number of ratifications necessary for it to enter into force. At its highest level ICAO encourages those of its member States who have not ratified aeronautical treaties to deposit their instruments of ratification so that the cycle of treaty making becomes complete and the world has globally applicable rules.

The following discussion defines treaties and agreements and the various categories the latter fall into, while discussing their general principles in an aeronautical context. It also discusses the various nuances and complex principles relating to modern treaty law and practice, delving into the approaches taken by State legislatures and judiciaries in incorporating treaties into domestic legislation. There is also a discussion with regard to the manner in which ICAO performs its depository and registration functions.

---

[27]The Bureau also provides advice and assistance to the Secretary General and through him to Council of ICAO and other bodies of the Organization and to ICAO Member States on constitutional, administrative and procedural matters, on problems of international law, air law, commercial law, labour law and related matters. Additionally, the Bureau conducts research and studies in the field of private and public international air law, prepares documentation for, and serves as the Secretariat of the Legal Committee: relevant bodies of the Assembly; and Diplomatic Conferences which adopt multilateral treaties on international air law.

[28]The Legal Committee of ICAO was established by the Interim Council on 24th June 1946 and approved by the First Assembly on 23rd May 1947.

Aeronautical treaties and other agreements pertaining to civil aeronautics are subject to the law and practice applicable to treaties in general and come under the purview of ICAO particularly with regard to the registration of such documents. Additionally, ICAO functions as the depository of treaties in instances where such treaties provide that ICAO be the depository. The designation of the depositary of a treaty may be made by the negotiating States, either in the treaty itself or in some other manner. The depositary may be one or more States, an international organization or the chief administrative officer of the organization. The functions of the depositary of a treaty are international in character and the depositary is under an obligation to act impartially in its performance. In particular, the fact that a treaty has not entered into force between certain of the parties or that a difference has appeared between a State and a depositary with regard to the performance of the latter's functions does not affect that obligation. There are also attendant obligations of States Parties to treaties calling for them to register international treaties with the United Nations. Within this context, this article intends to clarify the law and practice pertaining to aeronautical treaties and agreements in the field of civil aviation.

8.2 Treaties

A treaty is an international agreement concluded between States[29] in written form and governed by international law, whether embodied in a single instrument or in two or more related instruments and whatever its particular designation.[30] The above notwithstanding, a treaty can be concluded between a State and another subject of international law such as an international Organization. An example is the Headquarters Agreement between ICAO and the Government of Canada.[31] When a State places its signature on a treaty it merely means that the State has agreed to the text in the instrument. It comes into effect for that State when it is

[29] A State has been defined in Article 1 of the Montevideo Convention of 1933 as having the following characteristics: a permanent population; a defined geographic territory; a government; and the legal capacity to enter into relations with other States. See Montevideo Convention on the Rights and Duties of States, Signed at Montevideo, 26 December 1933. The Convention entered into Force, 26 December 1934. At http://www.taiwandocuments.org/montevideo01.htm.

[30] Vienna Convention on the Law of Treaties, 1969, Done at Vienna on 23c May 1969, United Nations General Assembly Document *A/CONF.39/27*, 23 May 1969, Article 2(a). The Convention entered into force on 27 January 1980. UNTS Vol. 1155, p. 331.

[31] Headquarters Agreement between Canada and ICAO of 14 April 1951, which paraphrased the 1947 Convention on the Privileges and Immunities of the Specialized Agencies. On 20 February 1992, the 1951 Agreement was terminated and superseded by a new Agreement that entered into force the same day. A new Supplementary Agreement was signed on 28 May 1999 superseding the Supplementary Agreement signed in 1980 in order to reflect the relocation of the Organization's Headquarters to a new location on 999 University Street on November 1, 1996. See Supplementary Agreement Between the International Civil Aviation Organization and the Government of Canada Regarding the Headquarters of the International Civil Aviation Organization, Doc 9591.

ratified[32] by the State. At the time of ratification a State can record a reservation to a part of the treaty.[33] These generic principles and those discussed below also apply to aeronautical treaties and agreements.

It must be noted that a State can sign a treaty in two ways. The first is called attestation by "simple signature" which corresponds to the above statement—that such a signature merely denotes that a State agrees with the text of an instrument and a simple signature is subject to ratification, acceptance or approval. However, if a State attaches to the instrument what is called a "definitive signature" it means that the State has agreed to be bound by the treaty. Therefore a definitive signature obviates the need for that State to later ratify the treaty, as it has the same force as ratification.

The process of ratification usually goes through two phases. The first is the internal procedure where the State concerned has to attend to its constitutional provisions by sending the text of the instrument it has signed through its national legislature or parliament. Once parliament adopts the text as its internal law, the State then has to proceed with its international procedure of depositing its notice of ratification with the depository. In formal terminology this process is called the doctrine of incorporation where customary international law as incorporated in a treaty that has been signed by a State is recognized as the internal law of the land on the common law practice based on a presumption that the legislature does not intend to commit a breach of international law.[34]

As the Vienna Convention[35] provides, a treaty need not be signed. According to Article 12 of the Convention the consent of a State to be bound by a treaty is expressed by the signature of the representative of that State only in certain circumstances.[36] Article 13 goes on to say that the consent of States to be bound by a treaty constituted

[32]"Ratification", "acceptance", "approval" and "accession" mean in each case the international act so named whereby a State establishes on the international plane its consent to be bound by a treaty. See Vienna Convention, including meteorological information, aeronautical telecommunications, search and rescue services, charts and the distribution of information. Para. 2.4.5 at Article 2(b).

[33]"Reservation" means a unilateral statement, however phrased or named, made by a State, when signing, ratifying, accepting, approving or acceding to a treaty, whereby it purports to exclude or to modify the legal effect of certain provisions of the treaty in their application to that State. Id. 2(d).

[34]See *West Rand Central Gold Mining Company* v. *R.* [1905] 2. K.B. 391 per Lord Alverstone who stated in his judgment: "whatever has received the common consent of civilized nations must have received the assent of our country, and that to which we assented along with other nations in general may properly be called international law: and as such will be acknowledged and applied by our municipal tribunals to decide questions to which doctrines of international law may be relevant". *Id.* 397.

[35]*Supra*, note 30.

[36]Article 12 (1) states that the consent of a State to be bound by a treaty is expressed by the signature of its representative when: (*a*) the treaty provides that signature shall have that effect; (*b*) it is otherwise established that the negotiating States were agreed that signature should have that effect; or (*c*) the intention of the State to give that effect to the signature appears from the full powers of its representative or was expressed during the negotiation. Article 12.2 provides that for the purposes of paragraph 1: (*a*) the initialling of a text constitutes a signature of the treaty when it is established that the negotiating States so agreed; (*b*) the signature ad referendum of a treaty by a representative, if confirmed by his State, constitutes a full signature of the treaty.

by instruments exchanged between them is expressed by that exchange when: the instruments provide that their exchange shall have that effect; or it is otherwise established that those States were agreed that the exchange of instruments should have that effect. States may also contract with each other under their domestic laws.

A treaty enters into force when the number of ratifications as specified in that treaty is received by the depository. When a treaty enters into force it is in force for only those States who have consented to be bound by it which are called "Parties".[37] However, an expression by a State that it consents to be bound by a particular treaty does not mean that *ipso facto* that treaty enters into force for that State. Either, the treaty must already be in force at that time, or as already mentioned the number of ratifications must be deposited. The Vienna Convention (1969) is more specific when it says that a treaty enters into force in such manner and upon such date as it may provide or as the negotiating States may agree.[38] There are three ways in which a treaty may enter into force. They are: on a date specified in the treaty; on signature only, as agreed by the negotiating States; or on ratification by all or a specified number of States. A treaty may be considered to apply to a State provisionally when the treaty itself so provides; or the negotiating States have in some other manner so agreed. Unless the treaty otherwise provides or the negotiating States have otherwise agreed, the provisional application of a treaty or a part of a treaty with respect to a State will be considered as terminated if that State notifies the other States between which the treaty is being applied provisionally of its intention not to become a party to the treaty.[39]

Treaties, conventions, agreements, protocols, exchanges of notes and other synonyms all mean one and the same thing at international law—that they are international transactions of a legal character. Treaties are concluded between States in written form and governed by international law, whether embodied in a single instrument or in two or more related instruments and whatever its particular designation.[40] Each treaty has four constituent elements: the capacity of the parties thereto to conclude agreement of the provisions of the treaty under international law; the intention of the parties to apply principles of international law when concluding agreement under a treaty; *consensus ad idem* or a meeting of the minds of the parties[41]; and, the parties must have the intention to create legal obligations among themselves. These four

---

[37]Vienna Convention, *supra* note 30 at Article 2 (1) (g). It should be noted that such States should not be called "signatories" as some refer to them erroneously.

[38]*Id.* Article 24 (1).

[39]*Id.* Article 25.

[40]*Id* Article 2 (1) (a).

[41]There are instances where States may record their reservation on particular provisions of a convention while signing the document as a whole. The International Court of Justice in its examination of the *Genocide Convention* has ruled:

> The object and purpose of the Convention. . .limit both the freedom of making reservations and that of objecting to them. It follows that it is the compatibility of a reservation with the object and purpose of the Convention that must furnish the criterion for the attitude of a State in objecting to the reservation. 1 *I.C.J Rep.* 1951, at 15.

elements form a composite regulatory process whereby a treaty becomes strong enough at international law to enable parties to settle their differences within the parameters of the treaty, make inroads into customary international law if necessary, and, transform an unorganized international community into one which may be organized under a uniform set of rules. Treaties are based on three fundamental principles of international law: good faith; consent; and fundamental international responsibility.[42] Since international customary law does not prescribe any particular form for consensual agreements and requirements that would make them binding, the parties to a treaty could agree upon the form of treaty they intend entering into and make it binding among them accordingly. Legal bonds are established between nations because they wish to create them and, as is seen in the *Preamble* to the Chicago Convention, a statement to this effect is reflected in the treaty itself.[43] The main feature of a multilateral international agreement is that absolute rights that may have existed within States before the entry into force of such treaty would be transformed into relative rights in the course of a balancing process in which considerations of good faith and reasonableness play a prominent part. However, treaty provisions must be so written and construed as best to conform to accepted principles of international customary law.[44]

Great reliance is placed on treaties as a source of international law. The international Court of Justice, whose function it is to adjudicate upon disputes of an international character between States, applies as a source of law, international conventions which establish rules that are expressly recognized by the States involved in a dispute.[45] The Court also has jurisdiction to interpret a treaty at the request of a State.[46]

The Vienna Convention[47] while recognizing treaties as a source of law, accepts free consent, good faith and the *pacta sunt servanda* as universally recognized elements of a treaty.[48] Article 11 of the Vienna Convention provides that the consent of a State to be bound by a treaty may be expressed by signature, exchange of instruments constituting a treaty, ratification, acceptance, approval or accession, or by any other means agreed upon. "Ratification", "acceptance", "approval", and

---

[42]Schwarzenberger and Brown (1976), at 118.

[43]The *Preamble* to the Chicago Convention states:

> … the undersigned governments having agreed on certain principles and arrangements in order that international civil aviation may be developed in a safe and orderly manner and that international air transport services may be established on the basis of equality of opportunity and soundly and economically; have accordingly concluded this Convention to that end.

[44]Greig (1976), at 8.

[45]*Statute of the International Court of Justice, Charter of the United Nations and Statute of the International Court of Justice*, United Nations: New York, Article 38. 1(a).

[46]*Id.* Article 36.2 (a).

[47]*Supra* note 30.

[48]*Id.*, Preamble.

"accession" generally mean the same thing, i.e. that in each case the international act so named indicates that the State performing such act is establishing on the international plane its consent to be bound by a treaty. A State demonstrates its adherence to a treaty by means of the *pacta sunt servanda*, whereby Article 26 of the Vienna Convention reflects the fact that every treaty in force is binding upon the parties and must be performed by them in good faith. The validity of a treaty or of the consent of a State to be bound by a treaty may be impeached only through the application of the Vienna Convention[49] which generally requires that a treaty could be derogated upon only in circumstances the treaty in question so specifies[50]; a later treaty abrogates the treaty in question[51]; there is a breach of the treaty[52]; a *novus actus interveniens* or supervening act which makes the performance of the treaty impossible[53]; and the invocation by a State of the *Clausula Rebus Sic Stantibus*[54] wherein a fundamental change of circumstances (when such circumstances constituted an essential basis of the consent of the parties to be bound by the treaty) which has occurred with regard to those existing at the time of the conclusion of the treaty, and which was not foreseen by the parties, radically changes or transforms the extent of obligations of a State. A State may not invoke the fact that its consent to be bound by a treaty has been expressed in violation of a provision of its internal law regarding competence to conclude treaties and seek to invalidate its consent unless such violation was manifest and concerned a rule of its internal law of fundamental importance.[55]

States or international organizations which are parties to such treaties have to apply the treaties they have signed and therefore have to interpret them. Although the conclusion of a treaty is generally governed by international customary law to accord with accepted rules and practices of national constitutional law of the signatory States, the application of treaties is governed by principles of international law. If however, the application or performance of a requirement in an international treaty poses problems to a State, the constitutional law of that State would be applied by courts of that State to settle the problem. Although Article 27 of the Vienna Convention requires States not to invoke provisions of their internal laws as justification for failure to comply with the provisions of a treaty, States are free to choose the means of implementation they see fit according to their traditions and political organization.[56] The overriding rule is that treaties are juristic acts and have to be performed.

Every international treaty is affected by the fundamental dichotomy where on the one hand, the question arises whether provisions of a treaty are enforceable at law, and

---

[49]*Id.* Article 42. 1.

[50]*Id.* Article 57.

[51]*Id.* Article 59.

[52]*Id.* Article 60.

[53]*Id.* Article 61.

[54]*Id.* Article 62.

[55]*Id.* Article 46.

[56]Reuter (1989), at 16.

on the other, whether the principles of State sovereignty, which is *jus cogens* or mandatory law, would pre-emt the provisions of a treaty from being considered by States as enforceable. Article 53 of the Vienna Convention addresses this question and provides that where treaties, which at the time of their conclusion conflict with a peremptory norm of general international law or *jus cogens* are void. A peremptory norm of general international law is a norm accepted and recognized by the international community of States as a whole as a norm from which no derogation is permitted and which can be modified only by a subsequent norm of general international law having the same character. The use of the words "as a whole" Article 53 effectively precludes individual States from considering on a subjective basis, particular norms as acceptable to the international community.[57] According to this provision therefore, a treaty such as the Chicago Convention could not have derogated from principles of accepted international legal norms when it was being concluded. The Vienna Convention has, by this provision, implicitly ensured the legal legitimacy of international treaties, and established the principle that treaties are in fact *jus cogens* and therefore are instruments containing provisions, the compliance with which is mandatory.

Once a State ratifies a treaty it has to deposit instruments of ratification with the United Nations or with the specialized agency of the United Nations as prescribed in the treaty concerned. Such Instruments must emanate from and be signed by the Head of State, Head of Government or Minister for Foreign Affairs or a person exercising, *ad interim*, the powers of one of the above authorities; clearly identify the treaty concerned and the type of action consistent with the provisions of the treaty, i.e., ratification, acceptance, approval, accession, consent to be bound, etc.; contain an unambiguous expression of the will of the Government, acting on behalf of the State, to recognize itself as being bound by the treaty concerned and to undertake faithfully to observe and implement its provisions (a simple reference to a domestic statutory provision will be inadequate); indicate the title of the signatory. In the case of a person exercising, *ad interim*, the powers of the Head of State, Head of Government or Minister for Foreign Affairs, the title must indicate that the person is exercising such powers *ad interim*. In this respect, the depositary accepts the following formulations: Acting Head of State, Acting Head of Government, Acting Minister for Foreign Affairs, Head of State *ad interim*, Head of Government ad interim and Minister for Foreign Affairs ad interim; indicate the date and place where the instrument was issued;

If required, the instrument of deposit must specify the scope of its application in accordance with the provisions of the relevant treaty; and contain all mandatory declarations and notifications in accordance with the provisions of the relevant treaty. Where reservations are intended, such reservations since reservations must be signed by the Head of State, Head of Government or Minister for Foreign Affairs or a person exercising, ad interim, the powers of one of the above authorities.

---

[57]See Frans G. von der Dunk, Jus Cogens Sive Lex Ferenda: Jus Cogendum, Masson-Zwaan and De Leon (1992), 219, at 223–224.

8.3 Other Agreements

A significant aspect of treaty law is the difference between a treaty and a memorandum of understanding (MOU). The fundamental difference between the two instruments lies in the terminology, and the fact that a treaty establishes legally binding obligations whereas an MOU or any other agreement such as a memorandum of cooperation (MOC), arrangement or exchange of letters does not create such obligations *stricto sensu*. A treaty would contain such language as "shall", "undertake", and "rights", whereas an MOC or other agreement would use the word "will" and avoid "shall" and "undertake". Treaties enter into force (usually when the required number of ratifications is received) whereas instruments such as MOUs and MOCs become applicable. However, a delicate nuance in the practice of treaty law is that merely because a particular instrument contains typical words used in a treaty, it may not necessarily be a treaty. Conversely, the use of the word "will" may not always mean that the Parties to a treaty did not intend to create a legally binding obligation.

The distinction between instruments that are conventionally recognized as treaties in accordance with the criteria discussed above, and other documents, has been blurred in practice. In the 1994 case of *Qatar* v. *Bahrain* the International Court of Justice went on to identify as a treaty a double exchange of letters i.e. between Bahrain and Saudi Arabia on the one hand and between Qatar and Saudi Arabia on the other, together with minutes of a meeting (held in 1990) between representatives of the three States and signed by the foreign Ministers of each State.

As will be discussed later in this article in some detail, registration with the United Nations (whether it be with the UN in New York or with a specialized agency such as ICAO) raises the presumption that the instrument so registered is a treaty,[58] although aeronautical agreements concluded between States that are not treaties in the strict sense are also registered at ICAO. MOUs are not registered in the United Nations (nor at ICAO), thus raising the presumption that they are not treaties.[59] In the aeronautical context, while States send their bilateral air transport services agreements to ICAO for registration, they do not send the Confidential Memoranda of Understanding (CMOUs) attached to such treaties for registration. The reason for this is the confidential nature of these documents which cannot be disclosed by any entity to third parties, unlike the main bilateral air services agreement which contains general principles and which ICAO discloses to all its 191 member States. These CMOUs contain confidential details of agreement between the two States concerned relating to what routes shall be operated with what equipment and the capacity and frequencies allowed for each carrier.

[58] Aust (2000), at 42.

[59] *Ibid.*

8.4 Aeronautical Treaties and Agreements

The Chicago Convention requires that all aeronautical agreements which were in existence on the coming into force of the Convention, and which were between a Contracting State and any other State or between an airline of a Contracting State and any other State or the airline of any other State, be forthwith registered[60] with the Council[61] of ICAO. There is a corresponding provision in Article 102 of the Charter of the United Nations[62] which provides that every treaty and every international agreement entered into by any Member of the United Nations after the Charter comes into force shall as soon as possible be registered with the Secretariat and published by it.[63] The provision goes on to say that no party to any such treaty or international agreement which has not been registered in accordance with the provisions of paragraph 1 of Article 102 may invoke that treaty or agreement before any organ of the United Nations.

This generic United Nations principle brings to bear an interesting dichotomy which can only be resolved by the fact that the Charter which was signed on 26 June 1945 and came into force on 24 October 1945 (earlier than the Chicago Convention did) as the earlier treaty will, under the legitimacy accorded to it by Article 30 of the Vienna Convention, prevail. The Charter goes on authoritatively to say in Article 103 that in the event of a conflict between the obligations of the

---

[60]Chicago Convention, Preamble *supra* note 1, Article 81.

[61]The ICAO Council is a permanent body responsible to the Assembly. It is composed of 36 Member States elected by the Assembly. In electing the members of the Council, the Assembly gives adequate representation to States of chief importance to air transport; States not otherwise included which make the largest contribution to the provision of facilities for international air navigation; and States not otherwise included whose designation will ensure that all the major geographic areas of the world are represented on the Council. The mandatory and permissive functions of the Council are stipulated in Articles 54 and 55 of the Convention on International Civil Aviation respectively. The Council has its genesis in the Interim Council of the Provisional International Civil Aviation Organization (PICAO). PICAO occupied such legal capacity as may have been necessary for the performance of its functions and was recognised as having full juridical personality wherever compatible with the Constitution and the laws of the State concerned. See Interim Agreement on International Civil Aviation, opened for signature at Chicago, December 7 1944, Article 3. Also in Hudson, *International Legislation*, Vol. 1X, New York: 1942–1945, at 159.

[62]The United Nations Charter is the constituting instrument of the Organization, setting out the rights and obligations of Member States, and establishing the United Nations organs and procedures. An international treaty, the Charter codifies the major principles of international relations, from sovereign equality of States to the prohibition of the use of force in international relations.

[63]The objective of article 102, which can be traced back to article 18 of the *Covenant of the League of Nations*, which provided that every treaty or international engagement entered into by any member of the League of Nations was required to be registered with the Secretariat and that any treaty or agreement not registered in such manner would be considered legally binding, is to ensure that all treaties and international agreements remain transparent in the public domain and thus assist in eliminating secret diplomacy. The *Charter of the United Nations* was drafted in the aftermath of the Second World War. At that time, secret diplomacy was believed to be a major cause of international instability.

Members of the United Nations under the Charter and their obligations under any other international agreement, their obligations under the Charter shall prevail. Furthermore, the Vienna Convention in Article 80 prescribes that treaties shall, after their entry into force, be transmitted to the Secretariat of the United Nations for registration or filing and recording, as the case may be, and for publication.

Technically, this creates confusion where three instruments say different things that *ex facie* would require any aeronautical treaty or agreement to be submitted for registration both in the Council of ICAO and in the Secretariat of the United Nations. One can find some respite in the fact that the Vienna Convention goes on to say in Article 80 that the designation of a depository constitutes authorization for it to perform registration or filing and recording as the case might be. Arguably, therefore, with regard to treaties and agreements which designate ICAO as the depository, ICAO could ascribe to itself the responsibility of registering such instruments.

The Chicago Convention also provides that any Contracting State may make arrangements not inconsistent with the provisions of the Convention and that any such arrangement shall be forthwith registered with the Council, which shall make it public as soon as possible.[64] Furthermore it requires that Contracting States accept the Convention as abrogating all obligations and understandings between them which are inconsistent with its terms, and that they undertake not to enter into any such obligations and understandings. A Contracting State which, before becoming a member of the Organization has undertaken any obligations toward a non-Contracting State or a national of a Contracting State or of a non-Contracting State inconsistent with the terms of this Convention, is required to take immediate steps to procure its release from the obligations. If an airline of any Contracting State has entered into any such inconsistent obligations, the State of which it is a national is obligated to use its best efforts to secure their termination forthwith and to cause them to be terminated in any event as soon as such action can lawfully be taken after the coming into force of the Convention.[65]

Taking a cue from the legal provisions enshrined in the Charter of the United Nations as well as the Chicago Convention, the ICAO Assembly,[66] at its first session in 1947, resolved that uniform rules pertaining to the registration and publication of aeronautical treaties and agreements should be adopted along the lines of Articles 81, 82 and 77[67] of

[64]Article 83 of the Chicago Convention.

[65]Article 82 of the Chicago Convention.

[66]The ICAO Assembly, comprised of the Organization's 191 Member States, meets once every 3 years. An extraordinary meeting of the Assembly may be convened by the Council at any time. The powers and duties of the Assembly are stated in Article 49 of the Chicago Convention.

[67]Articles 81 and 82 have already been discussed. Article 77 provides: "Nothing in this Convention shall prevent two or more contracting States from constituting joint air transport operating organizations or international operating agencies and from pooling their air services on any routes or in any regions, but such organizations or agencies and such pooled services shall be subject to all the provisions of this Convention, including those relating to the registration of agreements with the Council. The Council shall determine in what manner the provisions of this Convention relating to nationality of aircraft shall apply to aircraft operated by international operating agencies".

the Convention, so as to provide Contracting States clear guidelines in this regard.[68] The resultant Resolution of the Assembly gave rise to ICAO's *Rules for Registration with ICAO of Aeronautical Agreements and Arrangements.*[69]

8.5 Depository Functions of ICAO

Consistent with the similarity of procedure between the United Nations procedure and ICAO procedure for registration, ICAO's depository functions also follow that of the United Nations. The treaties that are deposited by States with ICAO are those that specify in the text that ICAO is the depository. The Secretary General, who is the depositary of a treaty, is responsible for ensuring the proper execution of all treaty actions relating to that treaty. The depositary's duties are international in character, and the depositary is under an obligation to act impartially in the performance of those duties. Article 77(1) of the Vienna Convention lists the functions of the depository to include (a) keeping custody of the original text of the treaty and of any full powers delivered to the depositary; (b) preparing certified copies of the original text and preparing any further text of the treaty in such additional languages as may be required by the treaty and transmitting them to the parties and to the States entitled to become parties to the treaty; (c) receiving any signatures to the treaty and receiving and keeping custody of any instruments, notifications and communications relating to it; (d) examining whether the signature or any instrument, notification or communication relating to the treaty is in due and proper form and, if need be, bringing the matter to the attention of the State in question; (e) informing the parties and the States entitled to become parties to the treaty of acts, notifications and communications relating to the treaty; (f) informing the States entitled to become parties to the treaty when the number of signatures or of instruments of ratification, acceptance, approval or accession required for the entry into force of the treaty has been received or deposited; (g) registering the treaty with the Secretariat of the United Nations; and (h) performing the functions specified in other provisions of the present Convention.

An instrument sent to ICAO for deposit must contain: (a) the title of the treaty concerned and identification of the type of action, consistent with the provisions of the treaty, i.e., ratification, acceptance, approval, accession, consent to be bound, etc.; (b) the title of the signatory. In the case of a person acting, *ad interim*, as the Head of State, Head of Government or Minister for Foreign Affairs, the title must indicate that the person is exercising such powers *ad interim.* In this respect, the depositary accepts the following formulations: Acting President, etc., Acting Prime Minister, etc., Acting Minister for Foreign Affairs, President *ad interim*, Prime Minister *ad interim* and Minister for Foreign Affairs *ad interim*; (c) an unambiguous expression of the

[68] Assembly Resolution A1-45, ICAO Doc. 7670 (1947).

[69] Doc 6685 –C/676. These rules were adopted by the Council on 1 April 1949 and amended on 22 November 1965, 16 May 1974, 16 March 1977, 25 November 1985, and 19 November 2003. The current edition is the Second Edition, published in 2004.

will of the Government, acting on behalf of the State, to recognize itself as being bound by the treaty concerned and to undertake faithfully to observe and implement its provisions (a simple reference to a domestic statutory provision will be inadequate); (d) if required, a clear identification of the scope of the treaty in conformity with the provisions of the relevant treaty; (e) if required, all mandatory declarations and notifications in accordance with the provisions of the relevant treaty; (f) the date and place where the instrument was issued must be specified; (g) the signature of the Head of State, Head of Government or Minister for Foreign Affairs or a person acting, *ad interim*, as one of the above authorities must be placed on the instrument;(h) the official seal. This is optional and cannot replace the signature of one of the authorities of State; and (i) where reservations are intended, signatures on such reservations of the Head of State, Head of Government or Minister for Foreign Affairs or a person acting, *ad interim*, as one of the above authorities. Reservations may either be included in the instrument or, if not, separately signed by one of the authorities of State.

The instrument of ratification, acceptance, approval or accession becomes effective only when it is deposited with the Secretary-General. Delivery of such instruments to the Treaty Section directly ensures prompt processing of the action. Instruments may be faxed to the Treaty Section, provided that the original promptly follows. The depositary will also accept a scanned copy of a document transmitted by electronic mail.

On receipt of deposit of ratification, the Secretary General of ICAO writes to the depositing State referring to the instrument of ratification by the Government title of the relevant Treaty and its date of signature by the Parties. The purpose of the Secretary General's letter is to acknowledge officially the deposit of ratification on the date of deposit and to advise of the date that the treaty would become effective for the depositing State, which usually is some days after the deposit, as stipulated in the Treaty. The Secretary General also encloses, for information, an up-to-date list of parties to the treaty as well as a form indicating the current status of the depositing State with regard to international air law instruments.

8.6 Expectation of Compliance with Treaties

Traditionally, States which follow the common law have adopted the dualist approach to treaties, in that they have considered treaties, which are customary international law, as being separate from domestic law.[70] A fundamental principle of the law applicable in the United Kingdom is that unless an international treaty has been adopted into English Law by an Act of Parliament it has no legal status under domestic law. The legal legitimacy ascribed to this principle lies in a statement by Lord Atkin in the 1937 case of *Attorney General of Canada* v. *Attorney General of Ontario*[71] where his Lordship said that the making of a treaty

[70]See *Cook* v. *Sprigg* [1899] AC 572. Also *Secretary of State in Council of India* v. *Kamachee Boye Sahaba* [1859] 13 Moo PCC 22 (Privy Council).

[71](1937) AC 326.

is an executive act whereas its compliance, if its principles are at variance with existing domestic law, would require legislative action so that existing domestic law could be brought into consistency with that of the treaty concerned.[72] Later, in 1971, Lord Denning in considering whether the then E.E.C. Treaty could be relied upon as principles of English law stated:

> We have no notice of treaties until they are embodied in laws enacted in Parliament.[73]

This somewhat rigid dualistic approach of the British judiciary established certain inarticulate principles which were entrenched in the rule that the courts will not take cognizance of nor apply an international treaty to which Britain was a Party unless and until it was adopted by an Act of Parliament and that even after such adoption, the language of the treaty cannot be used as interpreting a statute in which that treaty was incorporated, particularly in instances of ambiguity in the language of the statute.[74] Also, in accordance with a decision in 1950[75] the courts are not at liberty, when construing a treaty which has been incorporated into English law to have recourse to the *travaux préparatoires* (preparatory treaty work of the conference which adopted the treaty). This judicial approach absolutely obviated the applicability and relevance in domestic law of an unincorporated treaty, making such instrument irrelevant and inapplicable in grounding a cause of action for individuals.

A notable exception has been in the area of human rights where British courts have relied on the *European Convention on the Protection of Human Rights and Fundamental Freedoms* in construing the meaning of the *English Immigration Act* in instances affecting the rights of immigrants.[76] This trend, which seemingly recognizes the fact that individual human rights should be based on global norms that should be seen as universally applicable irrespective of domestic legislation, has received strength in a continuing manner from the British judiciary. Lord Scarman in 1975 extended the principle to all English courts by stressing their duty to take into consideration the *European Convention on Human Rights* when they were adjudicating the rights of individuals in Britain.[77]

On balance, one could argue, in view of the *cursus curiae*, that in general terms and as a matter of law treaties do not give rise to private rights[78] and the principles discussed above relating to the dualist approach would prevail unless there are compelling circumstances for the British judiciary to apply the exception. For instance, in a 1979 decision, Megary J. held in circumstances involving a claim by a plaintiff who alleged breach of confidentiality and privacy invoking Article 8 of the European

[72] *Id.* 347.

[73] (1971) 1 WLR 1087 at 1092.

[74] *Ellerman Lames Ltd.* v. *Murray* (1931) AC 126.

[75] *Porter* v. *Freudenbvurg* (1950) 1 KB 876.

[76] See *Aliamed* v. *Inner London Educational Authority* (1878) 1 Al E.R. 574 and *R.* v. *Secretary of State for Home Affairs* Ex Parte Bajan Singh (1975) 2 All. E. R. 1081.

[77] *R.* v. *Secretary of State for Home Affairs* Ex Parte Phanse Fkar (1975) 2 All. E.R. 497 at 511.

[78] *British Airways* v. *Laker Airways* (1983) 3 All. E.R. 375.

Convention that the Convention had not been ratified by the United Kingdom and as such was inapplicable under British law.[79] In 1991 a decision followed in the House of Lords reaffirming this general approach.[80]

Elsewhere in common law jurisdictions, the trend established by the British courts in separating treaties from domestic law has prevailed. In Australia, courts have held that an individual cannot enforce individual rights accruing from an international treaty unless such treaty was incorporated in and formed part of Australian law.[81] Once again, in this region of common law jurisdictions, an exception has been made in the instance of fundamental human rights, as was seen in the 1994 New Zealand case of *Tavita* v. *Minister of Immigration*[82] where the New Zealand Court of Appeals upheld the provisions of the *International Covenant on Civil and Political Rights* of 1966 which guaranteed the right of a child to acquire a nationality although the *Immigration Act* (1987) of that country did not admit of the grant of nationality to a child of a person who was in the country under a temporary residence permit at the birth of the child.

The arguments of monism and dualism aside, and as already mentioned, a State is expected to be bound by a treaty it signs and later ratifies by the legal maxim *pacta sunt servanda* invoked in Article 27 of the Vienna Convention which provides that every treaty in force is binding upon the parties to it and must be performed by them in good faith. A State Party to a treaty can opt out of this obligation in two instances, the first being the recording of what is called "*an interpretative declaration*"[83] during the diplomatic conference that discusses the text of a Treaty in the context of a multilateral treaty or during bilateral discussions in the instance of negotiating a bilateral treaty with another State. These declarations are widely used and go back in history to 1815.[84] In essence such a declaration is made when a State has a difference of view with regard to the meaning of a particular provision. In such an instance the State concerned makes a formal statement expressing the interpretation favoured by it and that statement is usually reflected in the *Travaux Préaparatiores* or negotiating history (record of proceedings). It is not uncommon for a State to put forward an interpretative declaration even after a Treaty has been concluded and this occurrence

[79]*Malone* v. *Commissioner of Police* (1979) 2 All. E.R. 620.

[80]*R.* v. *Secretary for the Home Department* Ex Parte Brind (1991) 2 WLR 588.

[81]*Tasmania Wilderness Society* v. *Fraser* (1982) 153 CLR 270 at 274. Also, *Ditrich* v. *The Queen* 1992 67 ALGR 1 at 6. See also, *Minister of State for Immigration and Ethnic Affairs* v. *Ah Hin Teoh* 183 CLR 273, *Chow Hung Ching* v. *the King* (1948) 77 CLR 449, *Bradly* v. *The Commonwealth* (1973) 128 CLR 557, *Simsek* v. *Macphee* (1982) 148 CLR 636, *Koowarta* v. *Bjelke-Petersen* (1982) 153 CLR.

[82](1994) 2 NZLR 257.

[83]An interpretative declaration is "a unilateral declaration, however phrased or named, made by a State or by an international Organization whereby that State or that Organization purports to clarify the meaning or scope attributed by the declarant to the treaty or to certain of its provisions". See UN Doc, A/CN.4/491/Add4, para. 361.

[84]Interpretative Declaration of the United Kingdom in respect of an instrument adopted during the Congress of Vienna, 1815. See 64 CTS 454.

is seen mainly in instances where a State realizes subsequently that a provision of a treaty it has signed and/or ratified is inconsistent or contrary to its domestic law in whole or in part thereof. A watered down version of an interpretative declaration is a *political declaration* which does not *per se* address the legality of a treaty provision in the eyes of a State but rather clarifies the State's policy towards that provision. By making such a declaration a State may keep open a window of opportunity that would enable the State to make a reservation at the point of ratification.

The second instance wherein a State could opt out of its obligation from adhering to an entire treaty is when it records a *reservation* to any particular provision in the Treaty at the point of definitive signature or ratification. Article 19 of the Vienna Convention provides that a State may, when signing, ratifying, accepting, approving or acceding to a treaty,[85] formulate a reservation unless: the reservation is prohibited by the treaty; the treaty provides that only specified reservations, which do not include the reservation in question, may be made; or in cases not falling under the abovementioned conditions, the reservation is incompatible with the object and purpose of the treaty. A reservation is:

> a unilateral statement, however phrased or named, made by a State, when signing, ratifying, accepting, approving or acceding to a treaty, whereby it purports to exclude or to modify the legal effect of certain provisions of the treaty in their application to that State.[86]

A reservation, an express acceptance of a reservation and an objection to a reservation must be formulated in writing and communicated to the Contracting States and other States entitled to become parties to the treaty. If formulated when signing the treaty subject to ratification, acceptance or approval, a reservation must be formally confirmed by the reserving State when expressing its consent to be bound by the treaty. In such a case the reservation will be considered as having been made on the date of its confirmation. An express acceptance of, or an objection to a reservation made previously to confirmation of the reservation does not itself require confirmation.[87]

In practice a reservation need not necessarily be unilateral and two or more States can put forward the same reservation.[88] There is also a *derogation* which should be distinguished from a *reservation*, the former being a concession accorded by a treaty to States' Parties to derogate from a provision or provisions of that treaty to accommodate special exigencies such as a state of emergency.

---

[85]The Vienna Convention does not contain provision for States Parties to make reservations to a treaty subsequent to their ratifying a treaty. However, The Secretary General of the United Nations may circulate a reservation received subsequently with a note that, unless he hears otherwise from other States' Parties any objections to the reservation within 90 days the reservation will deem to have been accepted. This same practice may be applied when a State wishes to modify a reservation previously made. See Aust (2000), at 129.

[86]Vienna Convention. Including meteorological information, aeronautical telecommunications, search and rescue services, charts and the distribution of information. Para. 2.4.5 at pp. 2–6, Article 2 (1) (d).

[87]*Id.* Article 23.

[88]Aust (2000), at 105.

A reservation established with regard to another party modifies for the reserving State in its relations with that other party the provisions of the treaty to which the reservation relates to the extent of the reservation; and modifies those provisions to the same extent for that other party in its relations with the reserving State. The reservation does not modify the provisions of the treaty for the other parties to the treaty *inter se*. When a State objecting to a reservation has not opposed the entry into force of the treaty between itself and the reserving State, the provisions to which the reservation relates do not apply as between the two States to the extent of the reservation.[89]

A reservation can be withdrawn at any stage and such an instance usually occurs when the situation prevailing in a State at the time the reservation is made ceases to exist. The Vienna Convention expressly provides that unless the treaty otherwise provides, a reservation may be withdrawn at any time and the consent of a State which has accepted the reservation is not required for its withdrawal.[90] The withdrawal of a reservation or of an objection to a reservation must be formulated in writing.[91]

The *pacta sunt servanda* element of a treaty (consent to be bound) may be adversely affected in instances of State succession when one State succeeds wholly or in part to the legal personality of another State, both of whom are parties to the same treaty. Instances of war and armed conflict, although they do not automatically sever treaty relationships between States may affect them. A treaty is void if its conclusion has been procured by the threat or use of force in violation of the principles of international law embodied in the Charter of the United Nations.[92] The basic principle, following the Charter of the United Nations is that treaties are no less binding in instances of war. However, termination of a treaty may be by consent of the parties, express or implied. Article 54 of the Vienna Convention prescribes that the termination of a treaty or the withdrawal of a party may take place either in conformity with the provisions of the treaty; or at any time by consent of all the parties after consultation with the other Contracting States. A treaty does not terminate merely because the number of Parties to that treaty falls below the required number to enter into force. The termination of a treaty under its provisions or in accordance with the Convention releases the parties from any obligation further to perform the treaty but does not affect any right, obligation or legal situation of the parties created through the execution of the treaty prior to its termination.[93]

There is no room for doubt that common law jurisdictions follow the dualist approach mainly because of their adherence to the doctrine of separation of

---

[89]Vienna Convention on the Law of Treaties, 1969, Done at Vienna on 23c May 1969, United Nations General Assembly Document A/CONF.39/27, 23 May 1969, Article 2(a). The Convention entered into force on 27 January 1980. UNTS Vol. 1155, p. 331, Article 21.

[90]*Id.* Article 22.

[91]Id. Article 23 (4).

[92]*Id.* Article 52.

[93]*Id*, Article 70.

powers.[94] A corollary to this approach is that the rights of a citizen lie primarily in domestic legislation and if a treaty were to be applicable within a State, it has to be formally included in the Statute books. This brings to bear the perennial debate on the primacy of international law over domestic law in instances where a State not only places its signature on a treaty but ratifies it subsequently.

There have been many different theories on which takes precedence over which—whether international law has primacy over municipal or domestic law or is it the other way around. In essence therefore it is a "toss up" between the dualist (or pluralistic) theory and the monist theory. The former considers that international law is law between sovereign States and municipal law is between citizens and the executive and therefore they are mutually exclusive and neither legal order has the power to create or amend the principles or rules of the other. Therefore, in the dualist view the doctrine of incorporation is a rule of construction that allows international law to apply within the municipal context if it has been formally adopted within a jurisdiction. The latter (monism) espouses the principle that international law is supreme even within the municipal arena. Monism implies that the State is but an abstraction of insularity which has no right to override the norms of international law that protect such fundamental legal entitlements as human rights, particularly when treaties incorporating such rights have been ratified by that State.

Whatever may be the judgment of a municipal court on this issue, from a wider perspective one wonders whether a State could hide behind its internal laws to ignore a rule of an international treaty it has ratified on the ground that its internal law prescribes the opposite or different rule from the international rule. The law on this is quite clear. The Vienna Convention provides that a State party may not invoke the provisions of its internal law as justification for its failure to perform a treaty.[95] Furthermore, a State may not invoke the fact that its consent to be bound by a treaty has been expressed in violation of a provision of its internal law regarding competence to conclude treaties as invalidating its consent unless that violation was manifest and concerned a rule of its internal law of fundamental importance. A violation is manifest if it would be objectively evident to any State conducting itself in the matter in accordance with normal practice and in good faith.[96]

---

[94]The doctrine of separation of powers ascribes separate powers to the three limbs of a State: the legislature; judiciary and the executive. The judiciary dispenses justice based on applicable domestic law and as such a treaty would be considered by the judiciary as part of domestic law only if it is adopted as a legislative enactment by the legislature and incorporated into the law of the land.

[95]Vienna Convention, *supra* note 19, Article 27.

[96]*Id.* Article 46.

It is submitted that from a logical perspective, monism triumphs. How else could one justify or legitimize the act of a State which is represented with full powers,[97] sitting in an international arena at a diplomatic conference, signing off on global principals which it then ratifies, if it can say later on, "I really did not mean what I did when I accepted the principles in this treaty"; or "I really did not mean to be bound by this treaty"; or, more importantly, "I meant that the principles contained in this treaty should apply to all others except the citizens of my country". In the aeronautical context, this logicality becomes significant in that there would be no meaning and purpose for a member State of ICAO to ratify a treaty by which, say, that State undertakes to mark plastic explosives or prosecute a person who hijacks an aircraft if its local laws prevent such prosecution on subjective grounds. It is for the State to decide either way using its legislative clout rather than let a hapless judiciary address the issue, the only duty of which is to dispense justice in accordance with municipal law.

References

Aust A (2000) Modern treaty law and practice. Cambridge University Press, New York

Greig DW (1976) International law, 2nd edn. Butterworths, London

Hymens M (1948) Results of a meeting. Interavia 3(8):422

MacKenzie D (2008) ICAO, a history of the international civil aviation organization. University of Toronto Press, Toronto

Masson-Zwaan TL, De Leon PMJM (eds) (1992) Air and space law: De Lege Ferenda. Essays in Honour of Henri A Wassenbergh. Martinus Nijhoff, Dordrecht

Milde M (1992) New headquarters agreement between ICAO and Canada. Ann Air Space Law (Part II):305–322

Reuter P (1989) Introduction to the law of treaties. Pinter Publishers, London

Schenkman J (1955) International civil aviation organization. Librairie E. Droz, Geneve, p 121

Schwarzenberger G, Brown ED (1976) A manual of international law, 6th edn. Professional Books Limited, Oxon

Vivian J (1973) ICAO assists developing nations in $ million annual programme. ICAO Bull 3:10–12

[97] Article 2(1) (c) of the Vienna Convention defines full powers as "a document emanating from the competent authority of a State designating a person or persons to represent the State for negotiating, adopting or authenticating the text of a treaty, for expressing the consent of the State to be bound by a treaty, or for accomplishing any other act with respect to a treaty". Granting full powers does not mean that a person who is given such powers by his State is empowered to do as he likes. He would need to obtain specific instructions from his government before acceding to any principle of the text being negotiated or performing any act in relation to a treaty. General full powers are deposited with the Secretariat of the Organization which convenes the diplomatic conference that discusses the text of a treaty.

Article 44
Objectives

The aims and objectives of the Organization are to develop the principles and techniques of international air navigation and to foster the planning and development of international air transport so as to:

(a) Insure the safe and orderly growth of international civil aviation throughout the world;
(b) Encourage the arts of aircraft design and operation for peaceful purposes;
(c) Encourage the development of airways, airports, and air navigation facilities for international civil aviation;
(d) Meet the needs of the peoples of the world for safe, regular, efficient and economical air transport;
(e) Prevent economic waste caused by unreasonable competition;
(f) Insure that the rights of contracting States are fully respected and that every contracting State has a fair opportunity to operate international airlines;
(g) Avoid discrimination between contracting States;
(h) Promote safety of flight in international air navigation;
(i) Promote generally the development of all aspects of international civil aeronautics

Contents

1 ICAO's Aims and Objectives

It is interesting to note that ICAO has aims and objectives, and not a mandate. Another interesting feature is that, while ICAO can develop principles and techniques of air transport (through adopting Annexes to the Chicago Convention) it can only "foster" the development of air transport (issue guidelines on air transport economics). The Council, at the 8th meeting of its 196th Session in June 2012, approved ICAO's the revised Vision and Mission Statements, as well as the new set

R. Abeyratne, *Convention on International Civil Aviation*,
DOI 10.1007/978-3-319-00068-8_45,

of five Strategic Objectives of the Organization for 2014–2015–2016.[1] Accordingly, ICAO's new vision statement is "Achieve the sustainable growth of the global civil aviation system". Its Mission Statement is

> The International Civil Aviation Organization is the global forum of States for international civil aviation. ICAO develops policies, standards, undertakes compliance audits, performs studies and analyses, provides assistance and builds aviation capacity through the cooperation of Member States and stakeholders.

Under these two statements come ICAO's new Strategic Objectives:

- Safety: Enhance global civil aviation safety
- Air Navigation Capacity and Efficiency: Increase capacity and improve efficiency of the global civil aviation system
- Security and Facilitation: Enhance global civil aviation security and facilitation
- Economic Development of Air Transport: Foster the development of a sound and economically-viable civil aviation system
- Environmental Protection: Minimize the adverse environmental effects of civil aviation activities

For ICAO's work on a global scheme on aircraft engine emissions the Strategic Objectives D and E are relevant and the priority in this regard would be to minimize the adverse environmental effects of civil aviation activities through the fostering of a sound and economically-viable civil aviation system. In other words, the global scheme to be proposed by ICAO and accepted by its 191 member States should contribute towards fostering the development of a sound and economically-viable civil aviation system while at the same time going towards the adverse environmental effects of civil aviation activities.

The ICAO Council has to be commended for extending ICAO's scope in air transport issues from "fostering the development of air transport" as prescribed in Article 44 of the Chicago Convention to "fostering the development of a sound and economically-viable civil aviation system". With this quantum leap, the Council has ascribed to ICAO the responsibility of fostering the progress of the entire civil aviation system. The civil aviation system, which is not defined in any known air law instrument or document, could be taken to comprise all elements of civil aviation from the manufacture and operation of aircraft (at least with regard to how many aircraft should serve the needs of the people and their descriptions according to the demography and economy of a particular society) and parts to ground handling which, prior to this new mandate were not the purview of ICAO. The Council has also deftly obviated the requirements of amendment of the Chicago Convention as prescribed therein, by interpreting one of the main aims and objectives of the Convention in a liberal manner. It will be interesting to learn how the

[1] Earlier, ICAO had four Strategic Objectives. They were: Safety—Enhance global civil aviation safety; Security—Enhance global civil aviation security; Environmental Protection—Minimize the adverse effect of global civil aviation on the environment; and Sustainable Development of Air Transport.

38th Session of the Assembly will view this positive but potentially contentious measure of the Council and endorse the two separate and distinctly different objectives of the Organization now existing side by side in two separate documents.

As discussed, ICAO's functions in the technical field are straightforward and clear. However, in the economic field the Organization's functions are somewhat obfuscated by the words "foster the development of air transport", ICAO is seemingly empowered to exercise economic oversight in a limited way.

The Conference of ICAO on the Economics of Airports and Air Navigation Services (CEANS), which has already been discussed under Article 15 agreed to submit to the Council of ICAO crucial recommendations for international civil aviation which will take cooperation between the air transport, airport and air navigation services industries to a higher level and increase the efficiency and cost-effectiveness in the provision and operation of airports and air navigation services around the world. These recommendations are calculated to serve the aviation industry expeditiously in coping with the current challenges that air transport faces.

Furthermore, it was the view of CEANS that the recommendations will make ICAO's policies on charges, which regulate the relationship between airports and air navigation services providers (ANSPs) on the one hand, and airlines and other airport and airspace users on the other, more authoritative in practice. The enhanced cooperation suggested by these recommendations would strengthen policies on States' economic oversight responsibility, requirements on implementation of performance management systems by all airports and ANSPs, and the establishment of a clearly defined, regular consultation process by all airports and ANSPs. At the same time, they recommend that States enshrine the main principles of non-discrimination, cost-relatedness, transparency and consultation with users in their national legislation, regulations or policies as well as all air services agreements between States.[2]

One of the fundamental premises addressed by CEANS is that the protection of users against the potential abuse of dominant position by airports and air navigation services providers is the primary responsibility of the State and could be discharged by the exercise of economic oversight. It was suggested during the discussions that such oversight could be effectively carried out by diligent monitoring by a State of the commercial and operational practices of service providers.

2 The ICAO Perspective

The ICAO Secretariat, in submitting its views on economic oversight to the Conference, commenced with the fundamental postulate that the State is ultimately responsible for protecting the interests of users through economic oversight defined

[2]Other essential features of the recommendations of the Conference are: more flexibility for commercialized airports and ANSPs in setting charges; support for separation of regulation from service provision; the application of good governance through best practices; and the efficient and cost-effective implementation of the global Air Traffic Management (ATM) concept.

the term "economic oversight" as monitoring by a State of the commercial and operational practices of service providers.[3] It was suggested that economic oversight may take several different forms, from a light-handed approach (such as the reliance on competition law) to more direct regulatory interventions in the economic decisions of service providers. It is interesting that the Secretariat took a direct and clear position that States may perform their economic oversight function through economic regulation, either through legislation or rule-making, and/or the establishment of a regulatory mechanism.[4]

It was also argued that the objectives of economic oversight could include: ensuring that there is no abuse of dominant position by service providers; ensuring non-discrimination and transparency in the application of charges; providing incentives for service providers and users to reach agreements on charges; ensuring that appropriate performance management systems are developed and implemented by service providers and assuring investments in capacity to meet future demand. The priority for each objective may vary depending on the specific circumstances in each State, and there should be a balance between such public policy objectives and the efforts of the autonomous/private entities to obtain the optimal effects of commercialization or privatization.

The Conference was advised that there were already several modalities in Doc 9082,[5] paragraph 15 of which recommends that States establish an independent mechanism for the economic regulation of airports and air navigation services. This provision suggests that such a mechanism would oversee economic, commercial and financial practices and its objectives could be drawn or adopted from, but need not be limited to certain principles.[6] The Secretariat also drew the attention of the Conference to the *Manual on Air Navigation Services Economics*[7] and the *Airports Economics Manual*[8] which suggest such modalities of economic oversight as (a) application of competition law; (b) fall-back regulation, whereby regulatory interventions are limited to situations when the behaviour of the regulated entity breaches publicly-stated acceptable bounds; (c) institutional arrangements such as requirements on consultation with users (often supplemented by arbitration/dispute resolution procedures), information disclosure, and a particular ownership, control

[3]Economic Oversight, CEANS-WP/4, 16/4/08.

[4]*Id.* 1.

[5]ICAO's Policies on Charges for Airports and Air Navigation Services Doc 9082/7 Seventh Edition-2004.

[6]The principles alluded to in paragraph 15 are: ensure non-discrimination in the application of charges; ensure there is no overcharging or other anti-competitive practice or abuse of dominant position; ensure transparency as well as the availability and presentation of all financial data required to determine the basis for charges; assess and encourage efficiency and efficacy in the operation of providers; establish and review standards, quality and level of services provided; monitor and encourage investments to meet future demand; and ensure user views are adequately taken into account.

[7]*Manual on Air Navigation Services Economics*, Doc 9161/3, Third Edition, 1997.

[8]*Airport Economics Manual*, Doc 9562, Second Edition: 2006.

and financial structure; (d) a third-party advisory commission, whereby a group of interested parties reviews pricing, investment and service levels proposals; (e) contract regulation, whereby the State grants a contract, or concession, to provide airport or air navigation services under certain conditions; (f) incentive-based or price-cap regulation; and (g) cost of service or rate of return regulation.

ICAO's conferred powers enable the Organization to adopt policy by majority decision (which is usually unnecessary as most of ICAO policy is adopted through consensus). However, States could opt out of these policies or make reservations thereto, usually before such policy enters into force. This is because States have delegated power to ICAO to make decisions on the basis that they accept such decisions on the international plane. In such cases States could contract out and enter into binding agreements outside the purview of ICAO even on subjects on which ICAO has adopted policy. The only exception to this rule lies in the adoption of Standards in Annex 2 to the Chicago Convention on Rules of the Air, in particular navigation over the high seas and other overflight areas where freedom of flight prevails which all Contracting States are bound to follow in order to maintain global safety.

It is an established fact that States retain the powers to act unilaterally and they are not bound to comply with obligations flowing from the Organization's exercise of conferred powers. A State could also distance itself from the State practice of other Contracting States if such activity is calculated to form customary international law that could in turn bind the objecting State if it does not persist in its objections.[9] However, It is implicitly recognized by the international community that every State has the duty to carry out in good faith its obligations arising from treaties and other sources of international law, and it may not invoke provisions in its constitution or in its laws as an excuse for failure to perform this duty. Also, every State has the duty to conduct its relations with other States in accordance with international law and with the principle that sovereignty of each State is subject to the supremacy of international law.[10] In this context, ICAO remains the only binding link of international obligations between its 190 member States that would forge mutual agreement and will continue to provide a global forum to these States in its triennial Assembly.

3 The Doctrine of Incorporation

The only way in which States could be considered as having incorporated principles into their legislation is through customary international law, where customary rules could be considered part of the national law. However, such custom should not be

---

[9]See Sarooshi (2005) at p. 110.

[10]Draft Declaration of Rights and Duties of States of the International Law Commission (1949), Articles 13 and 14. See http://untreaty.un.org/ilc/texts/instruments/english/draft%20articles/2_1_1949.pdf.

inconsistent with already existing legislation or judicial decisions of last instance or final authority.[11] There is a long *cursus curiae* in support of this principle.[12]

The doctrine of incorporation is essentially a rule of construction and, as already discussed, is a product of the British judiciary of the eighteenth century. Perhaps the most compelling thrust of the incorporation of custom into national law through the judiciary is contained in the dictum of Lord Alverstone in *West Rand Central Gold Mining Company* v. *R*[13] where his Lordship observed:

Whatever has received the common consent of civilized nations must have received the assent of our country, and that to which we assented along with other nations in general may properly be called international law, and as such will be acknowledged and applied by our municipal tribunals to decide questions to which doctrines of international law may be relevant.[14]

Accordingly, rights and duties flowing from customary law will *ipso facto*, and without any formal adoption become the law of the land. The advantage of incorporation lies in the empowerment of the judiciary to develop and administer the law taking into consideration contemporary international norms and jurisprudence. It must be stressed however, that the doctrine of incorporation largely applies to principles of international treaties, and therefore Articles 15 and 28 of the Chicago Convention which have already been referred to and are recognized by a significant body of nations and practiced accordingly, could be considered as law in a common law jurisdiction.

The application of the doctrine can be seen in countries such as Australia, South Africa and the United Kingdom. In Australia, courts have construed as legitimate the development of new rationales that are founded on the principle of legitimate expectation in administrative law. This is based on the fact that a legitimate expectation is created in the citizen based on the fact that the State would act in accordance with its commitments on the international arena. Judicial approval and acceptance of this principle can be seen in the 1995 decision of *Minister for Immigration and Ethnic Affairs* v. *Ah Hin Teoh*.[15] The expectation is tied in with the doctrine of attribution or State responsibility in honouring its international obligations on the domestic plane.[16]

---

[11]Blackstone, *Commentaries*, iv, ch 5; Oppenheim, i. 39–40; See also the dictum of Lord Finley in the *Lotus* (1927) PCIJ Ser. A, no. 10, p. 54.

[12]*Triquet* v. *Bath*, (1764), 3 Burr. 1478; per Mansfield LJ; *Buvot* v. *Barbuit* 91737) Cases t. Talb. 281. *Dolder* v. *Lord Huntingfield* (1805), 11 Ves. 283; *Vivesh* v. *Becker* (1814), 3 M&S. 284; *Wolff* v. *Oxholm* (1817), 6 M. &S 92, 100–106; *Novello* v. *Toogood* (183), 1, I.B. & C, 554; *De Wutz* v. *Hendricks* (1984), 2 Bing. 314, 315; *Emperor of Austria* v. *Day*, (1861), 30 LJ, Ch. 690, 70; *Trendtex Trading Corporation* v. *Central Bank of Nigeria* [1977] 1 QB 529 CA, Cf.; *R.* v. *Secretary of State*, *ex parte* Thakrar [1974] 1 QB 694, CA; *International Tin Council Appeals* [1989] 3 WLR 969, HL.

[13][1905] 2 KB. 391.

[14]*Id.* 396.

[15][1995] 183 CLR 273.

[16]Jan Arno Hessbruegge, The Historical Development of the Doctrine of Atribution and Due Diligence in International law, http://www1.law.nyu.edu/journals/jilp/issues/36/36_2_3_Hessbruegge.pdf.

The United Kingdom has, through statutory provision, extended the doctrine of incorporation to its national legislation. Section 3 of the *Human Rights Act* of 1998 explicitly stipulates that customary law could be considered part of the law in the United Kingdom. South Africa, taking a different route, incorporates a constitutional interpretation clause in Article 39 (1) of its 1996 Constitution. The aforementioned aims and objectives are accomplished by ICAO on a global basis, particularly through its policy on regional cooperation.

4 ICAO's Policy on Regional Cooperation

The Council of ICAO, at its 188th Session in October 2009, adopted a Policy and Framework for Regional Cooperation with a view to enhancing the Organization's interaction and cooperation with regional organizations and regional civil aviation bodies.

4.1 The Policy

The Policy, which is a first in the Organizations 65 year old history, is timely and necessary, and calls for a harmonious blend of leadership and partnership. It is essentially focused on ICAO's rendering assistance, advice and any other form of support, to the extent possible in the technical and civil aviation policy aspects of international civil aviation to Contracting States, and is based on the fact that regional cooperation should involve the triumvirate of ICAO, the Contracting States and regional organizations and regional civil aviation bodies. The Policy goes on to affirm that ICAO will promote regional cooperation through close partnerships with such organizations and bodies, and that, in implementing the Policy, ICAO will optimally use its resources both at Headquarters and its Regional Offices and apply the principles enunciated in the relevant ICAO Assembly Resolutions, guidance and policy.

In developing the Policy, which conforms to the Vision and Mission Statements of ICAO and will be implemented by the Secretary General, the Council was mindful that globalization confronts smaller and disadvantaged countries with many challenges that can be handled more effectively through cooperation with other States, and that emerging trends suggest that the world economy in this century is likely to consist of a network of various forms of regional cooperation. It was also noted that developments in regionalism, particularly in Europe and the Americas, make it clear that States of other regions may need to speed up their aviation activities in the fields of safety, security, sustainability and efficiency of air transport or risk becoming marginalized in an increasingly competitive global landscape.

The Policy aims at promoting cooperation through the expanded use of best practices and better utilization of existing capabilities and resources within the regions; improving services and making best use of resources, taking into account the different levels of competence that exist in States. It also takes into account relevant provisions of the *Convention on International Civil Aviation* (Chicago Convention) and relevant ICAO Assembly Resolutions. The Policy is aimed at enabling States to understand ICAO policy and implementing Standards and Recommended Practices (SARPs) contained in the 18 Annexes to the Chicago Convention; identifying existing opportunities for sub-regional cooperation and promote potential bodies for new partnerships; assisting with advice on capacity building; assessing the gaps in knowledge and capabilities to meet requirements; and improving existing practices.

In practical terms the Policy would also apply to cooperation in technical and/or policy matters, as appropriate, with technical bodies (such as ACAC, AFCAC, LACAC, and regional safety oversight organizations) as well as between ICAO and regional organizations (such as the African Union and the European Union).

Important vehicles for implementation of this Policy are ICAO's Regional Offices. Therefore these offices will take into account in their inputs to strategic planning the regional needs and opportunities for cooperation with regional civil aviation bodies, regional organizations and other stakeholders, with a view to assisting States in ensuring harmonization in adherence to ICAO policy.

Implicit in the Policy is the statement that, while ICAO encourages the activities of States, regional civil aviation bodies and regional organizations in facilitating, among others, the development of civil aviation infrastructure and implementation of SARPs and ICAO policy, States ultimately remain responsible for their obligations under the Chicago Convention, notwithstanding whatever arrangements States may conclude with their regional organizations and regional civil aviation bodies.

The objective of the Policy is to avoid duplication and achieve harmonization in all regions on improvements in the technical and/or policy areas by strengthening cooperation between ICAO, the regional civil aviation bodies and regional organizations. It also aims at ensuring adequate expertise and resources for aviation infrastructure development and for carrying out oversight functions. Sharing information and data and ensuring specialized training and expertise in the development of national/regional plans are also notable objectives, along with enacting civil aviation legislation as necessary.

In implementing the Policy, ICAO will enhance its cooperation with regional civil aviation bodies and regional organizations and ensure that cooperation with States which do not belong to regional organizations and regional civil aviation bodies is not jeopardized or compromised. It will also encourage States to direct their respective regional civil aviation bodies and regional organizations to closely cooperate with ICAO and to assign them tasks in the context of that cooperation and invite regional civil aviation bodies, pursuant to their rules of procedure, to give sympathetic consideration to the possibility of inviting ICAO Contracting States not members of the regional civil aviation body in question to participate as observers in its meetings. ICAO will meet periodically with regional civil aviation bodies including at an annual high level meeting with such bodies and define as necessary the role to be played by the Regional Offices in coordinating ICAO cooperation with regional civil aviation bodies.

4.2 The Framework of Cooperation

The Framework for Regional Cooperation, which is the driver of the Policy, is essentially a Strategic Plan of Action drawn in accordance with the ICAO Policy on Regional Cooperation and the Business Plan of the Organization. The objective of this Plan is to formulate and implement regional cooperation activities to enhance ICAO's role as the global forum for international civil aviation as well as further strengthen ICAO's regional activities with regional civil aviation bodies and regional organizations. The Framework is calculated to prepare ICAO to develop a bilateral mechanism of regional cooperation between ICAO on the one hand, and the regional civil aviation bodies and/or regional organizations (e.g. agreement with EC) on the other.

The Strategic Plan of Action on Regional Cooperation will be developed by the Secretary General in consultation with the Council of ICAO, and will be implemented by regional operational plans which are consistent with the Business Plan of the Organization. They will establish tasks, accountability and timelines and will be measured by performance indicators.

The regional operational plans will be drawn in accordance with the needs and priorities of the different regions, and tasks will be clearly identified and assigned both at Headquarters and the Regional Offices. The strategic thrusts of the Strategic Plan of Action are:

- Common efforts at harmonizing, between States, operational regulations, requirements and procedures based on ICAO SARPs implementation;
- Understanding each others' roles and responsibilities;
- Establishment of improved mechanisms for consultation and cooperation, including electronic information sharing;
- Coordinated programme planning and implementation between ICAO and the regional civil aviation bodies;
- Periodic review of regional issues;
- Maximising the effective use of resources at ICAO; and
- Benefiting from each other's competence and expertise; and joint training and capacity building.

The creation of a security and safety culture, and awareness of the adverse effects of aviation on the environment among ICAO's 190 member States is the most compelling need at present. In this context, enhanced cooperation between ICAO, the regional organizations and regional civil aviation is a critical factor. The Policy and Framework for Regional Cooperation serves as a vital tool that could ensure global and regional harmonization in facing current issues and consolidating cooperation and mutual assistance. At a time when ICAO is shifting its focus from pure standardization to assistance and implementation, the basic principles of such a shift should necessarily involve consolidation of responsibilities and assurances of accountability and partnerships between ICAO and regional organizations and civil aviation bodies.

ICAO is at the defining crossroads of its continuing path towards achieving its aims and objectives as set out in the Chicago Convention. With a view to setting its course in line with rapidly evolving trends of globalization and regionalization, the Organization has embarked on implementing an aggressive Business Plan that calls for a cultural transition and change of mind-set that accords with the dynamics of an evolving aviation industry. New leadership and new thinking have been catalysts in this process, and, through a fog of rhetoric which in the past tended to obfuscate the role of the Organization, a flight path has cleared that enables the Organization to steer towards a more relevant role at present and in the years to come. In this context, the ICAO Policy and Framework for Regional Cooperation is an integral tool that will enable the Organization to realize its Mission and Vision Statements.

Reference

Sarooshi D (2005) International organizations and their exercise of sovereign powers. Oxford University Press, Oxford

Article 45

Permanent Seat

The permanent seat of the Organization shall be at such place as shall be determined at the final meeting of the Interim Assembly of the Provisional International Civil Aviation Organization set up by the Interim Agreement on International Civil Aviation signed at Chicago on December 7, 1944. The seat may be temporarily transferred elsewhere by decision of the Council, and otherwise than temporarily by decision of the Assembly, such decision to be taken by the number of votes specified by the Assembly. The number of votes so specified will not be less than three-fifth of the total number of contracting States.

R. Abeyratne, *Convention on International Civil Aviation*,

DOI 10.1007/978-3-319-00068-8_46,

Article 46
First Meeting of Assembly

The first meeting of the Assembly shall be summoned by the Interim Council of the above-mentioned Provisional Organization as soon as the Convention has come into force, to meet at a time and place to be decided by the Interim Council.

DOI 10.1007/978-3-319-00068-8_47,

Article 47
Legal Capacity

The Organization shall enjoy in the territory of each contracting State such legal capacity as may be necessary for the performance of its functions. Full juridical personality shall be granted wherever compatible with the constitution and laws of the State concerned.

Contents

1 Legal Status of ICAO

ICAO is primarily governed by international law, being recognized by the United Nations Charter as a specialized agency of the United Nations. It is also governed by two major agreements, one between the United Nations and ICAO and the other between the Government of Canada and ICAO. The Headquarters Agreement between ICAO and Canada,[1] in Article 2, explicitly provides that ICAO shall possess juridical personality and shall have the legal capacities of a body corporate including the capacity to contract; to acquire and dispose of movable and immovable property; and to institute legal proceedings in Canada Article 3 of the Agreement stipulates that ICAO, its property and its assets, wherever located and by whomsoever held, shall enjoy the same immunity from suit and every form of judicial processes as is enjoyed in foreign States. (Assets include funds administered by the Organization in furtherance of its constitutional functions.)

The headquarters premises of ICA is inviolable and is given the same protection by the Government of Canada as is given to diplomatic missions in Canada.[2] The organization, its assets, income and property, owned or occupied in Canada are exempt from taxes[3] as well as goods purchased under appropriate certificates from manufacturers or wholesalers who are licenced under the Excise Act.[4]

A seminal judicial decision relating to the powers of international organizations was handed down by the Permanent Court of International Justice in 1922 in a

[1]Headquarters Agreement Between the International Civil Aviation Organization and the Government of Canada, ICAO Doc 9591. For further information on the agreement see Milde (1992).

[2]Headquarters Agreement id, Article 4 (1) and (2).

[3]*Id.* Article 6.

[4]*Id.* Article 7.

DOI 10.1007/978-3-319-00068-8_48,

case[5] relating to the issue as to whether the International Labour Organization (set up to regulate international labour relations) was competent to regulate labour relations in the agricultural sector. The court proceeded on the basis that the competence of an international organization with regard to a particular function lay in the treaty provisions applicable to the functions of that organization and that the determination of such competence would be based on interpretation. However, the principle of implied extension should be carefully applied, along the fundamental principle enunciated by Judge Green Hackworth in the 1949 *Reparation for Injuries Case*[6]—that powers not expressed cannot freely be implied and that implied powers flow from a grant of express powers, and are limited to those that are necessary to the exercise of powers expressly granted.[7]

The universal solidarity of ICAO Contracting States that was recognized from the outset at the Chicago Conference brings to bear the need for States to be united in recognizing the effect of ICAO policy and decisions. This principle was given legal legitimacy in the *ERTA* decision[8] handed down by the Court of Justice of the European Community in 1971. The court held that the competence of the European Community to conclude an agreement on road transport could not be impugned since the member States had recognized Community solidarity and that the Treaty of Rome which governed the Community admitted of a common policy on road transport which the Community regulated.

It should be noted that ICAO does not only derive implied authority from its Contracting States based on universality but it also has attribution from States to exercise certain powers. The doctrine of attribution of powers comes directly from the will of the founders, and in ICAO's case, powers were attributed to ICAO when it was established as an international technical organization and a permanent civil aviation agency to administer the provisions of the Chicago Convention. In addition, ICAO could lay claims to what are now called "inherent powers" which give ICAO power to perform all acts that the Organization needs to perform to attain its aims not due to any specific source of organizational power but simply because ICAO inheres in organizationhood. Therefore, as long as acts are not prohibited in ICAO's constituent document (the Chicago Convention), they must be considered legally valid.[9]

States retain the powers to act unilaterally and they are not bound to comply with obligations flowing from the Organization's exercise of conferred powers. States which have delegated powers on ICAO have the legal right under public international law to take measures against a particular exercise by ICAO of

[5]*Competence of the ILO to regulate the Conditions of Labour of Persons Employed in Agriculture*, Advisory Opinion [1922] Publ. PCIJ Series B, nos. 2&3.

[6]*Reparation for Injuries Suffered in the Service of the United Nations*, advisory opinion, [1949] ICJ Reports 174.

[7]*Id.* at p. 198.

[8]Case 22/70, *Commission* v. *Council* (European Road Transport Agreement) [1971] ECR 273.

[9]Seyersted (1963), at p. 28.

conferred powers which is considered to be *detournement de pouvoir*, *ultra vires* or an internationally wrongful act with which the objecting States do not wish to be associated. A State could also distance itself from the State practice of other Contracting States within the Council if such activity is calculated to form customary international law that could in turn bind the objecting State if it does not persist in its objections.[10]

ICAO's identity before courts having national jurisdiction would strictly be restricted to the nature of the organization and the type of work it carries out. Any special privilege accorded to ICAO by agreement or treaty would therefore be applicable only in relation to ICAO's scope of work.[11] Conceptually, it has been argued that in an instance of national litigation involving an international organization, courts would, in the event the litigious issue pertains to the work of that organization, apply the "functional theory" in an *acta jure gestionis* (commercial act), which means that the organization concerned will not be viewed as having special immunities or privileges. In the 1953 case of *Re International Bank for Reconstruction and Development and International Monetary Fund* v. *All America Cables and Radio Inc., and other cable companies*[12] the US Federal Communications Commission was confronted with the argument of the plaintiffs—the World Bank and the IMF—that the purpose of granting privileges and immunities to organizations located in the jurisdiction of a State where national law applied to contracts is to protect such organizations from unfair and undue interference including excessively high rates. The defendant (radio and cable) companies argued that there was no evidence or reason to allow the banks lower-than-commercial rates. The rationale that can be drawn from this case is that the purpose of immunity will be destitute of effect if courts were asked to determine the legality of an organization's work if such inquiry were to obstruct the work of that organization.

A question arises as to what extent or within what parameters must a court apply the principle of functional immunity to commercial acts of an international organization. Courts have veered from one extreme, coming close to recognizing absolute immunity as in the case of *Broadbent* v. *Organization of American States*[13] to linking key activities of an organization, such as its interpretation and translation services to *acta jure imperii* (sovereign act) on the basis that language services were integral to the main functions of an organization.[14]

[10]See Sarooshi (2005) at p. 110.

[11]In *United States* v. *Malekh et al.*, 32 ILR 308–334 (1960) where the defendant, a United Nations employee, was charged with espionage, the US District Court for the Southern District of New York held that neither the defendant nor his employer should have any claims to immunity as espionage was not a part of the functions of the United Nations.

[12]22 *ILR* 705–712.

[13]63 *ILR* 162–163: US District Court for the District of Columbia, 28 March 1978.

[14]*Iran-United States Claims Tribunal* v. *A.S.*, 94 ILR 321–330.

2 Status of ICAO Staff

A fundamental principle of public international law and the law of the international civil service bestows a certain degree of immunity and some privileges to members of the international civil service (those serving in the United Nations system). Therefore it follows that negligence of a member of the international civil service cannot be judged at the same level as that of a member of the public, particularly in relation to professional duties discharged. This is because the international civil service is granted immunity from liability for acts committed and opinions given in the course of their employment, provided such acts related to the performance of official duties. The special position occupied both by an international Organization of nations and its employees in the national courts is due to an explicit recognition of "rootlessness" and international character of both the Organization and its international civil service which, if brought into subjugation by national jurisdictions and legislation, would be rendered destitute of independence in their work for the international community. This article discusses the nature of international organizations and their staff, the types and degree of immunity they enjoy, the difficulties posed by the grant of such immunities and instances of waiver of such immunity.

Organizations themselves, as well as their employees have been called up by national courts from time to time as parties to litigation. In this respect, besides the inherent legal personality of the international organization, the employees of the Secretariat of that organization also have en "essential novelty"[15] where men and women of various nationalities form the international civil service of that organization, mostly as internationally recruited staff. International civil servants so recruited also have a somewhat different standing in national courts in relation to any issue arising from the discharge of their professional duties within the scope of their employment. Article 100 of the United Nations Charter provides that in the performance of their duties, the Secretary General and the staff shall not seek or receive instructions from any government or from any other authority external to the Organization. They are required to refrain from any action which might reflect on their position as international officials responsible only to the Organization. Article 103 of the Charter identifies as the paramount consideration in the employment of staff the securing of the highest standards of efficiency, competence and integrity. The special position occupied both by an Organization and its employees in the national courts is due to an explicit recognition of "rootlessness" and international character of both the Organization and its international civil service which, if brought into subjugation by national jurisdictions, would be rendered destitute of independence in their work for the international community.

International civil servants are recruited for their superior skills and knowledge and are usually expected to perform tasks that are normally beyond the capabilities

[15] James (1970), 53.

of the ordinary person. This imputes to the international civil servant an elevated standard of care as much as is attributed to members of particular professions such as medicine, law and accountancy. However, here the distinction ends, as unlike the other categories mentioned; they are accorded immunity from judicial process in respect of professional duties performed. There is a line drawn, however, precluding this category of employee from shielding himself absolutely from the law. The delicate balance between immunity and liability was brought to bear in the 1976 decision of the Criminal Court of the City of New York in *People* v. *Mark S. Weiner*[16] where the court held that, in an instance where a United Nations security officer used undue force on the plaintiff, immunity from suit would be so unconscionable that it violated on its face the concepts of fundamental fairness and equal treatment of all persons who sought judicial determination of a dispute.[17] In the early French case of *Avenol* v. *Avenol*[18] involving the Secretary General of the League of Nations who claimed diplomatic immunity from a suit for maintenance filed by his former spouse, the court held that immunity of League officials was functionally and territorially limited to the exercise of functions performed for the League and within the territory of the country in which such official duties were performed.

Members of the international civil service are protected in their official correspondence through the Vienna Convention on Diplomatic Relations, Article 27(2) of which states that the official correspondence of the mission shall be inviolable. Official correspondence means all correspondence relating to the mission and its functions. Article 31 of the Convention which states that a diplomatic agent shall enjoy immunity from the criminal jurisdiction of the receiving State and also enjoy immunity from its civil and administrative jurisdiction, makes some exceptions except in the case of a real action relating to private immovable property situated in the territory of the receiving State, unless he holds it on behalf of the sending State for the purposes of the mission; an action relating to succession in which the diplomatic agent is involved as executor, administrator, heir or legatee as a private person and not on behalf of the sending State; and an action relating to any professional or commercial activity exercised by the diplomatic agent in the receiving State outside his official functions. The operative question is whether an international Organization or its staff must wholly be at the mercy of a national court. The argument has been adduced that domestic courts should not have absolute jurisdiction or adjudicatory authority over international organizations since such exercise of authority might cause damage or adversely affect that organization's independence.[19] The rationale of this argument was accepted by

---

[16]378 NYS 2d 966.

[17]*Id*. 975.

[18]*Juge de Paix* Paris 8 March 1935.

[19]August Reinisch, *International Organizations Before National Courts*, Cambridge Studies in International and Comparative Law: 2000 at 388.

the Quebec Superior Court in 2003[20] where, in an instance where a former employee of ICAO sued the Organization et al. for wrongful dismissal from his position at ICAO, the court recognized the need to grant immunities to international organizations so that they could sustain their independence and freedom.

The court drew a parallel between freedom and independence of the Organization with the notion of immunity, recognizing that neither an international organization nor a State should be subject to the laws and conditions of the courts of another State. The Court acknowledged the bifurcation of immunity into absolute and functional immunity and concluded that ICAO has quasi-absolute immunity[21] in this particular case. According to the Court, functional immunity would be conferred regarding acts performed by officials of an international organization in the course of their duties and within the scope of their employment.

Article 29 of the Vienna Convention declares inviolable the person of a diplomatic agent against arrest or detention. The United Nations has endorsed this principle in Resolution 53/97 of January 1999 by strongly condemning acts of violence against diplomatic missions and agents. This Resolution followed condemnation in the Security Council of the murder of nine Iranian diplomats in Afghanistan.[22] In the 1988 case of *Boos* v. *Barry*[23] the US Supreme Court handed down its decision that diplomatic immunity is reciprocal among States based on mutual interest founded on functional requirements and reciprocity.[24] This effectively precludes the punishment of a member of the international civil service in a general sense, where the only remedy available to the host State against alleged offences of diplomat or a member of the international civil service to declare him *persona non grata*.[25]

---

[20]*Trempe* c. *Association du personel de l'OACI et Wayne Dixon* C.S. 500-05-061028-005, *Trempe* c. *OACI et Dirk Jan Goosen*, C.S. 500-05 063492-019.

[21]The court cited a line of cases to support its view on quasi-absolute immunity of An international organization, comprising: *Miller* v. *Canada* [2001] 1. S.C.R. 407; *Re Canadian Labour Code* [1992] 2. S.C.R. 50 and *Canada* v. *Lavigne* [1997] R.J.Q. 405 (CA).

[22]SC/6573 (15 September 1998). See also the Statement of the Secretary General of the United Nations SG/SM/6704 (14 September 1998).

[23]99 LEd 2d 333, 346 (1988); 121 ILR at 678.

[24]It should be noted that official recognition of this principle can be seen I the 1973 *United Nations Convention on the Prevention and Punishment of Crimes Against Internationally Protected Persons*.

[25]Under Article 9 of the Vienna Convention on Diplomatic relations, a receiving State may "at any time and without having to explain its decision" declare any member of a diplomatic staff persona non grata. A person so declared is considered unacceptable and is usually recalled to his or her home nation. If not recalled, the receiving State "may refuse to recognize the person concerned as a member of the mission." While diplomatic immunity protects mission staff from prosecution for violating civil and criminal laws, depending on rank, under Articles 41 and 42 of the Vienna Convention, they are bound to respect national laws and regulations (amongst other issues). Breaches of these articles can lead to persona non grata being used to "punish" erring staff. See 94 AJIL 2000 at 534 where it is reported that an attaché of the Russian embassy was declared persona non grata for suspected bugging of the State Department of the United States.

While this principle is seemingly reasonable, given the service and contribution provided by the international civil service, and the detrimental effect of interference by States of the provision of such services, an absolute application of this principle could tip the balance to the disadvantage of the public. In this context, specific problems have surfaced with regard to the conduct of the members of the international civil service which results in instances of criminal liability such as when a diplomat or member of the international civil service causes motor accidents and injury to third parties through their negligence.[26] Immunity from civil and administrative jurisdiction of the State in which international civil servants serve in an absolute sense could also cause inconsistencies of the administration of justice. Article 31 (1) of the Vienna Convention addresses this issue effectively by having three exemptions where liability would ensue: where the action relates to private immovable property situated within the host State; in matters of succession and litigation related thereto involving the diplomat as a private person; and with respect to unofficial and professional or commercial activity engaged in by the diplomat concerned. A compelling practical example of these exemptions lies in the United Kingdom. The Memorandum on Diplomatic Privileges and Immunities in the United Kingdom of 1987 takes a stringent stand against any reliance on diplomatic immunity which is calculated to evade a legal obligation.

It must be noted that diplomatic immunity afforded to international civil servants, such as exemptions from social security provisions in force in the host State (as per Article 33 of the Vienna Convention), exemptions from taxes and dues regional or municipal (except for indirect taxes), exemptions from dues regarding personal or public services (as per Article 35 of the Vienna convention) and from customs duties and inspection [as per Article 36(1)] of personal belongings and baggage, extends to members of the family of a diplomatic agent forming part of his/her household (as per Article 37).[27]

Such immunities start from the moment the diplomatic agent (or member of the international civil service) and his family enter the territory of the host State,[28] and last till the persons concerned leave the host country.

The immunity so afforded to diplomatic agents and members of the international civil service does not bind third nations. In a case involving a former ambassador of Syria to the German Democratic Republic, A German Federal Court ruled that benefits of the *persona non grata* rule applied only to the host State and not to other States such as the Federal Republic of Germany in that case.[29]

---

[26]A diplomat or member of the international civil service who is a national or permanent resident of the receiving State will only enjoy immunity from jurisdiction and inviolability in respect of official acts performed in the execution of his professional duties. See Article 38 of the *Vienna Convention on Diplomatic Relations*.

[27]Members of technical and administrative staff of a diplomatic agent may also benefit from such privileges.

[28]See *R.v. Secretary of State for the Home Department ex parte* Bagga, [1991] 1 QB 485; 88 ILR at 404.

[29]121 *ILR* 352.

Another point of contention arising from the broad principle of diplomatic immunity pertains to contracts of employment. Although generally, States and instrumentalities of State come within the purview of local legislation with regard to the hiring and firing of employees, this principle does not apply to diplomatic missions.[30] A point of concern is that such a principle may give rise to absolute discretion being bestowed on a diplomatic mission in disregarding established community rights such as racial, religious, gender and social equality.

3 Waiver of Immunity

The answer to the problem of according undue flexibility to diplomatic agents and members of the international civil service may lie in the practice of waiver of immunity. There are instances where the courts might deem immunity granted by treaty or other agreement to be waived. Waiver of immunity might result either from express agreement between the parties to a contract or by implied acquiescence of the party purporting to enjoy immunity through overt or covert acts. The leading case in this area concerns a 1967 decision[31] where the District of Columbia Circuit Court ruled that the Inter- American Development Bank did not enjoy immunity as any immunity given to the bank had been waived by the Bank by virtue of Article XI(3) of its Articles of Agreement with a Brazilian Corporation who was the other party to the action. An advance waiver, incorporated in a commercial agreement, even though it is calculated to apply only to a particular situation, cannot be deemed invalid and will be generally applicable according to the merits of the case. In *Standard Chartered Bank* v. *International Tin Council and others*.[32] The Queen's Bench in England rejected the claim that an advance waiver is inapplicable to a dispute if it were meant specifically in the contract to apply to "a particular case", which was interpreted by the court as a particular transaction and not a whole dispute. A choice of forum clause in a specific agreement could also be interpreted as a waiver of immunity from suit that could be effectively performed in advance.[33]

Usually, in the case of diplomatic agents and members of the international civil service, only the sending State can waive immunity.[34] In the case of the international civil service the immunity is granted by the host State and can only be waived by the Secretary General or CEO of the Organization served

[30]*Sengupta* v. *Republic of India*, 64 *ILR* 372.

[31]*Lutcher SA Celulose e Papel* v. *Inter-American Development Bank*, 382 F.2d. 454 (DC Cir. 1967).

[32][1986] 2 All ER 257; [1987] 1 WLR 641(1988) 77 ILR 16.

[33]See *Arab Banking Corporation* v. *International Tin Council and Algemene Bank Nederland and Others* (*Interveners*) *and Holo Trading Company Ltd.* (*Interveners*) (1988) 77 ILR 1–8.

[34]*Fayed* v. *Al-Tajir* [1987] All. E.R. 396. Article 32 of the Vienna Convention provides that the sending State may waive the immunity of diplomatic agents and others granted immunity under the Convention.

by the staff member concerned. The General Convention on the privileges and Immunities of the United Nations of 1946 sets out the immunities of the United Nations and its personnel and emphasizes the inviolability of its premises, archives and documents. The privileges and immunities blend with the concept of accountability of an international organization which is broader than principles of responsibility and liability for internationally wrongful acts.[35] The latter acts as a harmonious balance between impunity and answerability of the international civil service.

As already discussed, privileges and immunities are guaranteed by Article 105 of the Charter of the United Nations which provides that an international organization shall enjoy in the territory of each of its members such privileges and immunities as are necessary for the fulfilment of its purposes and that representatives of the members of the United Nations and officials of the Organization enjoy the privileges and immunities that are necessary for the independent exercise of their functions in connection with their official duties in the Organization.[36] This fundamental principle is usually enshrined in the Headquarters Agreement between the host State and the United Nations specialized agency concerned. In the 1988 opinion of the International Court of Justice, the Court opined that the United States was obliged to respect an obligation contained in Section 22 of the United Nations Headquarters Agreement with the United States that admitted of arbitration in the determination of domestic legislative power to close an observer mission of the Palestine Liberation Organization.[37] In an earlier case decided in 1983 the United States Court of Appeals held that immunities and privileges were granted to the United Nations by a host State specifically to preclude State intervention in the execution of duties by the United Nations in that jurisdiction.[38]

Diplomatic immunity and privileges are crucial to the harmonious inter relationships between States. Despite the inherent disadvantages of their abuse, which has sometimes resulted in harm to members of the public and business enterprise, it must be noted that the origins of diplomacy date back to the period of darkness preceding the dawn of history.[39] It is claimed that anthropoid apes living in caves practised a form of diplomacy in reaching understandings with their neighbours on territorial boundaries pertaining to their own hunting grounds.

---

[35]See Recommended Rules and Practices Drafted by the International Law Association's Committee on the Accountability of International Organizations. Report of the Seventeenth Conference, New Delhi 2002 at 774.

[36]This provision is supplemented by the General Convention on the Privileges and Immunities of the United Nations, 1946 and the Convention on Privileges and Immunities of the United Nations.

[37]*ICJ* Reports 1988 at 12. Also at 82 *ILR* at 225.

[38]*Mendaro* v.*World Bank, US Court of Appeals, 27 September 1983, 717 F.2d. 610 (DC Cir. 1983)*. See also *Iran-US Claims Tribunal* v. *AS* 94 ILR 321 at 329 where a Dutch court held that immunities and privileges guaranteed that an international organization may carry out its duties without let or hindrance.

[39]Nicolson (1953), at p. 2.

The compelling need to ensure the preservation of life of an emissary, on the ground that no negotiation could take place if emissaries, however hostile, were murdered on arrival, gave rise to the practice of diplomatic immunity, which is attributed to Australian aborigines, and is mentioned in the Institutes of Manu and in Homeric poems.[40] In the modern world, the institution of the permanent diplomatic mission is the cornerstone of international diplomacy and comity and the diplomat[41] carries out the function of diplomacy which is generally termed "diplomatic practice".[42] These privileges are extremely important if diplomacy is to be effective. The overall aim and objective of diplomacy is to ensure that peace and justice prevails throughout the world and to this end, the institution of diplomacy is a pre-eminent example of the growth of modern civilization. For these reasons the advantages of diplomatic immunities and privileges override their disadvantages.

References

James R (1970) The concept of the international civil servant. In: Jordan R (ed) International administration: its evolution and contemporary applications. Oxford University Press, Oxford

Milde M (1992) New headquarters agreement between ICAO and Canada. Ann Air Space Law (Part II):305–322

Nicolson H (1953) The evolution of diplomatic method. Constable & Co, London

Sarooshi D (2005) International organizations and their exercise of sovereign powers. Oxford University Press, Oxford

Sen B (1950) International law relating to diplomatic practice. Metropolitan Book Co. Ltd., Delhi

Seyersted F (1963) Objective international personality of intergovernmental organizations: do their capacities really depend upon the conventions establishing them? Krohns Bogtnykker, Copenhagen

[40] *Id.* p. 3.

[41] The terms "diplomat" or "diplomatic" are essentially modern and derive their origin from the Latin word "diplomas" meaning official documents emanating from the princes. See Sen (1950). Introduction. See also, Vienna Convention on Diplomatic Relations, United Nations: New York, 1961, *A/CONF.20/14/Add.1* Article 1 for various definitions pertaining to categories of diplomats.

[42] *Ibid.*

Article 48

Meetings of the Assembly and Voting

(a) The Assembly shall meet not less than once in three years[1] and shall be convened by the Council at a suitable time and place. An extraordinary meeting of the Assembly may be held at any time upon the call of the Council or at the request of not less than one-fifth of the total number of contracting States addressed to the Secretary General.

(b) All contracting States shall have an equal right to be represented at the meetings of the Assembly and each contracting State shall be entitled to one vote. Delegates representing contracting States may be assisted by technical advisers who may participate in the meetings but shall have no vote.

(c) A majority of the contracting States is required to constitute a quorum for the meetings of the Assembly. Unless otherwise provided in this Convention, decisions of the Assembly shall be taken by a majority of the votes cast.

Contents

1 The Supreme ICAO Body

The Assembly, at its 14th Session (Rome, 21 August–15 September 1962) adopted Resolution A14-5 (Protocol relating to the Amendment of Article 48 (a) of the Chicago Convention on International Civil Aviation) whereby the Assembly approved on the fourteenth day of September 1962 in accordance with provisions

[1]The expression "not less than once in three years" was introduced into clause (a) of Article 48 by an amendment which was adopted by the Assembly in Resolution A8-1, that expression replacing the word annually" which appears in the original text adopted at Chicago (the original being still in force for States who have not ratified the Amendment).

DOI 10.1007/978-3-319-00068-8_49,

of Article 94 (a)[2] of the Convention resolved that in Article 48 (a) of the Convention, the second sentence be deleted and substituted by

> An extraordinary meeting of the Assembly may be held at any time upon the call of the Council or at the request of not less than one-fifth of the total number of Contracting States addressed to the Secretary General.

The Assembly, at its 16th session (Buenos Aires, 3–26 September 1968) adopted Resolution A16-13 (Frequency and Site of Ordinary Sessions of the Assembly) where the Assembly resolved that the triennial arrangement of ordinary sessions of the Assembly should be regarded as the normal practice of the Organization, with the understanding that intervening ordinary sessions may be convened by decisions of the Assembly or Council. The same resolution provided for the flexibility of having assembly sessions away from the Headquarters of the Organization, taking into account the attendant benefits to the Organization and Contracting States.

At its 21st Session (Montreal, 24 September–15 October 1974) the Assembly adopted Resolution A21-15 (Study of a System of Rotation of Sites for Assembly Sessions) where the Assembly directed the Council to study the feasibility of establishing a system of rotation of sites for the ordinary sessions of the Assembly in the various regions, taking into consideration the facilities available in the regions concerned.

The Assembly is governed by Standing Rules, Rule 28 of which provides that a majority of Contracting States shall constitute a quorum for the plenary meetings of the Assembly.

2 Article 49: Powers and Duties of the Assembly

The Powers and duties of the Assembly, as provided by Article 49 of the Chicago Convention are: (a) Elect at each meeting its President and other officers; (b) Elect the Contracting States to be represented on the Council, in accordance with the provisions of Chapter IX; (c) Examine and take appropriate action on the reports of the Council and decide on any matter referred to it by the Council; (d) Determine its own rules of procedure and establish such subsidiary commissions as it may consider to be necessary or desirable; (e) Vote annual budgets and determine the financial arrangements of the Organization, in accordance with the provisions of Chapter XII; (f) Review expenditures and approve the accounts of the

---

[2]Article 94 (a) states: Amendment of Convention (a) Any proposed amendment to this Convention must be approved by a two-thirds vote of the Assembly and shall then come into force in respect of States which have ratified such amendment when ratified by the number of contracting States specified by the Assembly. The number so specified shall not be less than two thirds of the total number of contracting States.

(b) If in its opinion the amendment is of such a nature as to justify this course, the Assembly in its resolution recommending adoption may provide that any State which has not ratified within a specified period after the amendment has come into force shall thereupon cease to be a member of the Organization and a party to the Convention.

Organization; (g) Refer, at its discretion, to the Council, to subsidiary commissions, or to any other body any matter within its sphere of action; (h) Delegate to the Council the powers and authority necessary or desirable for the discharge of the duties of the Organization and revoke or modify the delegations of authority at any time; (i) Carry out the appropriate provisions of Chapter XIII; (j) Consider proposals for the modification or amendment of the provisions of this Convention and, if it approves of the proposals, recommend them to the Contracting States in accordance with the provisions of Chapter XXI; (k) Deal with any matter within the sphere of action of the Organization not specifically assigned to the Council.

3 The 37th Session

To have some familiarity with what goes on in an Assembly, it is useful to have a discussion as to what the 37th Session of the Assembly (Montreal, September/October 2010) accomplished.

The 37th Session of the ICAO Assembly was held from 28 September to 8 October 2010 at ICAO Headquarters in Montréal. Many significant milestones were achieved by its 1,588 participants, from 176 Member States and 40 international organizations involved in civil aviation. Major inroads were made in the areas of safety, security and environmental protection, and emphasis was given on the safety and security side on assuring even greater safety performance levels in what is already the safest and most secure mode of mass transport in the world.

The Assembly adopted a comprehensive resolution to reduce the impact of aviation emissions on climate change, an agreement which provides a roadmap for action through 2050 for the 190 Member States of the Organization. ICAO's leadership role in the field of environmental protection was reaffirmed by the Assembly, and the Organization, strengthening its global influence, signed numerous international agreements, including cooperation agreements with regional civil aviation organizations and bodies from all the world's regions.

The Assembly endorsed a proactive safety strategy based on the sharing of critical safety information among governments and industry stakeholders. Further, the Assembly endorsed ICAO's plan to establish a multi-disciplinary approach to address the critical issue of runway safety, which will bring together representatives from airlines, airports, air navigation service providers and regulatory authorities. The Assembly built on the achievements of a diplomatic Conference in Beijing in September 2010 by recognizing the need to strengthen aviation security worldwide. In a Declaration on Aviation Security, unanimously adopted by participants, international commitment was reaffirmed to enhance aviation security collaboratively and proactively through screening technologies to detect prohibited articles, strengthening of international standards, improvement of security information-sharing and provision of capacity-building assistance to States in need. The Assembly also put its full support behind the new ICAO Comprehensive Aviation Security Strategy.

Given the Resolution adopted by the Assembly on climate change, ICAO is now the first United Nations agency to lead a sector in the establishment of a globally harmonized agreement for addressing its CO_2 emissions. The Resolution was adopted with some States expressing reservations and calling upon the Council of ICAO to continue its work on specific aspects of the agreement. This remarkable accomplishment came only 2 months before negotiations are again undertaken by these very same States at the 16th Conference of the Parties of the United Nations Framework Convention on Climate Change (UNFCCC) meeting, scheduled to be held in December in Mexico.

Increased momentum at ICAO in the first half of 2010 reflected the culmination of an energetic and active triennium (2008–2010) during which the Organization endeavoured to attain global recognition as a performance and results-driven and values-based Organization. This period also brought to bear ICAO's transition from the past several decades of acting as an overall bureaucratic international body to becoming one that also implements policies and assists its Member States[3] as a priority in the areas of safety, security, environmental protection and the sustainable development of air transport. In particular, the activities of the first half of 2010, as a countdown to the 37th Session of the Assembly, were a fitting precursor leading to the success of Assembly, which made progressive inroads in these three areas, as well as in the legal field.

4 Background and Assembly Outcomes: Safety

From 29 March to 1 April, ICAO held a High-level Safety Conference attended by 551 participants, including Ministers and Directors General of Civil Aviation from 117 Member States, as well as representatives from 32 international organizations. The Conference called on ICAO to facilitate the collection, analysis and dissemination of safety information provided by States and industry partners and the Organization was given a strong mandate to create a strategy to further reduce the global accident rate, through the sharing of safety-related information among Member States and the air transport industry.

The comprehensive systems approach (CSA) was continued with regard to the Universal Safety Oversight Audit Programme (USOAP). Fourteen States received CSA audits in the first half of 2010, bringing the total number of completed safety oversight audits to 159. With regard to the continuation of the USOAP beyond 2010, development began of a continuous monitoring approach (CMA) under the direction of the Assembly and the Council. Consistent with safety management principles, ICAO's Integrated Safety Trend Analysis and Reporting System (iSTARS), a system which offers analysis capability for monitoring the achievement of global safety objectives through the assessment of numerous criteria, was further developed throughout 2010.

[3] See Abeyratne (2009), 529–544.

In accordance with Article 21 of the Chicago Convention, an ICAO Online Aircraft Safety Information System (OASIS) was developed. OASIS contains pertinent information concerning all aircraft habitually involved in international civil aviation, including registration, ownership and control, with the capability of establishing identifiers for aircraft using two fields of data that, when combined, uniquely identify all aircraft entered into the database, regardless of their current registration marks. This repository of information will also contain a history of aircraft ownership and greater control will be made possible.

An international register of Air Operator Certificates (AOCs) has been initiated by ICAO to facilitate the surveillance of foreign operators. The project was begun in 2010 and will take place in two development phases. ICAO continued its close collaboration with the International Air Transport Association (IATA) on work related to training and qualifications initiatives including, specifically, the development of guidance material for inclusion in the *Procedures for Air Navigation Services—Training*[4] on competency-based training and the assessment of maintenance personnel, evidence-based training for flight crew, and instructor and examiner qualifications. A symposium on the next generation of aviation professionals was held in March 2010, whose theme was "*Looking beyond the economic crisis: mobilizing the aviation community to recruit, educate, train and retain the next generation of aviation professionals*."[5]

A dangerous goods training programme was launched in 2010, consisting of a training manual and course that will assist States in complying with the broad principles governing the international transport of dangerous goods by air outlined in Annex 18 to the Chicago Convention and detailed in the *Technical Instructions for the Safe Transport of Dangerous Goods by Air*.[6] With a view to promoting performance-based navigation (PBN), and assisting States in their PBN implementation, ICAO planned PBN airspace design workshops and operational approval courses as well as continuous descent operations workshops.

As a first step, the 37th Session of the Assembly considered safety management systems (SMS) and safety data, including a report on the evolution of ICAO's proactive safety management approach that provided an overview of ICAO's safety analysis strategy. This strategy includes the eventual integration of operational data generated through future implementation of the State Safety Programme (SSP) and SMS. ICAO's leadership role in SMS was also addressed, as well as the development of common safety metrics, analysis methods and interoperable database systems to support safety performance measurement and ensure the effective sharing of safety information among States. Further discussions focused on

[4] *PANS-TRG*, Doc 9868.

[5] The event attracted 403 participants from 71 States and 14 international organizations. Especially noteworthy was the participation of over 80 students involved in aviation-related university and college programmes. Following the symposium, the Next Generation Of Aviation Professionals Task Force focused its work on the development of competencies for flight crew, air traffic management professionals and maintenance personnel.

[6] Doc 9284.

common methods and processes related to SMS implementation, acceptance, performance measurement and oversight, as well as the need to: educate senior management regarding their respective roles in support of SMS implementation and to develop skills within States and aviation organizations to support safety risk management activities, particularly the ability to investigate safety-related events of low consequence; define global safety metrics necessary to support a harmonized approach to safety analysis, citing the work of the Civil Air Navigation Services Organization (CANSO) in the development of leading and lagging safety indicators; develop an international standard for SMS terms and definitions, risk forecasting techniques and computer systems to support proactive safety analysis; and develop a new Annex to the Chicago Convention that would address safety management.

A Resolution on global planning for safety and sustainability was adopted by the Assembly, which recognizes *inter alia*, the importance of a global framework to support ICAO's Strategic Objectives and of basing regional and national plans and initiatives on the global framework for effective implementation, as well as the fact that further progress in improving the global safety and efficiency of civil aviation can best be achieved through a cooperative, collaborative and coordinated approach in partnership with all stakeholders, under the leadership of ICAO. The Resolution calls upon ICAO to implement and keep current the Global Aviation Safety Plan (GASP) and the Global Air Navigation Plan (GANP) to support the relevant Strategic Objectives; calls upon States and invites other stakeholders to cooperate in the development and implementation of regional, sub-regional and national plans based on these global plans; instructs the Council to report to future regular sessions of the Assembly on the implementation and evolution of the global plans; and instructs the Secretary General to promote, make available and effectively communicate the GASP, GANP and associated Global Aviation Safety Roadmap (GASR).

Another Resolution adopted by the Assembly pertains to the GASP, and reaffirms that the Organization's primary objective continues to be the improvement of safety, with an associated reduction in the number of accidents and related fatalities in the international civil aviation system. The Resolution recognizes that safety is a shared responsibility involving ICAO, Contracting States and all other stakeholders and urges Contracting States to support GASP objectives by: implementing the SSP; expeditiously implementing SMS across the aviation industry to complement the existing regulatory framework; sharing operational safety intelligence among States and relevant aviation stakeholders; ensuring that the travelling public has access to easily understandable safety-related information to enable informed decisions; creating an environment in which the reporting and sharing of information is encouraged and facilitated and in which remedial action is undertaken in a timely fashion when deficiencies are reported; and reporting accident and incident data as required to ICAO.

Inter alia, this Resolution also urges Contracting States, regional safety oversight organizations and concerned international organizations to work with all stakeholders to implement GASP and GASR methodology objectives in order to reduce the number and rate of aircraft accidents. Contracting States are called upon to demonstrate the political will necessary for taking remedial actions to address deficiencies, including those identified by USOAP audits and the ICAO regional planning process; to fully exercise safety oversight of their operators in full compliance with applicable Standards and Recommended Practices (SARPs); and to assure themselves that foreign operators flying in their territory receive adequate oversight from their own State and take appropriate action when necessary to preserve safety. For this purpose, States should develop sustainable safety solutions to fully exercise their safety oversight responsibilities, which may be achieved by sharing resources and utilizing internal and/or external resources, such as regional and sub-regional safety oversight organizations and expertise available from other States.

The Assembly discussed ICAO's Runway Safety Programme and adopted a Resolution which recognizes that runway accidents constitute a large portion of all accidents and have resulted in a great number of fatalities, as well as the fact that runway excursions are the highest single occurrence category of all accidents over the last 10 years for all commercial and general aviation operations of fixed-wing aircraft above 5,700 kg certified maximum take-off mass. The Resolution further recognizes that there are several areas of technological developments underway in the aviation industry that show promise in the prevention and mitigation of runway accidents and other serious incidents. The Resolution declares that ICAO shall actively pursue runway safety using a multidisciplinary approach, and urges States to take measures to enhance runway safety, including the establishment of runway safety programmes using a multidisciplinary approach that includes, at the least, regulators, aircraft operators, air navigation services providers, aerodrome operators and aircraft manufacturers, to prevent and mitigate the effects of runway excursions and incursions and other occurrences related to runway safety. States are invited to monitor runway safety events and related precursors as part of the safety data collection and processing system established under their SSP.

Another Resolution adopted by the Assembly concerned the development of an up-to-date consolidated statement of continuing ICAO policies and practices related to a global air traffic management system and communications, navigation and surveillance/air traffic management (CNS/ATM) systems. This Resolution calls upon States and regional safety oversight organizations to establish a framework for joint planning and cooperation at the sub-regional level for the development of CNS/ATM systems. A further Resolution, on the development of an up-to-date consolidated statement of continuing ICAO policies and associated practices related specifically to air navigation, contains Appendices on such areas as the: provision of adequate aerodromes; cooperation among Contracting States in investigations of aircraft accidents; and coordination and cooperation between civil and military air traffic.[7] On the latter, the Resolution requires the common use by civil and military aviation of airspace and certain facilities and for services to be arranged so as to ensure the safety, regularity and efficiency of civil aviation and

[7]For more information and a discussion on this subject, see Abeyratne (2010a), 129–144.

that the requirements of military air traffic are met, while also ensuring that the regulations and procedures established by Contracting States to govern the operation of their State aircraft over the high seas assure that these operations do not compromise the safety, regularity and efficiency of international civil air traffic and that, to the extent practicable, these operations comply with Annex 2—*Rules of the Air* to the Chicago Convention. The Secretary General is required to provide guidance on best practices for civil and military coordination and cooperation, and Contracting States may include, when appropriate, representatives of military authorities in their delegations to ICAO meetings. A significant pronouncement of the Resolution is that ICAO serves as an international forum for facilitating improved civil and military cooperation, collaboration and the sharing of best practices, and in providing the necessary follow-up activities that build on the success of the Global Air Traffic Management Forum on Civil/Military Cooperation (2009) with the support of civil and military partners.

Another Resolution, on performance based navigation global goals, urges all States to implement Area Navigation (RNAV)[8] and Required Navigation Performance (RNP)[9] air traffic services (ATS) routes and approach procedures in accordance with the ICAO concept laid down in the *Performance Based Navigation (PBN)* manual.[10] A further Resolution, on ICAO global planning for safety and sustainability, instructs the Council to amend the GANP to include a framework that will allow ICAO to easily analyze the impact of States' air navigation modernization plans on the global system and then take appropriate action as needed to ensure global harmonization. It also calls upon States, planning and implementation regional groups and the aviation industry to utilize the guidance provided in the GANP for planning and implementation activities and urges Contracting States, industry and financing institutions to provide the necessary support for coordinated implementation of the GANP, avoiding duplication of effort. The Resolution calls upon States that are developing new generation plans for their own air navigation modernization to share their plans in a timely manner with ICAO to ensure global compatibility and harmonization, and instructs the Council to ensure that the GANP

---

[8]RNAV may be defined as a method of navigation that permits aircraft operation on any desired course within the coverage of station-referenced navigation signals or within the limits of a self-contained system capability, or a combination of these. RNAV was developed to provide more lateral freedom and thus a more complete use of available airspace. This method of navigation does not require a track directly to or from any specific radio navigation aid, and has three principal applications: a route structure can be organized between any given departure and arrival point to reduce flight distance and traffic separation; aircraft can be flown into terminal areas on varied pre-programmed arrival and departure paths to expedite traffic flow; and instrument approaches can be developed and certified at certain airports, without local instrument landing aids at that airport.

[9]RNP is a type of PBN that allows an aircraft to fly a specific path between two 3-dimensionally defined points in space. RNAV and RNP systems are fundamentally similar. However, the key difference lies in the fact that an RNP specification includes a requirement for on-board navigation performance monitoring and alerting. A navigation specification that does not include such a requirement is referred to as an RNAV specification.

[10]Doc 9613.

is continuously maintained up to date in light of further operational and technical developments, in close collaboration with States and other stakeholders. A significant guideline issued to the Council in this Resolution is to organize a Twelfth Air Navigation Conference in 2012, with a view to developing longer-term planning for ICAO based on an update of the GANP.

Another current issue discussed by the Assembly was the prevention of communicable diseases through air travel and a Resolution was adopted that, linked to Article 14[11] to the Chicago Convention, urges Contracting States and regional safety oversight organizations to ensure that the public health sector and the aviation sector collaborate to develop a national preparedness plan for aviation which addresses public health emergencies of an international concern and which is integrated with a general national preparedness plan. The Resolution also urges Contracting States to develop a national preparedness plan for aviation that is in compliance with the World Health Organization International Health Regulations (2005) and based on scientific principles as well as guidelines from ICAO and the World Health Organization. Contracting States, and regional safety oversight organizations, as appropriate, are also urged to establish requirements for the involvement of stakeholders such as airport operators, aircraft operators and air navigation service providers in the development of a national preparedness plan for aviation. Contracting States are further requested to join and participate in the Cooperative Arrangement for the Prevention of the Spread of Communicable Disease through Air Travel (CAPSCA) project, where available, to ensure that its goals are achieved, unless equivalent measures are already in place.

On a regional basis, the Assembly adopted two Resolutions, one on a Comprehensive Regional Implementation Plan for Aviation Safety in Africa and the other on Regional Safety Oversight Organizations. The former, *inter alia*, urges Contracting States of the AFI Region to commit to and accelerate the establishment of regional safety oversight organizations and regional accident investigation agencies, where required, and otherwise strengthen cooperation across the region in order to make the most efficient use of available resources. The Council is instructed to notify States, industry and donors of the priority projects arising from a gap analysis performed in accordance with the GASP. Other requirements of this Resolution are that: States, industry and donors implement priority projects identified by the gap analysis and make contributions in cash and kind towards the implementation of the AFI Plan and that the Council recognize all such contributions; African States, ICAO and the African Civil Aviation Commission (AFCAC) jointly address deficiencies identified through the USOAP audits and implement the

[11] Article 14 provides: "Each Contracting State agrees to take effective measures to prevent the spread by means of air navigation of cholera, typhus (epidemic), smallpox, yellow fever, plague, and such other communicable diseases as the contracting States shall from time to time decide to designate, and to that end contracting States will keep in close consultation with the agencies concerned with international regulations relating to sanitary measures applicable to aircraft. Such consultation shall be without prejudice to the application of any existing international convention on this subject to which the contracting States may be parties."

recommendations made by the ICAO/AFCAC joint meeting on aviation safety in Africa and that the Council monitor the implementation of the recommendations of the joint meeting; the Council ensure a stronger ICAO leadership role in coordinating activities, initiatives and implementation strategies aimed specifically at implementing priority projects to achieve the sustainable improvement of flight safety in the AFI Region and to allocate resources to the relevant regional offices accordingly; and the Council monitor and measure the status of implementation in the AFI Region throughout the triennium and report to the next ordinary session of the Assembly on progress made.

The Resolution on Regional Safety Oversight Organizations directs the Council to promote the concept of regional cooperation for the purpose of enhancing safety and safety oversight, including the establishment of regional safety oversight organizations. The Council is also directed to continue to partner with Contracting States, industry and other stakeholders for coordinating and facilitating the provision of financial and technical assistance to States and sub-regional and regional safety and safety oversight bodies, including regional safety oversight organizations, in order to enhance safety and strengthen safety oversight capabilities. Further, the Resolution directs the Council to continue the Implementation Support and Development—Safety Programme and the analysis of relevant safety-critical information for determining effective means of providing assistance to States and sub-regional and regional safety and safety oversight bodies, including regional safety oversight organizations. Contracting States are urged to develop and further strengthen regional and sub-regional cooperation in order to promote the highest degree of aviation safety, and encouraged to foster the creation of regional or sub-regional partnerships to collaborate in the development of solutions to common problems, to build State safety oversight capability, and to participate in, or provide tangible support for, the strengthening and furtherance of sub-regional and regional aviation safety and safety oversight bodies, including safety oversight organizations.

Finally, the Assembly adopted a Resolution on proficiency in the English language used for radiotelephony communications which, *inter alia*, urges Contracting States that have not complied with the language proficiency requirement by the applicability date to post on the ICAO website their language proficiency implementation plans, including their interim measures to mitigate risk, as required, for pilots, air traffic controllers and aeronautical station operators involved in international operations, outlined in accordance with the associated practices and in ICAO guidance material.

4.1 Background and Assembly Outcomes: Security

In response to the attempted sabotage of Northwest Airlines Flight 253 on 25 December 2009,[12] ICAO used the Aviation Security Point of Contact (PoC) Network to communicate information and recommendations to participating States,

[12]For a discussion of this incident, see Abeyratne (2010b), 167–181.

numbering 134 as at 31 October. States were encouraged to conduct risk assessments and implement appropriate screening measures following the incident, and were reminded of the need for cooperation in all matters related to aviation security. The twenty-first meeting of the AVSEC Panel was held at ICAO Headquarters from 22 to 26 March 2010. The Panel considered the threat and risk environment in light of the attempted sabotage of 25 December 2009 and issued a number of recommendations. Provisions in Annex 17—*Security* to the Chicago Convention were updated and strengthened, and are expected to become applicable in 2011, following formal consultation with Member States and approval by the Council.

The sixth meeting of the Facilitation Panel, held at ICAO Headquarters from 10 to 14 May, recommended the introduction of a new Standard in Annex 9—*Facilitation*, obliging all States to adhere to internationally recognized requirements for the transmission of advance passenger information (API) data. The Facilitation Panel also agreed on a new set of guidelines for the passenger name record (PNR) data exchange that will serve to assist States in implementing their national PNR programmes. The Panel also agreed to commence work, on an urgent basis, on the development of new guidelines for advanced data exchange programmes in coordination with the World Customs Organization and IATA.

In considering ICAO's security policy, the Assembly adopted a consolidated statement on continuing ICAO policies related to the safeguarding of international civil aviation against acts of unlawful interference, which strongly condemns all acts of unlawful interference against civil aviation wherever and by whomsoever and for whatever reason they are perpetrated. The Resolution notes with abhorrence acts and attempted acts of unlawful interference aimed at the destruction in flight of civil aircraft in commercial service, including any misuse of civil aircraft as a weapon of destruction and the death of persons on board and on the ground, and reaffirms that aviation security must continue to be treated as a matter of highest priority and that appropriate resources should be made available by ICAO and its Member States. It calls upon all Contracting States to confirm their resolute support for established ICAO policy by applying the most effective security measures, individually and in cooperation with one another, to prevent acts of unlawful interference and to punish the perpetrators, planners, sponsors, and financiers of conspirators in any such acts.

The Assembly makes reference in the Resolution to legal instruments pertaining to aviation security[13] and calls upon Contracting States to give special attention to the adoption of adequate measures against persons committing, planning,

[13]*Convention on Offences and Certain Other Acts Committed on Board Aircraft* (Tokyo, 1963), *Convention for the Suppression of Unlawful Seizure of Aircraft* (The Hague, 1970), *Convention for the Suppression of Unlawful Acts Against the Safety of Civil Aviation* (Montréal, 1971), *Protocol for the Suppression of Unlawful Acts of Violence at Airports Serving International Civil Aviation, Supplementary to the Convention for the Suppression of Unlawful Acts Against the Safety of Civil Aviation* (Montréal, 1988), *Convention on the Marking of Plastic Explosives for the Purpose of Detection* (Montréal, 1991), *Convention for the Suppression of Unlawful Acts Relating to International Civil Aviation* (Beijing, 2010), and *Protocol Supplementary to the Convention for the Suppression of Unlawful Seizure of Aircraft* (Beijing, 2010).

sponsoring, financing or facilitating acts of unlawful seizure of aircraft, acts of sabotage or attempted sabotage or other acts or attempted acts of unlawful interference against civil aviation, and particularly the inclusion, in their legislation, of rules for the severe punishment of such persons. The Resolution also calls upon Contracting States to take adequate measures relating to the extradition or prosecution of persons committing such acts by adopting appropriate provisions in law or treaty for that purpose or by strengthening existing arrangements, and by concluding appropriate agreements for the suppression of such acts, which would provide for the extradition of persons committing criminal attacks against civil aviation.

On the subject of technical security measures, in this Resolution the Assembly urges all States, on an individual basis and in cooperation with other States, to take all possible measures for the prevention of acts of unlawful interference, in particular, those required or recommended in Annex 17 and those recommended by the Council. The Resolution also urges Contracting States to intensify their efforts for the implementation of existing SARPs and procedures relating to aviation security, to monitor such implementation, to take all necessary steps to prevent acts of unlawful interference against international civil aviation and to give appropriate attention to the guidance material contained in the ICAO *Security Manual for Safeguarding Civil Aviation Against Acts of Unlawful Interference*[14] and available on the ICAO restricted website. Finally, the Resolution encourages Contracting States to promote aviation security as a fundamental component of national, social and economic priorities, planning and operations.

Recognizing, *inter alia*, that acts of unlawful interference continue to compromise the safety, regularity and efficiency of international civil aviation, the Assembly urges Contracting States to cooperate for the purpose of providing a joint response in connection with an act of unlawful interference, as well as utilizing, if necessary, the experience and capabilities of the State of the operator, the State of manufacture and the State of registration of an aircraft which has been subjected to an act of unlawful interference, while taking measures in their territory to free the passengers and crew members of that aircraft. The Resolution also condemns any failure by a Contracting State to fulfil its obligations to return without delay an aircraft which is being illegally detained and to submit to competent authorities or extradite without delay the case of any person accused of an act of unlawful interference with civil aviation, along with the reporting of false threats to civil aviation and calls upon Contracting States to prosecute the perpetrators of such acts in order to prevent the disruption of civil aviation operations. Finally, Contracting States are called upon to continue to assist in the investigation of such acts and in the apprehension and prosecution of those responsible.

Within the Resolution, in connection with the Universal Security Audit Programme (USAP), the Assembly urges all Member States to give full support to ICAO by: accepting USAP audit missions as scheduled by the Organization, in

[14]Doc 8973.

coordination with relevant States; facilitating the work of the audit teams; preparing and submitting to ICAO the required pre-audit documentation; and preparing and submitting an appropriate corrective action plan to address deficiencies identified during an audit, as well as other post-audit documentation. The Resolution also urges all Member States, if requested by another State, to share the results of an audit carried out by ICAO and the corrective actions taken by the audited State, as appropriate and consistent with their sovereignty. The Resolution requests that the Council report to the next ordinary session of the Assembly on the overall implementation of the USAP, including its decision with regard to the study to assess the feasibility of extending the CMA to the USAP after the conclusion of the current audit cycle in 2013.

The Assembly adopted a Declaration on Aviation Security that urges Contracting States to, *inter alia*: strengthen and promote the effective application of ICAO SARPs, with particular focus on Annex 17, and to develop strategies to address current and emerging threats; strengthen security screening procedures, enhance human factors and utilize modern technologies to detect prohibited articles and support the research and development of technology for the detection of explosives, weapons and prohibited articles in order to prevent acts of unlawful interference; develop enhanced security measures to protect airport facilities and improve in-flight security, with appropriate enhancements in technology and training; and develop and implement strengthened and harmonized measures and best practices for air cargo security, taking into account the need to protect the entire air cargo supply chain.

5 Background and Assembly Outcomes: Environmental Protection

The eighth meeting of the Committee on Aviation Environmental Protection (CAEP/8) was held at ICAO Headquarters from 1 to 12 February and was attended by 184 participants nominated by 22 Member States and 13 international organizations. The meeting considered alternatives for reducing and limiting the environmental impact of aviation, and standards, policies and guidance material on measures to address aircraft noise and engine emissions were developed, which included: technological improvements; operating procedures; proper organization of air traffic; appropriate airport and land-use planning; and the use of market based options.

ICAO held its third Environmental Colloquium from 11 to 14 May 2010 at ICAO Headquarters, with a broad regional participation from representatives of ICAO's Member States, international organizations, aviation industries and academic/research institutions. The objective of the Colloquium was to provide the most up-to-date information to form the basis for discussions and high-level decisions at the 37th Session of the Assembly. The first day featured a tutorial to help familiarize participants with the vocabulary and concepts used in the description, measurement, regulation, and management of aviation greenhouse gas (GHG) emissions. The Colloquium addressed the latest developments on the assessment of aviation emissions and highlighted various solutions for the related environmental impacts. As well, the Colloquium considered key developments from the following meetings: ICAO High-level Meeting on International Aviation and Climate Change, October 2009; ICAO Conference on Aviation and Alternative Fuels, held in Rio de Janeiro in November 2009; United Nations Framework Convention on Climate Change (UNFCCC) 15th Conference of Parties (COP/15), held in Copenhagen in December 2009; and CAEP/8.

The most topical and arguably contentious issue under discussion at the Assembly was climate change and it was a significant achievement for ICAO Member States to adopt a Resolution on the issue, given the reservations recorded by some States. The Assembly noted that, if the global community is to stabilize greenhouse gas emissions in the atmosphere and maintain emissions at a level that would prevent dangerous anthropogenic interference with the earth's climate, the increase in global temperature would have to be maintained below 2 °C. In order to achieve this target, sizeable cuts in global emissions will be needed and all sectors of the economy are looked to for their contribution, including international aviation, which is well known as representing a significant and growing source of emissions.

The Assembly considered a proposal[15] from the Secretary General of ICAO, an unusual step—for a proposal to come from the Chief Executive of the ICAO Secretariat rather than from the Council that reports to the Assembly—as it reflected unequivocally that the Council had unprecedentedly failed to reach consensus on a comprehensive approach to aviation and climate change. The underlying reason for this impasse was that developing States could not agree to the ambitious emissions reductions suggested by developed States. The proposal suggested that the Assembly adopt a resolution to: require ICAO to exercise continuous leadership on environmental issues relating to international civil aviation, including greenhouse gas emissions; continue to study policy options to limit or reduce the environmental impact of aircraft engine emissions and develop concrete proposals and provide advice as soon as possible to the Conference of the Parties of the UNFCCC, encompassing technical solutions and market-based measures, taking into account the potential implications of such measures for developing as well as developed States; and continue to cooperate with organizations involved in policy-making in this field, notably the UNFCCC.

The proposed Resolution, which was subsequently adopted by the Assembly, also suggests, *inter alia*, that States and relevant organizations work through ICAO to achieve a global annual average fuel efficiency improvement of 2 % until 2020 and an aspired global fuel efficiency improvement rate of 2 % per annum from 2021 to 2050, calculated on the basis of volume of fuel used per revenue tonne kilometre

[15] A37-WP/262 EX/53.

performed. ICAO and its Member States, with relevant organizations, are encouraged to work together to strive to achieve a collective medium term global aspired goal of keeping the global net carbon emissions from international aviation from 2020 at the same level, taking into account the special circumstances and respective capabilities of developing States, the maturity of aviation markets and the sustainable growth of the international aviation industry. In this regard, the proposed resolution suggests that the Council consider a *de minimis* exception for States that do not have substantial international aviation activity levels, in the submission of action plans and regular reports on aviation CO_2 emissions to ICAO. The Resolution also invites the Assembly to recognize that in the short term, voluntary carbon offsetting schemes constitute a practical way to offset CO_2 emissions, and invites States to encourage those operators wishing to take early action to use carbon offsetting, particularly through the use of credits generated from internationally recognized schemes such as the Clean Development Mechanism.[16]

In addition to the 2 % annual improvement in fuel efficiency discussed above, the 37th Session of the Assembly also considered a proposal to further explore the feasibility of more ambitious medium and long term goals, including carbon neutral growth and emissions reductions. Three States proposed that a more ambitious goal be set, of carbon neutral growth by 2020 compared to 2005 levels. In response, a developing State took the position that ICAO should be guided by the principle of common but differentiated responsibilities under the UNFCCC, and that the next task for ICAO is to assist States to achieve the goal of 2 % annual fuel efficiency improvement, while the goal of carbon neutral growth is not realistic and not fair for developing States and that no State should be allowed to take unilateral actions on market-based measures. The latter suggestion was given some support from other developing States.

The main argument of the developing States present at the Assembly was that since the larger quantity of greenhouse gas emissions has been caused by developed States, developing States should not be called upon to pay for ambitious emissions reduction levels at the same level as developed States. Furthermore, developing States claimed that stabilizing the climate should be based on the principles of equity, and common but differentiated responsibilities and those obligations detailed under the framework of the UNFCCC. They concluded that any measure taken should not unduly curb the development of aviation in developing States.

The challenge faced by the Assembly during its discussions was to achieve consensus on establishing guiding principles when designing new, and implementing existing, market based measures for international aviation, and to engage in constructive bilateral and/or multilateral consultations and negotiations with other States to reach an agreement on issues such as carbon neutral growth and market

[16]The **Clean Development Mechanism** allows a developed State with an emission-reduction or emission-limitation commitment under the Kyoto Protocol to implement an emission-reduction project in developing countries. Such projects can earn saleable certified emission reduction credits, each equivalent to one tonne of CO_2, which can be counted towards meeting Kyoto targets. See http://www.icao.int/icao/fr/env2010/ClimateChange/Finance_f.htm.

based measures, as well as on a *de minimis* threshold of international aviation activity, consistent with 1 % of total revenue tonne kilometres as follows:

- Commercial aircraft operators of States below the threshold should qualify for exemption from the application of market based measures that are established on national, regional and global levels; and
- States and regions implementing market based measures may wish to also consider an exemption for small aircraft operators.

It is significant that, notwithstanding the divergence of views between States on the abovementioned issues, and the reservations of some developing States, the Assembly was successful in adopting a Resolution which sets the way forward to more understanding and progress in the years to come.

6 Background and Assembly Outcomes: Law

ICAO held a Diplomatic Conference on Aviation Security in Beijing from 30 August to 10 September 2010, at which 76 States and four international organizations participated. The Conference adopted the *Convention for the Suppression of Unlawful Acts Relating to International Civil Aviation* (Beijing Convention) and the *Protocol Supplementary to the Convention for the Suppression of Unlawful Seizure of Aircraft* (Beijing Protocol). These treaties criminalize, *inter alia*, the act of using civil aircraft as a weapon, and of using dangerous materials to attack aircraft or other targets on the ground. The unlawful transport of biological, chemical or nuclear weapons and their related material has also been made punishable. Moreover, the criminal liability of directors and organizers of an offence under the treaties is specifically covered. Making a threat to commit an offence under the treaties may also trigger criminal liability, when the circumstances indicate that the threat is credible. Each of the two treaties requires 22 ratifications to bring it into force, and as of 1 October 2010, the Convention had been signed by 20 States and the Protocol by 21 States.

The Assembly adopted a Resolution on the Beijing Diplomatic Conference and the Convention and Protocol issuing from the Conference, which recalls Resolution A36-26, Appendix C, relating to the ratification of instruments which have been developed and adopted under the auspices of the Organization, and recognizes the importance of broadening and strengthening the global aviation security regime to meet new and emerging threats. On this basis, the Resolution urges all States to support and encourage the universal adoption of the *Convention for the Suppression of Unlawful Acts Relating to International Civil Aviation* (Beijing, 2010) and the *Protocol Supplementary to the Convention for the Suppression of Unlawful Seizure of Aircraft* (Beijing, 2010). The Resolution urges all States to sign and ratify the Beijing Convention and Beijing Protocol as soon as possible; and directs the Secretary General to provide assistance with the ratification process, as appropriate, and if so requested by a Member State.

The 37th Session of the Assembly will be remembered as a forum at which ICAO's leadership in international civil aviation as the only global forum that can assist States was reiterated in the key areas of safety, security and environmental protection. The Assembly also implicitly recognized that ICAO stands at the threshold of a renewed vision under a new leadership inspired by innovative thinking, and that the Organization is at the defining crossroads of its continuing path towards achieving its aims and objectives as set out in the Chicago Convention. A flight path has been cleared through the fog of rhetoric that, in the past, tended to obfuscate the role of ICAO, and the Organization is now better able to steer towards a more relevant role in the twenty-first century.

From a legal perspective, and particularly with regard to the Resolution on climate change, it appears that ICAO Member States ascribe to Assembly Resolutions a binding nature that they do not necessarily have. ICAO, as a specialized agency of the United Nations, and the Resolutions issued by the Assembly, are governed in general by the laws applicable to the United Nations.[17] The record of the United Nations over its six decades of history is that Member States have on occasion, but in a consistent manner, refused to automatically comply with the corporate will of the Organization.[18] Brownlie has expressed the view that decisions by international conferences and organizations can in principle only bind those States accepting them.[19] Shaw, referring to the binding force of United Nations General Assembly Resolutions states:

> . . .one must be alive to the dangers in ascribing legal value to everything that emanates from the Assembly. Resolutions are often the results of political compromises and arrangements and, comprehended in that sense, never intended to constitute binding norms. Great care must be taken in moving from a plethora of practice to the identification of legal norms.[20]

With regard to the practice of international organizations, the issuance of resolutions may create a custom and, indeed, non binding instruments form a special category that is sometimes referred to as "soft law", which is definitely not law in the sense of enforceability.[21]

---

[17]Article 57 of the United Nations Charter provides that the various specialized agencies, established by intergovernmental agreement and having wide international responsibilities, as defined in their basic instruments, in economic, social, cultural, educational, health, and related fields, shall be brought into relationship with the United Nations in accordance with the provisions of Article 63, which provides that the Economic and Social Council may enter into agreements with any of the agencies referred to in Article 57, defining the terms on which the agency concerned shall be brought into relationship with the United Nations. Such agreements shall be subject to approval by the General Assembly and it may coordinate the activities of the specialized agencies through consultation with and recommendations to such agencies and through recommendations to the General Assembly and to the Members of the United Nations.

[18]Zoller (1987), at 32.

[19]Brownlie (1990), 691.

[20]Shaw (2003), 110.

[21]*Id.* 111. See also Tammes (1958), 265.

ICAO's conferred powers enable the Organization to adopt binding regulations by majority decision (which is usually unnecessary as most of ICAO policy is adopted through consensus). However, States may opt out of these policies or make reservations thereto, usually before such policy enters into force. This is because States have delegated power to ICAO to make decisions on the basis that they accept such decisions internationally. In such cases, States could contract out and enter into binding agreements outside the purview of ICAO even on subjects on which ICAO has adopted policy. The only exception to this rule lies in the adoption of Standards in Annex 2, in particular navigation over the high seas and other overflight areas where freedom of flight prevails, which all Contracting States are bound to follow in order to maintain global safety.

7 Treaties and the Assembly

The Assembly, at its 7th Session (Brighton, 16 June–6 July 1953) adopted Resolution A7-6 (Procedure for Approval of Draft Conventions on International Air Law) whereby the Assembly resolved that any draft convention which the ICAO Legal Committee considered ready for presentation to the States as a final draft shall be transmitted to the Council with an accompanying report of the Committee. The Council was given the flexibility of taking action as it deemed fit, including the circulation of the draft to the Contracting States and to such other States and international Organizations.

Resolution A7-6 also provided that such draft convention shall be considered, with a view to its approval, by a conference which may be convened in conjunction with a session of the Assembly, the opening date of such Conference to be not less than 6 months after the date of transmission of the draft Convention.

References

Abeyratne R (2009) The role of ICAO in the twenty-first century. Ann Air Space Law XXXIV:529–544

Abeyratne R (2010a) Compromises in the use of airspace in civil and military aviation, Eur Transp Law XLV(2):129–144

Abeyratne R (2010b) The NW 253 flight and the global framework of aviation security. Air Space Law 35(2):167–181

Brownlie I (1990) Principles of public international law, 4th edn. Clarendon Press, Oxford

Shaw MN (2003) International law, 5th edn. Cambridge University Press, Cambridge

Tammes AJP (1958) Decisions of international organs as a source of international law. HR 94:265

Zoller E (1987) The corporate will of the United Nations and the rights of the minority contributors. Am J Int Law 81(3):32

Article 49
Powers and Duties of Assembly

The powers and duties of the Assembly shall be to:

(a) Elect at each meeting its President and other officers;
(b) Elect the contracting States to be represented on the Council, in accordance with the provisions of Chapter IX;
(c) Examine and take appropriate action on the reports of the Council and decide on any matter referred to it by the Council;
(d) Determine its own rules of procedure and establish such subsidiary commissions as it may consider to be necessary or desirable;
(e) Vote annual budgets and determine the financial arrangements of the Organization, in accordance with the provisions of Chapter XII;
(f) Review expenditures and approve the accounts of the Organization;
(g) Refer, at its discretion, to the Council, to subsidiary commissions, or to any other body any matter within its sphere of action;
(h) Delegate to the Council the powers and authority necessary or desirable for the discharge of the duties of the Organization and revoke or modify the delegations of authority at any time;
(i) Carry out the appropriate provisions of Chapter XIII;
(j) Consider proposals for the modification or amendment of the provisions of this Convention and, if it approves of the proposals, recommend them to the contracting States in accordance with the provisions of Chapter XXI;
(k) Deal with any matter within the sphere of action of the Organization not specifically assigned to the Council.

R. Abeyratne, *Convention on International Civil Aviation*,
DOI 10.1007/978-3-319-00068-8_50, © Springer International Publishing Switzerland 2014

Article 50
Composition and Election of Council

(a) The Council shall be a permanent body responsible to the Assembly. It shall be composed of thirty-six contracting States elected by the Assembly. An election shall be held at the first meeting of the Assembly and thereafter every three years, and the members of the Council so elected shall hold office until the next following election.

(b) In electing the members of the Council, the Assembly shall give adequate representation to (1) the States of chief importance in air transport; (2) the States not otherwise included which make the largest contribution to the provision of facilities for international civil air navigation; and (3) the States not otherwise included whose designation will insure that all the major geographic areas of the world are represented on the Council. Any vacancy on the Council shall be filled by the Assembly as soon as possible; any contracting State so elected to the Council shall hold office for the unexpired portion of its predecessor's term of office.

(c) No representative of a contracting State on the Council shall be actively associated with the operation of an international air service or financially interested in such a service.

Contents

1 The Resident Governing Body

The Council now has 36 Members—13 States in Part I cited in Article 50 (a), 12 States in Part II cited in Article 50 (b) and 11 in Part III. Membership in the Council has gradually increased over the years.[1] At its 37th Session, the Assembly in 2010 elected its Council. The 36-member Council is the governing body of the Organization and was elected for a 3-year term. The election process was divided into three parts, with the following States elected:

[1]At the 13th Session, the Assembly, by Resolution A13-1, increased Council membership to twenty seven States, and at its 17th Session, by Resolution A17-1, the Assembly increased this number to 30. At its 21st Session in 1974, by Resolution A21-2, the membership of the Council was increased to 33. At its 28th Session in 1990, the Assembly increased this number to 36.

DOI 10.1007/978-3-319-00068-8_51,

Part I
(States of chief importance in air transport)—Australia*, Brazil*, Canada*, China*, France*, Germany*, Italy*, Japan*, Russian Federation*, United Kingdom*, and the United States*.

Part II
(States which make the largest contribution to the provision of facilities for international civil air navigation)—Argentina*, Belgium, Colombia, Denmark, Egypt*, India*, Mexico*, Nigeria*, Saudi Arabia*, Singapore*, South Africa* and Spain*.

Part III
(States ensuring geographic representation)—Burkina Faso, Cameroon*, Cuba, Guatemala, Malaysia*, Morocco, Paraguay, Peru, Republic of Korea*, Slovenia, Swaziland, Uganda*, United Arab Emirates*.
(*indicates re-election)

The voting process is governed by established rules of Procedure of the Assembly[2]:

Rule 54
Each contracting State which intends to stand for election to the Council may at any time so inform, in writing, the Secretary General who shall, at the opening of the session, publish a list showing the names of all the States which have so notified him. This list shall serve the purpose of information only. The official notification of candidacy may be given only at the times specified in Rules 56 and 58 and the official lists of candidatures shall be only those specified in Rules 56 b) and 58 b).

Rule 55

A. *The election of the Council shall be so conducted as to enable adequate representation on the Council to be given to the contracting States described in Article 50 b) of the Convention and shall be held in three parts as follows:*

   i. *The first part—election of States of chief importance in air transport—shall be held within four days of the opening of the session.*
   ii. *The second part — election of States not already elected in the first part but which make the largest contribution to the provision of facilities for international civil air navigation — shall be held immediately after the first part of the election.*
   iii. *The third part—election of States not elected in either the first or the second part, and whether or not they were candidates in either of those parts, and whose designation will ensure that all the major geographical areas of the world are represented on the Council — shall be held as soon as possible after the expiry of twenty-four hours following the publication of the list of candidates mentioned in Rule 58 b).*

B. *As early as possible after the opening of the session, the Assembly shall fix the maximum number of contracting States to be elected in each part of the election and fix also the day on which the first two parts of the election shall be held.*

[2]*Voting on Election of the Council*, Doc 7600 Section IX.

2 Notification and Announcement of Candidates for Parts 1 and 2

In order to comply with Rule 56 (a), reproduced below, candidacies for Parts 1 and 2 must be submitted between approximately 1,200 h on Tuesday, 28 September 2010 and 1,200 h on Thursday, 30 September 2010. They must be dated within that time frame (28, 29, or 30 September 2010) and be addressed to the Secretary General of ICAO, making reference to Rule 56 (a) of the Assembly's Rules of Procedure and indicating under which part of the election the candidate wishes to be considered.

The letter of notification should be signed by a national of the State presenting its candidacy. Examples of officials who would sign the letter would include a Minister, a Director General of Civil Aviation, the Chief of the Delegation of that State at the Assembly, or, in the case of a Council Member State submitting its candidacy for re-election, by the current Representative of that State.

Information on candidates for Parts 1 and 2 will be issued early in the afternoon of Thursday, 30 September in accordance with Rule 56 (b).

Rule 56

A. *Each contracting State which desires to stand for election in either the first or the second part shall so notify the Secretary General in writing during the period of forty-eight hours following the opening of the session.*
B. *At the end of the period of forty-eight hours mentioned above, the Secretary General shall publish a list of the States which have notified him, in accordance with paragraph a) above, of their candidacy for the first or the second part of the election.*
C. *All States entered in the aforesaid list shall be deemed to be available for consideration for the first part as well as for the second part, if necessary, of the election unless a contracting State notifies the Secretary General that it does not wish to be considered in the first part or the second part of the election. Accordingly, and subject to the foregoing, any contracting State included in the said list and not elected in the first part of the election will automatically be included amongst those to be considered in the second part of the election.*

3 Notification and Announcement of Candidates for Part 3

Candidates for Part 3 of the election are addressed in Rules 57 and 58 of the Assembly's Rules of Procedure.

In order to comply with Rule 58 (a), reproduced below, Candidacies for Part 3 must be submitted during the 48 h following the election of Part 2 (between approximately 1,200 h on Saturday, 2 October 2010 and 1,200 h on Monday, 4 October 2010). They must be dated within that time frame (2, 3 or 4 October 2010) and be addressed to the Secretary General of ICAO, making reference to Rule 58 (a) of the Assembly's Rules of Procedure.

The letter of notification should be signed by a national of the State presenting its candidacy. Examples of officials who would sign the letter would include a Minister, a Director General of Civil Aviation, the Chief of the Delegation of that

State at the Assembly, or, in the case of a Council Member State submitting its candidacy for re-election, by the current Representative of that State.

Information on candidates for Part 3 will be issued early in the afternoon of Monday, 4 October 2010 in accordance with Rule 58 (b).

Rule 57

After the second part of the election the President of the Assembly shall declare an interval of approximately forty-eight hours specifying the hour at which that interval will expire, in order that candidatures may be presented for the third part of the election.

Rule 58

A. *Any contracting State not elected in the first or the second part of the election, and whether or not it was a candidate in either of those parts, shall, if it wishes to be a candidate for the third part, so notify the Secretary General in writing after the commencement, but before the expiry, of the interval mentioned in Rule 57.*
B. *A list showing the names of the States which are candidates in accordance with this Rule for the third part of the election shall be published at the end of the aforesaid interval.*

4 Electronic Voting Procedure

The Council has decided that with the understanding that manual votes using voting slips and ballot boxes will remain in place as a fall back option, the voting system developed by the International Labour Organization, currently using three working languages (English, French and Spanish), will be used at the 37th Session of the Assembly for the Council elections. If the introduction of the electronic voting system using the three languages proves to be a success, the Organization may work towards having the system adapted to include all six languages at the time of the next ordinary session of the Assembly in 2013.

The voting system involves wireless technology that were being used in 2010 by the International Labour Organization and the World Meteorological Organization for the same purpose, and will allow Delegates to cast their votes from their seats using tablets while ensuring complete anonymity.

Delegations of States whose voting privileges have been suspended because of long-standing arrears will not be provided with the access information required to activate their voting tablets during the distribution phase of the election procedures.

5 President of the Council

The Council is presided over by its President, who by virtue of Article 51 is elected for a period of 3 years. He serves a maximum of two terms Article 51 provides that the Council shall elect from among its members one or more Vice Presidents who shall retain their right to vote when serving as acting President. The President need not be selected from among the representatives of the members of the Council but, if a representative is elected, his seat shall be deemed vacant and it shall be filled by the State which he represented. The duties of the President shall be to: (a) Convene meetings of the Council, the Air Transport Committee, and the Air Navigation Commission; (b) Serve as representative of the Council; and (c) Carry out on behalf of the Council the functions which the Council assigns to him.

Milde States:

> It is a historic anomaly not seen in any other organization of the UN system that ICAO has two leading officials – the President of the Council and the Secretary General – without a distinct delimitation of their jurisdictions that would convincingly justify this duality. The common practice is that the agencies have one single executive official named Secretary-General, Director or Director General.[3]

The President of the Council, who is elected by that body, receives a salary from ICAO as the highest ranking employee of ICAO. In ICAO's history of 64 years only four elected persons holding the position as Presidents of the Council. *Milde* goes on to say:

> Each of them impressed his character and personality on the position and the office of the President became what the incumbent wished it to be.[4]

Noteworthy is the President's function, which is not one explicitly mentioned in Article 51, of mediation. A glaring example of the intervention of the President in bringing about a compromise between two polarized groups was his successful mediation in the noise issue that sprang up in the 28th Session of the ICAO Assembly. At the 28th Assembly Sessions held in October 1990, the ICAO Assembly observed that while certification standards for subsonic jet aircraft noise levels are specified in Volume 1, Chapter 2 and Chapter 3 of Annex 16 and that environmental problems due to aircraft noise continued to exist in the neighbourhood of many international airports, some States were consequently considering restrictions on the operations of aircraft which exceed the noise levels in Volume I, Chapter 3 of Annex 16. The Assembly also recognized that the noise standards in Annex 16 were not intended to introduce operating restrictions on aircraft and that operating restrictions on existing aircraft would increase the costs of airlines and would impose a heavy economic burden, particularly on those airlines which do not have the financial resources to re-equip their fleets. Therefore, considering that resolution of problems due to aircraft noise must be based on the mutual recognition of the difficulties encountered by States and a balance among their different concerns, the Assembly, by Resolution A28-3, urged States not to introduce any new operating restrictions on aircraft which exceed the noise levels in Volume I, Chapter 3 of Annex 16 before considering:

---

[3]Milde (2008) at 146. Also, Secretary-General in the UN and IMO; Director-General in the FAO, ILO, UNESCO and WHO.

[4]Milde (2008), *ibid*. The first President was Dr Edward Warner (USA) in 1947–1956, The second President was Mr. Walter Binaghi (Argentina) in 1956–1976, The third President was Dr Assad Kotaite (Lebanon) in 1976–2006. On 21 November 2007 the Council elected Mr. Roberto Kobe Gonzales as its fourth President for a period of three years, which was extended for three years by the Council in 2010. In 2013 the Council will elect its fifth President.

- Whether the normal attrition of existing fleets of such aircraft will provide the necessary protection of noise climates around their airports;
- Whether the necessary protection can be achieved by regulations preventing their operators from adding such aircraft to their fleets through either purchase, or lease/charter/interchange, or alternatively by incentives to accelerate fleet modernization;
- Whether the necessary protection can be achieved through restrictions limited to airports and runways the use of which has been identified and declared by them as generating noise problems and limited to time periods when greater noise disturbance is caused; and,
- The implications of any restrictions for other States concerned, consulting these States and giving them reasonable notice of intention.

The Assembly further urged States:

- To frame any restrictions so that Chapter 2 compliant aircraft of an individual operator which are presently operating to their territories may be withdrawn from these operations gradually over a period of not less than 7 years;
- Not to begin the above phase-in period for any restrictions before I April 1995;
- Not to restrict before the end of the phase-in period the operations of any aircraft less than 25 years after the date of issue of its first individual certificate of airworthiness;
- Not to restrict before the end of the phase-in period the operations of any presently existing wide-body aircraft or of any fitted with high by-pass ratio engines;
- To apply any restrictions consistently with the non-discrimination principle in Article 15 of the Chicago Convention so as to give foreign operators at least as favourable treatment as their own operators at the same airports; and,
- To inform ICAO, as well as the other States concerned, of all restrictions imposed.

The Assembly also strongly encouraged States to continue to cooperate bilaterally, regionally and inter-regionally with a view to:

- Alleviating the noise burden on communities around airports without imposing severe economic hardship on aircraft operators; and
- Taking into account the problems of operators of developing countries with regard to Chapter 2 aircraft presently on their register, where they cannot be replaced before the end of the phase-in period, provided that there is proof of a purchase order or leasing contract placed for a replacement Chapter 3 compliant aircraft and the first date of delivery of the aircraft has been accepted;

The Assembly, while urging States, if and when any new noise certification standards are introduced which are more stringent than those in Volume I, Chapter 3 of Annex 16, not to impose any operating restrictions on Chapter 3 compliant aircraft, urged the Council to promote and States to develop an integrated approach to the problem of aircraft noise, including land-use planning procedures around international airports, so that any residential, industrial or other land-use that might be adversely affected by aircraft noise is minimal. The Assembly further urged States to assist aircraft operators in their efforts to accelerate fleet modernization and thereby prevent obstacles and permit all States to have access to lease or purchase aircraft compliant with Chapter 3, including the provision of multilateral technical assistance where appropriate. This Resolution superseded Resolution A23-10, which was discussed above.

Resolution A28-3 represented a cautious balance between the concerns of the aircraft manufacturers, the airline industry and developing States who do not wish to lose in the near future, the services of Chapter 2 aircraft which are already in use and service. Although aircraft manufactured prior to October 1977 that are included in Chapter 2 of Annex 16 and called "Chapter 2 aircraft" are required to be phased out, the compromise in Resolution A28-3 allows States that have noise problems at airports to start phasing out operations by Chapter 2 aircraft from the year 1995 and to have all of them withdrawn by the year 2002, with some exceptions. The Resolution envisages that by the year 2002 only aircraft manufactured after October 1977 and described in Chapter 3 of Annex 16 (called "Chapter 3 aircraft") would be in operation. Following this resolution, a number of developed States have already started to phase out Chapter 2 aircraft, while giving due recognition to the compromise reached in Resolution A28-3.

At its 32nd Assembly, held in September 1998, Assembly Resolution A32-8[5] containing a consolidated statement of continuing ICAO policies and practices related to environmental protection was adopted, making current the regulatory policies relating to aviation and the environment. Appendix B to the Resolution cites Annex 16 Volume 1 as comprising, *inter alia,* noise certification standards for future subsonic aircraft and mentions that aircraft manufacturers and operators need to note that future generations of aircraft have to be so designed as to operate efficiently and with the least possible environmental disturbance. Appendix C calls upon Contracting States and international Organizations to recognize the leading role of ICAO in dealing with aircraft noise and requests the former to work closely together to ensure the greatest harmonization of work in the area of environmental protection as related to air transport. In Appendix G, which relates to the problem of sonic boom, the Assembly reaffirms the importance attached to ameliorating problems caused to the public by sonic boom as a result of supersonic flight, invites States involved in the manufacture of supersonic aircraft to furnish ICAO with proposals that would meet specifications established by ICAO on the subject.

The most topical issue addressed by Resolution A32-8, is in its Appendix D, which, whilst reiterating the time limits specified for the phasing out of Chapter 2 Aircraft and related dates, strongly encourages States to continue to co-operate bilaterally, regionally and inter-regionally with a view to alleviating the noise burden on communities and also to take into account the problems that may be faced by some operators in phasing out their Chapter 2 aircraft before the end of the

[5] *Assembly Resolutions in Force* (As of 2 October 1998), *ICAO Doc 9730*, ICAO: Montreal, at I-36.

period specified. The Resolution also urges States, if and any noise certification standards are introduced new which are more stringent than those in Volume 1, Chapter 3 of Annex 16 not to impose any operating restrictions on Chapter 3 compliant aircraft. More importantly, States are urges to assist operators in their efforts of fleet modernization with a view to preventing obstacles and permit all States to have access to lease or purchase aircraft compliant with Chapter 3.

The qualification in Resolution A32-8 seemingly admits of Chapter 2 aircraft which were reconverted to be compliant with Chapter 3 noise levels being considered for operation at least until 1 April 2002. The Resolution urges States to consider the difficulties faced by operators of Chapter 2 aircraft who are unable to make them Chapter 3 compliant by the given date, implying that it would be in the economic interests of such operators to be given additional time in order to make the necessary replacements. Chapter 2 aircraft could be made Chapter 3 compliant whereby the aircraft can be re-certified to Chapter 3 standards through re-engining or hush kitting. Chapter 2 aircraft which are likely to be re-engined or hush-kitted are Boeing 727s and 737s, DC-9s, BAC1-11s and some Boeing 747-100s that need hush-kitting.

There was an attempt on the part of the European Union to limit and eventually eliminate Chapter 3 compliant aircraft from operating within countries of the European Union. This ban would also be calculated to affect the importation of such aircraft into the region. Legislation passed by the Union in April 1999 was intended to bar Chapter 3 compliant aircraft from European registries from 4 May 2000 (originally 1 April 1999, which is 3 years before the date specified in Resolution A32-8, namely, 1 April 2002) and to prohibit their operation into the countries of the European Union after 1 April 2002. Such action has been reportedly criticized by the Air Transport Association which claims that the inflexibility of such a deadline "will severely undercut, if not destroy entirely—ICAO's efforts to address environmental issues on a uniform international basis".[6]

The action of the European Union sought justification on the basis that the exponential air traffic growth in Europe would increase noise around European airports, requiring stringent noise standards. A related fear is reportedly that hush-kitted aircraft, which are rare in Europe, will find a new home in the Continent. Unlike in Europe, airlines in the United States have been somewhat prolific in the use of hush-kits in aircraft[7] and many United States carriers operate Stage 2 hush-kitted aircraft into Europe and even have based equipment in the Continent. At the time of writing, although the U.S. State Department had reacted forcefully to the European Union's hush-kitted aircraft ban, the application of which was extended by the Union until May 2000, it was reported that both the United States and

[6]Flint (1999), at p. 29.

[7]It was reported that American Airlines is installing the Raisbeck System on 52 B 727 aircraft and hush-kitting 20 more. United is installing hush-kits on 75 B 727 aircraft and 24 B 737-200s. Delta is hush-kitting 104 727s and 54 737-200s. Southwest, TWA, Alaska Airlines and US Airways are other carriers who plan to hush-kit their Chapter 2 aircraft. See Flint (1999), *Id.* at p. 34.

the European Union were working together on a possible new ICAO Standard (presumably to be called Chapter 4).[8]

By this time, the aircraft noise issue had entered a phase where trading and environmental issues were at a delicate balance. On the one hand, competition issues pertaining to the sale of hush-kits and other equipment calculated to reduce aircraft engine noise to levels prescribed by ICAO were quite significant from the perspective of international trade, and on the other hand, environmental issues that had been addressed by the ICAO Council through CAEP and also by the Assembly should also be given careful consideration. As discussed earlier, ICAO Assembly Resolution A32-8 urged States to give consideration to the economic difficulties that some States may have to face in phasing out Chapter 2 aircraft by the year 2002. However, the Resolution did not leave room for States to claim that pure economic factors would effectively preclude them from phasing out Chapter 2 aircraft by the date stipulated in the Resolution.

The issue was a "double edged sword" involving two distinct disciplines. As discussed earlier, the European contention is based on the strictly legal issue of noise pollution and overtones of the tort of nuisance committed by operators whose aircraft are not compliant with Chapter 3 standards to the satisfaction of the European Union. Others who oppose what they claim to be a premature enforcement of ICAO standards, as contained in Resolution A32-8, argue that the European hush-kit rule would cost the manufacturing industry significant losses. It was reported that the United States industry would lose $ 2 billion if the ban were to be enforced in Europe as scheduled and a compromise was reached with the intervention of the good offices of the President of the Council.[9]

References

Esler D (1999) The latest noise about noise. Bus Commercial Aviat 52

Flint P (1999) Breaking the sound barrier. Air Transp World 3:29

Milde M (2008) In: Benkoe M (ed) International air law and ICAO. Eleven International Publishing, The Hague

[8]Esler (1999) at p. 53.

[9]Europe Considers Delaying Hushkit Ban, *Aviation Daily*, Thursday October 7 1999 at p. 1.

Article 51

President of Council

The Council shall elect its President for a term of three years. He may be reelected. He shall have no vote. The Council shall elect from among its members one or more Vice Presidents who shall retain their right to vote when serving as acting President. The President need not be selected from among the representatives of the members of the Council but, if a representative is elected, his seat shall be deemed vacant and it shall be filled by the State which he represented. The duties of the President shall be to:

(a) Convene meetings of the Council, the Air Transport Committee, and the Air Navigation Commission;
(b) Serve as representative of the Council; and
(c) Carry out on behalf of the Council the functions which the Council assigns to him.

R. Abeyratne, *Convention on International Civil Aviation*,
DOI 10.1007/978-3-319-00068-8_52, © Springer International Publishing Switzerland 2014

Article 52

Voting in Council

Decisions by the Council shall require approval by a majority of its members. The Council may delegate authority with respect to any particular matter to a committee of its members. Decisions of any committee of the Council may be appealed to the council by any interested contracting State.

DOI 10.1007/978-3-319-00068-8_53,

Article 53
Participation Without a Vote

Any contracting State may participate, without a vote, in the consideration by the Council and by its committees and commissions of any question which especially affects its interests. No member of the Council shall vote in the consideration by the Council of a dispute to which it is a party.

R. Abeyratne, *Convention on International Civil Aviation*,
DOI 10.1007/978-3-319-00068-8_54,

Article 54
Mandatory Functions of Council

The Council shall:

(a) Submit annual reports to the Assembly;
(b) Carry out the directions of the Assembly and discharge the duties and obligations which are laid on it by this Convention;
(c) Determine its organization and rules of procedure;
(d) Appoint and define the duties of an Air Transport Committee, which shall be chosen from among the representatives of the members of the Council, and which shall be responsible to it;
(e) Establish an Air Navigation Commission, in accordance with the provisions of Chapter X;
(f) Administer the finances of the Organization in accordance with the provisions of Chapters XII and XV;
(g) Determine the emoluments of the President of the Council;
(h) Appoint a chief executive officer who shall be called the Secretary General, and make provision for the appointment of such other personnel as may be necessary, in accordance with the provisions of Chapter XI;
(i) Request, collect, examine and publish information relating to the advancement of air navigation and the operation of international air services, including information about the costs of operation and particulars of subsidies paid to airlines from public funds;
(j) Report to contracting States any infraction of this Convention, as well as any failure to carry out recommendations or determinations of the Council;
(k) Report to the Assembly any infraction of this Convention where a contracting State has failed to take appropriate action within a reasonable time after notice of the infraction;
(l) Adopt, in accordance with the provisions of Chapter VI of this Convention, international standards and recommended practices; for convenience, designate them as Annexes to this Convention; and notify all contracting States of the action taken;
(m) Consider recommendations of the Air Navigation Commission for amendment of the Annexes and take action in accordance with the provisions of Chapter XX;
(n) Consider any matter relating to the Convention which any contracting State refers to it.

R. Abeyratne, *Convention on International Civil Aviation*,
DOI 10.1007/978-3-319-00068-8_55, © Springer International Publishing Switzerland 2014

Contents

1 The Powers of the Council

There are a few home truths one has to consider with regard to the Council. The first is that the 36 members of the Council are voted in by the Assembly of 191 States. Therefore the Council members, who are not sent in by their States on the States' own accord, are obliged to act in the best interests of their constituents, which are all States comprising ICAO. It would therefore be morally reprehensible for a member of the Council to act in the interests of its own. A good example of the Council's representation of the interests of all States is the adoption of the Annexes to the Convention, particularly those relating to safety and security. As regards environmental protection, be verified by an independent and accredited verifier prior to submission.

On 2 November 2011 the Council of ICAO adopted a declaration unanimously urging the EU not to subject non-EU carriers to its ETS. This Declaration was sponsored by India and 25 other member States of the Council. The entire aviation community, including the EU—which has stated that it is willing to amend its scheme if an ICAO sponsored alternative is finalized—supports the view that market based measures should be global in application and should be developed through ICAO.

Much of the mandatory functions of the Council are straightforward and require no discussion. However, there are a few that deserve some inquiry. Article 54 (j), which requires the Council to *Report to Contracting States any infraction of this Convention, as well as any failure to carry out recommendations or determinations of the Council* is one such provision. As discussed earlier, the Council is bound by this provision to advise Contracting States of the failure to carry out recommendations or determinations of the Council. When applied to the corrective recommendations to States with serious safety and security concerns based on results of audits conducted by ICAO, the relevance of this requirement becomes significant.

2 Article 54 (j)

Article 54 (j) makes it a mandatory function of the Council to report to any Contracting State any infraction of the Chicago Convention as well as any failure to carry out recommendations and determinations of the Council. There are various dimensions to this provision in the context of Resolution A36-2. Firstly, it is surprising that the Assembly Resolution does not also request the Council to perform its mandatory function in Article 54 (k), which is to report to the Assembly any infraction of the Convention where a Contracting State fails to take appropriate action within a reasonable time after notice of the infraction. This would have arguably been a more coercive and effective tool than the measure prescribed in Article 54 (j) in that States would be quite concerned if their shortcomings were to be aired out in front of 190 Contracting States at an ICAO Assembly.

The Chicago Convention bestows neither the ability nor the power on the Council to investigate and determine on its own initiative whether there has been an infraction of the Convention. There is also no specific provision which entitles the Council to notify the State concerned that an infraction has taken place. However, Article 54 (n) provides that the Council can consider any matter relating to the Convention which any Contracting State refers to it, giving the Council the capacity to make its own determination and recommendations pertaining to a matter referred to it. It is also noteworthy that both Article 15 of the Convention, which allows the Council to report and make recommendations resulting from a review by the Council of charges imposed for the airports and other facilities, and Article 69, which gives the Council competency to make recommendations to member States for the improvement of air navigation facilities, are two instances of specific provision being made within the Convention where the Council can make recommendations for the consideration of ICAO member States.

Clearly, non compliance with SARPs and shortcomings or deficiencies in safety and security cannot be classified as infractions of the Convention. An infraction is a violation and arguably applicable to the Chicago Convention itself and not to the Annexes which only contain SARPs that are not strictly legally binding so as to constitute a violation if not followed. Therefore, the Assembly, in A36-2 quite clearly meant the reportage of failure to carry out recommendations and determinations of the Council with regard to SARPs. This is clearly an administrative function and not a judicial function, since an administrative act is usually referred to as similar or related activities regarding the handling and processing of information.

The second dimension to the Resolution is that it is the function of the Council in this case, to use the words of operative clause 6 of Resolution A36-2 to

> apply and review... the procedure to inform Contracting States within the scope of Article 54j) of the Chicago Convention, in the case of a State having significant shortcomings with respect to ICAO safety related SARPs in order for other Contracting States to take action in an adequate and timely manner.

Surprisingly the Council is asked by the Assembly to restrict itself to determining the adherence to SARPs and report its findings thereof, which is already a function handed down in the Convention to the Council in Article 38.[1] Again, it is not clear as to why the Assembly refrained from applying the rest of Article 54 (j) to its

[1]Article 38 provides: *inter alia* that any Contracting State can file a difference to a standard and notify the Council which in turn is required to make immediate notification to all other States of the difference which exists between one or more features of an international standard and the corresponding national practice of that State.

Resolution, which makes it incumbent upon the Council to report the failure to carry out recommendations or determinations of the Council. This application would have served the purpose of the Assembly better than the mere restriction to the SARPs in the Annexes.

The third dimension is that the Council, under the Convention, has only functions (which are in essence duties) and no powers.[2] On the other hand the Assembly has powers and duties accorded to it in the Chicago Convention,[3] one of which is to delegate to the Council the powers and authority necessary or desirable for the discharge of the duties of the Organization and revoke or modify the delegations of authority at any time.[4] However, in this instance there is no indication that the Assembly exercised its powers to delegate its authority or power to the Council to apply and review the procedure in Article 54 (j). If this had been the case, the Council would have had the same right and the authority of the Assembly to take appropriate action as deemed necessary in the manner in which the information derived from safety audits would be disseminated and reported to other States.

From the above discussion it becomes clear that, while on the one hand the ICAO Assembly, which in essence is the representative voice of the 190 member States comprising ICAO, has directed the Council to apply and review procedures to inform member States within the scope of Article 54 (j) of shortcomings, on the other hand, the overriding separate and individual memoranda signed by ICAO with its member States in the area of safety would have to be revised in terms of the confidentiality clause. Additionally, the Council would have to set in place an understanding with States and appropriate mutually agreed guidelines on the content of such information and the manner in which it is to be divulged.

Article 54 (j) brings to bear the inevitable question as to whether the Council would be disposed toward putting into effect what has legitimately been prescribed by the Chicago Convention as a mandatory function. If the Council does not carry out this function, will the international aviation community consider the Council to be in dereliction of its legal duty? Could it be argued that, if the Council performs all of its other mandatory functions under Article 54 [from 54 (a) to 54 (n)] it cannot abdicate nor can it ignore one particular function?

In such instances the ICAO audit process has a structured approach where, through the intervention of the ICAO Council, the States concerned are requested to remedy deficiencies that are discovered during the audits. If States do not heed the call of ICAO to remedy such deficiencies, the only mechanism available to ICAO is the recourse offered in Article 54 (j) of the Chicago Convention which makes it mandatory function of the Council to report to Contracting States any

[2]Although Jacob Schenkman, in his well documented and logically reasoned treatise on ICAO states that "The Council has been entrusted with duties, powers and functions..." he does not give a single example of such a power. See Capt. Schenkman (1955) at 158.

[3]Article 49 of the Convention.

[4]Article 49 (h).

infraction of the Convention, as well as any failure to carry out recommendations or determinations of the Council.

The answer to the above questions would *ex facie* be that, since the Chicago Convention explicitly requires the Council to report to Contracting States its recommendations and determinations, the Council is bound to report to States if its recommendations regarding deficiencies found in safety oversight audits are not followed and carried out by the States concerned. In addition to the fact that it is the duty of the Council to implement its mandatory function under Article 54 (j), it can be argued that ICAO has been empowered by its member States to carry out its functions and duties under Chicago Convention.

3 Functions of the Council in Ensuring Safety

Article 54 (j) alludes to a mandatory function of the Council of ICAO. This brings one to the distinction between a function and a power. While a power is the capacity to direct the decisions and actions of others, a function on the other hand is to perform, execute or administer.[5] A power is also defined as an ability on the part of a person to produce a change in a given legal relation by doing or not doing a certain act.[6] In this context the Council only has a function to report to States shortcomings of other States detected during the course of safety audits with regard to adherence by the ICAO member States of SARPs. It is therefore incontrovertible that Assembly Resolution A36-2 merely hands over to the Council the function to report an infraction of the Chicago Convention as well as shortcomings with regard to SARPs and recommendations and determinations of the Council in that regard.

The Chicago Convention bestows neither the ability nor the power on the Council to investigate and determine on its own initiative whether there has been an infraction of the Convention. There is also no specific provision which entitles the Council to notify the State concerned that an infraction has taken place. However, Article 54 (n) provides that the Council can consider any matter relating to the Convention which any Contracting State refers to it, giving the Council the capacity to make its own determination and recommendations pertaining to a matter referred to it. It is also noteworthy that both Article 15 of the Convention, which allows the Council to report and make recommendations resulting from a review by the Council of charges imposed for the airports and other facilities, and Article 69, which gives the Council competency to make recommendations to member States for the improvement of air navigation facilities, are two instances of specific provision being made within the Convention where the Council can make recommendations for the consideration of ICAO member States.

Clearly, non-compliance with SARPs and shortcomings or deficiencies in safety cannot be classified as infractions of the Convention. An infraction is a violation

[5]Deluxe Black's Law Dictionary, Sixth Edition, St. Paul. Minn: 1990, at 673.

[6]*Id.* at 1189.

and arguably applicable to the Chicago Convention itself and not to the Annexes which only contain SARPs that are not strictly legally binding so as to constitute a violation if not followed. Therefore, the Assembly, in A36-2 quite clearly meant the reportage of failure to carry out recommendations and determinations of the Council with regard to SARPs. This is clearly an administrative function and not a judicial function, since an administrative act is usually referred to as similar or related activities regarding the handling and processing of information.

Another important dimension to the Council's role as per A36-2 in divulging safety information is that ICAO has already entered into memoranda of understanding with the States audited—that audit reports will be confidential and made available to the State audited and relevant ICAO staff on a need-to-know basis. These agreements also require that, concurrently with the preparation of the report, a non-confidential audit activity report limited to the name of the audited State, the identity of airports visited during the audit, and the completion date of the audit will be developed for release to all Contracting States. Reports to the Council are required to be in a form that maintains the confidentiality of the audit report in relation to the State concerned. Accordingly, ICAO has restricted itself for purposes of confidentiality to giving only limited and non specific details of audits to its member States. This raises a legal issue as to ICAO's right to contravene its agreement with member States in deference to an Assembly Resolution. This issue also seemingly goes to the root of ICAO's empowerment by its member States and ICAO's accreditation to such States.

International organizations can generally only work on the basis of legal powers that are attributed to them. Presumably, these powers emanate from the sovereign States that form the membership of such organizations.[7] Therefore, the logical conclusion is that if international organizations were to act beyond the powers accorded to them, they would be presumed to act *ultra vires*.[8] It should be noted that ICAO does not only derive implied authority from its Contracting States based on universality but it also has attribution from States to exercise certain powers. The doctrine of attribution of powers comes directly from the will of the founders, and in ICAO's case, powers were attributed to ICAO when it was established as an international technical organization and a permanent civil aviation agency to administer the provisions of the Chicago Convention. In addition, ICAO could lay claims to what are now called "inherent powers" which give ICAO power to perform all acts that the Organization needs to perform to attain its aims not due to any specific source of organizational power but simply because ICAO inheres in organizationhood. Therefore, as long as acts are not prohibited in ICAO's constituent document (the Chicago Convention), they must be considered legally valid.[9]

A seminal judicial decision relating to the powers of international organizations was handed down by the Permanent Court of International Justice in 1922 in a

[7]See de Witte (1998) at pp. 277–304.

[8]Klabbers (2002) at p. 60.

[9]Seyersted (1963), at p. 28.

case[10] relating to the issue as to whether the International Labour Organization (set up to regulate international labour relations) was competent to regulate labour relations in the agricultural sector. The court proceeded on the basis that the competence of an international organization with regard to a particular function lay in the treaty provisions applicable to the functions of that organization and that the determination of such competence would be based on interpretation. However, the principle of implied extension should be carefully applied, along the fundamental principle enunciated by Judge Green Hackworth in the 1949 *Reparation for Injuries Case*[11]—that powers not expressed cannot freely be implied and that implied powers flow from a grant of express powers, and are limited to those that are necessary to the exercise of powers expressly granted.[12]

The universal solidarity of ICAO Contracting States that was recognized from the outset at the Chicago Conference brings to bear the need for States to be united in recognizing the effect of ICAO policy and decisions. This principle was given legal legitimacy in the *ERTA* decision[13] handed down by the Court of Justice of the European Community in 1971. The court held that the competence of the European Community to conclude an agreement on road transport could not be impugned since the member States had recognized Community solidarity and that the Treaty of Rome which governed the Community admitted of a common policy on road transport which the Community regulated.

It should be noted that ICAO does not only derive implied authority from its Contracting States based on universality but it also has attribution from States to exercise certain powers. The doctrine of attribution of powers comes directly from the will of the founders, and in ICAO's case, powers were attributed to ICAO when it was established as an international technical organization and a permanent civil aviation agency to administer the provisions of the Chicago Convention. In addition, ICAO could lay claims to what are now called "inherent powers" which give ICAO power to perform all acts that the Organization needs to perform to attain its aims not due to any specific source of organizational power but simply because ICAO inheres in organizationhood. Therefore, as long as acts are not prohibited in ICAO's constituent document (the Chicago Convention), they must be considered legally valid.[14]

States retain the powers to act unilaterally and they are not bound to comply with obligations flowing from the Organization's exercise of conferred powers. States which have delegated powers on ICAO have the legal right under public international law to take measures against a particular exercise by ICAO of conferred

---

[10]*Competence of the ILO to regulate the Conditions of Labour of Persons Employed in Agriculture*, Advisory Opinion [1922] Publ. PCIJ Series B, nos. 2&3.

[11]*Reparation for Injuries Suffered in the Service of the United Nations*, advisory opinion, [1949] ICJ Reports 174.

[12]*Id.* at p. 198.

[13]Case 22/70, *Commission* v. *Council* (European Road Transport Agreement) [1971] ECR 273.

[14]Seyersted (1963), at p. 28.

powers which is considered to be *detournement de pouvoir*, *ultra vires* or an internationally wrongful act with which the objecting States do not wish to be associated. A State could also distance itself from the State practice of other Contracting States within the Council if such activity is calculated to form customary international law that could in turn bind the objecting State if it does not persist in its objections.[15]

The above notwithstanding, a significant issue in the determination of ICAO's effectiveness as an international organization is the overriding principle of universality and global participation of all its 191 Contracting States in the implementation of ICAO policy. This principle, which has its genesis in the Chicago Conference of 1944, has flowed on gaining express recognition of legal scholars. This is what makes ICAO unique as a specialized agency of the United Nations and establishes without any doubt that ICAO is not just a tool of cooperation among States.

4 Adoption of Annexes

4.1 Annex 19: Safety Management

Article 54 (l) provides that the Council is required to adopt, in accordance with the provisions of Chapter VI of this Convention, international standards and recommended practices; for convenience, designate them as Annexes to this Convention; and notify all Contracting States of the action taken. As already discussed, the Council has already adopted 18 Annexes and, at the time of writing a new Annex—Annex 19 on Safety Management—was coming up before the Council for adoption.

Safety Management Systems (SMS) are processes which proactively manage the projected increase in aircraft incidents and accidents brought about by the increase in air traffic movements. SMS require vigilance in the liberalization of air transport and the correspondent increase in capacity. At the Directors General of Civil Aviation Conference on a Global Strategy for Aviation Safety, convened by the International Civil Aviation Organization in Montreal from 20 to 22 March 2006, Canada defined a Safety Management System as a business-like approach to safety. An SMS is a systematic, explicit and comprehensive process for the management of safety risks that integrates operations and technical systems with financial and human resource management, for all activities related to an air operator as an approved maintenance organization's certificate holder.[16]

Any management system, including that which involves aviation safety, would necessarily entail planning, goal setting, performance measurement and accountability. This process is systemic, in that there cannot be one factor without the other

[15] See Sarooshi (2005) at p. 110.

[16] Management of Aviation Safety, Presented by Canada, DGCA/06-WP/15, 4/2/06, at 2.

to complete the entire system of the safety management process. It also requires a symbiotic relationship between operator and regulator, who are jointly responsible for the determination of the parameters of regulations and their intent. While the regulator's role is to give a clear set of instructions to the operator reflecting the expectations of the regulator, the operator has to ensure their compliance as well as maintain close coordination with the regulator.

References

de Witte B (1998) Sovereignty and European integration: the weight of tradition. In: Slaughter A-M et al (eds) The European court and national courts: doctrine and jurisprudence. Oxford University Press, Oxford

Klabbers J (2002) An introduction to international institutional law. Cambridge University Press, Cambridge

Sarooshi D (2005) International organizations and their exercise of sovereign powers. Oxford University Press, Oxford

Schenkman J (1955) International civil aviation organization. Librairie E. Droz, Geneve

Seyersted F (1963) Objective international personality of intergovernmental organizations: do their capacities really depend upon the conventions establishing them? Krohns Bogtnykker, Copenhagen

Article 55
Permissive Functions of the Council

The Council may:

(a) Where appropriate and as experience may show to be desirable, create subordinate air transport commissions on a regional or other basis and define groups of states or airlines with or through which it may deal to facilitate the carrying out of the aims of this Convention;
(b) Delegate to the Air Navigation Commission duties additional to those set forth in the Convention and revoke or modify such delegations of authority at any time;
(c) Conduct research into all aspects of air transport and air navigation which are of international importance, communicate the results of its research to the contracting States, and facilitate the exchange of information between contracting States on air transport and air navigation matters;
(d) Study any matters affecting the organization and operation of international air transport, including the international ownership and operation of international air services on trunk routes, and submit to the Assembly plans in relation thereto;
(e) Investigate, at the request of any contracting State, any situation which may appear to present avoidable obstacles to the development of international air navigation; and, after such investigation, issue such reports as may appear to it desirable.

Of some contention is Article 55 (c) which requires the council to

Conduct research into all aspects of air transport and air navigation which are of international importance, communicate the results of its research to the Contracting States, and facilitate the exchange of information between Contracting States on air transport and air navigation matters.

This could be tied to the objective of ICAO to meet the needs of the people of the world for safe, regular, efficient and economical air transport. In this regard the Council should initiate studies that involve research into all aspects of air transport which are of international importance. This is provided for in the Chicago Convention. Such studies, taking into account global, regional and national economic trends, could analyse their effects on the demand for air transport and how such demands could be met. This could result in a compendium of planning for States, aircraft and component manufacturers, environmentalists and service providers.

One of the forgotten issues in air transport—the right of the consumer (particularly the passenger) should be a key issue for study. The October 2011 crew strike of QANTAS as well as the August/September 2012 strike by Lufthansa crew left thousands of passengers stranded at airports. Although these grave consequences to passengers were the result of industrial actions, States must take responsibility for

DOI 10.1007/978-3-319-00068-8_56, © Springer International Publishing Switzerland 2014

providing regular, unbroken air transport that is not arbitrarily hindered by unmet demands of the service provider. This is also emphasized in the Chicago Convention. There should be core guidance to States with a view to ensuring regularity of service. This would meet the objective that air transport meets the needs of the people for air travel.

The Council should discuss justifications for and against free trade in aviation when considered against national concerns on aviation safety and security; aircraft manufacturing and commercial aspects concerned therewith including government subsidies and the manner in which civil aviation would affect trade surpluses and deficits. Another aspect worthy of in depth study is the effect on the global and national economies of open skies.

Another critical area is reflected in Article 55 (d) which calls on the Council to:

> *study any matters affecting the organization and operation of international air transport, including the international ownership and operation of international air services on trunk routes, and submit to the Assembly plans in relation thereto.*

Studies on the operation of air transport, ownership and control issues and more importantly the significance of air services operations on the trunk routes are critical to the development of air transport. *A fortiori,* the need for these are even more compelling now, with the new strategic objective of the Council dedicated to air transport economics. In this regard, the Council should produce a comprehensive annual or triennial report on global trends and their effect on air transport, with guidance to key stakeholders in the provision of air services that meet the needs of the consumer while ensuring sustainable development of air transport. This will defragment the current process whereby each key player has his own forecast—often based on a "predict and provide" model rather than a strategic management model.

An early initiative in this regard was the adoption by the Assembly, at its sixteenth session of Resolution A16-32 (Registration of Agreements and Arrangements) whereby the Assembly resolved that in accordance with Articles 54, 55 (c) and (d) of the Chicago Convention the Council request Contracting States to submit to ICAO preferably on an ad hoc basis, such agreements and arrangements as may be necessary to the Organization for the furtherance of specific studies undertaken in accordance with the Convention. The Resolution also encouraged States to provide as much information as possible to ICAO in order to facilitate and add value to such studies.

Article 56
Nomination and Appointment of Commission

The Air Navigation Commission shall be composed of nineteen members appointed by the Council from among persons nominated by contracting States. These persons shall have suitable qualifications and experience in the science and practice of aeronautics. The Council shall request all contracting States to submit nominations. The President of the Air Navigation Commission shall be appointed by the Council.

R. Abeyratne, *Convention on International Civil Aviation*,

DOI 10.1007/978-3-319-00068-8_57,

Article 57
Duties of Commission

The Air Navigation Commission shall:

(a) Consider, and recommend to the Council for adoption, modifications of the Annexes to this Convention;
(b) Establish technical subcommissions on which any contracting State may be represented, if it so desires;
(c) Advise the Council concerning the collection and communication to the contracting States of all information which it considers necessary and useful for the advancement of air navigation.

R. Abeyratne, *Convention on International Civil Aviation*,
DOI 10.1007/978-3-319-00068-8_58,

Article 58
Appointment of Personnel

Subject to any rules laid down by the Assembly and to the provisions of this Convention, the Council shall determine the method of appointment and of termination of appointment, the training, and the salaries, allowances, and conditions of service of the Secretary General and other personnel of the Organization, and may employ or make use of the service of nationals of any contracting State.

R. Abeyratne, *Convention on International Civil Aviation*,
DOI 10.1007/978-3-319-00068-8_59,

Article 59
International Character of Personnel

The President of the Council, the Secretary General, and other personnel shall not seek or receive instructions in regard to the discharge of their responsibilities from any authority external to the Organization. Each contracting State undertakes fully to respect the international character of the responsibilities of the personnel and not to seek to influence any of its nationals in the discharge of their responsibilities.

R. Abeyratne, *Convention on International Civil Aviation*,
DOI 10.1007/978-3-319-00068-8_60,

Article 60

Immunities and Privileges of Personnel

Each contracting State undertakes, so far as possible under its constitutional procedure, to accord to the President of the Council, the Secretary General, and the other personnel of the Organization, the immunities and privileges which are accorded to corresponding personnel of other public international organizations. If a general international agreement on the immunities and privileges of international civil servants is arrived at, the immunities and privileges accorded to the President, the Secretary General, and the other personnel of the Organization shall be the immunities and privileges accorded under that general international agreement.

R. Abeyratne, *Convention on International Civil Aviation*,
DOI 10.1007/978-3-319-00068-8_61,

Article 61
Budget and Apportionment of Expenses

The Council shall submit to the Assembly annual budgets, annual statements of accounts and estimates of all receipts and expenditures. The Assembly shall vote the budgets with whatever modification it sees fit to prescribe, and with the exception of assessments under Chapter XV to States consenting thereto, shall apportion the expenses of the Organization among the contracting States on the basis which it shall from time to time determine.

R. Abeyratne, *Convention on International Civil Aviation*,
DOI 10.1007/978-3-319-00068-8_62,

Article 62
Suspension of Voting Power

The Assembly may suspend the voting power in the Assembly and in the Council of any contracting State that fails to discharge within a reasonable period its financial obligations to the Organization.

R. Abeyratne, *Convention on International Civil Aviation*,
DOI 10.1007/978-3-319-00068-8_63,

Article 63
Expenses of Delegations and Other Representatives

Each contracting State shall bear the expenses of its own delegation to the Assembly and the remuneration, travel, and other expenses of any person whom it appoints to serve on the Council, and of its nominees or representatives on any subsidiary committees or commissions of the Organization.

R. Abeyratne, *Convention on International Civil Aviation*,
DOI 10.1007/978-3-319-00068-8_64,

Article 64
Security Arrangements

The Organization may, with respect to air matters within its competence directly affecting world security, by vote of the Assembly enter into appropriate arrangements with any general organization set up by the nations of the world to preserve peace.

Contents

1 ICAO and Its UN Mission

At its first Session (Montreal, 6–27 May 1947), the Assembly recognized that the Interim Council of the Provisional International Civil Aviation Organization (PICAO) had negotiated a draft agreement of relationship between ICAO and the United Nations, and adopted Resolution A1-2 (Approval of Agreement with the United Nations) whereby it was resolved that the Council be authorized to enter into such supplementary arrangements with the Secretary General of the United Nations for the implementation of the agreement, in accordance with Article XIX thereof, as may be found desirable in the light of operating experiences of the two Organizations.

The Resolution also authorized the Council of ICAO to enter into negotiations with the United Nations for the conclusion of further appropriate arrangements between the United Nations and ICAO with respect to air matters within the competence of ICAO. To this end, the President of the ICAO Council was authorized to sign with the appropriate official of the United Nations a protocol bringing the agreement of relationship between the United Nations and ICAO. This agreement came into force on 13 May 1947.

The Assembly, at its 5th Session (Montreal, 5–18 June 1951) adopted Resolution A5-5 (Coordination of Activities Between the United Nations and ICAO Relating to Emergency Action to Assist in the Maintenance of International Peace and Security) which declared that ICAO agreed to cooperate with and render all possible assistance to the principal organs of the United Nations with respect to matters within the competence of the Organization directly affecting international peace and security, as contemplated in the Chicago Convention , account being taken of the special position of the members of ICAO who are not members of the United Nations.

R. Abeyratne, *Convention on International Civil Aviation*,
DOI 10.1007/978-3-319-00068-8_65, © Springer International Publishing Switzerland 2014

United Nations Security Council Resolution 1269,[1] adopted by the Security Council on 19 October 1999 reflects the concern of the world community with regard to the increase of international terrorism which endangers the lives and well-being of individuals worldwide as well as the peace and security of all States. The Resolution goes on to condemn all acts, methods and practices of terrorism as criminal and unjustifiable, regardless of their motivation, and in all their forms and manifestations, wherever and by whomever committed, in particular those which could threaten international peace and security. With this declaration, the United Nations Security Council has widened the scope for combating terrorism, particularly to encompass such instances as the 11 September events, which could expand to economic paralysis of global commercial activity through attacks aimed at the aviation industry.

The principle of State Responsibility with regard to world peace and security lies primarily in Article 24 of the United Nations Charter[2] which calls upon all members to refrain in their international relations from the threat or use of force against the territorial integrity or political independence of any state, or in any other manner inconsistent with the purposes of the United Nations. Furthermore, Article 51 of the Charter preserves the right of individual or collective self-defence if an armed attack occurs against a member of the United Nations, notwithstanding any right granted by the Charter that would preclude any member State from interfering in the affairs of another member State, particularly with regard to matters of State sovereignty.

Although the United Nations Charter involves action between States, it would not be incorrect to assume that international terrorism, purportedly committed by private individuals, could nonetheless be brought within the preview of the above-mentioned provisions of the United Nations Charter, particularly from the neo post modernist approach of collective involvement. This assumption is embodied in recent work of the International Law Commission, through Article 2 of the Draft Articles on Responsibility of States for Internationally Wrongful Acts[3] which provides that international responsibility of a State, which is referred to in Article 1, is attributable to that State if conduct of the State constitutes a breach of an international obligation of that State. The document also provides that the wrongfulness of an act of a State is precluded if the act constitutes a lawful measure of self defence taken in conformity with the Charter of the United Nations.[4] The State responsible for an internationally wrongful act is under an obligation to compensate for damage caused, including reparation for financially assessable damage including loss of profits.[5]

---

[1]S/RES/1269 (1999), 19 October 1999.

[2]Charter of the United Nations and Statute of the International Court of Justice, United Nations, New York.

[3]Draft Articles on Responsibility of States for Internationally Wrongful Acts, Adopted by the International Law Commission (53rd Session, 2001).

[4]*Id.* Article 21.

[5]*Id.* Article 36.

In addition to State responsibility for conduct attributable to that State, the International Law Commission has established that a crime against the peace and security of mankind entails individual responsibility, and in a crime of aggression.[6] A further link drawing civil aviation to the realm of international peace and security lies in the Rome Statute of the International Criminal court, which defines a war crime, *inter alia*, as intentionally directing attacks against civilian objects; attacking or bombarding, by whatever means, towns, villages, dwellings or buildings which are undefended and which are not military objects; employing weapons, projectiles, and material and methods of warfare that cause injury.[7] The Statute also defines as a war crime, any act which is intentionally directed at buildings, material, medical units and transport, and personnel using the distinctive emblems of the Geneva Conventions in conformity with international law.[8]

The vulnerability of civil aviation against acts of terrorism may be further affected in instances where potential offenders are domiciled in certain States which may avail themselves of acts of self defence by States when threatened. An explicit political interpretation of the role of the United Nations Charter in this regard was given in 1986 by the then United States Secretary of State.

George Shultz said:

> The Charter's restrictions on the use or threat of force in international relations include a specific exception on the right of self-defence. It is absurd to argue that international law prohibits us from capturing terrorists in international waters or airspace....[9]

The inherent nature of civil aviation, in its dependence on national and international regulation and control, calls for ineluctable State involvement and responsibility in ensuring world peace and security in relation to matters pertaining to civil aviation. In other words, if States were to take action in the field of civil aviation through which self defence can be achieved, any threat to civil aviation must be linked to a State to which can be attributed or imputed some relationship with actual or possible offenders. In the Nicaragua Case,[10] the International Court of Justice in 1986, accepted the premise that self defence could involve responses to counter groups of persons who have either been sent by a State or are acting on behalf of that State with its explicit or tacit acquiescence. Therefore, an "armed attack" could be definitively sustained within the text of the United Nations Charter only when the link between the State and non State factor is sufficiently close, and the attack could tantamount to an attack by that State.

---

[6]Draft Code of Crimes Against the Peace and Security of Mankind, International Law Commission Report, 1996, Chapter II Article 2.

[7]Rome Statute of the International Criminal Court, Article 8.2 (b) (ii), (v) and (xx).

[8]*Id.* Article 8.2 (b) (XXIV).

[9]Shultz, Low-Intensity Warfare: The Challenge of Ambiguity, Address to the National Defence University, Washington D.C., 15 Jan 1986 reproduced in (1986) 25 International Legal Materials 204 at 206.

[10](1986) ICJ Reports 14.

It must be mentioned at the outset that any agreement or arrangement between ICAO and the United Nations under this provision has to relate to two exclusive aspects: civil aviation; and peace. Article 89 of the Chicago Convention enables Contracting States to have freedom of action irrespective of the provisions of the Convention in case of war, whether belligerents or neutrals. It also allows a State which has declared a state of national emergency (and notifies the ICAO Council of such) to have the same freedom of action notwithstanding the provisions of the Convention. Therefore, unless a State is at war (which the Convention does not define) or has declared a state of national emergency, it would be bound by the provisions of the Convention. War is conventionally defined as a behavior pattern of organized violent conflict typified by extreme aggression, societal disruption, and high mortality. This behavior pattern involves two or more organized groups.

ICAO has also to be mindful of the fact that The United Nations Charter lists the achievement of international cooperation in solving international problems of an economic, social, cultural or humanitarian character, as one of the purposes of the United Nations. The problems that the United Nations is mandated by its Charter to solve should therefore be necessarily of an international nature. Article 2(7) of the Charter expands the scope of this philosophy further when it provides that the United Nations is not authorized to intervene in matters which are essentially within the domestic jurisdiction of any State, without prejudice to the right of the United Nations to intervene in matters which are within the domestic jurisdiction of any State, and apply enforcement measures where there is an occurrence of acts of aggression, a threat to the peace or breach thereof. Therefore *stricto sensu*, the United Nations cannot intervene in instances where natural disasters such as famine, drought or earthquakes render the citizens of a State homeless, destitute and dying of starvation unless invited by the States concerned. The principle however cannot be too strictly interpreted, as natural disasters may usually lead to breaches of the peace. In such instances the United Nations Security Council may take such actions by air, sea or land as may be necessary to maintain or restore international peace and security.

Peace is achieved through global friendship and understanding, which in turn is achieved through global connectivity. Aviation connects cities more than any other mode of transport, but there has to be peace to achieve this objective. There are instances where attacks against aviation have caused global friction. One such instance was when in 1991, PAN AM flight 103 over Lockerbie was destroyed in mid air. States retaliated against Libya, whose nationals were considered to have caused the terrorist act. This collective retaliation came through the United Nations in Security Council Resolution 731 of 1992. Which condemned the destruction of Pan Am flight 103 and UTA flight 772 and the resultant loss of hundreds of lives; and strongly deplored the fact that the Libyan Government had not responded effectively to the requests to cooperate fully in establishing responsibility for the terrorist acts referred to above against Pan Am flight 103 and UTA flight 772; The Resolution urged the Libyan Government immediately to provide a full and effective response to those requests so as to contribute to the elimination of international terrorism and requested the Secretary-General to seek the cooperation of the Libyan

Government to provide a full and effective response to those requests. It also urged all States individually and collectively to encourage the Libyan Government to respond fully and effectively to those requests.

Earlier, The United Nations had adopted Resolution 286 on 9 September 1970 which expressed grave concern at the threat to innocent civilian lives from the hijacking of aircraft and any other interference in international travel, and appealed to all parties concerned for the immediate release of all passengers and crews without exception, held as a result of hijackings and other interference in international travel. It also called on States to take all possible legal steps to prevent further hijackings or any other interference with international civil air travel.

2 The Lockerbie Case

On 3 March 1992, the Registry of the ICJ received an application by the Socialist Peoples' Libyan Arab Jamahiriya (hereafter referred to as Libya), instituting proceedings against the United States of America. The application referred to a dispute between Libya and the United States, which arose as a result of the destruction of Pan Am flight 103 on December 1988 over Lockerbie, Scotland. On 14 November 1991, a Grand Jury of the United States District Court for the District of Columbia, indicted two Libyan nationals, charging them *inter alia* that they had caused a bomb to be placed on board Pan Am flight 103 which bomb had exploded causing the aeroplane to crash. Consequently, the British and American Governments declared that Libya must:

> surrender for trial all three charged with the crime; and accept responsibility for the actions of Libyan officials; disclose all it knows of this crime, including the names of all those responsible, and allow access to all witnesses, documents and other material evidence, including all the remaining timers; pay appropriate compensation.[11]

It is interesting to note that the United Kingdom and United State Governments had demanded the payment of compensation from Libya, even before the two accused Libyan nationals had been tried or their guilt otherwise determined.

The United Nations Security Council considered this declaration and on 21 January 1992 adopted Resolution 731 (1992), strongly deploring the fact that the Libyan Government did not effectively respond to the requests contained in the declaration of the British and the American Governments, and urging Libya to provide a full and effective response to those requests so as to contribute to the elimination of international terrorism.[12]

In the course of the oral proceedings before the ICJ, reference had been made by both the United Kingdom and the United States to the possibility of sanctions being

[11]Questions of Interpretation and Application of the 1971 Montreal Convention Arising From the Aerial Incident at Lockerbie (Libyan Arab Jamahiriya v. The United States of America) Provisional Measures, Order of 14 April 1992, I.C.J. Reports, 1992, 114 at 122.

[12]Id. 123–124.

imminently imposed by the Security Council on Libya in order to require it to extradite the accused to the United States or the United Kingdom. Libya's application before the ICJ was therefore to invoke the jurisdiction of the Court for provisional measures that would *inter alia* preclude the United States and the United Kingdom from taking any initiative within the Security Council for the purpose of impairing Libya's right to exercise its own jurisdiction over the accused.

The Security Council adopted a further Resolution—Resolution 748 of 1992[13]—expressing its deep concern that the Libyan Government had not provided a full and effective response to the requests of its earlier Resolution (731 of 1992) and deciding that the Libyan Government must commit itself to ceasing all forms of terrorist action and by concrete actions, demonstrate its renunciation of terrorism. The Security Council, acting under authority of Chapter VII of the United Nations Charter, which, under Article 39 thereof grants the Security Council powers to determine the existence of any threat to international peace, also decided by Resolution 748 that all States adopt appropriate measures against Libya. The Security Council Resolution also called upon all States, including States that were not members of the United Nations, and all international organizations, to act strictly with the provisions of the Resolution, notwithstanding the existence of any rights or obligations conferred or imposed by any international agreement or any contract entered into or any licence or permit granted prior to 15 April 1992.

The Agent of the United States, by letter of 2 April 1992, drew the ICJ's attention to the adoption of Security Council Resolution 748 of 1992. In his letter, the United States Agent stated:

> That resolution, adopted pursuant to Chapter V11 of the United Nations Charter, decides that the Libyan Government must now comply without any further delay with paragraph 3 of Resolution 731 (1992) of 21 January 1992 regarding the requests contained in documents s/23306, S/23308 and S/23309. It will be recalled that the reference requests include the request that Libya surrenders the two Libyan suspects in the bombing of Pan Am flight 103 to the United States or to the United Kingdom. For this additional reason, the United States maintains its submission of 28 March 1992 that the request of the Government of the Great Socialist Peoples' Libyan Arab Jamahiriya for the indication of provisional measures of protection should be denied, and that no such measures should be indicated.[14]

Libya in reply claimed that the risk of contradiction between Resolution 748 of 1992 and the provisional measures requested of the Court by Libya did not render the Libyan request inadmissible, since there is in law no competition or hierarchy between the Court and the Security Council, both being equal organs of the United Nations, exercising their own competence.[15] Libya also recorded that it regarded the decision of the Security Council as contrary to international law and one which had been taken to avoid applicable law in the guise of the Security Council's powers under Chapter VII of the United Nations Charter. Libya claimed that the foundation

[13]Ibid.

[14]*Id.* 125.

[15]*Id.* 126.

of the Court's jurisdiction lay in the *Convention for the Suppression of Unlawful Acts Against the Safety of Civil Aviation*, signed in Montreal on 23 September 1971[16] (hereafter referred to as the Montreal Convention).

Article 14(1) of the Montreal Convention stipulates:

> Any dispute between two or more Contracting States concerning the interpretation or application of this Convention which cannot be settled through negotiation, shall, at the request of one of them, be submitted to arbitration. If within six months from the date of the request for arbitration the parties are unable to agree on the organization of the arbitration, any one of those Parties may refer the dispute to the International Court of Justice by request in conformity with the Statute of the Court.

There is no doubt that this provision has been drafted in imperative terms, and any dispute that required the interpretation or application of the provisions of the Convention could be submitted to the Court for arbitration. Therefore, the contentious issue of the extradition of the Libyan nations was a matter clearly within the jurisdiction of the Court since Article 7 of the Montreal Convention provides:

> The Contracting State in the territory of which the alleged offender is found shall, if it does not extradite him, be obliged, without exception whatsoever and whether or not the offence was committed in its territory, to submit the case to its competent authorities for the purpose of prosecution. . .

It was Libya's claim that according to this provision, it had the right to try its own nationals and was not obligated to extradite them to either the United States or the United Kingdom as was required by the Security Council Resolution. It was also Libya's contention that The ICJ had explicit jurisdiction by virtue of Article 14(1) of the Montreal Convention to interpret Article 7.

The United States claimed that irrespective of the Montreal Convention, Libya was bound as a member of the United Nations to abide by the United Nations Charter which in Chapter VII (Article 39) granted absolute power to the Security Council to decide on the measures to be taken to restore and maintain international peace and security. The United States also invoked Article 25 of the United States Charter which provides that:

> The members of the United Nations agree to accept and carry out the decisions of the Security Council in accordance with the present Charter.

The Court, in its decision agreed with the contention of the United States that both Libya and the United States, as members of the United Nations, are obliged to accept and carry out the decisions of the Security Council in accordance with Article 25 of the United Nations Charter. This obligation, according to the Court, extended to Resolution 748 of 1992. The Court also cited Article 103 of the United Nations Charter which stipulates:

[16] *Ibid.*

> In the event of a conflict between the obligations of the members of the United Nations under the present Charter and their obligations under any other international agreement, their obligations under the present Charter shall prevail.

The Court held therefore that the provisions of the Montreal Convention cited by Libya were subservient to the Security Council Resolution and accordingly denied the application of Libya.

One of the areas in which ICAO could link itself to the meaning and purpose of the United Nations is in the operation of relief flights. On July 27, 2011, the flight bringing first aid to famine stricken Somalia landed in Mogadishu. It carried 10 tonnes of plumpy nuts—enough to offer 3,500 children suffering from starvation a respite from death. According to the *Economist*, famine is declared when 30 % of the children are actually malnourished, 20 % of the population is without food and deaths are running at 10,000 adults or four per 10,000 children every day. This is the peace side of aviation.

In 1969 the XXIst International Conference of the Red Cross adopted a Resolution whereby States were requested to exercise their sovereign and legal rights so as to facilitate the transit, admission and distribution of relief supplies provided by impartial international humanitarian organizations for the benefit of civilian populations in disaster areas when disaster situations imperil the life and welfare of such populations. The United Nations subsequently announced that this Resolution would also apply to situations arising from armed conflict.

The above discussion is not intended to obfuscate the fact that the provisions of the Geneva Conventions and the United Nations Charter do not admit of the operation of relief flights at the will and pleasure of the benefactor, without the permission of the recipient State. It is also not intended to circumvent the fact that States are primarily responsible for organizing relief. Relief societies such as the Red Cross and the Red Crescent Organizations are merely called upon to play a supplementary role by assisting the authorities of the States concerned in their task.

Since it is clear that the intervention of the United Nations Security Council in a matter lying within the domestic jurisdiction of a State can only be justified in instances where there is a threat to international peace and security, a breach of the peace within a State or an act of aggression, a question which arises when a relief flight is operated as a part of a humanitarian project is whether the operation of such a flight could be considered a legitimate unilateral action by States. The Question would essentially be ground in a legal analysis of the principles of humanitarian law and State sovereignty.[17] On the one hand, everyone has the right to life, liberty and security of person and the right to a standard of living adequate for the health and well being of himself and of his family, including food, clothing, housing and medical care.

On the other, there is overall recognition of the fact that every State has complete and exclusive sovereignty over the airspace above its territory. Except for the Paris Convention of 1956, which provides for civil aircraft registered in a member State

---

[17]See Ruwantissa Abeyratne, Legal and Aeronautical Issues Concerning the Earthquake in Haiti, Air & Space Law Vol 35(2), 183–193.

of the European Civil Aviation Conference (ECAC) to fly freely into member States for the purposes of discharging or taking on traffic where such aircraft are engaged *inter alia* in non-scheduled flights for the purpose of meeting humanitarian or emergency needs, there is no multilateral or bilateral agreement that admits of unilateral intervention of a State in another for humanitarian purposes, where the intervening State does not obtain permission of the recipient State. In fact, Resolution 46/182 explicitly provides in the Annex to the Resolution that the sovereignty, territorial integrity and national unity of States must be fully respected in accordance with the Charter of the United Nations and that in this context, humanitarian assistance should be provided with the consent of the affected country and in principle on the basis of an appeal by the affected country. These conflicting principles, although not bestowing legal authority on the United Nations to intervene in a State with relief flights, at least give some degree of justification to the United Nations' efforts to mediate with States concerned in the promotion of relief operations and to seek the support of other States, with the concurrence of affected States.

If humans are dying, one has got to help at all costs.

As was seen in the severe humanitarian crisis that occurred as a result of the Earthquake in early January 2010, and more recently in mid 2011 resulting from the famine in the Horn of Africa, in the first stage of a disaster, airlift of persons and relief material is the first line of response. Humanitarian assistance has to be provided with speed, flexibility and mobility. The World Food Programme (WFP) of the United Nations pioneers such efforts but, as was seen in the aftermath of the Haitian disaster, the proliferation of aircraft carrying out relief flights causes bottlenecks which require management and planning of airspace. United Nations aviation standards depend on Standards adopted by States under the aegis of the International Civil Aviation Organization, the Federal Aviation Administration (FAA) of the United States and the European Joint Aviation Agency (JAA). Base on these criteria, WFP carried out several airlifts to Cote d'Ivoire and Liberia in April 2011 as well as to Somalia in July 2011.

Annex 9 to the Convention on International Civil Aviation, which is on facilitation of air transport provides that Contracting States shall facilitate the entry into, departure from and transit through their territories of aircraft engaged in relief flights performed by or on behalf of inter-national organizations recognized by the UN or by or on behalf of States themselves and shall take all possible measures to ensure their safe operation. Such relief flights are those undertaken in response to natural and man-made disasters which seriously endanger human health or the environment, as well as similar emergency situations where UN assistance is required. Such flights shall be commenced as quickly as possible after obtaining agreement with the recipient State.

According to its Internationally Agreed Glossary of Basic Terms, the United Nations Department of Humanitarian Affairs considers an emergency to be "a sudden and usually unforeseen event that calls for immediate measures to minimize its adverse consequences", and a disaster to be

> a serious disruption of the functioning of society, causing widespread human, material or environmental losses which exceed the ability of the affected society to cope using only its own resources.

The Annex also stipulates that ICAO member States shall ensure that personnel and articles arriving on relief flights are cleared without delay. Therefore, The operation of relief flights, either by States or such bodies as the United Nations, to alleviate human suffering in times of war, natural or manmade catastrophe, is yet another area in which the role of civil aviation is brought to bear in securing peace and security. There is a specific provision in Annex 9 to the Chicago Convention for provision by State of relief flights. Contracting States are required, by Standard 8.8 of Chapter 8 of the Annex, to facilitate the entry into, departure from and transit through their territories of aircraft engaged in relief flights performed by or on behalf of international organizations recognized by the United Nations or by or on behalf of States themselves and to take all possible measures to ensure their safe operation. The relief flights referred to should be undertaken to respond to natural and man-made disasters which seriously endanger human health or the environment. An emergency is acknowledged in the Annex as "a sudden and usually unforeseen event that calls for immediate measures to minimize its adverse consequences". A disaster is described in the Annex as

> a serious disruption of the functioning of society, causing wide spread human, material or environmental losses which exceed the ability of the affected society to cope using its own resources.

Article 65

Arrangements with Other International Bodies

The Council, on behalf of the Organization, may enter into agreements with other international bodies for the maintenance of common service and for common arrangements concerning personnel and, with the approval of the Assembly, may enter into such other arrangements as may facilitate the work of the Organization.

R. Abeyratne, *Convention on International Civil Aviation*,
DOI 10.1007/978-3-319-00068-8_66,

Article 66
Functions Relating to Other Agreements

(a) **The Organization shall also carry out the functions placed upon it by the International Air Services Transit Agreement and by the International Air Transport Agreement drawn up at Chicago on December 7, 1944, in accordance with the terms and conditions therein set forth.**

(b) **Members of the Assembly and the Council who have not accepted the International Air Services Transit Agreement or the International Air Transport Agreement drawn up at Chicago on December 7, 1944 shall not have the right to vote on any questions referred to the Assembly or Council under the provisions of the relevant Agreement.**

R. Abeyratne, *Convention on International Civil Aviation*,
DOI 10.1007/978-3-319-00068-8_67, © Springer International Publishing Switzerland 2014

Part III
International Air Transport

Chapter XIV. Information and Reports

Article 67. File Reports with Council

Chapter XV. Airports and Other Air Navigation Facilities

Article 68. Designation of Routes and Airports
Article 69. Improvement of Air Navigation Facilities
Article 70. Financing of Air Navigation Facilities
Article 71. Provision and Maintenance of Facilities by Council
Article 72. Acquisition or Use of Land
Article 73. Expenditure and Assessment of Funds
Article 74. Technical Assistance and Utilization of Revenues
Article 75. Taking Over of Facilities from Council
Article 76. Return of Funds

Chapter XVI. Joint Operating Organizations and Pooled Services

Article 77. Joint Operating Organizations Permitted
Article 78. Function of Council
Article 79. Participation in Operating Organizations

Article 67
File Reports with Council

Each contracting State undertakes that its international airlines shall, in accordance with requirements laid down by the Council, file with the Council traffic reports, cost statistics and financial statements showing among other things all receipts and the sources thereof.

R. Abeyratne, *Convention on International Civil Aviation*,
DOI 10.1007/978-3-319-00068-8_68,

Article 68
Designation of Routes and Airports

Each contracting State may, subject to the provisions of this Convention, designate the route to be followed within its territory by any international air service and the airports which any such service may use.

R. Abeyratne, *Convention on International Civil Aviation*,
DOI 10.1007/978-3-319-00068-8_69, © Springer International Publishing Switzerland 2014

Article 69
Improvement of Air Navigation Facilities

If the Council is of the opinion that the airports or other air navigation facilities, including radio and meteorological services, of a contracting State are not reasonably adequate for the safe, regular, efficient, and economical operation of international air services, present or contemplated, the Council shall consult with the State directly concerned, and other States affected, with a view to finding means by which the situation may be remedied, and may make recommendations for that purpose. No contracting State shall be guilty of an infraction of this Convention if it fails to carry out these recommendations.

R. Abeyratne, *Convention on International Civil Aviation*,

DOI 10.1007/978-3-319-00068-8_70,

Article 70

Financing of Air Navigation Facilities

A contracting State, in the circumstances arising under the provisions of Article 69, may conclude an arrangement with the Council for giving effect to such recommendations. The State may elect to bear all of the costs involved in any such arrangement. If the State does not so elect, the Council may agree, at the request of the State, to provide for all or a portion of the costs.

R. Abeyratne, *Convention on International Civil Aviation*,

DOI 10.1007/978-3-319-00068-8_71,

Article 71

Provision and Maintenance of Facilities by Council

If a contracting State so requests, the Council may agree to provide, man, maintain, and administer any or all of the airports and other air navigation facilities including radio and meteorological services, required in its territory for the safe, regular, efficient and economical operation of the international air services of the other contracting States, and may specify just and reasonable charges for the use of the facilities provided.

Contents

1 ICAO Assistance to Air Navigation

As stated earlier, responsibility of States for the provision of air navigation services in their territories is founded in principles contained in Article 28 of the Chicago Convention of 1944.[1] It must be noted that this is not an absolute obligation as the State is called upon to provide such services only in so far as it finds practicable to do so. In order to cover an eventuality of a State not being able to provide adequate air navigation services, the Convention imposes an overall obligation on the Council of ICAO in Article 69 to the effect that the Council shall consult with a State which is not in a position to provide reasonably adequate air navigation services for the safe, regular, efficient and economical operations of aircraft. Such consultations will be with a view to finding means by which the situation may be remedied. Article 70 of the Chicago Convention even allows for a State to conclude an arrangement with the Council regarding the financing of air navigation facilities and the Council is given the option in Article 71 of agreeing to provide, man, maintain and administer such services at the request of a State.

Joint financing of air navigation services under the Chicago Convention is based on the fundamental concept enunciated in the Convention—that the future development of civil aviation can greatly help to create understanding and friendship among nations. The operation of joint financing in civil aviation becomes necessary to coordinate non-stop transatlantic flights north of the 45th Parallel, which are under the jurisdiction of the flight information region of Iceland. Air navigational facilities provided by Iceland are indispensable for aircraft flying within this region.

---

[1] *Id.* Article 28 (a).

R. Abeyratne, *Convention on International Civil Aviation*,
DOI 10.1007/978-3-319-00068-8_72,

Because major storm tracks converge very near Iceland and that area of the Atlantic is favorable to the redevelopment of certain types of storms and to the formation of secondary depressions, it is common during unfavorable meteorological conditions that air traffic over the North Atlantic becomes congested, requiring full utilization of the air traffic control and flight information services offered by Iceland.

At its 14th Session (Rome, 21 August–15 September 1962) the ICAO Assembly made further provision for increased participation by "user States" in joint financing agreements than were contemplated among the Council and the States involved. At its 16th Session (Buenos Aires, 3–26 September 1968) the ICAO Assembly issued guidelines for the implementation of joint financing agreements. Since its inception, ICAO has held several conferences on joint financing: in Geneva, June 8–16 1948 on the subject of Iceland; in London, April 20–May 12, 1949 on Greenland and the Faroe Islands; and in Geneva, September 6–24, 1956 for the revision of the Danish and Icelandic Arrangements.

At the North Atlantic Route Service Conference, held in Dublin in March 1946, a recommendation was made *inter alia* that Iceland should provide an area control center in Reykjavik as well as certain telecommunications and meteorological services for the North Atlantic Region. This recommendation was approved subsequently by the PICAO Council on April 17 1946 and May 9 1946. During the Conference, the Delegation of Iceland made a statement to the effect that Iceland would not be able to provide the services recommended by the conference owing to the magnitude of aircraft crossings that required services. On May 16 1947, as a follow up to its Statement, Iceland submitted to ICAO a request for financial and technical aid in regard to the air traffic control, meteorological and telecommunications services in Iceland in accordance with Articles 69, 70 and 71 of the Chicago Convention. The ICAO Council, on June 25 1947, concluded that the request of Iceland constituted prima facie grounds for aid to be rendered in the manner sought.[2]

The Assembly, at its First Session (Montreal, 6–27 May 1947) adopted Resolution A1-65 (Joint Support Policy) which resolved that financial and technical aid through ICAO for furthering the provision of air navigation facilities and services adequate for the safe, regular, efficient and economical operation of international air services will be rendered, under the terms of Chapter XV of the Chicago Convention, in accordance with the basic principles and general policy laid down in Annex 1[3] to Resolution A1-65. The ensuing Joint Financing Agreement of 1948 has since been replaced by the Agreements of 1956[4] which was amended by the Montreal Protocol of 1982.[5] This agreement requires Iceland and Denmark to

[2]Report on the Conference on Air Navigation Services in Iceland, ICAO Doc 7000-JS/550 (June 8–26, 1948) at 8.

[3]Annex 1 to the Resolution set forth in some detail the general policy of ICAO relating to the joint support of air navigation services.

[4]Agreement on the Joint Financing of Certain Air Navigation Services in Iceland, Doc 9586-JS/682 (1956).

[5]Agreement on the Joint Financing of Certain Air Navigation Services in Iceland as Amended by the Montreal Protocol 1982, Doc 9586-JS/682 (June 1982).

operate and maintain air navigation services without interruption[6] and provides for Iceland to be reimbursed by the Contracting States for 95 % of the costs incurred.

2 Joint Support

Currently, two Joint Financing Agreements cover the operation and financing of facilities and services provided by Denmark and Iceland respectively for civil aircraft flying across the North Atlantic, north of the 45°N latitude. These services comprise air traffic control, communications and meteorology. Today, the services continue to be provided and financed in accordance with these two Agreements, as amended and updated in 1982 and 2008. At present, 24 States with civil aircraft flying across the North Atlantic are parties to the Agreements, including the two Provider States, Denmark and Iceland. All States whose aircraft make a significant number of North Atlantic crossings are invited to adhere to these Agreements.

Most of the costs are recovered from the users. The United Kingdom National Air Traffic Services Limited serves as billing and collecting agent. The small portion of the costs which is not allocable to international civil aviation is recovered from the States parties to the Agreements based on the percentage of North Atlantic crossings performed by aircraft operators from those States.

With effect from 27 March 1997, the first phase of reduced vertical separation minimum (RVSM) was implemented in the North Atlantic minimum navigation performance specifications (MNPS) airspace, in accordance with regional supplementary procedures approved by Council. This enables RVSM-approved aircraft, between flight level 290 and 410 inclusive, to operate with 1,000 ft vertical separation instead of 2,000 ft, as was previously required. Height-keeping performance needs to be demonstrated as part of the RVSM approval process, and is being monitored on an on-going basis by the Central Monitoring Agency on behalf of the North Atlantic Systems Planning Group (NAT SPG). RVSM almost doubles the airspace capacity in the flight level band concerned.

The new system is being financed under a new joint financing arrangement inspired by the experience of the Danish and Icelandic Joint Financing Agreements.

Under the arrangement, Canada, Iceland, Ireland, Portugal and the United Kingdom fund and own the facilities for a height monitoring project on the basis of their share of the North Atlantic traffic. The United States obligations include the provision of GMU equipment, a number of reference stations and post-processing equipment. Operational, maintenance and depreciation costs, as well as ICAO's administrative costs, are met through user charges. As with the Danish and Icelandic Joint Financing Agreements, the United Kingdom Civil Aviation Authority serves as billing and collecting agent.

The 35th Session of the ICAO Assembly adopted Resolution A35-15, by which participating States recognized that ICAO is the only international organization in a

---

[6] *Id.* at Art. III(1).

position to effectively coordinate global CNS/ATM activities and that the ICAO CNS/ATM systems should be utilized to serve the interests and the objectives of civil aviation throughout the world. The Resolution also recognized that Contracting States should have equal rights to benefit from global systems incorporated within the ICAO CNS/ATM systems. Referring to a Statement of ICAO Policy on CNS/ATM Systems Implementation and Operation developed and adopted by the ICAO Council on 9 March 1994, the Assembly resolved that nothing should deprive a Contracting State from its right to benefit from the ICAO CNS/ATM systems or cause discrimination between provider and user States and that States' sovereignty and borders should not be affected by the ICAO CNS/ATM systems implementation. The Resolution urged that provisions and guidance material relating to all aspects of the ICAO CNS/ATM systems should be sought and developed through the convening of adequate meetings, conferences, panels and workshops with the participation of Contracting States and called upon Contracting States, the Planning and Implementation Regional Groups (PIRGs) and the aviation industry to use the ICAO Global ATM Operational Concept as the common framework to guide planning and implementation of CNS/ATM systems and to focus all such development work on the Global ATM Operational Concept, while urging the Council to ensure that ICAO develop the transition strategies, ATM requirements and SARPs necessary to support the implementation of a global ATM system and to continue considering without delay the economic, institutional, legal and strategic aspects related to the implementation of the ICAO CNS/ATM systems.

3 SADIS, SCAR and SCRAG

On 6 March 2000, the Council of ICAO decided that, effective 1 January 2001, all States receiving the service provided by the Satellite distribution system for information relating to air navigation (SADIS) shall participate in the SADIS Cost Allocation and Recovery (SCAR) arrangement, thereby rendering null and void the already existing mechanism implemented through the *Agreement on the Voluntary Sharing of Costs of the Satellite Distribution System for Information relating to Air Navigation* as of that date. The new arrangement was given effect through the Agreement on the Sharing of Costs of the Satellite Distribution System for Information relating to Air Navigation (SADIS Agreement)[7] which provided *in limine* that the United Kingdom, as the provider State, shall provide, operate and maintain the SADIS and do so in conformity with all relevant ICAO standards and recommended practices and in accordance with relevant recommendations and decisions approved by the ICAO Council or other authorized ICAO body. Each Party receiving the SADIS

[7]The objective of this Agreement is for the Parties to establish and administer a mechanism to share in an equitable and fair manner the costs of providing, operating and maintaining the SADIS as approved by the Council, the services of which are described in Annex I to the Agreement. The Agreement, and its Annexes which form an integral part thereof, entered into force on 1 January 2001.

service, including the United Kingdom, but excluding the Parties referred to under Article IV, are required to pay its share of the costs of providing, operating and maintaining the SADIS as attributable to it in accordance with Article XI.[8] The Agreement also provides that any Party receiving the SADIS service and which falls within the group of States defined by the United Nations as "least developed countries (LDCs)" shall, unless it chooses not to, be exempt from paying its share of the respective costs as far as it remains in that situation.

Article VI of the SADIS Agreement exempts parties to the Agreement from liability for any damages or losses, physical or financial, inflicted as the consequence of failures and/or omissions in the provision, operation and maintenance of the SADIS. Article IX provides that the costs to be shared among the Parties shall be the full costs to the United Kingdom of employing fully or in part the facilities and personnel listed in Annex II to the Agreement for the purpose of providing, operating and maintaining the SADIS, including depreciation of assets and cost of capital and an appropriate amount for administration..

The SCAR arrangement is administered by a group, the *SADIS Cost Recovery Administrative Group* (SCRAG), which assesses the annual cost-share attributable to each Party, including reassessments arising from new Parties adhering to the Agreement, and audit the costs of the SADIS provision and any related financial activities, incurred by the provider State and subject to cost sharing. The SCRAG is composed of one Party from the European Region nominated by the European Air Navigation Planning Group (EANPG), one Party from the AFI Region nominated by the AFI Planning and Implementation Regional Group (APIRG), one Party from the MID Region nominated by the MID Air Navigation Planning and Implementation Regional Group (MIDANPIRG) and one Party from the Asia Region nominated by the ASIA/PAC Air Navigation Planning and Implementation Regional Group (APANPIRG). An additional member has been nominated by the planning and implementation regional group for the region wherein Parties are located which in the aggregate are responsible for more than 50 % of the total current assessments. The representative from the Party so nominated shall be chairman of the SCRAG. If none of the regions includes Parties which in the aggregate are responsible for more than 50 % of the total current assessments, SCRAG shall elect its chairman from among its members. Only those Parties which participate in the SCAR arrangement are eligible to serve on the SCRAG. The United Kingdom, in its capacity as the SADIS provider State, participates in the SCRAG as an observer. Furthermore, the International Air Transport Association (IATA), as a representative of user interests, is invited to participate as an observer. The Chairman of the SADIS Operations Group (SADISOPSG) is invited to participate as an observer as needed to provide information on the technical efficacy of the SADIS services provided and on the inventory of the facilities and services falling under the SCAR arrangement. Each member Party of the SCRAG has one vote, and when voting is required, decisions by the Group are

[8] SADIS Agreement, Article III 2.

arrived at by simple majority; however, when the votes are equally divided, the Chairman's vote prevails.

The costs to be shared among the Parties are the full costs to the United Kingdom of employing fully or in part the facilities and personnel listed in Annex II to the Agreement for the purpose of providing, operating and maintaining the SADIS, including depreciation of assets and cost of capital and an appropriate amount for administration. Each Party as encompassed by Article III, paragraph 2, is assessed a share of the total costs of the SADIS arrangement in proportion to the total number of available tonne-kilometres (ATKs) in scheduled services (international and domestic) performed by air carriers based in the territory of the State of that Party. The share of each Party is calculated from the total number of ATKs performed by all air carriers based in the territory of the State of that Party as a percentage of the total number of such ATKs performed by all air carriers of all the Parties participating in the arrangement. The total costs to be shared include the costs attributable to the Parties exempted from paying.

On or before 1 November of each year, the Secretary General of ICAO furnishes the SCRAG with the total number of ATKs performed in scheduled services (international and domestic) in the preceding calendar year by air carriers of each party based in the territory of the State of that Party. For example, the assessments for year n are calculated on the basis of the cost estimates for that year as approved by the SCRAG and ATKs as provided with regard to each Party by the Secretary General for year $n-2$. The cost basis for the assessments in year n , however, are first adjusted upwards or downwards as the case may be by the amount by which the total estimated costs for year $n-2$ were below or above the approved actual costs for that year. Likewise the assessment of each Party is adjusted to take into account any difference between the amounts paid by it under this Agreement as advances for year $n-2$ and its share as determined on the basis of actual ATKs and approved actual costs in year $n-2$. Any under-recovery of costs for year n arising from the failure of a Party to pay the SADIS cost share attributable to it for that year is added to the total SADIS costs to be shared for year $n+2$. Any subsequent remittance by the Party concerned to offset the consequential debt is deducted from the total costs to be shared for the year following that in which the remittance was received.

Article XII of the Agreement provides that the SCRAG communicates to the United Kingdom as the SADIS provider State on or before 1 December each year the approved estimated assessments for each Party adjusted as provided for in Article XI and authorize their collection by the United Kingdom, which thereupon may proceed to issue the invoices to each Party for its respective assessment as adjusted. Failure by a Party receiving the SADIS service to pay its share of the costs of providing the service (other than a Party exempted in accordance with Article IV) would lead to the service to that Party being withdrawn at the end of the calendar year in which payment was due. The service is not re-instated until the Party concerned fully settles its debt. It is the prerogative of each Party to decide whether or not to recover the assessment it has paid under the Agreement from users (aircraft operators). Such cost recovery by a Party should, however, in so far as it applies to international civil aviation, be in conformity with the principles and practices set out in the *Chicago*

Convention and *ICAO's Policies on Charges for Airports and Air Navigation Services.*[9]

Any dispute relating to the interpretation or application of the Agreement which is not settled by negotiation between the Parties involved, could be referred to the Council of ICAO for its recommendation upon request of any of these Parties. The Agreement is open to accession by the civil aviation administration or other such designated entity of any State being served by the SADIS. Accession is effected by notice in writing to that effect given to the Secretary General of ICAO by the head of the civil aviation administration or other such designated entity in the State concerned. Any Party may withdraw from participation in the Agreement on 31 December in any year by notice in writing to that effect given to the Secretary General not later than 1 January of that year by the Party concerned. The Agreement may be terminated by the United Kingdom as the SADIS provider State on 31 December in any year by notice in writing given to the Secretary General not later than 1 January of that year.[10]

[9]*Doc 9082.*

[10]Article XIX (b) provides that If at any time it proves impossible for the United Kingdom to perform the services within the limit determined pursuant to the provisions of Article X, the United Kingdom shall immediately notify the Secretary General in writing of such fact and shall furnish to the SCRAG through the Secretary General a detailed estimate of the additional amount required.

Article 72
Acquisition or Use of Land

Where land is needed for facilities financed in whole or in part by the Council at the request of a contracting State, that State shall either provide the land itself, retaining title if it wishes, or facilitate the use of the land by the Council on just and reasonable terms and in accordance with the laws of the State concerned.

R. Abeyratne, *Convention on International Civil Aviation*, DOI 10.1007/978-3-319-00068-8_73,

Article 73

Expenditure and Assessment of Funds

Within the limit of the funds which may be made available to it by the Assembly under Chapter XII, the Council may make current expenditures for the purposes of this Chapter from the general funds of the Organization. The Council shall assess the capital funds required for the purposes of this Chapter in previously agreed proportions over a reasonable period of time to the contracting States consenting thereto whose airlines use the facilities. The Council may also assess to State that consent any working funds that are required.

R. Abeyratne, *Convention on International Civil Aviation*,
DOI 10.1007/978-3-319-00068-8_74,

Article 74

Technical Assistance and Utilization of Revenues

When the Council, at the request of a contracting State, advance funds or provides airports or other facilities in whole or in part, the arrangement may provide, with the consent of that State, for technical assistance in the supervision and operation of the airports and other facilities, and for the payment, from the revenues derived from the operation of the airports and the other facilities, and of interest and amortization charges.

R. Abeyratne, *Convention on International Civil Aviation*,
DOI 10.1007/978-3-319-00068-8_75,

Article 75

Taking Over of Facilities from Council

A contracting State may at any time discharge any obligation into which it has entered under Article 70, and take over airports and other facilities which the Council has provided in its territory pursuant to the provisions of Articles 71 and 72, by paying to the Council an amount which in the opinion of the Council is reasonable in the circumstances. If the State considers that the amount fixed by the Council is unreasonable it may appeal to the Assembly against the decision of the Council and the Assembly may confirm or amend the decision of the Council.

R. Abeyratne, *Convention on International Civil Aviation*,

DOI 10.1007/978-3-319-00068-8_76,

Article 76
Return of Funds

Funds obtained by the Council through reimbursement under Article 75 and from receipts of interest and amortization payments under Article 74 shall, in the case of advances originally financed by States under Article 73, be returned to the States which were originally assessed in the proportion of their assessment, as determined by the Council.

R. Abeyratne, *Convention on International Civil Aviation*,
DOI 10.1007/978-3-319-00068-8_77,

Article 77

Joint Operating Organizations Permitted

Nothing in this Convention shall prevent two or more contracting States from constituting joint air transport operating organizations or international operating agencies and from pooling their air services on any routes or in any regions, but such organizations or agencies and such pooled services shall be subject to all the provisions of this Convention, including those relating to the registration of agreements with the Council. The Council shall determine in what manner the provisions of this Convention relating to nationality of aircraft shall apply to aircraft operated by international operating agencies.

R. Abeyratne, *Convention on International Civil Aviation*,
DOI 10.1007/978-3-319-00068-8_78,

Article 78
Function of Council

The Council may suggest to contracting States concerned that they form joint organizations to operate air services on any routes or in any region.

R. Abeyratne, *Convention on International Civil Aviation*,
DOI 10.1007/978-3-319-00068-8_79,

Article 79

Participation in Operating Organizations

A State may participate in joint operating organizations or in pooling arrangements, either through its government or through an airline company or companies designated by its government. The companies may, at the sole discretion of the State concerned, be state-owned or partly state-owned or privately owned.

R. Abeyratne, *Convention on International Civil Aviation*,
DOI 10.1007/978-3-319-00068-8_80,

Part IV
Final Provisions

Chapter XVII. Other Aeronautical Agreements and Arrangements

Article 80. Paris and Habana Conventions
Article 81. Registration of Existing Agreements
Article 82. Abrogation of Inconsistant Arrangements
Article 83. Registration of New Arrangements
Article 83 *bis*. Transfer of Certain Functions and Duties

Chapter XVIII. Disputes and Default

Article 84. Settlement of Disputes
Article 85. Arbitration Procedure
Article 86. Appeals
Article 87. Penalty for Non-conformity of Airline
Article 88. Penalty for Non-conformity by State $

Chapter XIX. War

Article 89. War and Emergency Conditions

Chapter XX. Annexes

Article 90. Adoption and Amendment of Annexes

Chapter XXI. Ratifications, Adherences, Amendments, and Denunciations

Chapter XXII. Definitions

Article 80

Paris and Habana Conventions

Each contracting State undertakes, immediately upon the coming into force of this Convention, to give notice of denunciation of the Convention relating to the Regulation of Aerial Navigation signed at Paris on October 13, 1919 or the Convention on Commercial Aviation signed at Habana on February 20, 1928, if it is a party to either. As between contracting States, this Convention supersedes the Conventions of Paris and Habana previously referred to.

R. Abeyratne, *Convention on International Civil Aviation*,

DOI 10.1007/978-3-319-00068-8_81,

Article 81
Registration of Existing Agreements

All aeronautical agreements which are in existence on the coming into force of this Convention, and which are between a contracting State and any other State or between an airline of a contracting State and any other State or the airline of any other State, shall be forthwith registered with the Council.

R. Abeyratne, *Convention on International Civil Aviation*,
DOI 10.1007/978-3-319-00068-8_82,

Article 82
Abrogation of Inconsistent Arrangements

The contracting States accept this Convention as abrogating all obligations and understandings between them which are inconsistent with its terms, and undertake not to enter into any such obligations and understandings. A contracting State which, before becoming a member of the Organization has undertaken any obligations toward a non-contracting State or a national of a contracting State or of a non-contracting State inconsistent with the terms of this Convention, shall take immediate steps to procure its release from the obligations. If an airline of any contracting State has entered into any such inconsistent obligations, the State of which it is a national shall use its best efforts to secure their termination forthwith and shall in any event cause them to be terminated as soon as such action can lawfully be taken after the coming into force of this Convention.

R. Abeyratne, *Convention on International Civil Aviation*,
DOI 10.1007/978-3-319-00068-8_83,

Article 83
Registration of New Arrangements

Subject to the provisions of the preceding Article, any contracting State may make arrangements not inconsistent with the provisions of this Convention. Any such arrangement shall be forthwith registered with the Council, which shall make it public as soon as possible.

Contents

1 Registration and Depositary Functions of ICAO

The Assembly, at its 16th Session (Buenos Aires, 3–26 September 1968) adopted Resolution A 16-32 (Registration of Agreements and Arrangements), whereby the Assembly resolved that each Contracting State comply with Article 83 of the Chicago Convention by registering with the Council all arrangements relating to international civil aviation between such Contracting State and any other State and between such Contracting States and an airline of another State.

2 ICAO's Rules for Registration

Registration at ICAO of aeronautical agreements follows the same principles as registration under Article 102 of the United Nations Charter. Except for CMOUs, which are confidential to the States Parties concerned, every treaty and every international agreement is required to be registered and made public. For purposes of registration, ICAO recognizes that an "aeronautical agreement or arrangement" means any agreement or arrangement, whatever its form and descriptive name, relating to international civil aviation.[1] These instruments are divided into two categories for registration with the ICAO Council. The first category[2] is any aeronautical agreement or arrangement in existence on the coming into force of the Chicago Convention, i.e. on April 4th, 1947, between a Contracting State and any other State; a Contracting State and an airline of any other State; an airline of a Contracting State and a non-Contracting State; or an airline of a Contracting State and an airline of any other State. This category is governed by Article 81 of the

[1] *Id.*, Article 1.
[2] *Id.* Article 2.

DOI 10.1007/978-3-319-00068-8_84,

Chicago Convention.[3] The second category, in accordance with Article 83 of the Convention, includes any aeronautical agreement or arrangement coming into force after April 4th, 1947, between a Contracting State and any other State; a Contracting State and an airline of any other State; or a Contracting State and a national (physical person or corporation) of any other State if it relates to the ownership or operation of any international air service, aerodrome or air navigation service.[4]

Each Contracting State is responsible for the registration of any aeronautical agreement or arrangement registrable under ICAO's rules of registration to which it is a party.[5] Any Contracting State responsible for the registration of an aeronautical agreement or arrangement, which receives written confirmation of registration of that agreement from ICAO, is relieved of the obligation of registration. Registration of an agreement is usually effected by the State concerned transmitting to the Secretary General of ICAO a true copy thereof duly certified by the appropriate authority of the registering party.[6] The registering party is also required to, at the same time or as soon as possible thereafter, notify ICAO of the date of the coming into force of any such agreement or arrangement if it is not evident from the terms thereof and the method of entry into force (signature, ratification, approval, acceptance, exchange of letters, etc.).

The date of receipt by the Secretary General of ICAO is considered to be the date of registration.[7] Any aeronautical agreement or arrangement registered *ex officio* by ICAO pursuant to Article 7[8] of these Rules is deemed to be registered on the date on which such agreement or arrangement first came into force between two or more of the parties thereto. The Secretary General issues the State which registers with

---

[3] C-WP/212 (1949) at 2.

[4] The text of aeronautical agreements or arrangements concluded subsequent to April 4th, 1947 and registered with ICAO are to be made public immediately after such registration. Doc 6685, the Third Restatement constitutes a comprehensive revision of the earlier (1965) Restatement, covering many more subjects, and reflecting important developments in the intervening decades. This Restatement consists of international law as it applies to the United States, and domestic law that has substantial impact on the foreign relations of the United States or has other important international consequences, Article 12. If a party to an aeronautical agreement or arrangement registered with ICAO transmits to ICAO a decision or an advisory opinion thereon, rendered by a judicial or arbitral body or other person authorized or agreed by the parties to such agreement or arrangement, the Secretary General is required to indicate that fact on an electronic database of aeronautical agreements and arrangements.

[5] *Id.* Article 6.

[6] *Id.* Article 8. It is a requirement that such certified copy reproduces the original text in the language or languages in which the said agreement or arrangement was concluded and be accompanied by one additional copy. If the original text is not in one of the following languages—English, Arabic, Chinese, French, Russian, Spanish—it shall also be accompanied by two copies of a translation into one of these languages.

[7] *Id.* Article 9.

[8] Article 7 provides: "Any aeronautical agreement or arrangement registrable hereunder shall be registered, *ex officio*, by ICAO in any case where the Organization is a party thereto or where, by the terms thereof, it is entrusted with the registration thereof; or where it is the custodian thereof. Any party thereto other than ICAO shall be relieved of the obligation of registration".

ICAO any aeronautical agreement or arrangement and to any other party thereto a written confirmation of registration of any aeronautical agreement or arrangement signed by the Secretary General or his representative.[9] The Secretary General is required to maintain a register containing, in respect of each agreement or arrangement registered, a record of:

- The serial registration number assigned to it;
- The title of the agreement or arrangement and a summarized statement of its purpose or effect;
- The names of the parties thereto;
- The dates of signature; the date of ratification, acceptance, exchange of ratification, accession or adherence; and the date of entry into force;
- The duration;
- The language or languages used;
- The name of the party registering and the date of registration;
- A reference, where available, to the Contracting States' own electronic databases of agreements or arrangements.[10]

Article 83 *bis*. Transfer of Certain Functions and Duties

(a) Notwithstanding the provisions of Articles 12, 30, 31 and 32 (a), when an aircraft registered in a contracting State is operated pursuant to an agreement for the lease, charter or interchange of the aircraft or any similar arrangement by an operator who has his principal place of business or, if he has no such place of business, his permanent residence in another contracting State, the State registry may, by agreement with such other State, transfer to it all or part of its functions and duties as State registry in respect of that aircraft under Articles 12, 30, 31 and 32 (a). The State of registry shall be relieved of responsibility in respect of the functions and duties transferred.

(b) The transfer shall not have effect in respect of other contracting States before either the agreement between States in which it is embodied has been registered with the Council and made public pursuant to Article 83 or the existence and scope of the agreement have been directly communicated to the authorities of the other contracting State or States concerned by a State party to the agreement.

(c) The provisions of paragraph (a) and (b) above shall also be applicable to cases covered by Article 77.

[9] *Id.* Article 10.

[10] *Id.* Article. 11.

Article 84
Settlement of Disputes

If any disagreement between two or more contracting States relating to the interpretation or application of this Convention and its Annexes cannot be settled by negotiation, it shall, on the application of any State concerned in the disagreement, be decided by the Council. No member of the Council shall vote in the consideration by the Council of any dispute to which it is a party. Any contracting State may, subject to Article 85, appeal from the decision of the Council to an ad hoc arbitral tribunal agreed upon with the other parties to the dispute or to the Permanent Court of International Justice. Any such appeal shall be notified to the Council within sixty days of receipt of notification of the decision of the Council.

Contents

1 Is ICAO a Judicial Body?

The Assembly, at its first session (Montreal, 6–27 May 1947) adopted Resolution A1-23 (Authorization to the Council to Act as an Arbitral Body) whereby the Assembly resolved that, pending further discussion and ultimate decision by the Organization as to the methods of dealing with international disputes in the field of civil aviation, the Council be authorized to act as an arbitral body on any differences arising among Contracting States relating to international civil aviation matters submitted to it, when expressly requested to do so by all Parties to such differences, and that the Council, on such occasions, be authorized to submit an advisory report, or a decision binding upon the Parties, if the Parties expressly decide to bind themselves in advance to accept the decision of the Council as binding.

A significant feature of this provision is that the Council of ICAO has jurisdiction to decide on a dispute relating *to the interpretation or application of this Convention and its Annexes*. This provision reflects two significant points: the first is that Contracting States should first attempt to resolve their disputes by themselves, through negotiation[1]; the second is that the word **shall** in this provision infuses into the decision making powers of the Council an unquestionably mandatory character. Furthermore, a decision taken by the Council is juridically dignified

[1]Hingorani (1959), at 16. See also, *Rules of Procedure for the Council*, Fifth Edition 1980, Article 14.

DOI 10.1007/978-3-319-00068-8_85, © Springer International Publishing Switzerland 2014

by Article 86 of the Convention, when the Article states that unless the Council decides otherwise, any decision by the Council on whether an international airline is operating in conformity with the provisions of the Convention shall remain in effect unless reversed in appeal. The council also has powers of sanction granted by the Convention, if its decision is not adhered to.[2]

In April 1952, an application was made by the Government of India to the Council in regard to a disagreement between that Government and the Government of Pakistan on the interpretation and application of Articles 5, 6 and 9 of the Chicago Convention and the International Air Services Transit Agreement. This did not go into the dispute settlement process under Article 84 of the Convention as the two governments had settled their dispute by negotiation.[3]

In September 1967 the United Kingdom filed an application under Article 84 of the Convention, relating to the establishment of a prohibited area by Spain near Gibraltar, involving questions of application and interpretation of Article 9 of the Convention. At the Council's Eleventh Meeting of its Sixty Eighth Session, consideration of the matter was deferred *sine die* at the request of the United Kingdom and Spain.

In March 1971, the Government of Pakistan filed an application before the Council, under Article 2 of the Rules of Settlement of Differences, seeking redress against India in regard to suspension of flights of Pakistani aircraft over Indian territory, in derogation of Article 5 of the Convention, among other grounds. The substantive issue relating to the application of Article 5 of the Convention, thus bringing the dispute under Article 84 of the Convention. The matter was resolved by the Council in favour of Pakistan on 29 July 1971.

The Council has therefore the power under the Chicago Convention to adjudicate disputes between the member States of ICAO on matters pertaining to international civil aviation. As already mentioned, the Council is a permanent body responsible to the ICAO Assembly and is composed of 36 Contracting States elected by the Assembly. It has its genesis in the Interim Council of the Provisional International Civil Aviation Organization (PICAO).[4] PICAO occupied such legal capacity as may have been necessary for the performance of its functions and was recognised as having full juridical personality wherever compatible with the Constitution and the laws of the State concerned.[5] The definitive word "juridical" attributed to PICAO a mere judicial function, unequivocally stipulating that the organization and its component bodies, such as the Interim Council were obligated

---

[2]Article 87.

[3]Doc 7388-C/860.

[4]Hereafter referred to as PICAO. See Interim Agreement on International Civil Aviation, opened for signature at Chicago, December 7 1944, Article 3. Also in Hudson, *International Legislation*, Vol 1X (1942–1945, New York) at 159.

[5]*Id.* Article 1 Section 4. It is interesting to note that PICAO was established as a provisional organization of a technical and advisory nature for the purpose of collaboration in the field international civil aviation. *Vide* Article 1 Section 1.

to remain within the legal parameters allocated to them by the Interim Agreement[6] and that PICAO was of a purely technical and advisory nature. A legislative or quasi-legislative function could not therefore be imputed to the Interim Council of PICAO. It could mostly study, interpret and advise on standards and procedures[7] and make recommendations with respect to technical matters through the Committee on Air Navigation.[8] The International Civil Aviation Organization (ICAO) which saw the light of day on April 4, 1947 derived the fundamental postulates of its technical and administrative structure from its progenitor—PICAO—and it would seem reasonable to attribute a certain affinity *ipso facto* between the two organizations and hence, their Councils. One of the Council's functions is to consider any matter relating to the Convention which any Contracting State refers to it.[9] Since one of the distinctive features of the ICAO Council is its ability to make rules for international civil aviation, it follows incontrovertibly that the Council's dispute resolution powers are compelling.

The ICAO Council has played a signal role in dispute resolution in the 1990s up to date over the past two decades. One of the best examples of ICAO's role in the international community was seen in The Iranair Incident—IR 655 (Iran, United States 1998). This concerned the shooting down of an Iran Air Airbus A300 (IR655) carrying commercial passengers on a scheduled flight from Bandar-Abbas (Iran) to Dubai. The aircraft was brought down by the *U.S. S. Vincennes* over the Persian Gulf, resulting in the death of all 290 persons on board the aircraft.

One of the emergent features of the ICAO Council which became clear at its deliberations was the Council's resolve to address its deliberations to purely technical issues pertaining to the incident, while stringently avoiding political issues and diplomatic pitfalls. This is certainly true of all incidents discussed above, where the Council restricted its scope to technical issues as applicable to the principles embodied in the Chicago Convention.

2 Some Clashes of Regulation

There seems to be an unfortunate dichotomy in terminology in the Convention since on the one hand, Article 54 (n) makes it mandatory that the Council shall merely consider any matter relating to the Convention which any Contracting State refers to it, while on the other, Article 84 categorically states that any disagreement between two or more States relating to the interpretation or application of the Convention and

---

[6]*Op. cit.*

[7]Interim Agreement, *supra* note 4, Section 6.4.b(1).

[8]*Id.* Section 6.4.b(6). Also, Buergenthal (1969) at 4, where the author states that PICAO's functions were merely advisory, which precludes any imputation of legislative or quasi-legislative character to its Interim Council.

[9]Chicago Convention, Preamble *supra* note 1, Article 54.

its Annexes, that cannot be settled by negotiation shall. . .be decided by the Council. The difficulty arises on a strict interpretation of Article 54 (n) where even a disagreement between two States as envisaged under Article 84 could well be considered as 'any matter' under Article 54 (n). In such an instance, the Council could well be faced with the dilemma of choosing between the two provisions. It would not be incorrect for the Council to merely consider a matter placed before it, although a decision is requested by the applicant State, since, Article 54 (n) is perceived to be comprehensive as the operative and controlling provision that lays down mandatory functions of the Council. It is indeed unfortunate that these two provisions obfuscate the issue which otherwise would have given a clear picture of the decision making powers of the Council. A further thread in the fabric of adjudicatory powers of the Council is found in Article 14 of the Rules of Settlement promulgated by the Council in 1957[10] which allows the Council to request the parties in dispute to engage in direct negotiations at any time.[11] This emphasis on conciliation has prompted the view that the Council, under article 84 would favour the settling of disputes rather than adjudicating them.[12] This view seems compatible with the proposition that the consideration of a matter under Article 54 (n) would be a more attractive approach *in limine* in a matter of dispute between two States.

Milde noted in 1979:

The Council of ICAO cannot be considered a suitable body for adjudication in the proper sense of the word—*i.e.* settlement of disputes by judges and solely on the basis of respect for law. The Council is composed of States (not independent individuals) and its decisions would always be based on policy and equity considerations rather than on pure legal grounds. . .truly legal disputes. . .can be settled only by a true judicial body which can bring into the procedure full judicial detachment, independence and expertise. The under-employed ICJ is the most suitable body for such types of disputes.[13]

The perceived inadequacies of the ICAO Council in being ethically unsuitable to decide on disputes between States can only be alleviated by the thought that the members of the Council are presumed to be well versed in matters of international civil aviation and therefore would be deemed to be better equipped to comprehend the issues placed before them than the distinguished members of the International Court of Justice, some of whom may not be experts of international air law. Nonetheless, there is no doubt that the ICAO Council possess juridical powers[14] and that as one commentator said:

---

[10]Rules for the Settlement of Differences, ICAO Doc 7782/2 (2ed. 1975).

[11]*Id.* Article 14 (a).

[12]Buergenthal (1969), 121 at 136.

[13]Milde (1979) 87, at 88.

[14]de Lacerda (1978) at 219.

If ICAO did not exist, it would have to be invented; otherwise, international civil aviation would not function with the safety, efficiency and regularity that it has achieved today.[15]

International organizations can generally only work on the basis of legal powers that are attributed to them. Presumably, these powers emanate from the sovereign States that form the membership of such organizations. Therefore, the logical conclusion is that if international organizations were to act beyond the powers accorded to them, they would be presumed to act *ultra vires* or beyond the scope of their mandate. The universal solidarity of UN Contracting States that was recognized from the outset at the establishment of the Organization brings to bear the need for States to be united in recognizing the effect of UN policy and decisions. This principle was given legal legitimacy in the 1971 decision concerning the European Road Transport Agreement handed down by the Court of Justice of the European Community. The court held that the competence of the European Community to conclude an agreement on road transport could not be impugned since the member States had recognized Community solidarity and that the Treaty of Rome which governed the Community admitted of a common policy on road transport which the Community regulated.

It should be noted that the United Nations does not only derive implied authority from its Contracting States based on universality but it also has attribution from States to exercise certain powers. The doctrine of attribution of powers comes directly from the will of the founders, and in the UN's case, powers were attributed to the Organization when it was established as an international organization that would administer the provisions of the United Nations Charter. In addition, the UN could lay claims to what are now called "inherent powers" which give it power to perform all acts that the Organization needs to perform to attain its aims not due to any specific source of organizational power but simply because the United Nations inheres in organizationhood. Therefore, as long as acts are not prohibited in the UN's constituent document (the UN Charter), they must be considered legally valid.

Over the past two decades the inherent powers doctrine has been attributed to the United Nations Organization and its specialized agencies on the basis that such organizations could be stultified if they were to be bogged down in a quagmire of interpretation and judicial determination in the exercise of their duties. The advantages of the inherent powers doctrine is twofold. Firstly, inherent powers are functional and help the organization concerned to reach its aims without being tied by legal niceties. Secondly, it relieves the organization of legal controls that might otherwise effectively preclude that organization from achieving its aims and objectives. The ability to exercise its inherent powers has enabled the UN to address issues on promoting self determination and independence, strengthening international law, handing down judicial settlements of major international disputes, providing humanitarian aid to victims of conflict, alleviating chronic hunger and rural poverty in developing countries and promoting women's rights, just to name a few.

---

[15]Fitzgerald (1976) 47 at 50.

References

de Lacerda JCS (1978) A study about the decisions of the ICAO council. Ann Air Space Law 111:219

Fitzgerald GF (1976) ICAO now and in the coming decades. In: Matte NM (ed) International air transport law, organization and politics for the future. McGill: Montreal

Hingorani (1959) Dispute settlement in international civil aviation. Arb J 14:14

Milde M (1979) Dispute settlement in the framework of the international civil aviation organization (ICAO). Settlement Space Law Disputes

Buergenthal T (1969) Law making in the international civil aviation organization. Syracuse University Press, Syracuse

Article 85
Arbitration Procedure

If any contracting State party to a dispute in which the decision of the Council is under appeal has not accepted the Statute of the Permanent Court of International Justice and the contracting States parties to the dispute cannot agree on the choice of the arbitral tribunal, each of the contracting States parties to the dispute shall name a single arbitrator who shall name an umpire. If either contracting State party to the dispute fails to name an arbitrator within a period of three months from the date of the appeal, an arbitrator shall be named on behalf of that State by the President of the Council from a list of qualified and available persons maintained by the Council. If, within thirty days, the arbitrators cannot agree on an umpire, the President of the Council shall designate an umpire from the list previously referred to. The arbitrators and the umpire shall then jointly constitute an arbitral tribunal. Any arbitral tribunal established under this or the preceding Article shall settle its own procedure and give its decisions by majority vote, provided that the Council may determine procedural questions in the event of any delay which in the opinion of the Council is excessive.

R. Abeyratne, *Convention on International Civil Aviation*,

DOI 10.1007/978-3-319-00068-8_86,

Article 86
Appeals

Unless the Council decides otherwise any decision by the Council on whether an international airline is operating in conformity with the provisions of this Convention shall remain in effect unless reversed on appeal. On any other matter, decisions of the Council shall, if appealed from, be suspended until the appeal is decided. The decisions of the Permanent Court of International Justice and of an arbitral tribunal shall be final and binding.

R. Abeyratne, *Convention on International Civil Aviation*,
DOI 10.1007/978-3-319-00068-8_87,

Article 87

Penalty for Non-conformity of Airline

Each contracting State undertakes not to allow the operation of an airline of a contracting State through the airspace above its territory if the Council has decided that the airline concerned is not conforming to a final decision rendered in accordance with the previous Article.

R. Abeyratne, *Convention on International Civil Aviation*,
DOI 10.1007/978-3-319-00068-8_88,

Article 88

Penalty for Non-conformity by State$

The Assembly shall suspend the voting power in the Assembly and in the Council of any contracting State that is found in default under the provisions of this Chapter.

R. Abeyratne, *Convention on International Civil Aviation*, 675
DOI 10.1007/978-3-319-00068-8_89, © Springer International Publishing Switzerland 2014

Article 89
War and Emergency Conditions

In case of war, the provisions of this Convention shall not affect the freedom of action of any of the contracting States affected, whether as belligerents or as neutrals. The same principle shall apply in the case of any contracting State which declares a state of national emergency and notifies the fact to the Council.

Contents

1 Limitations of the Convention

On the occasion of the adherence of Israel to the Chicago Convention, The Government of Egypt advised that in view of the considerations of fact and of law which still affect Egypt's special position with regard to Israel, and in pursuance of Article 89 of the Chicago Convention, Israeli aircraft may not claim the privilege of flying over the territory of Egypt.[1] In a separate event, the Government of Iraq also informed the Council that a state of emergency had been declared on 14 May 1948 and therefore Article 89 of the Chicago Convention was applicable and all Israeli aircraft were denied the privilege of flying over the territory of Iraq.

This provision has ominous nuances in that the compelling protection afforded to civil aircraft is effectively removed by it, opening it up to States in times of war to engage in intervention of civil aircraft. Although the Charter contains no provision which deals directly with the security of civil aviation, it is one of the most salutary international legal documents in the area of civil aviation security. The Preamble to the Charter stipulates that citizens of the member States of the United Nations will practice tolerance and live together in peace with one another as good neighbours. The principle of security is embodied in several articles of the Charter. Article 1 (2) provides that the purpose of the United Nations is to pursue the development of friendly relations among nations based on respect for the principle of equal rights and self-determination of peoples, and to take other appropriate measures to strengthen universal peace.

[1] Letter dated 16 October 1949, reproduced I Annex A to Doc 6922-C/803 at 125.

R. Abeyratne, *Convention on International Civil Aviation*,

DOI 10.1007/978-3-319-00068-8_90,

2 Ensuring Civilian Protection

As civil aircraft are by definition presumed to transport civilians, the principles of the Chicago Convention should ensure the protection of civilians and their property from dangers affecting civil aircraft in flight. The United Nations Charter can therefore be regarded as imputing to the international community a duty to protect the human being and his property in relation to flight:

There is a mandatory obligation implied in article 55 of the Charter that the United Nations "shall promote respect for, and observance of, human rights and fundamental freedoms"; or, in terms of article 13, that the Assembly shall make recommendations for the purpose of assisting in the realization of human rights and freedoms. There is a distinct element of legal duty in the understanding expressed in article 56 in which all members pledge themselves to take joint and separate action in co-operation with the organization for the achievement of the purpose set forth in article 55.[2]

A civil aircraft, when identified as such cannot be attacked.[3] *The United Nations Charter opposes the use of force against civilian aircraft. Article 2(4) of the charter prohibits the use of force in any manner inconsistent with the purposes of the Charter. There is also provision for the settlement of disputes by peaceful means.*[4]

An armed attack against an aircraft is a special kind of aggression[5] *and is protected by the right of self-defence which is recognized against an such an attack, by Article 51 of the Charter. This provision narrows the field of the exercise of self-defence to circumstances involving an armed attack. An unauthorized entry into the airspace of a State by an unarmed aircraft does not constitute an armed attack, even if such entry is effected for the purposes of espionage or provocation. Although no authoritative definition of an armed attack has ever been adopted internationally, it is generally presumed that an armed attack would constitute belligerence endangering the safety of those affected by such attack when it is carried out by an offender(s) wielding weapons.*

Reference

Kunz JL (1948) The inter-American treaty of reciprocal assistance. Am J Int Law 42:111, 115

[2]H. Lauterpact, *International Law and Human Rights* (1950), p. 149.

[3]I.A. Vlasic, *Casebook on International Air Law* (1982), p. 161.

[4]Art. 33 of the U.N. Charter.

[5]Kunz (1948), pp. 111, 115.

Article 90
Adoption and Amendment of Annexes

(a) The adoption by the Council of the Annexes described in Article 54, subparagraph (l), shall require the vote of two-thirds of the Council at a meeting called for that purpose and shall then be submitted by the Council to each contracting State. Any such Annex or any amendment of an Annex shall become effective within three months after its submission to the contracting States or at the end of such longer period of time as the Council may prescribe, unless in the meantime a majority of the contracting States register their disapproval with the Council.

(b) The Council shall immediately notify all contracting States of the coming into force of any Annex or amendment thereto.

Contents

1 Procedures for Enforcing SARPs

This is a purely procedural provision and the historical and legal bases of the Annexes have already been discussed under Articles 37 and 38. The ICAO Assembly, at its 2nd Session (Geneva, 1–21 June 1948) adopted Recommendation 8 which recommended that the Council, when adopting further Annexes and establishing a date by which States may notify their disapproval of them, take fully into account the time needed for transmitting the Annexes, so as to allow for their effective study during the full period provided in Article 90. In Resolution A7-9 (adopted at the 7th Session of the Assembly—Brighton, 16 June–6 July 1953) the assembly resolved that the Council, in fixing the dates for the application by Contracting States of International Standards, allow sufficient time to enable States to complete their arrangements for implementation thereof.

There are two operative phrases of importance in Article 90 which speak of an Annex being "effective" within three months after its submission to the Contracting States and the Annex "coming into force". The issue is: "what is the significance of these two practices?"

2 Difference Between Standards and Recommended Practices

Firstly one must start with the two main components of the Annexes i.e. Standards on the one hand, and Recommended Practices on the other. A Standard is defined as any

R. Abeyratne, *Convention on International Civil Aviation*,

DOI 10.1007/978-3-319-00068-8_91,

specification for physical characteristics, configuration, material, performance, personnel or procedure, the uniform application of which is recognized as necessary for the safety or regularity of international air navigation and to which Contracting States will conform in accordance with the Convention; in the event of impossibility of compliance, notification to the Council is compulsory under Article 38 of the Convention.

A Recommended Practice is any specification for physical characteristics, configuration, material, performance, personnel or procedure, the uniform application of which is recognized as desirable in the interest of safety, regularity or efficiency of international air navigation, and to which Contracting States will endeavour to conform in accordance with the Convention. States are invited to inform the Council of non-compliance.

SARPs are formulated in broad terms and restricted to essential requirements. For complex systems such as communications equipment, SARPs material is constructed in two sections: core SARPs—material of a fundamental regulatory nature contained within the main body of the Annexes, and detailed technical specifications placed either in the Appendices to Annexes or in manuals.

The differences to SARPS notified by States are published in Supplements to Annexes. After the Council adopts an Annex it is sent to ICAO member States for their comments and notification of disapproval of Standards with a date identified by the Council of the Annexes effective date which is within three months of submission. This is an interim edition of the Annex, referred to as the "Green Edition", which is dispatched to States with a covering explanatory letter. This covering letter also gives the various dates associated with the introduction of the Annex including its effective date. Once the Annex becomes effective, the States have three months to indicate disapproval of adopted amendments to SARPs. Unless a majority of the States indicate their disapproval on or before the time allocated to them to respond with their disapprovals, the Council declares the Annex to have come into force at a particular date.

Article 91
Ratification of Convention

(a) This Convention shall be subject to ratification by the signatory States. The instruments of ratification shall be deposited in the archives of the Government of the United States of America, which shall give notice of the date of the deposit to each of the signatory and adhering States.

(b) As soon as this Convention has been ratified or adhered to by twenty-six States it shall come into force between them on the thirtieth day after deposit of the twenty-sixth instrument. It shall come into force for each State ratifying thereafter on the thirtieth day after the deposit of its instrument of ratification.

(c) It shall be the duty of the Government of the United States of America to notify the government of each of the signatory and adhering States of the date on which this Convention comes into force.

R. Abeyratne, *Convention on International Civil Aviation*,
DOI 10.1007/978-3-319-00068-8_92,

Article 92

Adherence to Convention

(a) This Convention shall be open for adherence by members of the United Nations and States associated with them, and States which remained neutral during the present world conflict.

(b) Adherence shall be effected by a notification addressed to the Government of the United States of America and shall take effect as from the thirtieth day from the receipt of the notification by the Government of the United States of America, which shall notify all the contracting States.

R. Abeyratne, *Convention on International Civil Aviation*,
DOI 10.1007/978-3-319-00068-8_93,

Article 93
Admission of Other States

States other than those provided for in Articles 91 and 92 (a) may, subject to approval by any general international organization set up by the nations of the world to preserve peace, be admitted to participation in this Convention by means of a four-fifths vote of the Assembly and on such conditions as the Assembly may prescribe: provided that in each case the assent of any State invaded or attacked during the present war by the State seeking admission shall be necessary.

Contents

1 Entering into Force and State Responsibilities

Article 91 is straightforward and clear. The first consideration in Article 92 is the link between the applicability of the Chicago Convention to a Contracting State and its status as a member of the United Nations. As discussed earlier, the Chicago Convention established ICAO in Article 43. By virtue of Article 1 of an agreement entered into between ICAO and the United Nations,[1] the United Nations recognizes ICAO as the specialized agency responsible for taking such action as may be appropriate under its basic instrument (Chicago Convention) for the accomplishment of the purposes set forth therein. By virtue of Article 57 of the United Nations Charter specialized agencies, established by inter-governmental agreement and having wide international responsibilities as defined in their basic instruments in economic, social, cultural, educational, health and related fields, are brought into relationship with the United Nations. Article 64 of the Chicago Convention provides that the ICAO may, with respect to air matters within its competence, directly affecting world security, enter into appropriate arrangements with any general organization set up by the nations of the world to preserve peace. Article 65 of the Convention provides that ICAO may enter into agreements with international

[1]The Economic and Social Council on 21 June 1946 directed its Committee on Negotiations with Specialized Agencies to enter into negotiations with the Provisional International Civil Aviation Organization for the purpose of bringing it into relationship with the United Nations and to submit a report of the negotiations to the third session of the Council, including therein a draft preliminary agreement based on such negotiations. See *Protocol Concerning the Entry Into Force of the Agreement Between the United Nations and the International Civil Aviation Organization*, Signed at New York, on 1 October 1947, United Nations Treaty Series No. 45: 1947 at 316–343.

DOI 10.1007/978-3-319-00068-8_94,

bodies for the maintenance of common service, for common arrangements concerning personnel and for the facilitation of its work.

In accordance with Article 65, ICAO entered into the aforementioned agreement with the United Nations, Article II of which stipulates that any application submitted to ICAO by States other than those provided for in articles 91 and 92 (a) of the Chicago Convention to become parties to the Convention, shall be immediately transmitted by the secretariat of ICAO to the General Assembly of the United Nations. The General Assembly may recommend the rejection of such application, and any such recommendation shall be accepted by the Organization. If no such recommendation is made by the General Assembly at the first session following receipt of the application, the application shall be decided upon by the Organization in accordance with the procedure established in article 93 of the Convention.

Article 92 is unequivocal that member States of the United Nations (or States associated with them) should deposit their instruments of ratification or adherence with the United States. ICAO has nothing to do with such deposit, and the United States would advise ICAO subsequent to the fact.

At the time of writing, the United Nations General Assembly had, on 29 November 2012, endorsed an upgraded U.N. status for the Palestinian Authority. The resolution elevates their status from "non-member observer entity"[2] to "non-member observer state,"[3] the same category as the Vatican, which Palestinians hope will provide new leverage in their dealings with Israel. This brings up an issue

[2]Historically it can be observed that national liberation movements have been given some recognition by the United Nations which has granted observer status to them. Policies of decolonization, particularly in regions such as Africa have been a compelling precursor to this trend. However, economic and social perspectives and issues have also propelled the United Nations to admit observers in addition to national liberation. Such entities admitted have been looked at as future authoritative governments that will be responsible for the social and economic well-being of their people. This was the commencement in the United Nations of the "proto-state "approach. Hence Resolution 3237 (XXIX) adopted in 1974, granted the Palestine Liberation Organization (PLO) observer status in the United Nations. The Resolution, inter alia, admitted of the right of the PLO to participate in the sessions and the work of the General Assembly in the capacity of observer and invited the PLO to participate in the sessions and the work of all international conferences convened under the auspices of the General Assembly in the capacity of observer. The resolution also gave the PLO entitlement to participate as an observer in the sessions and the work of all international conferences convened under the auspices of other organs of the United Nations.

[3]The Charter of the United Nations does not address the issue of observer status. The issue has been dealt with purely on the basis of practice which has been ascribed an inarticulate legal basis through discussions and decisions in the General Assembly. There are various types of observer, including non-member states (such as Palestine now); intergovernmental organizations; national liberation movements. Observers have the right to speak at United Nations General Assembly meetings, but no vote. Various other rights (e.g., to participate in debates, to submit proposals and amendments, the right of reply, to raise points of order and to circulate documents, etc.) are given selectively to some observers only. The only international organisation to be accorded these rights is the European Union. There is a distinction between state and non-state observers. Non-Member States of the United Nations, which are members of one or more specialized agencies, can apply for the status of Permanent Observer State. The non-state observers are the international organizations and other entities.

from an aviation perspective for Palestine. Article 91 speaks of "signatory States" to the Chicago Convention. Palestine is not in that category. Article 92 refers to "members of the United Nations and States associated with them". This effectively rules out Palestine which is a "nonmember". Is it a State associated with a member of the United Nations? There is no such official recognition. For that matter, is Palestine a "State" within the legal definition of the Montevideo Convention of 1933 which defines a State as being required to have a geographically defined territory; a permanent population; a government; and the ability to enter into arrangements with other States? The main issue would be the first one concerning land boundaries which have been subject to some contention. It is submitted that although 138 members voted for the recognition of Palestine as a "nonmember observer State" if the legal requirement for the recognition of Palestine as a State is not satisfied, there would be some contention at law, since membership in the United Nations and Statehood are two different things. It is entirely possible for a sovereign State to be a non-member of the United Nations as was Switzerland for a considerable number of years; and for a State which was not fully independent, as was India then, to be a member of the United Nations.

2 UN and ICAO Membership

Articles 3 to 6 of the United Nations Charter govern membership in the United Nations. As stipulated in Article 4(1) of the Charter, UN membership is open to all peace-loving states which accept the obligations contained in the Charter and, in the judgment of the United Nations, are able and willing to carry out such obligations. The admission of any such state to membership is effected by a decision of the General Assembly upon the recommendation of the Security Council.

The recommendation of the Security Council is essential for a state to gain admission to the United Nations. China, France, the Russian Federation, the United Kingdom, and the United States, as permanent members of the Council, each wield a veto, and any one veto can effectively preclude admission. In an instance where the Security Council recommends admission, it is then up to the General Assembly to decide whether to admit the candidate as a Member State. Article 18(2) of the Charter prescribes that the admission of new Members to the United Nations must be decided by the General Assembly by a two-thirds majority of the members present and voting each of the UN's 193 Member States gets one vote In the General Assembly and no Member State has veto power.

One commentator says:

If Palestine is accepted as a State by the UN General Assembly, then UN agencies such as the ILO, WHO, FAO, and ICAO would also regard Palestine as a state, and no US veto power could prevent Palestine's acceptance as a full member of such organizations. However accepting Palestine would mean an automatic cut-off of US funding to the organization, as occurred with UNESCO, which lost some $ 60 million in US contributions when it agreed to accept Palestine as a full member. Other organizations will presumably be very reluctant to commit financial suicide in order to satisfy Palestinian political aims.[4]

This statement, although seemingly acceptable in principle, has to be viewed with some caution with regard to Palestine vying for membership of ICAO. ICAO membership is acquired by a State through adherence to the Chicago Convention. Palestine's only recourse would be under Article 93 which provides that States other than those provided for in Articles 91 and 92 (a) may, subject to approval by any general international organization set up by the nations of the world to preserve peace, be admitted to participation in this Convention by means of a four-fifths vote of the Assembly and on such conditions as the Assembly may prescribe: provided that in each case the assent of any State invaded or attacked during the present war by the State seeking admission shall be necessary.

Article 93 *bis*

(a) Notwithstanding the provisions of Articles 91, 92 and 93 above:

(1) A State whose government the General Assembly of the United Nations has recommended be debarred from membership in international agencies established by or brought into relationship with the United Nations shall automatically cease to be a member of the International Civil Aviation Organization;

(2) A State which has been expelled from membership in the United Nations shall automatically cease to be a member of the International Civil Aviation Organization unless the General Assembly of the United Nations attaches to its act of expulsion a recommendation to the contrary.

(b) A State which ceases to be a member of the International Civil Aviation Organization as a result of the provisions of paragraph (a) above may, after approval by the General Assembly of the United Nations, be readmitted to the International Civil Aviation Organization upon application and upon approval by a majority of the Council.

(c) Members of the Organization which are suspended from the exercise of the rights and privileges of membership in the United Nations shall, upon the request of the latter, be suspended from the rights and privileges of membership in this Organization.

[4]http://www.canadafreepress.com/index.php/article/51293.

Article 94

Amendment of Convention

(a) Any proposed amendment to this Convention must be approved by a two-thirds vote of the Assembly and shall then come into force in respect of States which have ratified such amendment when ratified by the number of contracting States specified by the Assembly. The number so specified shall not be less than two-thirds of the total number of contracting States.

(b) If in its opinion the amendment is of such a nature as to justify this course, the Assembly in its resolution recommending adoption may provide that any State which has not ratified within a specified period after the amendment has come into force shall thereupon cease to be a member of the Organization and a party to the Convention.

R. Abeyratne, *Convention on International Civil Aviation*,
DOI 10.1007/978-3-319-00068-8_95,

Article 95
Denunciation of Convention

(a) **Any contracting State may give notice of denunciation of this Convention three years after its coming into effect by notification addressed to the Government of the United States of America, which shall at once inform each of the contracting States.**
(b) **Denunciation shall take effect one year from the date of the receipt of the notification and shall operate only as regards the State effecting the denunciation.**

R. Abeyratne, *Convention on International Civil Aviation*,

DOI 10.1007/978-3-319-00068-8_96,

Article 96

For the purpose of this Convention the expression:

(a) "Air service" means any scheduled air service performed by aircraft for the public transport of passengers, mail or cargo.
(b) "International air service" means an air service which passes through the air space over the territory of more than one State.
(c) "Airline" means any air transport enterprise offering or operating an international air service.
(d) "Stop for non-traffic purposes" means a landing for any purpose other than taking on or discharging passengers, cargo or mail.

Signature of Convention

IN WITNESS WHEREOF, the undersigned plenipotentiaries, having been duly authorized, sign this Convention on behalf of their respective governments on the dates appearing opposite their signatures.

DONE at Chicago the seventh day of December 1944, in the English language. A text drawn up in the English, French and Spanish languages, each of which shall be of equal authenticity, shall be open for signature at Washington, D.C. Both texts shall be deposited in the archives of the Government of the United States of America, and certified copies shall be transmitted by that Government to the governments of all the States which may sign or adhere to this Convention

R. Abeyratne, *Convention on International Civil Aviation*,
DOI 10.1007/978-3-319-00068-8_97,

Conclusion

The final consideration in this Commentary is whether the Chicago Convention is adequate for the current aviation scene. Clearly, the Chicago Convention cannot be superseded by a brand new instrument as many of its provisions are still relevant and useful. However, there has been at least one article on this subject, which recommends amendments to the Convention.[1] There is one provision that seriously requires review by the international community and that is Article 6 which states that

> no scheduled international air service may be operated over or into the territory of a contracting State, except with the special permission or other authorization of that State, and in accordance with the terms of such permission or authorization.

A detailed discussion on this issue has already been embarked on in this book. In addition, what could be said is that the Council of ICAO should firstly embark on a detailed study of whether open skies as a global concept would generally benefit the travelling public. Such a study could at least give some direction to the States on how to review their policies under Article 6.

As far as trends in liberalization go, privatization may continue steadily although the pace of privatization may be relatively slow unless there are turnarounds in strategic terms for the betterment of air transport and the environment remains conducive to privatization. Long-term traffic over the years to 2015 will average 5 % growth but growth rates will differ between key markets. Liberalization of existing regulatory constraints in air transport will also affect areas and markets hitherto unaffected and restrictions on market access, capacity, pricing and ownership and control will be progressively eroded making the airline industry truly global. There will be a blurring of boundaries between open skies and clear skies and liberalization will accelerate the process of industrial concentration within global alliances. A handful of very large airlines will carry 60–70 % of global air carriage and governments and regulatory authorities will attempt to control

[1] See Milde (1994), 401–452.

R. Abeyratne, *Convention on International Civil Aviation*,

DOI 10.1007/978-3-319-00068-8,

possible abusers of such dominance. Continued growth of traffic will bring immense pressure on airports and air traffic systems and environmentalists will bring pressure for reduction of airport noise and aircraft engine emissions. It is in this backdrop that the world aviation community gathered in Montreal in March 2003 in order to deliberate on future liberalization of Air transport.

The Council may wish to consider in the course of this study that, while the open skies policy sounds economically expedient, its implementation would undoubtedly phase out smaller carriers who are now offering competition in air transport and a larger spectrum of air transport to the consumer. Lower fares, different types of services and varied inflight service profiles are some of the features of the present system. It is desirable that a higher level of competitiveness prevails in the air transport industry, and to achieve this objective, preferential measures for carriers of developing countries would play a major role.

Another Consideration is that economic liberalization has implications for safety and security regulation, which need to be properly addressed at the national, bilateral, regional and global levels, as appropriate, in order to ensure continued safe, secure and orderly development of civil aviation. The Chicago Convention imposes responsibility on Contracting States for compliance with standards and practices related to safety and security. Therefore, irrespective of any change in economic regulation, safety and security must remain of paramount importance in the operation and development of air transport.

Convention on International Civil Aviation, Signed at Chicago, on 7 December 1944 (Chicago Convention)

Preamble

WHEREAS the future development of international civil aviation can greatly help to create and preserve friendship and understanding among the nations and peoples of the world, yet its abuse can become a threat to the general security; and

WHEREAS it is desirable to avoid friction and to promote that cooperation between nations and peoples upon which the peace of the world depends;

THEREFORE, the undersigned governments having agreed on certain principles and arrangements in order that international civil aviation may be developed in a safe and orderly manner and that international air transport services may be established on the basis of equality of opportunity and operated soundly and economically;

Have accordingly concluded this Convention to that end.

Part I. Air Navigation

Chapter I. General Principles and Application of the Convention

Article 1. Sovereignty

The contracting States recognize that every State has complete and exclusive sovereignty over the airspace above its territory.

Article 2. Territory

For the purposes of this Convention the territory of a State shall be deemed to be the land areas and territorial waters adjacent thereto under the sovereignty, suzerainty, protection or mandate of such State.

Article 3. Civil and State Aircraft

(a) This Convention shall be applicable only to civil aircraft, and shall not be applicable to state aircraft.

(b) Aircraft used in military, customs and police services shall be deemed to be state aircraft.

(c) No state aircraft of a contracting State shall fly over the territory of another State or land thereon without authorization by special agreement or otherwise, and in accordance with the terms thereof.

(d) The contracting States undertake, when issuing regulations for their state aircraft, that they will have due regard for the safety of navigation of civil aircraft.

Article 3 *bis*

(a) The contracting States recognize that every State must (a) refrain from resorting to the use of weapons against civil aircraft in flight and that, in case of interception, the lives of persons on board and the safety of aircraft must not be endangered. This provision shall not be interpreted as modifying in any way the rights and obligations of States set forth in the Charter of the United Nations.

(b) The contracting States recognize that every State, in the exercise of its sovereignty, is entitled to require the landing at some designated airport of a civil aircraft flying above its territory without authority or if there are reasonable grounds to conclude that it is being used for any purpose inconsistent with the aims of this Convention; it may also give such aircraft any other instructions to put an end to such violations. For this purpose, the contracting States may resort to any appropriate means consistent with relevant rules of international law, including the relevant provisions of this Convention, specifically paragraph a) of this Article. Each contracting State agrees to publish its regulations in force regarding the interception of civil aircraft.

(c) Every civil aircraft shall comply with an order given inconformity with paragraph b) of this Article. To this end each contracting State shall establish all necessary provisions in its national laws or regulations to make such compliance mandatory for any civil aircraft registered in that State or operated by an operator who has his principal place of business or permanent residence in that State. Each contracting State shall make any violation of such applicable laws or regulations punishable by severe penalties and shall submit the case to its competent authorities in accordance with its laws or regulations.

(d) Each contracting State shall take appropriate measures to prohibit the deliberate use of any civil aircraft registered in that State or operated by an operator who has his principal place of business or permanent residence in that State for any purpose inconsistent with the aims of this Convention. This provision shall not affect paragraph (a) or derogate from paragraph (b) and (c) of this Article.

Article 4. Misuse of Civil Aviation

Each contracting State agrees not to use civil aviation for any purpose inconsistent with the aims of this Convention.

Chapter II. Flight Over Territory of Contracting States

Article 5. Right of Non-Scheduled Flight

Each contracting State agrees that all aircraft of the other contracting States, being aircraft not engaged in scheduled international air services shall have the right, subject to the observance of the terms of this Convention, to make flights into or in transit non-stop across its territory and to make stops for non-traffic purposes without the necessity of obtaining prior permission, and subject to the right of the State flown over to require landing. Each contracting State nevertheless reserves the right, for reasons of safety of flight, to require aircraft desiring to proceed over regions which are inaccessible or without adequate air navigation facilities to follow prescribed routes, or to obtain special permission for such flights.

Such aircraft, if engaged in the carriage of passengers, cargo, or mail for remuneration or hire on other than scheduled international air services, shall also, subject to the provisions of Article 7, have the privilege of taking on or discharging passengers, cargo, or mail, subject to the right of any State where such embarkation or discharge takes place to impose such regulations, conditions or limitations as it may consider desirable.

Article 6. Scheduled Air Services

No scheduled international air service may be operated over or into the territory of a contracting State, except with the special permission or other authorization of that State, and in accordance with the terms of such permission or authorization.

Article 7. Cabotage

Each contracting State shall have the right to refuse permission to the aircraft of other contracting States to take on in its territory passengers, mail and cargo carried for remuneration or hire and destined for another point within its territory. Each contracting State undertakes not to enter into any arrangements which specifically grant any such privilege on an exclusive basis to any other State or an airline of any other State, and not to obtain any such exclusive privilege from any other State.

Article 8. Pilotless Aircraft

No aircraft capable of being flown without a pilot shall be flown without a pilot over the territory of a contracting State without special authorization by that State and in accordance with the terms of such authorization. Each contracting State undertakes to insure that the flight of such aircraft without a pilot in regions open to civil aircraft shall be so controlled as to obviate danger to civil aircraft.

Article 9. Prohibited Areas

(a) Each contracting State may, for reasons of military necessity or public safety, restrict or prohibit uniformly the aircraft of other States from flying over certain areas of its territory, provided that no distinction in this respect is made between the aircraft of the State whose territory is involved, engaged in international scheduled airline services, and the aircraft of the other contracting States likewise engaged. Such prohibited areas shall be of reasonable extent and location so as not to interfere unnecessarily with air navigation. Descriptions of such prohibited areas in the territory of a contracting State, as well as any subsequent alterations therein, shall be communicated as soon as possible to the other contracting States and to the International Civil Aviation Organization.

(b) Each contracting State reserves also the right, in exceptional circumstances or during a period of emergency, or in the interest of public safety, and with immediate effect, temporarily to restrict or prohibit flying over the whole or any part of its territory, on condition that such restriction or prohibition shall be applicable without distinction of nationality to aircraft of all other States.

(c) Each contracting State, under such regulations as it may prescribe, may require any aircraft entering the areas contemplated in subparagraphs (a) or (b) above to effect a landing as soon as practicable thereafter at some designated airport within its territory.

Article 10. Landing at Customs Airport

Except in a case where, under the terms of this Convention or a special authorization, aircraft are permitted to cross the territory of a contracting State without landing, every aircraft which enters the territory of a contracting State shall, if the regulations of that State so require, land at an airport designated by that State for the purpose of customs and other examination. On departure from the territory of a contracting State, such aircraft shall depart from a similarly designated customs airport. Particulars of all designated customs airports shall be published by the State and transmitted to the International Civil Aviation Organization established under Part II of this Convention for communication to all other contracting States.

Article 11. Applicability of Air Regulations

Subject to the provisions of this Convention, the laws and regulations of a contracting State relating to the admission to or departure from its territory of aircraft engaged in international air navigation, or to the operation and navigation of such aircraft while within its territory, shall be applied to the aircraft of all contracting States without distinction as to nationality, and shall be complied with by such aircraft upon entering or departing from or while within the territory of that State.

Article 12. Rules of the Air

Each contracting State undertakes to adopt measures to insure that every aircraft flying over or maneuvering within its territory and that every aircraft carrying its nationality mark, wherever such aircraft may be, shall comply with the rules and regulations relating to the flight and maneuver of aircraft there in force. Each contracting State undertakes to keep its own regulations in these respects uniform, to the greatest possible extent, with those established from time to time under this Convention. Over the high seas, the rules in force shall be those established under this Convention. Each contracting State undertakes to insure the prosecution of all persons violating the regulations applicable.

Article 13. Entry and Clearance Regulations

The laws and regulations of a contracting State as to the admission to or departure from its territory of passengers, crew or cargo of aircraft, such as regulations relating to entry, clearance, immigration, passports, customs, and quarantine shall be complied with by or on behalf of such passengers, crew or cargo upon entrance into or departure from, or while within the territory of that State.

Article 14. Prevention of Spread of Disease

Each contracting State agrees to take effective measures to prevent the spread by means of air navigation of cholera, typhus (epidemic), smallpox, yellow fever, plague, and such other communicable diseases as the contracting States shall from time to time decide to designate, and to that end contracting States will keep in close consultation with the agencies concerned with international regulations relating to sanitary measures applicable to aircraft. Such consultation shall be without prejudice to the application of any existing international convention on this subject to which the contracting States may be parties.

Article 15. Airport and Similar Charges

Every airport in a contracting State which is open to public use by its national aircraft shall likewise, subject to the provisions of Article 68, be open under uniform conditions to the aircraft of all the other contracting States. The like uniform conditions shall apply to the use, by aircraft of every contracting State, of all air navigation facilities, including radio and meteorological services, which may be provided for public use for the safety and expedition of air navigation.

Any charges that may be imposed or permitted to be imposed by a contracting State for the use of such airports and air navigation facilities

by the aircraft of any other contracting State shall not be higher,
(a) As to aircraft not engaged in scheduled international air services, than those that would be paid by its national aircraft of the same class engaged in similar operations, and
(b) As to aircraft engaged in scheduled international air services, than those that would be paid by its national aircraft engaged in similar international air services.
All such charges shall be published and communicated to the International Civil Aviation Organization: provided that, upon representation by an interested contracting State, the charges imposed for the use of airports and other facilities shall be subject to review by the Council, which shall report and make recommendations thereon for the consideration of the State or States concerned. No fees, dues or other charges shall be imposed by any contracting State in respect solely of the right of transit over or entry into or exit from its territory of any aircraft of a contracting State or persons or property thereon.

Article 16. Search of Aircraft

The appropriate authorities of each of the contracting States shall have the right, without unreasonable delay, to search aircraft of the other contracting States on landing or departure, and to inspect the certificates and other documents prescribed by this Convention.

Chapter III. Nationality of Aircraft

Article 17. Nationality of Aircraft

Aircraft have the nationality of the State in which they are registered.

Article 18. Dual Registration

An aircraft cannot be validly registered in more than one State, but its registration may be changed from one State to another.

Article 19. National Laws Governing Registration

The registration or transfer of registration of aircraft in any contracting State shall be made in accordance with its law and regulations.

Article 20. Display of Marks

Every aircraft engaged in international air navigation shall bear its appropriate nationality and registration marks.

Article 21. Report of Registrations

Each contracting State undertakes to supply to any other contracting State or to the International Civil Aviation Organization, on demand, information concerning the registration and ownership of any particular aircraft registered in that State. In addition, each contracting State shall furnish reports to the International Civil Aviation Organization, under such regulations as the latter may prescribe, giving such pertinent data as can be made available concerning the ownership and control of aircraft registered in that State and habitually engaged in international air navigation. The data thus obtained by the International Civil Aviation Organization shall be made available by it on request to the other contracting States.

Chapter IV. Measures to Facilitate Air Navigation

Article 22. Facilitation of Formalities

Each contracting State agrees to adopt all practicable measures, through the issuance of special regulations or otherwise, to facilitate and expedite navigation by aircraft between the territories of contracting States, and to prevent unnecessary delays to aircraft, crews, passengers and cargo, especially in the administration of the laws relating to immigration, quarantine, customs and clearance.

Article 23. Customs and Immigration Procedures

Each contracting State undertakes, so far as it may find practicable, to establish customs and immigration procedures affecting international air navigation in accordance with the practices which may be established or recommended from time to time, pursuant to this Convention. Nothing in this Convention shall be construed as preventing the establishment of customs-free airports.

Article 24. Customs Duty

(a) Aircraft on a flight to, from, or across the territory of another contracting State shall be admitted temporarily free of duty, subject to the customs regulations of the State. Fuel, lubricating oils, spare parts, regular equipment and aircraft stores on board an aircraft of a contracting State, on arrival in the territory of another contracting State and retained on board on leaving the territory of that State shall be exempt from customs duty, inspection fees or similar national or local duties and charges. This exemption shall not apply to any quantities or articles unloaded, except in accordance with the customs regulations of the State, which may require that they shall be kept under customs supervision.

(b) Spare parts and equipment imported into the territory of a contracting State for incorporation in or use on an aircraft of another contracting State engaged in international air navigation shall be admitted free of customs duty, subject to compliance with the regulations of the State concerned, which may provide that the articles shall be kept under customs supervision and control.

Article 25. Aircraft in Distress

Each contracting State undertakes to provide such measures of assistance to aircraft in distress in its territory as it may find practicable, and to permit, subject to control by its own authorities, the owners of the aircraft or authorities of the State in which the aircraft is registered to provide such measures of assistance as may be necessitated by the circumstances. Each contracting State, when undertaking search for missing aircraft, will collaborate in coordinated measures which may be recommended from time to time pursuant to this Convention.

Article 26. Investigation of Accidents

In the event of an accident to an aircraft of a contracting State occurring in the territory of another contracting State, and involving death or serious injury, or indicating serious technical defect in the aircraft or air navigation facilities, the State in which the accident occurs will institute an inquiry into the circumstances of the accident, in accordance, so far as its laws permit, with the procedure which may be recommended by the International Civil Aviation Organization. The State in which the aircraft is registered shall be given the opportunity to appoint observers to be present at the inquiry and the State holding the inquiry shall communicate the report and findings in the matter to that State.

Article 27. Exemption from Seizure on Patent Claims

(a) While engaged in international air navigation, any authorized entry of aircraft of a contracting State into the territory of another contracting State or authorized transit across the territory of such State with or without landings shall not entail any seizure or detention of the aircraft or any claim against the owner or operator thereof or any other interference therewith by or on behalf of such State or any person therein, on the ground that the construction, mechanism, parts, accessories or operation of the aircraft is an infringement of any patent, design, or model duly granted or registered in the State whose territory is entered by the aircraft, it being agreed that no deposit of security in connection with the foregoing exemption from seizure or detention of the aircraft shall in any case be required in the State entered by such aircraft.

(b) The provisions of paragraph (a) of this Article shall also be applicable to the storage of spare parts and spare equipment for the aircraft and the right to use and install the same in the repair of an aircraft of a contracting State in the territory of any other contracting State, provided that any patented part or equipment so stored shall not be sold or distributed internally in or exported commercially from the contracting State entered by the aircraft.

(c) The benefits of this Article shall apply only to such States, parties to this Convention, as either (1) are parties to the International Convention for the Protection of Industrial Property and to any amendments thereof; or (2) have enacted patent laws which recognize and give adequate protection to inventions made by the nationals of the other States parties to this Convention.

Article 28. Air Navigation Facilities and Standard Systems

Each contracting State undertakes, so far as it may find practicable, to:

(a) Provide, in its territory, airports, radio services, meteorological services and other air navigation facilities to facilitate international air navigation, in accordance with the

standards and practices recommended or established from time to time, pursuant to this Convention;
(b) Adopt and put into operation the appropriate standard systems of communications procedure, codes, markings, signals, lighting and other operational practices and rules which may be recommended or established from time to time, pursuant to this Convention;
(c) Collaborate in international measures to secure the publication of aeronautical maps and charts in accordance with standards which may be recommended or established from time to time, pursuant to this Convention.

Chapter V. Conditions to Be Fulfilled with Respect to Aircraft

Article 29. Documents Carried in Aircraft

Every aircraft of a contracting State, engaged in international navigation, shall carry the following documents in conformity with the conditions prescribed in this Convention:
(a) Its certificate of registration;
(b) Its certificate of airworthiness;
(c) The appropriate licenses for each member of the crew;
(d) Its journey log book;
(e) If it is equipped with radio apparatus, the aircraft radio station license;
(f) If it carries passengers, a list of their names and places of embarkation and destination;
(g) If it carries cargo, a manifest and detailed declarations of the cargo.

Article 30. Aircraft Radio Equipment

(a) Aircraft of each contracting State may, in or over the territory of other contracting States, carry radio transmitting apparatus only if a license to install and operate such apparatus has been issued by the appropriate authorities of the State in which the aircraft is registered. The use of radio transmitting apparatus in the territory of the contracting State whose territory is flown over shall be in accordance with the regulations prescribed by that State.
(b) Radio transmitting apparatus may be used only by members of the flight crew who are provided with a special license for the purpose, issued by the appropriate authorities of the State in which the aircraft is registered.

Article 31. Certificates of Airworthiness

Every aircraft engaged in international navigation shall be provided with a certificate of airworthiness issued or rendered valid by the State in which it is registered.

Article 32. Licenses of Personnel

(a) The pilot of every aircraft and the other members of the operating crew of every aircraft engaged in international navigation shall be provided with certificates of competency and licenses issued or rendered valid by the State in which the aircraft is registered.

(b) Each contracting State reserves the right to refuse to recognize, for the purpose of flight above its own territory, certificates of competency and licenses granted to any of its nationals by another contracting State.

Article 33. Recognition of Certificates and Licenses

Certificates of airworthiness and certificates of competency and licenses issued or rendered valid by the contracting State in which the aircraft is registered, shall be recognized as valid by the other contracting States, provided that the requirements under which such certificates or licenses were issued or rendered valid are equal to or above the minimum standards which may be established from time to time pursuant to this Convention.

Article 34. Journey Log Books

There shall be maintained in respect of every aircraft engaged in international navigation a journey log book in which shall be entered particulars of the aircraft, its crew and of each journey, in such form as may be prescribed from time to time pursuant to this Convention.

Article 35. Cargo Restrictions

(a) No munitions of war or implements of war may be carried in or above the territory of a State in aircraft engaged in international navigation, except by permission of such State. Each State shall determine by regulations what constitutes munitions of war or implements of war for the purposes of this Article, giving due consideration, for the purposes of uniformity, to such recommendations as the International Civil Aviation Organization may from time to time make.

(b) Each contracting State reserves the right, for reasons of public order and safety, to regulate or prohibit the carriage in or above its territory of articles other than those enumerated in paragraph (a): provided that no distinction is made in this respect between its national aircraft engaged in international navigation and the aircraft of the other States so engaged; and provided further that no restriction shall be imposed which may interfere with the carriage and use on aircraft of apparatus necessary for the operation or navigation of the aircraft or the safety of the personnel or passengers.

Article 36. Photographic Apparatus

Each contracting State may prohibit or regulate the use of photographic apparatus in aircraft over its territory.

Chapter VI. International Standards and Recommended Practices

Article 37. Adoption of International Standards and Procedures

Each contracting State undertakes to collaborate in securing the highest practicable degree of uniformity in regulations, standards, procedures, and organization in relation to aircraft, personnel, airways and auxiliary services in all matters in which such uniformity will facilitate and improve air navigation.

To this end the International Civil Aviation Organization shall adopt and amend from time to time, as may be necessary, international standards and recommended practices and procedures dealing with:

(a) Communications systems and air navigation aids, including ground marking;
(b) Characteristics of airports and landing areas;
(c) Rules of the air and air traffic control practices;
(d) Licensing of operating and mechanical personnel;
(e) Airworthiness of aircraft;
(f) Registration and identification of aircraft;
(g) Collection and exchange of meteorological information;
(h) Log books;
(i) Aeronautical maps and charts;
(j) Customs and immigration procedures;
(k) Aircraft in distress and investigation of accidents;

and such other matters concerned with the safety, regularity, and efficiency of air navigation as may from time to time appear appropriate.

Article 38. Departures from International Standards and Procedures

Any State which finds it impracticable to comply in all respects with any such international standard or procedure, or to bring its own regulations or practices into full accord with any international standard or procedure after amendment of the latter, or which deems it necessary to adopt regulations or practices differing in any particular respect from those established by an international standard, shall give immediate notification to the International Civil Aviation Organization of the differences between its own practice and that established by the international standard. In the case of amendments to international standards, any State which does not make the appropriate amendments to its own regulations or practices shall give notice to the Council within sixty days of the adoption of the amendment to the international standard, or indicate the action which it proposes to take. In any such case, the Council shall make immediate notification to all other states of the difference which exists between one or more features of an international standard and the corresponding national practice of that State.

Article 39. Endorsement of Certificates and Licenses

(a) Any aircraft or part thereof with respect to which there exists an international standard of airworthiness or performance, and which failed in any respect to satisfy that standard at the time of its certification, shall have endorsed on or attached to its airworthiness certificate a complete enumeration of the details in respect of which it so failed.

(b) Any person holding a license who does not satisfy in full the conditions laid down in the international standard relating to the class of license or certificate which he holds shall have endorsed on or attached to his license a complete enumeration of the particulars in which he does not satisfy such conditions.

Article 40. Validity of Endorsed Certificates and Licenses

No aircraft or personnel having certificates or licenses so endorsed shall participate in international navigation, except with the permission of the State or States whose territory is entered. The registration or use of any such aircraft, or of any certificated aircraft part, in any State other than that in which it was originally certificated shall be at the discretion of the State into which the aircraft or part is imported.

Article 41. Recognition of Existing Standards of Airworthiness

The provisions of this Chapter shall not apply to aircraft and aircraft equipment of types of which the prototype is submitted to the appropriate national authorities for certification prior to a date three years after the date of adoption of an international standard of airworthiness for such equipment.

Article 42. Recognition of Existing Standards of Competency of Personnel

The provisions of this Chapter shall not apply to personnel whose licences are originally issued prior to a date one year after initial adoption of an international standard of qualification for such personnel; but they shall in any case apply to all personnel whose licenses remain valid five years after the date of adoption of such standard.

Part II. The International Civil Aviation Organization

Chapter VII. The Organization

Article 43. Name and Composition

An organization to be named the International Civil Aviation Organization is formed by the Convention. It is made up of an Assembly, a Council, and such other bodies as may be necessary.

Article 44. Objectives

The aims and objectives of the Organization are to develop the principles and techniques of international air navigation and to foster the planning and development of international air transport so as to:

(a) Insure the safe and orderly growth of international civil aviation throughout the world;

(b) Encourage the arts of aircraft design and operation for peaceful purposes;

(c) Encourage the development of airways, airports, and air navigation facilities for international civil aviation;

(d) Meet the needs of the peoples of the world for safe, regular, efficient and economical air transport;

(e) Prevent economic waste caused by unreasonable competition;

(f) Insure that the rights of contracting States are fully respected and that every contracting State has a fair opportunity to operate international airlines;

(g) Avoid discrimination between contracting States;

(h) Promote safety of flight in international air navigation;

(i) Promote generally the development of all aspects of international civil aeronautics.

Article 45. Permanent Seat

The permanent seat of the Organization shall be at such place as shall be determined at the final meeting of the Interim Assembly of the Provisional International Civil Aviation Organization set up by the Interim Agreement on International Civil Aviation signed at Chicago on December 7, 1944. The seat may be temporarily transferred elsewhere by decision of the Council, and otherwise than temporarily by decision of the Assembly, such decision to be taken by the number of votes specified by the Assembly. The number of votes so specified will not be less than three-fifths of the total number of contracting States.

Article 46. First Meeting of Assembly

The first meeting of the Assembly shall be summoned by the Interim Council of the above-mentioned Provisional Organization as soon as the Convention has come into force, to meet at a time and place to be decided by the Interim Council.

Article 47. Legal Capacity

The Organization shall enjoy in the territory of each contracting State such legal capacity as may be necessary for the performance of its functions. Full juridical personality shall be granted wherever compatible with the constitution and laws of the State concerned.

Chapter VIII. The Assembly

Article 48. Meetings of the Assembly and Voting

(a) The Assembly shall meet not less than once in three years and shall be convened by the Council at a suitable time and place. An extraordinary meeting of the Assembly may be held at any time upon the call of the Council or at the request of not less than one-fifth of the total number of contracting States addressed to the Secretary General.

(b) All contracting States shall have an equal right to be represented at the meetings of the Assembly and each contracting State shall be entitled to one vote. Delegates representing contracting States may be assisted by technical advisers who may participate in the meetings but shall have no vote.

(c) A majority of the contracting States is required to constitute a quorum for the meetings of the Assembly. Unless otherwise provided in this Convention, decisions of the Assembly shall be taken by a majority of the votes cast.

Article 49. Powers and Duties of the Assembly

The powers and duties of the Assembly shall be to:

(a) Elect at each meeting its President and other officers;

(b) Elect the contracting States to be represented on the Council, in accordance with the provisions of Chapter IX;

(c) Examine and take appropriate action on the reports of the Council and decide on any matter referred to it by the Council;

(d) Determine its own rules of procedure and establish such subsidiary commissions as it may consider to be necessary or desirable;

(e) Vote annual budgets and determine the financial arrangements of the Organization, in accordance with the provisions of Chapter XII;

(f) Review expenditures and approve the accounts of the Organization;

(g) Refer, at its discretion, to the Council, to subsidiary commissions, or to any other body any matter within its sphere of action;

(h) Delegate to the Council the powers and authority necessary or desirable for the discharge of the duties of the Organization and revoke or modify the delegations of authority at any time;

(i) Carry out the appropriate provisions of Chapter XIII;

(j) Consider proposals for the modification or amendment of the provisions of this Convention and, if it approves of the proposals, recommend them to the contracting States in accordance with the provisions of Chapter XXI;

(k) Deal with any matter within the sphere of action of the Organization not specifically assigned to the Council.

Chapter IX. The Council

Article 50. Composition and Election of Council

(a) The Council shall be a permanent body responsible to the Assembly. It shall be composed of thirty-three contracting States elected by the Assembly. An election shall be held at the first meeting of the Assembly and thereafter every three years, and the members of the Council so elected shall hold office until the next following election.

(b) In electing the members of the Council, the Assembly shall give adequate representation to (1) the States of chief importance in air transport; (2) the States not otherwise included which make the largest contribution to the provision of facilities for international civil air navigation; and (3) the States not otherwise included whose designation will insure that all the major geographic areas of the world are represented on the Council. Any vacancy on the Council shall be filled by the Assembly as soon as possible; any contracting State so elected to the Council shall hold office for the unexpired portion of its predecessor's term of office.

(c) No representative of a contracting State on the Council shall be actively associated with the operation of an international air service or financially interested in such a service.

Article 51. President of Council

The Council shall elect its President for a term of three years. He may be reelected. He shall have no vote. The Council shall elect from among its members one or more Vice Presidents who shall retain their right to vote when serving as acting President. The President need not be selected from among the representatives of the members of the Council but, if a representative is elected, his seat shall be deemed vacant and it shall be filled by the State which he represented. The duties of the President shall be to:

(a) Convene meetings of the Council, the Air Transport Committee, and the Air Navigation Commission;

(b) Serve as representative of the Council; and

(c) Carry out on behalf of the Council the functions which the Council assigns to him.

Article 52. Voting in Council

Decisions by the Council shall require approval by a majority of its members. The Council may delegate authority with respect to any particular matter to a committee of its members. Decisions of any committee of the Council may be appealed to the Council by any interested contracting State.

Article 53. Participation Without a Vote

Any contracting State may participate, without a vote, in the consideration by the Council and by its committees and commissions of any question which especially affects its interests. No member of the Council shall vote in the consideration by the Council of a dispute to which it is a party.

Article 54. Mandatory Functions of Council

The Council shall:

(a) Submit annual reports to the Assembly;

(b) Carry out the directions of the Assembly and discharge the duties and obligations which are laid on it by this Convention;

(c) Determine its organization and rules of procedure;

(d) Appoint and define the duties of an Air Transport Committee, which shall be chosen from among the representatives of the members of the Council, and which shall be responsible to it;

(e) Establish an Air Navigation Commission, in accordance with the provisions of Chapter X;

(f) Administer the finances of the Organization in accordance with the provisions of Chapters XII and XV;

(g) Determine the emoluments of the President of the Council;

(h) Appoint a chief executive officer who shall be called the Secretary General, and make provision for the appointment of such other personnel as may be necessary, in accordance with the provisions of Chapter XI;

(i) Request, collect, examine and publish information relating to the advancement of air navigation and the operation of international air services, including information about the costs of operation and particulars of subsidies paid to airlines from public funds;

(j) Report to contracting States any infraction of this Convention, as well as any failure to carry out recommendations or determinations of the Council;

(k) Report to the Assembly any infraction of this Convention where a contracting State has failed to take appropriate action within a reasonable time after notice of the infraction;

(l) Adopt, in accordance with the provisions of Chapter VI of this Convention, international standards and recommended practices; for convenience, designate them as Annexes to this Convention; and notify all contracting States of the action taken;

(m) Consider recommendations of the Air Navigation Commission for amendment of the Annexes and take action in accordance with the provisions of Chapter XX;

(n) Consider any matter relating to the Convention which any contracting State refers to it.

Article 55. Permissive Functions of Council

The Council may:

(a) Where appropriate and as experience may show to be desirable, create subordinate air transport commissions on a regional or other basis and define groups of states or airlines with or through which it may deal to facilitate the carrying out of the aims of this Convention;

(b) Delegate to the Air Navigation Commission duties additional to those set forth in the Convention and revoke or modify such delegations of authority at any time;

(c) Conduct research into all aspects of air transport and air navigation which are of international importance, communicate the results of its research to the contracting States, and facilitate the exchange of information between contracting States on air transport and air navigation matters;

(d) Study any matters affecting the organization and operation of international air transport, including the international ownership and operation of international air services on trunk routes, and submit to the Assembly plans in relation thereto;

(e) Investigate, at the request of any contracting State, any situation which may appear to present avoidable obstacles to the development of international air navigation; and, after such investigation, issue such reports as may appear to it desirable.

Chapter X. The Air Navigation Commission

Article 56. Nomination and Appointment of Commission

The Air Navigation Commission shall be composed of fifteen members appointed by the Council from among persons nominated by contracting States. These persons shall have suitable qualifications and experience in the science and practice of aeronautics. The Council shall request all contracting States to submit nominations. The President of the Air Navigation Commission shall be appointed by the Council.

Article 57. Duties of Commission

The Air Navigation Commission shall:

(a) Consider, and recommend to the Council for adoption, modifications of the Annexes to this Convention;

(b) Establish technical subcommissions on which any contracting State may be represented, if it so desires;

(c) Advise the Council concerning the collection and communication to the contracting States of all information which it considers necessary and useful for the advancement of air navigation.

Chapter XI. Personnel

Article 58. Appointment of Personnel

Subject to any rules laid down by the Assembly and to the provisions of this Convention, the Council shall determine the method of appointment and of termination of appointment, the training, and the salaries, allowances, and conditions of service of the Secretary General and other personnel of the Organization, and may employ or make use of the services of nationals of any contracting State.

Article 59. International Character of Personnel

The President of the Council, the Secretary General, and other personnel shall not seek or receive instructions in regard to the discharge of their responsibilities from any authority external to the Organization. Each contracting State undertakes fully to respect the international character of the responsibilities of the personnel and not to seek to influence any of its nationals in the discharge of their responsibilities.

Article 60. Immunities and Privileges of Personnel

Each contracting State undertakes, so far as possible under its constitutional procedure, to accord to the President of the Council, the Secretary General, and the other personnel of the Organization, the immunities and privileges which are accorded to corresponding personnel of other public international organizations. If a general international agreement on the immunities and privileges of international civil servants is arrived at, the immunities and privileges accorded to the President, the Secretary General, and the other personnel of the Organization shall be the immunities and privileges accorded under that general international agreement.

Chapter XII. Finance

Article 61. Budget and Apportionment of Expenses

The Council shall submit to the Assembly annual budgets, annual statements of accounts and estimates of all receipts and expenditures. The Assembly shall vote the budgets with whatever modification it sees fit to prescribe, and, with the exception of assessments under Chapter XV to States consenting thereto, shall apportion the expenses of the Organization among the contracting States on the basis which it shall from time to time determine.

Article 62. Suspension of Voting Power

The Assembly may suspend the voting power in the Assembly and in the Council of any contracting State that fails to discharge within a reasonable period its financial obligations to the Organization.

Article 63. Expenses of Delegations and Other Representatives

Each contracting State shall bear the expenses of its own delegation to the Assembly and the remuneration, travel, and other expenses of any person whom it appoints to serve on the Council, and of its nominees or representatives on any subsidiary committees or commissions of the Organization.

Chapter XIII. Other International Arrangements

Article 64. Security Arrangements

The Organization may, with respect to air matters within its competence directly affecting world security, by vote of the Assembly enter into appropriate arrangements with any general organization set up by the nations of the world to preserve peace.

Article 65. Arrangements with Other International Bodies

The Council, on behalf of the Organization, may enter into agreements with other international bodies for the maintenance of common services and for common arrangements concerning personnel and, with the approval of the Assembly, may enter into such other arrangements as may facilitate the work of the Organization.

Article 66. Functions Relating to Other Agreements

(a) The Organization shall also carry out the functions placed upon it by the International Air Services Transit Agreement and by the International Air Transport Agreement drawn up at Chicago on December 7, 1944, in accordance with the terms and conditions therein set forth.

(b) Members of the Assembly and the Council who have not accepted the International Air Services Transit Agreement or the International Air Transport Agreement drawn up at Chicago on December 7, 1944 shall not have the right to vote on any questions referred to the Assembly or Council under the provisions of the relevant Agreement.

Part III. International Air Transport

Chapter XIV. Information and Reports

Article 67. File Reports with Council

Each contracting State undertakes that its international airlines shall, in accordance with requirements laid down by the Council, file with the Council traffic reports, cost statistics and financial statements showing among other things all receipts and the sources thereof.

Chapter XV. Airports and Other Air Navigation Facilities

Article 68. Designation of Routes and Airports

Each contracting State may, subject to the provisions of this Convention, designate the route to be followed within its territory by any international air service and the airports which any such service may use.

Article 69. Improvement of Air Navigation Facilities

If the Council is of the opinion that the airports or other air navigation facilities, including radio and meteorological services, of a contracting State are not reasonably adequate for the safe, regular, efficient, and economical operation of international air services, present or contemplated, the Council shall consult with the State directly concerned, and other States affected, with a view to finding means by which the situation may be remedied, and may make recommendations for that purpose. No contracting State shall be guilty of an infraction of this Convention if it fails to carry out these recommendations.

Article 70. Financing of Air Navigation Facilities

A contracting State, in the circumstances arising under the provisions of Article 69, may conclude an arrangement with the Council for giving effect to such recommendations. The State may elect to bear all of the costs involved in any such arrangement. If the State does not so elect, the Council may agree, at the request of the State, to provide for all or a portion of the costs.

Article 71. Provision and Maintenance of facilities by Council

If a contracting State so requests, the Council may agree to provide, man, maintain, and administer any or all of the airports and other air navigation facilities including radio and meteorological services, required in its territory for the safe, regular, efficient and economical operation of the international air services of the other contracting States, and may specify just and reasonable charges for the use of the facilities provided.

Article 72. Acquisition or Use of Land

Where land is needed for facilities financed in whole or in part by the Council at the request of a contracting State, that State shall either provide the land itself, retaining title if it wishes, or facilitate the use of the land by the Council on just and reasonable terms and in accordance with the laws of the State concerned.

Article 73. Expenditure and Assessment of Funds

Within the limit of the funds which may be made available to it by the Assembly under Chapter XII, the Council may make current expenditures for the purposes of this Chapter from the general funds of the Organization. The Council shall assess the capital funds required for the purposes of this Chapter in previously agreed proportions over a reasonable period of time to the contracting States consenting thereto whose airlines use the facilities. The Council may also assess to States that consent any working funds that are required.

Article 74. Technical Assistance and Utilization of Revenues

When the Council, at the request of a contracting State, advances funds or provides airports or other facilities in whole or in part, the arrangement may provide, with the consent of that State, for technical assistance in the supervision and operation of the airports and other facilities, and for the payment, from the revenues derived from the operation of the airports and other facilities, of the operating expenses of the airports and the other facilities, and of interest and amortization charges.

Article 75. Taking Over of Facilities from Council

A contracting State may at any time discharge any obligation into which it has entered under Article 70, and take over airports and other facilities which the Council has provided in its territory pursuant to the provisions of Articles 71 and 72, by paying to the Council an amount which in the opinion of the Council is reasonable in the circumstances. If the State considers that the amount fixed by the Council is unreasonable it may appeal to the Assembly against the decision of the Council and the Assembly may confirm or amend the decision of the Council.

Article 76. Return of Funds

Funds obtained by the Council through reimbursement under Article 75 and from receipts of interest and amortization payments under Article 74 shall, in the case of advances originally financed by States under Article 73, be returned to the States which were originally assessed in the proportion of their assessments, as determined by the Council.

Chapter XVI. Joint Operating Organizations and Pooled Services

Article 77. Joint Operating Organizations Permitted

Nothing in this Convention shall prevent two or more contracting States from constituting joint air transport operating organizations or international operating agencies and from pooling their air services on any routes or in any regions, but such organizations or agencies and such pooled services shall be subject to all the provisions of this Convention, including those relating to the registration of agreements with the Council. The Council shall determine in what manner the provisions of this Convention relating to nationality of aircraft shall apply to aircraft operated by international operating agencies.

Article 78. Function of Council

The Council may suggest to contracting States concerned that they form joint organizations to operate air services on any routes or in any regions.

Article 79. Participation in Operating Organizations

A State may participate in joint operating organizations or in pooling arrangements, either through its government or through an airline company or companies designated by its government. The companies may, at the sole discretion of the State concerned, be state-owned or partly stateowned or privately owned.

Part IV. Final Provisions

Chapter XVII. Other Aeronautical Agreements and Arrangements

Article 80. Paris and Habana Conventions

Each contracting State undertakes, immediately upon the coming into force of this Convention, to give notice of denunciation of the Convention relating to the Regulation of Aerial Navigation signed at Paris on October 13, 1919 or the Convention on Commercial Aviation signed at Habana on February 20, 1928, if it is a party to either. As between contracting States, this Convention supersedes the Conventions of Paris and Habana previously referred to.

Article 81. Registration of Existing Agreements

All aeronautical agreements which are in existence on the coming into force of this Convention, and which are between a contracting State and any other State or between an airline of a contracting State and any other State or the airline of any other State, shall be forthwith registered with the Council.

Article 82. Abrogation of Inconsistent Arrangements

The contracting States accept this Convention as abrogating all obligations and understandings between them which are inconsistent with its terms, and undertake not to enter into any such obligations and understandings. A contracting State which, before becoming a member of the Organization has undertaken any obligations toward a non-contracting State or a national of a contracting State or of a non-contracting State inconsistent with the terms of this Convention, shall take immediate steps to procure its release from the obligations. If an airline of any contracting State has entered into any such inconsistent obligations, the State of which it is a national shall use its best efforts to secure their termination forthwith and shall in any event cause them to be terminated as soon as such action can lawfully be taken after the coming into force of this Convention.

Article 83. Registration of New Arrangements

Subject to the provisions of the preceding Article, any contracting State may make arrangements not inconsistent with the provisions of this Convention. Any such arrangement shall be forthwith registered with the Council, which shall make it public as soon as possible.

Article 83 *bis*. Transfer of Certain Functions and Duties

(a) Notwithstanding the provisions of Articles 12, 30, 31 and 32 (a), when an aircraft registered in a contracting State is operated pursuant to an agreement for the lease, charter or interchange of the aircraft or any similar arrangement by an operator who has his principal place of business or, if he has no such place of business, his permanent residence in another contracting State, the State registry may, by agreement with such other State, transfer to it all or part of its functions and duties as State registry in respect of that aircraft under Articles 12, 30, 31 and 32 (a). The State of registry shall be relieved of responsibility in respect of the functions and duties transferred.

(b) The transfer shall not have effect in respect of other contracting States before either the agreement between States in which it is embodied has been registered with the Council and made public pursuant to Article 83 or the existence and scope of the agreement have been directly communicated to the authorities of the other contracting State or States concerned by a State party to the agreement.

(c) The provisions of paragraph (a) and (b) above shall also be applicable to cases covered by Article 77.

Chapter XVIII. Disputes and Default

Article 84. Settlement of Disputes

If any disagreement between two or more contracting States relating to the interpretation or application of this Convention and its Annexes cannot be settled by negotiation, it shall, on the application of any State concerned in the disagreement, be decided by the Council. No member of the Council shall vote in the consideration by the Council of any dispute to which it is a party. Any contracting State may, subject to Article 85, appeal from the decision of the Council to an ad hoc arbitral tribunal agreed upon with the other parties to the dispute or to the Permanent Court of International Justice. Any such appeal shall be notified to the Council within sixty days of receipt of notification of the decision of the Council.

Article 85. Arbitration Procedure

If any contracting State party to a dispute in which the decision of the Council is under appeal has not accepted the Statute of the Permanent Court of International Justice and the contracting States parties to the dispute cannot agree on the choice of the arbitral tribunal, each of the contracting States parties to the dispute shall name a single arbitrator who shall name an umpire. If either contracting State party to the dispute fails to name an arbitrator within a period of three months from the date of the appeal, an arbitrator shall be named on behalf of that State by the President of the Council from a list of qualified and available persons maintained by the Council. If, within thirty days, the arbitrators cannot agree on an umpire, the President of the Council shall designate an umpire from the list previously referred to. The arbitrators and the umpire shall then jointly constitute an arbitral tribunal. Any arbitral tribunal established under this or the preceding Article shall settle its own procedure and give its decisions by majority vote, provided that the Council may determine procedural questions in the event of any delay which in the opinion of the Council is excessive.

Article 86. Appeals

Unless the Council decides otherwise any decision by the Council on whether an international airline is operating in conformity with the provisions of this Convention shall remain in effect unless reversed on appeal. On any other matter, decisions of the Council shall, if appealed from, be suspended until the appeal is decided. The decisions of the Permanent Court of International Justice and of an arbitral tribunal shall be final and binding.

Article 87. Penalty for Non-conformity of Airline

Each contracting State undertakes not to allow the operation of an airline of a contracting State through the airspace above its territory if the Council has decided that the airline concerned is not conforming to a final decision rendered in accordance with the previous Article.

Article 88. Penalty for Non-conformity by State $

The Assembly shall suspend the voting power in the Assembly and in the Council of any contracting State that is found in default under the provisions of this Chapter.

Chapter XIX. War

Article 89. War and Emergency Conditions

In case of war, the provisions of this Convention shall not affect the freedom of action of any of the contracting States affected, whether as belligerents or as neutrals. The same principle shall apply in the case of any contracting State which declares a state of national emergency and notifies the fact to the Council.

Chapter XX. Annexes

Article 90. Adoption and Amendment of Annexes

(a) The adoption by the Council of the Annexes described in Article 54, subparagraph (l), shall require the vote of two-thirds of the Council at a meeting called for that purpose and shall then be submitted by the Council to each contracting State. Any such Annex or any amendment of an Annex shall become effective within three months after its submission to the contracting States or at the end of such longer period of time as the Council may prescribe, unless in the meantime a majority of the contracting States register their disapproval with the Council.

(b) The Council shall immediately notify all contracting States of the coming into force of any Annex or amendment thereto.

Chapter XXI. Ratifications, Adherences, Amendments, and Denunciations

Article 91. Ratification of Convention

(a) This Convention shall be subject to ratification by the signatory States. The instruments of ratification shall be deposited in the archives of the Government of the United States of America, which shall give notice of the date of the deposit to each of the signatory and adhering States.

(b) As soon as this Convention has been ratified or adhered to by twenty-six States it shall come into force between them on the thirtieth day after deposit of the twenty-sixth instrument. It shall come into force for each State ratifying thereafter on the thirtieth day after the deposit of its instrument of ratification.

(c) It shall be the duty of the Government of the United States of America to notify the government of each of the signatory and adhering States of the date on which this Convention comes into force.

Article 92. Adherence to Convention

(a) This Convention shall be open for adherence by members of the United Nations and States associated with them, and States which remained neutral during the present world conflict.

(b) Adherence shall be effected by a notification addressed to the Government of the United States of America and shall take effect as from the thirtieth day from the receipt of the notification by the Government of the United States of America, which shall notify all the contracting States.

Article 93. Admission of Other States

States other than those provided for in Articles 91 and 92 (a) may, subject to approval by any general international organization set up by the nations of the world to preserve peace, be admitted to participation in this Convention by means of a four-fifths vote of the Assembly and on such conditions as the Assembly may prescribe: provided that in each case the assent of any State invaded or attacked during the present war by the State seeking admission shall be necessary.

Article 93 *bis*

(a) Notwithstanding the provisions of Articles 91, 92 and 93 above:

(1) A State whose government the General Assembly of the United Nations has recommended be debarred from membership in international agencies established by or brought into relationship with the United Nations shall automatically cease to be a member of the International Civil Aviation Organization;

(2) A State which has been expelled from membership in the United Nations shall automatically cease to be a member of the International Civil Aviation Organization unless the General Assembly of the United Nations attaches to its act of expulsion a recommendation to the contrary.

(b) A State which ceases to be a member of the International Civil Aviation Organization as a result of the provisions of paragraph (a) above may, after approval by the General Assembly of the United Nations, be readmitted to the International Civil Aviation Organization upon application and upon approval by a majority of the Council.

(c) Members of the Organization which are suspended from the exercise of the rights and privileges of membership in the United Nations shall, upon the request of the latter, be suspended from the rights and privileges of membership in this Organization.

Article 94. Amendment of Convention

(a) Any proposed amendment to this Convention must be approved by a two-thirds vote of the Assembly and shall then come into force in respect of States which have ratified such amendment when ratified by the number of contracting States specified by the Assembly. The number so specified shall not be less than twothirds of the total number of contracting States.

(b) If in its opinion the amendment is of such a nature as to justify this course, the Assembly in its resolution recommending adoption may provide that any State which has not ratified within a specified period after the amendment has come into force shall thereupon cease to be a member of the Organization and a party to the Convention.

Article 95. Denunciation of Convention

(a) Any contracting State may give notice of denunciation of this Convention three years after its coming into effect by notification addressed to the Government of the United States of America, which shall at once inform each of the contracting States.

(b) Denunciation shall take effect one year from the date of the receipt of the notification and shall operate only as regards the State effecting the denunciation.

Chapter XXII. Definitions

Article 96

For the purpose of this Convention the expression:

(a) "Air service" means any scheduled air service performed by aircraft for the public transport of passengers, mail or cargo.

(b) "International air service" means an air service which passes through the air space over the territory of more than one State.
(c) "Airline" means any air transport enterprise offering or operating an international air service.
(d) "Stop for non-traffic purposes" means a landing for any purpose other than taking on or discharging passengers, cargo or mail.

Signature of Convention

IN WITNESS WHEREOF, the undersigned plenipotentiaries, having been duly authorized, sign this Convention on behalf of their respective governments on the dates appearing opposite their signatures.
DONE at Chicago the seventh day of December 1944, in the English language. A text drawn up in the English, French and Spanish languages, each of which shall be of equal authenticity, shall be open for signature at Washington, D.C. Both texts shall be deposited in the archives of the Government of the United States of America, and certified copies shall be transmitted by that Government to the governments of all the States which may sign or adhere to this Convention

Reference

Milde M (1994) The Chicago Convention - are major amendments necessary or desirable 50 years later? Ann Air Space Law XIX-I:401–452

Index

DOI 10.1007/978-3-319-00068-8,

B

C

T

MIX
Papier aus verantwortungsvollen Quellen
Paper from responsible sources
FSC® C105338

If you have any concerns about our products,
you can contact us on
ProductSafety@springernature.com

In case Publisher is established outside the EU,
the EU authorized representative is:
Springer Nature Customer Service Center GmbH
Europaplatz 3, 69115 Heidelberg, Germany

Printed by Libri Plureos GmbH
in Hamburg, Germany